BREAKTHROUGHS

A Chronology of Great Achievements in Science and Mathematics

1200–1930

BREAKTHROUGHS

A Chronology of Great Achievements in Science and Mathematics 1200–1930

CLAIRE L. PARKINSON
Foreword by L. PEARCE WILLIAMS

G.K.HALL &CO.
BOSTON

Parkinson, Claire L.
Breakthroughs : a chronology of great achievements in science and mathematics, 1200–1930.

Bibliography: p. 512
Includes index.
1. Science—History. 2. Mathematics—History.
3. Civilization, Occidental. I. Title.
Q125.P327 1985 509 85-7628
ISBN 0-8161-8706-1

Contents

Foreword vii
Preface ix
Chronology 1
Sources 512
Name Index 522
Subject Index 555

To my mother and father

Foreword

In the last fifty years, the history of science has grown from a subject practiced primarily by retired scientists and taught only on the periphery of the curriculum, to a major discipline that encompasses the growth of science and technology, the philosophy of science, and the mutual effects of science and society. Graduate degrees can be pursued in all the major universities of America and Britain, and undergraduate courses serve to bridge the gap between the sciences and the humanities in universities, colleges, and technical institutes. Recent modifications of undergraduate curricula have pushed the history of science closer and closer to the very center of what is increasingly considered to be the necessary core of higher education.

As the field has grown and prospered, various aids to its understanding and pursuit have been forthcoming. Beginning in 1970, quarto volume after quarto volume of that monument of modern scholarship, the *Dictionary of Scientific Biography* has rolled off the presses until, in 1980, the sixteenth and final volume (Index) appeared. No one who has not had to begin a new research project, or even make up weekly lectures without it, can fully appreciate the value of this work. Its cost and size, however, place it generally outside the study of the average undergraduate or graduate student. And its arrangement as biography makes it difficult to use for those students who are just becoming acquainted with the subject. Even if someone can put a name together with a law or a fact, the DSB is not always helpful. If you want to know what Ampere's law is, you will not find it in the article on Ampere because Ampere had never heard of it. It was Maxwell who gave it that name, long after Ampere had died. Other, smaller, dictionaries of scientific biography, of which there are a number, suffer from the same disability.

More recently, a scholarly and very useful *Dictionary of the History of Science* has been published. This is a history of scientific concepts in which general ideas are described and traced. Again, for the practitioner of the history of science who knows his way around the conceptual map, this work is invaluable. Like the DSB, it provides a good bibliography with which to begin to follow concepts both to their origins and to their consequences. But, like the DSB, it is a difficult work for the beginner to use. Concepts, and particularly scientific concepts, are difficult and subtle for those just introduced to them. For the neophyte, it is a little like coming in to the middle of the first act of a play. It is possible to figure out what is going on, but there is always the lingering sense that something important, even fundamental, has been missed.

What has been needed for years, and what Dr. Parkinson attempts to provide in this volume, is a most basic aid to the study of the history of science. It is, simply, a chronology of scientific discovery and, as such, it deals not with lives, nor concepts, but with much more simple facts. Who discovered what, when? This would seem to be about the simplest and most straightforward question that could be asked about the history of science, but it is not. One of the leading students of the history of science, Thomas S. Kuhn, has, indeed, made it a prominent point in much of his work on the philosophy of science that discoveries and laws and theories are not like continents, to be discovered, but like species that evolve over time. It is, therefore, impossible ever to assign a specific date for their appearance. For Kuhn and his disciples, then, this volume will be a futile exercise. But Kuhn and his disciples are already sophisticated students of the history of science who can easily adjust to this rather controversial notion; beginners are not, and surely it can do little harm to provide them some specific chronological guides as they start their studies. Perhaps later they will agree with Kuhn (not all of us do!) and fuzz their historical perceptions accordingly. For now, this work will permit students to find specific information easily and use it.

But, it may be objected, just how specific is the information? One of the things Dr. Parkinson discovered (as did I when I was asked to advise her) was that there is a good deal of scholarly disagreement over what should be simple matters. Authorities, in short, disagreed both with Dr. Parkinson and with each other, and sometimes the disagreements were quite violent. For example, Dr. Parkinson discovered that the estimates of modern scholars for the accuracy with which Tycho Brahe could observe heavenly bodies ranged from 8′ to 1″, which is quite a range! What should be done? Since there are few, if any, criteria for judging which of the experts is *right*, and since life is short and every such discrepancy could not be traced back to original sources, Dr. Parkinson has, I think rightly, gone ahead and published what seemed to her to be the best guess. Her authority for this course is Bacon's famous dictum that truth is more likely to emerge from error than from confusion. Experts who are offended by her entries in their fields may find them to be the starting points for new and interesting researches that may, perhaps, remove the startling inconsistencies among equally competent scholars on matters of fact.

I cannot close this short foreword without a word of praise for Dr. Parkinson's persistence and courage. The task she set herself was a difficult and tedious one, and she has continued with it in the face of considerable adverse reaction about both its possibility and desirability. Her reward, I am sure, will be the gratitude of generations of students who will find the book indispensable to them in both the study of the sciences and the study of its history. Even the scholars, I suspect, will ultimately come around and treat it as a standard handbook.

L. PEARCE WILLIAMS
Cornell University

Preface

Science has become a vital component of our civilization, influencing our work, our recreations, our goals, our perceptions of ourselves and the universe around us. Yet few of us have much understanding of the historical foundations of that science, and hence of the strengths and weaknesses upon which it and its individual disciplines will flourish or flounder. This volume is intended to convey a sense of the development of the science and mathematics of Western civilization by presenting a chronology of events important to that development from the preliminary introduction of Hindu-Arabic numerals through such varied events of the early twentieth century as the emergence of quantum mechanics and general relativity, the artificial transmutation of chemical elements, the discovery of vitamins and controlled experimentation on dietary deficiency diseases, the hypotheses of continental drift and the Big Bang explosion of the universe, and the determination of unexpected inherent limitations of Western science, as exemplified by the uncertainty principle in physics and Kurt Gödel's proofs of incompleteness in mathematics. The span of years covered is 1202 to 1930.

The central purpose of the book is to provide, in chronological order, concise, clear statements on events or accomplishments contributing to the development of Western science. In the hopes that the casual reader may obtain some understanding or insight as well as facts by browsing through the book, I have tried in some of the entries to include more than a simple statement of what happened when, by adding a flavor of how an accomplishment was made or how it was influenced by other scientific developments. All entries are brief, however, and the reader is encouraged to seek more detailed discussions, especially from primary sources, on those topics that interest him most.

Although I expect the book to be used chiefly as a reference or for random browsing, additional benefits can derive from reading it through as a whole or systematically following a specific topic. The sequence of entries demonstrates the uneven nature of scientific development, important interconnections within and among the various scientific disciplines, and the often extensive development of a subject before the emergence of a key individual providing crucial linking concepts. Examples of the latter include numerous theories of evolution developed before the mechanism of natural selection was suggested by Charles Darwin, with extensive supporting evidence, and the use of differentiation and integration techniques before the fundamental theorem of calculus was formulated and proved by Isaac Newton and Gottfried Leibniz.

A complete chronology of science never could be constructed and has not been attempted. A more realistic goal would be a very selective (and of necessity subjective) listing of the truly major scientific events; but even that has not been attempted. This book simply presents in chronological order a listing of numerous events in the history of Western science, with one aim being to include most of the major events in the disciplines covered and enough of the minor events to add interest and coherency.

The entries within each year are alphabetized according to nine main categories. These consist of the eight disciplines of Astronomy, Biology, Chemistry, Earth Sciences, Health Sciences, Mathematics, Meteorology, and Physics, plus a Supplemental category that contains entries on scattered related topics such as scientific methodology and a few philosophical, religious, and political issues directly connected to the contemporary or succeeding science. Although there had been some social science entries in an earlier version of the manuscript, these have by and large been eliminated, as have most entries regarding technology. In addition to the main category identification, the heading for each entry contains a more specific subtitle.

An important feature of the book for its use as a reference is the inclusion of detailed name and subject indexes. The name index provides a comprehensive list, by year and main subject category, of each time an individual is mentioned. The subject index is less comprehensive but should aid the reader in locating specific entries and determining, for instance, when a given event occurred or who was involved with it. For further convenience, the name index includes birth and death dates, nationalities, and middle and alternative names for each individual for whom that information was available. This basic biographical material is also generally included in the main text for the central person or persons discussed in each entry, except for entries supplemental to other entries in the same year.

This chronology has consumed much of my spare time over the past thirteen years. It began as an effort to facilitate my own understanding of the development of science, and it evolved into book form as its potential value to other people became apparent. For the first ten years it was exclusively a one-person task. I tried hard to keep the entries clear, compact, and accurate, but this was more difficult than might be expected, as time and time again both major and minor contradictions were found among published sources. In the past three years my publisher and I have had others read through the manuscript to pinpoint troublesome entries and suggest revisions. To the best of our abilities, errors and ambiguities have been eliminated.

In spite of the efforts expended, however, this book is certain to contain errors, and probably some significant ones. The state of the discipline of the history of science is such that unknowns abound in almost every aspect. Even the task of finding names and life-spans can be tremendously exasperating: in some cases four different books provided four different birth dates for the same scientist and two or three different spellings of his name. Obviously I had to make numerous subjective decisions on which source to accept as the most authoritative. On the advice of L. Pearce Williams, in the end I tended when possible to use the multivolume *Dictionary of Scientific Biography* to "resolve" such issues.

More significant than the contradictions among published sources regarding birth and death dates were the contradictions involving the main substance of a scientific event or achievement. Different histories of science often provide widely differing accounts of the same incident, many times containing outright contradictions and other times containing strong differences of interpretation on such issues as the importance of an event, the relative value of the contributions of different individuals, and the meaning of the reported results. Even when the event described centers on a published piece of research and the original source is read and reread, there still remain problems: how to summarize a substantial work in a few sentences; how to indicate fairly what might in hindsight appear to be the most important aspect of a work, although the author clearly had different emphases; how to word entries on theories that have long ago been discounted, while recognizing that our decade has no hold on scientific "truth" and that indeed discarded theories do on occasion reemerge with increased support.

By being concerned with resolving the contradictions, I have avoided repeating many of the errors published elsewhere, but probably not all. I apologize in advance especially for errors, but also for other imperfections within the entries, whatever the cause, and for any serious omissions and all other shortcomings. I ask the reader his forbearance in view of the impossibility of perfectly chronicling Western science. Although incomplete, the book contains a wealth of information and should help the reader to understand something about scientific development and to place given events within the context of other events. It is not meant to be all-inclusive, but it is meant to include enough events and with sufficient detail on the key events to be of interest and value to the general reader. I am aware of and terribly pained by the flaws, but the book covers a vibrant, fascinating subject, and I hope that there are readers who become informed and excited by it.

I thank the many people who have helped me through the final years of this endeavor. Most prominently this includes my brother William Parkinson, who not only has supported me throughout but who recommended an appropriate publisher, plus Janet Campbell, Myrtle Lane, Tony Busalacchi, George Demko, L. Pearce Williams, C. V. Parkinson, and David Atlas, all of whom gave freely of their time and advice at crucial moments during the lengthy review process. I thank those who read through all or portions of the manuscript and offered their suggestions, most notably Donald J. Cavalieri, C. V. Parkinson, L. Pearce Williams, Myrtle Lane, Aaron Ihde, Victor Thoren, Joseph Dauben, Lois Magner, Marshall Walker, Webb Dordick, Albert Carozzi, David Lindberg, and William Coleman. Finally, I thank my employer, Goddard Space Flight Center, and my family and friends who have helped me through the personal sacrifices necessitated when undertaking a project of this scope on top of an already-demanding full-time job. This includes many of the above individuals plus William Campbell, Thomas Heslin, Ann Parkinson, Jennifer Parkinson, Virginia Parkinson, Jean Ryan, and Warren Washington.

Claire L. Parkinson
Greenbelt, Maryland
September, 1985

Figure 1. *A woodcut contrasting the use of Hindu-Arabic computation methods (see 1202—*MATHEMATICS *entry) with the older use of a reckoning board. The satisfied expression on the face of the man on the left suggests the greater convenience of the new methods. (By permission of the Houghton Library, Harvard University.)*

Chronology

1202

MATHEMATICS (Arithmetic)

Leonardo Fibonacci (Leonardo of Pisa, c.1170–c.1250, Italian) writes *Liber abaci* [Book of the abacus], the major work awakening Europe to the advantages over the Roman numeral system of the Hindu numerals and the Hindu-Arabic computation methods. The book deals with arithmetic and elementary algebra, and is based largely on earlier Arabic and Greek sources. Individual merchants and bankers begin using Hindu-Arabic numerals over the next three centuries, but it is not until c.1500 that their use reasonably completely supersedes the use of Roman numerals.

c.1210

CHEMISTRY (Mineral Acids)

As a result of the discovery of mineral acids, a discovery made possible through improvements in stills, chemists become far better able to dissolve substances and create reactions in solution. Previously, weak organic acids had been used.

EARTH SCIENCES (Ocean Tides)

Wallingford (d.1213, English), in the first known recording of tide observations for the purpose of prediction, tabulates the occurrence of floods at London Bridge. For the next 500 years, tide prediction is the central objective of most observational oceanography.

1215

SUPPLEMENTAL (Religion and Science; Aristotle)

The church in Paris forbids the teaching of Aristotle's (384–322 B.C., Greek) metaphysical and scientific works, extending a prohibition on his philosophical works imposed in 1210.

1217

ASTRONOMY (Aristotle)

Through his translation of Alpetragius's (Abū Ishāq al-Bitrūjī al-Ishbīlī, fl. c.1190, Spanish) *Liber astronomiae* [Book of astronomy], Michael Scot (c.1175–c.1234, Scottish) awakens Europe to the Aristotelian system of astronomy and its contrasts with the better-known and more mathematical system of Ptolemy. In the Aristotelian system, planets revolve around the earth on concentric spheres, whereas in the Ptolemaic system the planets revolve on small circles (epicycles) which are centered on larger circles (deferents) centered near the earth. Although more complicated, the Ptolemaic system is better able to predict the positions of the planets and explain the changes in their apparent brightnesses. It is detailed in Ptolemy's *Almagest* and available in Latin translation since c.1160.

1220

MATHEMATICS (Geometry; Trigonometry)

In *Practica Geometriae,* Leonardo Fibonacci (Leonardo of Pisa, c.1170–c.1250, Italian) presents much of the material of Euclid's (c.330–c.275 B.C., Greek) *Elements* and of Greek trigonometry, as well as (1) introducing Fibonacci sequences (0, 1, 1, 2, 3, 5, 8, . . .), (2) showing that there exist irrationals other than those classified in Euclid's book 10, and (3) attempting to show that straightedge and compass are not sufficient to construct the roots of $x^3 + 2x^2 + 10x = 20$.

c.1220

BIOLOGY (Classification)

Michael Scot (c.1175–c.1234, Scottish) translates Aristotle's *History of Animals, Parts of Animals,* and *Generation of Animals* into Latin, precipitating Albertus Magnus's (c.1200–1280, German) classification of European fauna according to general Aristotelian principles.

EARTH SCIENCES/MATHEMATICS (Stereographic Projections)

Jordanus de Nemore (Nemorarius, fl. c.1220, French) projects a globe onto a plane tangent to it at the North Pole and describes the basic principles of stereographic projection, including the principle that a circle on the globe will retain its property of being a circle when projected onto the plane.

MATHEMATICS (Algebra)

In *De numeris datis,* Jordanus de Nemore (fl. c.1220, French) generalizes arithmetical problems by inserting letters in place of numbers, and examines problems later seen as forerunners to linear and quadratic equations. He accepts the Hindu numerals and Hindu-Arabic computation methods detailed earlier in the century by Leonardo Fibonacci (1202—MATHEMATICS).

PHYSICS (Optics; Lux Concept)

Robert Grosseteste (c.1168–1253, English), influenced by the writings of Saint Augustine, presents a conception of the universe based on light, or *lux,* as fundamental. In *De luce seu de inchoatione formarum* and *De motu corporali et luce* he suggests that light was the first form in matter, that lux propagated in all directions according to mathematical laws, thereby forming dimensionality and determining the structure of the universe, and that the lux concept can be used to explain transformations and differences in many spheres of study. Hence optics is seen as a study fundamental to many others.

PHYSICS (Statics)

Jordanus de Nemore (fl. c.1220, French), or followers of his using his name, writes several treatises on statics in which he includes many results from Aristotle, such as the law of the lever, and proceeds beyond Aristotle in the derivation of laws for bodies positioned on inclined planes, the latter being included in *Liber Jordani de Nemore de ratione ponderis.* In *Elementa super demonstrationem ponderis* [Elements for the demonstration of weights], Jordanus enunciates the principle that the strength required to lift a weight w to a height h is equivalent to that required to lift a weight kw to a height h/k.

1225

MATHEMATICS (Algebra)

In *Liber quadratorum* [Book of square numbers], Leonardo Fibonacci (Leonardo of Pisa, c.1170–c.1250, Italian) solves selected first, second, and third degree determinate and indeterminate equations. Like that of the Arabs, his algebra is based on methods adopted from arithmetic and employs almost no symbolism, although on occasion he generalizes his results by replacing numbers with letters. The work includes the general solution in rationals of a system of three equations in three unknowns, each having quadratic terms. Leonardo rejects negative roots of equations, but does interpret a negative answer to a financial problem as a loss rather than a gain.

c.1230

PHYSICS (Sound)

In *De generatione sonorum,* Robert Grosseteste (c.1168–1253, English) describes sound as a vibratory motion propagating through the air from the sound source to the receiving ear.

SUPPLEMENTAL (Scientific Methods)

In commentaries on the *Physics* and *Posterior Analytics* of Aristotle and in other treatises, Robert Grosseteste (c.1168–1253, English) presents views on proper scientific methodology, encouraging the search for general principles and causes, the testing of theories against experience, and the use of mathematics. He asserts that scientific inquiry should begin with facts (*quia*), either observed or experienced, and should aim at determining the reason behind the facts (*propter quid*), this involving the resolution of the phenomena into more elementary principles and the derivation from these principles of hypotheses which allow a reconstruction of the phenomena. The hypotheses should then be verified or rejected based on observation or on theory which has previously been verified. More specifically, he introduces the Principle of Falsification, asserting that a hypothesis should be rejected if experience proves its conclusions to be false.

1231

SUPPLEMENTAL (Religion and Science: Physics)

Pope Gregory IX (c.1143–1241, Italian) rescinds the general ban forbidding the study of physics, although those writings specifically viewed as heretical remain banned.

c.1231–1235

PHYSICS (Optics; Law of Refraction)

In his "De iride seu de iride et speculo" and other essays, Robert Grosseteste (c.1168–1253, English) describes the science of optics, concentrating especially on geometrical optics and the path of light rays, reflection and refraction. He claims that when a ray is refracted upon entering a denser medium, the angle between its refracted path and the perpendicular to the interface is half the angle the ray would have made if unrefracted. He thereby makes an early attempt at determining a quantitative law of refraction (see 1621—PHYSICS for the modern law of refraction). Although he claims to have verified experimentally specific results, the erroneous nature of some of these throws into question the claims of experimentation.

c.1235

METEOROLOGY (Rainbow)

Robert Grosseteste (c.1168–1253, English), in his works on optics, suggests that rainbows are caused by the refraction of sunlight inside clouds.

SUPPLEMENTAL (Encyclopedias)

Bartholomew Anglicus (fl. c.1250, English) writes *De proprietatibus rerum* [On the properties of things], an encyclopedic work on natural history, combining fact and fiction, and interspersing superstition with observation and theology with science. The influence of the stars on earthly affairs is considered major.

1238

BIOLOGY (Dissections)

In an effort to increase understanding of the human body, the Holy Roman Emperor Frederick II (1194–1250, Italian) is said to have ordered that every five years a corpse be dissected in Salerno. (Whether such an order was actually issued is uncertain.)

1240

HEALTH SCIENCES (Medical Regulations)

Frederick II (1194–1250, Italian) promulgates a set of laws regulating the study and practice of medicine, including the licensing of physicians, the preparation of drugs, and the role of apothecaries. A standard course of medical studies is to take five years, followed by a year of practicing medicine under the guidance of an experienced physician.

c.1240

MATHEMATICS (Arithmetic)

Using Al-Khwārizmī's *Treatise on Calculation with the Hindu Numerals* as a prototype and adding theory from Boethius's sixth-century paraphrase of Nicomachus of Gerasa's *Arithmetica,* John of Holywood (Sacrobosco, c.1200–1256, Irish) writes what is to become a standard university text for centuries: *Algorismus vulgaris.* The book becomes especially popular with the addition of a commentary on it by Peter of Dacia in 1291.

c.1245

BIOLOGY (Falcons)

The Holy Roman Emperor Frederick II (1194–1250, Italian) writes a detailed, meticulously illustrated treatise on falconry, *De arte venandi cum avibus* [On the art of hunting with birds]. In addition to discussing the habits and migrational tendencies of falcons, Frederick describes experiments on the artificial incubation of eggs, and presents an observation-based description of falcon anatomy, including such previously unrecorded details as the form of the sternum and the structure of the lungs.

1250

SUPPLEMENTAL (Encyclopedias)

Vincent of Beauvais (c.1190–c.1264, French) writes the *Speculum naturale, historiale, doctrinale,* a larger and less-conversational encyclopedia than that of Bartholomew Anglicus (c.1235—SUPPLEMENTAL), summarizing the knowledge of the time.

c.1250

BIOLOGY (Bird Migration)

Matthew Paris (c.1200–1259, English) details the migrations of crossbills.

BIOLOGY (Botany; Aristotle)

Albertus Magnus (c.1200–1280, German) helps introduce the work and methods of Aristotle to Europe and writes treatises on a variety of topics, including *De vegetabilibus* [On vegetables and plants] recording his botanical observations. This work classifies plants and vegetables into basic types and describes the structure and function of various parts of the plant, including root, stem, seed, leaf, and flower.

BIOLOGY (Vision)

Albertus Magnus (c.1200–1280, German) rejects the popular "extramission" theory that what makes sight possible is some emission from the eye, advocating instead a modified version of Aristotle's basic concept that the visible object alters the medium between the object and the eye and this alteration is propagated to the eye.

EARTH SCIENCES (Geography, Cartography)

Four maps of Great Britain drawn by Matthew Paris (c.1200–1259, English) locate major roads and towns.

EARTH SCIENCES (Glaciology)

Geographic details of the glaciers and icebergs of Iceland are described by an anonymous author in the Norwegian work *Konungs skuggsjä* [Mirror of the king].

MATHEMATICS (Geometry; Trigonometry)

Nasir al-Din al-Tūsī (Nasir Eddin, 1201–1274, Persian) attempts to determine whether Euclid's parallel postulate can be derived from the other Euclidean postulates, doing so in his *al-Risāla al-shāfiya,* and begins to separate the study of trigonometry from that of astronomy, doing so in his *Shakl al-qitā,* where, for instance, he presents and describes the law of sines. His works in these two areas may have influenced the further advances of Girolamo Saccheri (1733—MATHEMATICS) and Johannes Regiomontanus (1464—MATHEMATICS) respectively.

1252

HEALTH SCIENCES (Surgery)

A growing emphasis on surgery and human dissection at the Universities of Bologna and Padua leads to the *Chirurgia magna* [Major surgery] of Bruno of Longoburgo (fl. c.1250, Italian).

c.1255

MATHEMATICS (Geometry)

Johannes Campanus of Novara (d.1296, Italian) translates Euclid's *Elements* from Arabic into Latin, using various Arab sources plus an earlier Latin translation by Adelard of Bath. The Campanus translation becomes widely used, is first printed in 1482, and is reprinted over a dozen times over the succeeding century and a half.

1259

ASTRONOMY (General)

Nasir al-Din al-Tūsī (1201–1274, Persian) begins construction of a major observatory at Maragha. Among the instruments built for the observatory are a giant mural quadrant, an armillary sphere, a solstitial armill, and a parallactic ruler. After 12 years of observation, the Zīj-i īlkhānī (Īlkhānī astronomical tables) are completed in 1272. Al-Tūsī also writes a criticism of Ptolemaic astronomy (in *Tadhkira*) and includes suggested changes to insist more completely on uniform circular motions for the heavenly bodies. He notes that a reciprocating rectilinear motion can be produced by the combination of two uniform circular motions. His ideas and works probably become known to Nicolaus Copernicus and conceivably influence his further developments (1543—ASTRONOMY).

c.1260

BIOLOGY (Blood Circulation)

Ibn al-Nafis (1210–1288, Egyptian) rejects Galen's theory that the blood passes through pores in the septum and concludes that it instead passes through the lungs in moving from the right to the left ventricle of the heart. He thus arrives at the concept of the lesser circulation of the blood, but his work does not enter into the mainstream of European knowledge, and the concept is not rederived until almost three centuries later (1553—BIOLOGY).

EARTH SCIENCES (Minerals; Metals; Stones)

Albertus Magnus (c.1200–1280, German) writes a treatise on the origin of minerals, metals, and stones: *De Mineralibus et rebus metallicis.* He perceives these substances as arising through the effect of celestial influences on the material of the earth's crust, and as becoming compact and hardened by the heat and cold of the earth.

EARTH SCIENCES (Volcanoes; Mountains; Subterranean Winds)

Albertus Magnus (c.1200–1280, German) puts forward a theory of subterranean winds, in part to account for volcanic eruptions, and hypothesizes that there exist two major forces causing mountain formation: the pressure of subterranean winds attempting to escape their earth bondage, and the action of the sea piling sand into larger and larger mounds along its shore.

1266

HEALTH SCIENCES (Suppuration)

Theodoric Borgognoni of Lucca (1205–1298, Italian) completes the standard version of his most famous work, *Chirurgia* [Surgery], in which he presents the surgical knowledge and practices of his father Hugh of Lucca, supplemented by his own experiences and understandings. Notably, Theodoric advocates antiseptic surgery, denouncing the widespread belief in "laudable pus," adopted from the Arabs, that in treating wounds a first step is to encourage the formation of pus, doing so with unguents and poultices. Instead, he encourages thorough cleansing of the wound and using stitches where necessary. His aseptic surgery is practiced by his pupil Henri de Mondeville but not many others. Theodoric also encourages use of narcotics-impregnated soporific sponges held to the nose of a patient to induce sleep before surgery. He improves techniques for their preparation, drying them after initial preparation and then immersing them in water when ready to use them.

1267–1268

EARTH SCIENCES (Spherical Earth)

In his *Opus majus* (1267–1268—SUPPLEMENTAL), Roger Bacon (c.1219–c.1292, English) mentions the possibility of reaching India in a voyage of only a few days by traveling west from Spain, clearly implying adherence to the widely accepted belief in the basic spherical shape of the earth.

SUPPLEMENTAL (Scientific Methods)

Roger Bacon (c.1219–1292, English) writes his major work, *Opus majus* [Larger work], along with two summaries, *Opus minus* [Smaller work] and *Opus tertium*. *Opus majus* is divided into seven parts, the fourth, fifth and sixth concerning mathematics, optics, and experimental science respectively, and the remainder concerning ignorance and error, theology, language, and philosophy. Bacon encourages direct observation, at least some experimentation, and the application of mathematics to the various sciences. He describes experiments with lenses and with gunpowder, mentions of the latter being newly arrived in Europe from China.

1268

BIOLOGY (Vision)

Roger Bacon (c.1219–c.1292, English) writes an essay on vision, *De multiplicatione specierum* [On the multiplication of species], which, along with part 5 of his *Opus majus* (1267–1268—SUPPLEMENTAL), presents the thesis that the ability to see an object results from "species" originating from the object, species which pass through the intervening medium to the observer's eye not by moving from place to place but by sequentially producing likenesses of themselves in the adjacent medium. The actual stimulation of sight requires in addition the adjustment of the medium by the emanation of the species from the observer's eye. Bacon's explanation is based on those of many earlier authorities, particularly Ibn al-Haytham and Aristotle.

1269

PHYSICS (Magnetism)

Peter of Maricourt (Peter Peregrinus, Pierre de Maricourt, fl. c.1269, French) writes a small book, *Epistola de magnete* [Letter on the magnet], on the fundamental properties of magnetism, including the distinction between north and south poles and the procedure of magnetizing an iron rod by magnetic induction from a magnet moved above the rod. Peregrinus accomplishes two major improvements to the compass: he places the magnet on a pivot and he adds a direction scale. He explains magnetism celestially rather than terrestrially, the magnet's orientation being determined by the celestial poles.

SUPPLEMENTAL (Translations of Archimedes, Aristotle)

William of Moerbeke (c.1220–c.1286, Flemish) completes Greek-to-Latin translations of most of the works of Archimedes, including *Equilibrium of Planes* and *On Floating Bodies,* plus many of the works of Aristotle.

c.1270

HEALTH SCIENCES (Hydrocephalus)

William of Saliceto (1210–1277, Italian) describes treatment of hydrocephalus in children through use of a cautery to burn a small hole in the skull and thereby remove excess brain fluid.

PHYSICS (Optics)

Witelo (c.1230–c.1275, Silesian), in *Perspectiva,* treats a wide range of optical topics, beginning with a summary of the relevant geometry and proceeding to radiation, vision, perception, reflection, mirror images, and refraction, including meteorological phenomena such as the rainbow. His emphasis being on geometrical optics rather than the nature of light, Witelo includes detailed discussions of the images formed in various concave and convex mirrors. These discussions are based on the rectilinear propagation of light through uniform media and on the law of reflection, which he attempts to establish geometrically from the principle of minimum distance. Although he relates efforts to determine experimentally relationships involving angles of refraction, his results appear to be fabricated and based on a failure to understand the reciprocal nature of refraction, i.e., that light traveling from a more dense to a less dense medium will follow the reverse path of light traveling in the opposite direction.

SUPPLEMENTAL (Religion and Science)

St. Thomas Aquinas (1225–1274, Italian) completes his Christianization of the works of Aristotle, for which a major step is the substitution of "God" for "prime mover." St. Thomas encourages scientific research by emphasizing faith *through* understanding in preference to the earlier emphasis on faith without understanding. He also writes philosophical commentaries on the *Physics, Metaphysics, De anima,* and *Ethics* of Aristotle.

1272

ASTRONOMY (Planets)

Astronomers under Alfonso El Sabio (Alfonso X of Castile, 1221–1284, Spanish) write, in manuscript form, a revised edition of the Toledan Tables of al-Zarqālī. Known as the Alfonsine Tables, this compilation of planetary positions is used extensively for the succeeding three centuries. Predictions of celestial motions are based on Ptolemaic theory, trigonometric equations, and solar, lunar, planetary, and stellar observations, some from the decade 1262–1272.

1275

BIOLOGY (Anatomy)
HEALTH SCIENCES (Surgery)

William of Saliceto (1210–1277, Italian) describes his observations, as a practicing surgeon, in *Chirurgia* [Surgery], including descriptions of a human dissection, of the thoracic organs, and of a hernia patient's abdominal veins.

c.1276

BIOLOGY (Fetus)

Giles of Rome (1247–1316, Italian) discusses the development of the human fetus in the treatise *De formatione corporis humani in utero.* His considerations include the timing of the soul's entry and the relative importance of the two parents.

1277

SUPPLEMENTAL (Religion and Science)

The Bishop of Paris condemns 219 specific propositions of Aristotle and his commentators. The theological reaction against Aristotle throughout the thirteenth century is based in large part on the apparent contradictions with Biblical teachings, for instance the concept of the world's being eternal (and hence not emerging from some unique Creative act) and the implication that the processes of nature are regular and knowable (with no allowance for miracles). The condemnation works for as well as against further scientific advances, as it discourages blind acceptance of Aristotle, thereby encouraging critical examination of Aristotle's teachings, especially when they conflict with either experience or divine revelation.

1277–1279

PHYSICS (Optics; Vision)

John Pecham (c.1230–1292, English) writes what is to become the standard optical textbook for the next three centuries: *Perspectiva communis.* The book presents up-to-date understandings of the propagation of light, vision, reflection, refraction, rainbows, the eye's anatomy and physiology, and the psychology of vision. A central theme concerns the theory of direct vision. Pecham follows Ibn al-Haytham in arguing that the eye sees through receipt of rays from the visible object, but he also adopts concepts from theories of Aristotle, al-Kindī, and Robert Grosseteste, arguing that rays emitted from the eye are also essential, as they appropriately moderate the rays from the object. The latter rays, upon hitting the cornea can be separated to those hitting perpendicularly, which continue to the glacial humor, and those hitting nonperpendicularly, which he argues can be ignored. There is then a one-to-one correspondence between points of the object and points in the image on the glacial humor, allowing unconfused vision.

c.1280

PHYSICS (Impetus Theory)

Peter Olivi (1248–1298, French) further develops the impetus theory of motion outlined c.500 A.D. by John Philoponus. The theory asserts that after an object is placed into motion by an initial impetus it can continue to move even though the propellant force is no longer applied. This counters Aristotelian tradition, in which motion required the continued action of an external motive force. The continued force, still required, can now be internal to the body. For instance, from the perspective of the Aristotelian tradition at the time, an object thrown through the air is perceived as continually pushed forward by the air, the air or some other medium being essential to the object's motion. Here as elsewhere, the Aristotelian tradition was influenced by the interpretations of Aristotle's commentators, in this case particularly Averroës (Abū'l-Walīd Muhammad Ibn Ahmad Ibn Muhammad Ibn Rushd, 1126–1198, Arabic). Aristotle himself was indefinite regarding the precise manner by which the air moves the projectile forward, although he did suggest the possibility that the adjacent air receives a propellant power from the original motive force, with the propellant power then being passed on through the medium. Another possibility, suggested though not encouraged by Aristotle, was that the air rushes from the front to the rear of the object, pushing it forward. By contrast, from the perspective of the impetus theory, the object itself gains an internal motive force which propels it forward.

c.1281

ASTRONOMY (Planetary Motions)
METEOROLOGY (Rainbow)

Qutb al-Dīn al-Shīrāzī (1236–1311, Persian) completes his major astronomical work: *Nihāyat al-idrāk fī dirāyat al-aflāk,* in which he continues the work of Nasir al-Din al-Tūsī (1259—ASTRONOMY) in developing an alternative planetary model to that of Ptolemy, being more consistent than Ptolemy in the use of uniform circular motions. The work also includes sections on cosmography, geodesy, geography, mechanics, meteorology, and optics. Qutb al-Dīn's explanation of the rainbow is of particular note, as he concludes that the phenomenon results from the passage of light through the transparent sphere of the raindrop, where the ray of light undergoes two refractions and one reflection.

1282

EARTH SCIENCES (Stratigraphy; Fossils; Mountains)

In *La Composizione del mondo,* Ristoro d'Arezzo (c.1220–1282, Italian) describes various facets of the earth's composition, including observations of rock strata and fossilized mollusc shells in the Apennines. He attributes the presence of the fossils to Noah's Flood, understanding that they are most likely of marine origin and remains of living things. In book 6 he describes the origin of dry land and of mountains, contending that the primary cause for both phenomena resides in the stars. This accounts for the greater amount of dry land in the Northern than in the Southern Hemisphere, because the northern heavens have a greater number of stars. The topography of the earth reflects that of the heavens, as more distant stars pull mountains higher. Lesser forces leading to mountain formation include earthquakes, erosion from moving water, the deposition of sand and gravel by waves, and human activities. D'Arezzo further contends that celestial influences are necessary for the generation of any body on earth, whether animal, vegetable, or mineral. Scientific fact and fiction are intertwined, with many of the explanations containing an astrological bias.

1286

HEALTH SCIENCES (Postmortems)

The first known postmortem examination is undertaken as a physician in Cremona dissects a human corpse in an effort to find an explanation for an epidemic spreading through the city.

c.1287

BIOLOGY (Herbal)

In his *Herbal,* Rufinus (fl. 1250–1300, Italian) includes detailed physical descriptions of a wide variety of plants, supplementing the physical description with a commentary on the plant's medicinal benefits and a comparison of the plant with related plants.

1289

SUPPLEMENTAL (Block Printing)

The first recorded block printing in Europe is done at Ravenna, Italy, heralding technological developments over the next century and a half which lead to the development of printing from movable type by Johann Gutenberg (1440—SUPPLEMENTAL). Transmission of scientific and other knowledge is thereby greatly facilitated.

1290

ASTRONOMY (Ecliptic)

William of Saint-Cloud (fl. c.1295, English-French) calculates the obliquity of the ecliptic from the sun's altitude at the solstice, obtaining a value of 23°34′. (The correct value, by twentieth-century calculations, of this angle between the plane of the earth's equator and the plane of the earth's orbit around the sun, is 23°32′.)

EARTH SCIENCES (Latitude)

William of Saint-Cloud (fl. c.1295, English-French) correctly determines the latitude of Paris to be 48°50′.

c.1290

ASTRONOMY (Methodological Philosophy)

Bernard of Verdun (fl. late 13th century, French) and Giles of Rome (1247–1316, Italian) assert that the controversy between Aristotelian astronomy and Ptolemaic astronomy must be solved by the observational evidence. Giles further asserts that when two hypotheses are equally capable of explaining the observations, then the simplest should be accepted. Similar viewpoints are later stated by William of Ockham (c.1330—SUPPLEMENTAL).

HEALTH SCIENCES (Surgery)

Guido Lanfranchi (d.1315, Italian) discusses traditional and new surgical practices in his *Chirurgia parva.* Exiled from Milan due to political reasons, he spreads new surgical practices developed in Italy to France, where he becomes physician to the French king.

1296

HEALTH SCIENCES (Surgery)

Guido Lanfranchi (d.1315, Italian) writes his *Chirurgia magna,* providing detailed descriptions and recommendations on many conditions requiring surgical attention. He also warns against the common practice of leaving bloodletting and minor surgery to the barbers, stating the necessity that surgeons understand medicine, and also the necessity that physicians understand surgery.

1296–1297

EARTH SCIENCES (Descriptive Geography)

While a prisoner in Genoa, Marco Polo (1254–1324, Italian) dictates an account of his life and travels, providing Europeans with new information on the geography and customs of the Far East.

1299

HEALTH SCIENCES (Eyeglasses)

Clear reference is made in a Florentine manuscript to the use of spectacles to correct for farsightedness (for nearsightedness, see 1451). Although the inventor and precise date remain unknown, it appears that two lenses were first attached and set on the nose for viewing sometime in the late thirteenth century, spreading from Venice or Pisa to the rest of Italy and Europe.

MATHEMATICS (Hindu-Arabic Numerals)

A law passed in Florence, in the statutes of the *Arte del Cambio,* forbids the use of Hindu-Arabic numerals by Florentine bankers. Many banks elsewhere in Europe also forbid use of these numerals, regarding them as more easily forged than Roman numerals.

c.1300

CHEMISTRY (Atomic Theory)

Giles of Rome (1247–1316, Italian) asserts an atomic theory of matter, basing his ideas largely on the earlier work of the Spaniard Avicebron.

EARTH SCIENCES (Portolano Charts)

The first of the Spanish and Portuguese Portolano charts appear, helping navigators through inserting on a map a network of centers with lines radiating from each center in the directions of the 8 or 16 principal compass directions.

PHYSICS (Color)

Dietrich von Freiberg (Theodoric of Freiberg, c.1250–1311, German) reiterates the earlier speculation of Averroës that all colors are simply combinations of lightness and darkness, taken in different proportions.

SUPPLEMENTAL (Scientific Understanding)

Joannes Duns Scotus (c.1266–1308, Scottish), expressing the standard Aristotelian viewpoint, clearly distinguishes between scientific results understood through causal mechanisms versus those understood through empirical generalizations. Knowledge is considered most complete when the cause is known; for example, an eclipse of the sun occurs because the moon blocks the view. When no clear cause is known but the empirical evidence is strong that some particular happening always occurs after some other, then Dun Scotus regards extrapolation to a causal connection as legitimate.

1304

METEOROLOGY (Rainbow)
PHYSICS (Optics)

In one of three original works on optics, Dietrich von Freiberg (c.1250–1311, German) writes an extended explanation of the rainbow phenomenon (*On the Rainbow*) based on experiments in which he simulated a large raindrop with a glass globe filled with water. According to Freiberg, the rainbow arises due to two refractions and one reflection of solar rays, the first refraction occurring when the light ray first enters a raindrop, the reflection occurring within the raindrop, and the second refraction occurring as the ray exits the raindrop. Freiberg distinguishes primary and secondary arcs in the rainbow and asserts that the sequence of colors is the same for the primary arcs of all rainbows, and is reversed for the less-frequent secondary arcs. These latter result from the double reflection of light rays entering near the bottom of the raindrop.

1306

BIOLOGY (Practical Botany)

Pietro Crescenzi (1230–c.1310, Italian) writes the highly popular treatise *Ruralia commoda,* in which he describes a full range of topics of concern to the farmer, including cultivation of cereals, biology of plants, grafting of vines and trees, destruction of plants by insect larvae, problems concerning water supply, and positioning of farm buildings.

c.1310

BIOLOGY (Muscles)

Bernard of Gordon (c.1285–c.1320, French) theorizes on the cause of muscle movements, offering the explanation that muscles move due to a mechanical pull from the nerves.

CHEMISTRY (Alchemy)

Four works on alchemy are written by a Spaniard using the name Geber, the Latin form of a noted Arabic alchemist. The works include some theory but emphasize practical alchemy. Gold is considered the only perfect metal, other metals being theoretically convertible into gold through use of the philosopher's stone. Detailed descriptions are given of equipment, of methods for preparing substances, and of properties of individual metals. Others also attribute their writings to Geber, partly in anxiety over a formal condemnation issued against alchemy in 1307 by the church.

HEALTH SCIENCES (Epilepsy)

Bernard of Gordon (c.1285–c.1320, French) concurs with the ancient Greek explanation that epilepsy is caused by humors hindering the normal functioning of the brain and limiting the flow of air to the limbs.

1316

BIOLOGY (Dissection)

Mondino De' Luzzi (c.1275–1326, Italian) writes *Anatomia* [Anatomy], the first major western work devoted specifically to human anatomy. Mondino had introduced the instructional practice of public dissection of human corpses, and the book is essentially a manual for dissection, systematically arranged and written in a simple, concise style. It becomes highly popular as a text for the next two centuries.

1317

SUPPLEMENTAL (Religion and Alchemy)

A papal bull issued by Pope John XXII (1244–1334, French) forbids the practice of alchemy.

1320

EARTH SCIENCES (Continents)

In *De aqua et terra* [Of the water and the land], Dante Alighieri (1265–1321, Italian) discusses the issue of the existence of continents, concluding by a process of elimination that the power to raise a portion of the earth's surface above the level of the ocean to produce dry land derives from the realm of the fixed stars.

HEALTH SCIENCES (Surgery)

Henri de Mondeville (c.1260–c.1320, French) publishes *Chirurgia* [Surgery], in which he advocates use of sutures, total cleansing of wounds, limitation of suppuration, and a wine dressing for covering wounded areas.

c.1320

MATHEMATICS (Logic, Excluded Middle)

William of Ockham (c.1285–1349, English) creates a 3-valued logic, adding the excluded middle to the concepts of truth and falsity of Aristotle's 2-valued logic.

1321

MATHEMATICS (Probability; Mathematical Induction)

Levi ben Gerson (1288–1344, French) uses mathematical induction to obtain formulae for the number of permutations of n objects taken r at a time and for the number of combinations of n objects taken r at a time, doing so in his *Sefer ha mispar* [Book of numbers] on arithmetic and algebra.

1325–1330

EARTH SCIENCES (Cartography)

The detailed "Gough Map" of England is drawn. Distances along roads are shown, though there is no system of coordinates such as latitude and longitude.

1326–1335

EARTH SCIENCES (Tides)

Richard Wallingford (c.1292–1335, English) installs near the abby of St. Albans a mechanical clock to show the ebb and flow of the tide.

1328

PHYSICS (Mechanics)

In *Tractatus proportionum,* Thomas Bradwardine (c.1290–1349, English) attempts to describe motion mathematically, an emerging concern of Aristotelian commentators. Bradwardine relates a body's velocity v to the motive power p exerted upon it and the resistance to motion, r, provided by the medium through which the motion takes place. He argues that, as long as p exceeds r, velocity will be doubled by squaring the ratio p/r. In modern notation, Bradwardine's rule becomes $v = \log (p/r)$, a rule that produces zero velocity when motive power and resistance are equal. (In Newtonian mechanics, it is the acceleration of a body, not the velocity, that becomes zero when all forces on the body cancel [1687—ASTRONOMY/PHYSICS].) Bradwardine discusses the specific issue of the hypothetical free fall of bodies in a void, with zero resistance, and concludes that two homogeneous bodies of the same material will fall at the same speed even though they differ in size, contradicting the Aristotelian view that the heavier body falls faster. Bradwardine's speculations in these areas anticipate the much later, more complete work of Galileo Galilei (1609—PHYSICS).

1330

CHEMISTRY (Alchemy)

Petrus Bonus of Ferrara (fl. 1325, Italian) describes the current alchemical theories, largely Aristotelian in concept, in *Pretiosa margerita novella* [Precious new pearl].

c.1330

BIOLOGY (Vision)

William of Ockham (c.1285–1349, English) conjectures that an object can impress a likeness of itself upon the eye without affecting the space or medium between the two, hence eliminating the "species" of Roger Bacon (1268—BIOLOGY).

PHYSICS (Mechanics)

William of Ockham (c.1285–1349, English) criticizes the Aristotelian theory that motions require the continual contact of an external propellant force. He offers the sun's illumination of the earth and the attractive power of a magnet as evidence that action at a distance is a common occurrence in nature. Taking a previously stated concept and asserting it more rigorously, Ockham maintains that there exists no separate entity "motion" but that instead an object in motion should be examined through the successive places or forms it occupies.

SUPPLEMENTAL (Scientific Methods: Ockham's Razor)

In addition to encouraging a heavier emphasis on observation and the separation of science and theology, William of Ockham (c.1285–1349, English) calls for economy in scientific explanation, in particular for a minimizing of the number of explanatory entities. Although such a principle had been stated earlier by Robert Grosseteste and Joannes Duns Scotus, it gains wide recognition as "Ockham's Razor." Ockham presents it in his book *Summa totius logicae* summarizing the study of logic.

1337–1344

METEOROLOGY (Observational)

William Merle (d.1347, English) maintains a daily record of weather observations, providing the first such extant record in Western civilization. As no weather instruments are yet available, the observations are exclusively visual. Merle also uses his knowledge of weather changes to attempt some weather forecasts.

c.1340

MATHEMATICS (Fractions)

Johannes de Lineriis (French), in *Algorismus de minutiis,* describes sexagesimal and common fractions, including the practice of placing numerator over denominator, with a horizontal line dividing the two.

1344

MATHEMATICS (Infinity)

Gregory of Rimini (d.1358, Italian) addresses the issues of continuity and infinity in mathematics, discussing such problems as whether an infinite spiral can exist on a finite body. He notes that such phrases as "greater than" and "less than" must be interpreted differently when infinities are being considered than when all quantities are finite.

1347–1350

HEALTH SCIENCES (Plague)

Several preventive practices are encouraged to avoid the Black Death, which, having spread westward from India, is ravaging Europe, killing a third of the population. These practices include avoidance of heavy exercise, hot baths, and dampness; taking of antidotes against poison; burning of aromatic wood; bleeding; and, most effective, flight. Treatment for those having caught the plague includes bleeding, drugs, and cauterizing.

1348

HEALTH SCIENCES (Plague)

In *Consilium,* Gentile da Foligno (d.1348, Italian) describes various symptoms of the Black Death, including fever, side or chest pain, coughing, vomiting blood, inflammatory swellings in the groin or under the armpit, and a rapid pulse. Among other physicians also writing on the plague are Guy de Chauliac (c.1290–c.1368, French) and, later in the century, John of Glogau (c.1305–1377, Silesian). A particularly popular account is later written by writer and humanist Giovanni Boccaccio (see 1353—HEALTH SCIENCES).

c.1350

ASTRONOMY (Motions of Heavenly Bodies)

In an original application of a theory in mechanics (the Impetus Theory, c.1350—PHYSICS) to the heavenly bodies, Jean Buridan (c.1295–c.1358, French) rejects the belief that God or angels are needed constantly to guide these bodies in their orbits suggesting instead that an initial impetus from God is sufficient. In some respects this anticipates the much later development of the concept of inertia (PHYSICS entries for 1613 and 1644, plus 1687—ASTRONOMY). Buridan, along with others of his contemporaries, believes in the daily rotation of the earth about its axis.

EARTH SCIENCES (Ocean Tides)

Giacomo Dondi (1298–1359, Italian) suggests that, when near enough, the planet Venus and/or the planet Jupiter can affect the earth's tides.

HEALTH SCIENCES (Mineral Salts)

Giacomo Dondi (1298–1359, Italian) recommends the medical use of salts extracted from mineral waters.

PHYSICS (Mechanics)
MATHEMATICS (Graphical Representations)

Nicole Oresme (c.1320–1382, French) examines rates of change in his *De uniformitate et difformitate intensionum.* He plots time along a horizontal axis from which he draws vertical lines with lengths proportional to velocities, thereby anticipating coordinate geometry.

PHYSICS (Mechanics; Impetus Theory)

In a commentary on the *Physics* of Aristotle, Jean Buridan (c.1295–c.1358, French) refutes details of the Aristotelian theory of motion and further develops the impetus theory outlined in part by John Philoponus (c.500 A.D.) and Peter Olivi (see c.1280—PHYSICS). Although rejecting the Aristotelian notion that the propellant force must be external to the object, the impetus theory retains the basic Aristotelian requirement that for motion to continue a force must continue to act. The force is simply transferred into the body through the impetus. Buridan asserts that the impetus given an object at one moment can maintain the object in uniform motion indefinitely if no external forces, such as friction, restrict it. Buridan explains the increasing velocity of a falling body by allowing gravity to continually increase the impetus. Furthermore, he applies the impetus theory to the heavens, claiming that at the Creation God gave the heavenly spheres a rotation impetus, which, if encountering no resistance, will allow the rotations of the spheres to continue forever with no need of a Prime Mover.

1353

HEALTH SCIENCES (Plague)

Giovanni Boccaccio (1313–1375, Italian) publishes a popular, accessible first-hand account of the 1348 outbreak of the plague in Florence. The book, entitled the *Decamerone,* centers on 100 fictional tales told ten a day over a period of ten days by ten men and women attempting to cheer themselves during the plague outbreak. The tales have been described as vigorous, racy, and realistic, with strong anticlerical tendencies and colorful descriptions of the dress and customs of the medical men of the times. A prologue provides a factual description of the 1348 outbreak in Florence.

1360

HEALTH SCIENCES (Surgery)

Guy de Chauliac (c.1290–c.1368, French) completes *Chirurgia magna,* a systematic work that becomes a standard surgical text for a century and a half. Among the many topics treated in detail are fractures and hernias. De Chauliac suggests the use of pulleys and weights to extend fractured limbs and describes the loss of cerebrospinal fluid in cases of skull fractures.

c.1360

EARTH SCIENCES (Rivers)

Conrad von Megenburg (1309–1374, German) asserts that rivers result from rain.

MATHEMATICS (Fractional Exponents)

Nicole Oresme (c.1320–1382, French) introduces fractional exponents in his unpublished *Algorismus proportionum.* Although understood by Oresme and some later mathematicians, these are not widely used until the seventeenth century.

MATHEMATICS (Graphical Representation)

Nicole Oresme (c.1320–1382, French) writes *Tractatus de figuratione potentiarum et mensurarum* and the summary statement (perhaps written by a follower rather than Oresme himself) *Tractatus de latitudinibus formarum* on the "latitude of forms," in which independent and dependent variables are represented graphically. Expanding beyond the two-dimensional plots he had made in *De uniformitate et difformitate intensionum* (c.1350—MATHEMATICS), he suggests a three-dimensional extension, with a function of two independent variables plotted as a volume.

c.1370

ASTRONOMY (Earth's Rotation)

Although accepting the Ptolemaic system of the universe, Nicole Oresme (c.1320–1382, French) asserts in his commentary on the *De caelo et mundo* of Aristotle that we would still view from the earth the same sequence of stellar positions as we view now even if the earth were rotating daily on its axis and the sphere of the stars were fixed instead of vice versa. Furthermore, a daily rotation of the earth would not be incompatible with any other observed astronomical phenomena, such as eclipses of the sun and conjunctions and oppositions of the planets.

SUPPLEMENTAL (Astrological Prediction)

Nicole Oresme (c.1320–1382, French) rebuts the claim of astrologers that they are able to predict future events on earth, including specific rebuttal of claims for weather prediction.

1388

HEALTH SCIENCES (Sanitation)

Disturbed by the increasing pollution of the air and waterways, Parliament passes the first sanitary act applied to all of England. The act forbids dumping of garbage in ditches, rivers, or other water systems.

1391

ASTRONOMY (Astrolabe)

Geoffrey Chaucer (c.1343–1400, English) writes, in English, *A Treatise on the Astrolabe* describing the construction of the astrolabe and its use in computing the position of a star.

1392

ASTRONOMY (Planetary Positions)

Geoffrey Chaucer (c.1343–1400, English) writes *Equatorie of the Planetis,* describing an instrument for the computation of the positions of the planets and the moon according to the Ptolemaic theory of the universe. It, like the astrolabe for stellar positions (1391—ASTRONOMY), entails an analogue computation.

1410

BIOLOGY (Herbals)

The herbal *Liber de simplicibus* by Benedetto Rinio (Italian) includes not only individual paintings of 440 plants by the noted artist Andrea Amodio, but also a brief botanical description of each plant, including the appropriate season for collecting samples, the proper plant part for obtaining desired drugs, and the name of the plant in various languages.

EARTH SCIENCES (Earth Scale)

The influential astronomical compendium *Imago mundi* by Pierre d'Ailly (1350–1420, French) includes speculations that, with favorable winds, a ship could sail westward from Europe across the Atlantic to the Far East in only a few days.

EARTH SCIENCES (Geography)

Jacopo Angelo de Scarperia (Italian) translates into Latin from Greek the *Geography* of Ptolemy. The book is eagerly read, and is distinguished from most medieval works by its greater reliance on empirical evidence. Futhermore, it includes chapters on methods of spherical projection.

1416–1434

EARTH SCIENCES (Descriptive Geography)

Geographical knowledge is broadened as, under Henry the Navigator (1394–1460, Portuguese), the Portuguese undertake a scientific expedition to the Canaries in 1416, colonize Madeira in 1420, discover the Azores about 1430, voyage along the west coast of Africa, and round Cape Bojador in 1434.

c.1420

ASTRONOMY/EARTH SCIENCES (Navigation)

Henry the Navigator (1394–1460, Portuguese) establishes a school for the study of navigation and storage of maps and manuscripts plus an astronomical observatory. Scholars of many nationalities and specialties are brought together to teach Portugal's sea captains the methods of navigation and the relevant mathematics, geography, and astronomy. Among the works examined are those of Ptolemy.

1436

EARTH SCIENCES (Latitude; Longitude)

Andrea Bianco (Italian) draws a map of Europe in which, following a 1410 translation of a work of Ptolemy, he inserts lines of latitude and longitude.

1440

ASTRONOMY (General)

Nicholas Cusa (c.1401–1464, German) asserts that the heavenly bodies are made of the same elements as the earth, that there may be living beings on some of these bodies, and that the earth rotates daily on its axis, the motion having been provided once and for all from an initial impetus. He further suggests that an observer on the sun would view the earth as revolving around him, and that, from such considerations, the universe cannot be considered to have a unique center.

SUPPLEMENTAL (Printing from Movable Type)

Johann Gutenberg (c.1397–1468, German) and Lauren Janszoon Koster (c.1370–c.1440, Dutch), independently of each other and of the Chinese and Koreans preceding them, invent printing from movable type. A consequence is the more rapid spread of the Renaissance from Italy to the rest of Europe. Gutenberg, through much experience with the working of metals, creates type that does not wear down easily, is uniform in size, and produces an even line of print.

1442

EARTH SCIENCES (Descriptive Geography)

Upon returning from the Far East, Niccolo de Conti (d.1469, Italian) updates Marco Polo's 1296–1297 account of the customs and geography of eastern lands. In particular, he confirms Marco Polo on the nonexistence of the suspected land boundary blocking the Indian Ocean from surrounding oceans.

1443–1472

ASTRONOMY (Comets)

Paolo Toscanelli dal Pozzo (1397–1482, Italian) records observations of comets in 1443, 1449–1450, 1456, 1457, and 1472, the 1456 one presumably being the comet later named Halley's comet (1705—ASTRONOMY).

c.1450

EARTH SCIENCES (Speculative Geography)

Paolo Toscanelli dal Pozzo (1397–1482, Italian) proclaims the feasibility of sailing from Europe westward across the Atlantic Ocean to reach Asia. On a map of the Atlantic he estimates the distance between the two continents at roughly 3,000 miles, encouraging adventurers like Christopher Columbus to consider a westward voyage to reach the East.

HEALTH SCIENCES (Pulse)

Nicholas Cusa (c.1401–1464, German) recommends timing the pulse rate as an aid in medical diagnosis.

METEOROLOGY (Anemometer)

Leone Battista Alberti (1404–1472, Italian) describes the pressure-plate anemometer for measuring wind. The instrument resembles a weathervane with a small swinging plate attached to it. The plate swings along an arc with scaled gradations indicating the wind force.

METEOROLOGY (Hygrometer)

Nicholas Cusa (c.1401–1464, German) describes an instrument with which the humidity of the air is measured by balancing a mass of wool against small stones, the wool increasing in weight as the humidity increases. This is the first known description of a hygrometer in Western civilization.

SUPPLEMENTAL (Scientific Knowledge)

Nicholas Cusa (c.1401–1464, German) asserts that the truth can never be known in total, although it can be approached ever closer through scientific reasoning.

1451

HEALTH SCIENCES (Eyeglasses)

Concave lenses are aligned in spectacles to correct for nearsightedness (for farsightedness, see 1299—HEALTH SCIENCES).

c.1455

SUPPLEMENTAL (Printing; Gutenberg Bible)

Johann Gutenberg (c.1397–1468, German) issues the Gutenberg Bible, the first major type-printed book in the Western world.

1463

PHYSICS (Impetus Theory)

Nicholas Cusa (c.1401–1464, German) applies the impetus theory to the motion of a spherical ball in *De ludo globi.* The motive force gradually lessens following the initial impetus.

1464

MATHEMATICS (Trigonometry; Law of Sines)

De triangulis omnimodis [On triangles of all kinds], by Johannes Regiomontanus (Johannes Müller, 1436–1476, German), is the first systematic European work on trigonometry as a separate subject, divorced from astronomy (see also c.1250—MATHEMATICS). Regiomontanus includes in it a statement of the law of sines for planar triangles (in modern notation: $[\sin A]/[\sin B] = a/b$, where A and B are angles of a triangle and a and b are the sides opposite those angles) and statements of the law of sines and the law of cosines for spherical triangles.

1469

SUPPLEMENTAL (Platonic Academy)

Lorenzo de Medici (1449–1492, Italian) founds the Platonic Academy of Florence, dedicated to the study of Plato and the Pythagoreans, and selects Marsilio Ficino (1433–1499, Italian) as its first leader.

1473

CHEMISTRY (Atomic Theory)

The atomic theories of Democritus, Epicurus, and Lucretius are introduced to modern Europe through publication of Lucretius's poem *De rerum natura* [On the nature of things].

1476

SUPPLEMENTAL (Printing)

William Caxton (c.1421–1491, English) sets up a printing press in Westminster, England. As a result of this and other presses, by the end of the century Europeans possess an estimated nine million books.

1482

MATHEMATICS (Geometry; Euclid)

Euclid's *Elements* are printed, the version used being Johannes Campanus's Latin translation (c.1255—MATHEMATICS).

1482–1515

SUPPLEMENTAL (Translations; Printing)

Aldus Manutius (1450–1515, Italian) devotes himself to getting translated and then publishing the Greek and Roman classics, and to doing so in the form of small, inexpensive books. He employs translators, editors, and proofreaders to accomplish these tasks.

1483

ASTRONOMY (Planets)

Printing of the first Latin edition of the Alfonsine Tables for predicting the positions of the planets. These tables were compiled in the mid-thirteenth century by astronomers under Alfonso El Sabio, appearing in manuscript form at that time (1272—ASTRONOMY).

1489

MATHEMATICS (Symbolism: plus sign, minus sign)

With the usage of "+" and "−" to denote the operations of addition and subtraction, Johannes Widman (c.1462–c.1498, German) advances the operational symbolism of algebra beyond the works of Diophantus and the Hindus. Widman uses the symbolism in his work *Behend und hüpsch Rechnung uff allen Kauffmanschafften*.

1491

MATHEMATICS (Arithmetic; Long Division)

An arithmetic text by Filippo Calandri (fifteenth century, Italian) presents examples of long division in essentially the method still used in the twentieth century.

1492

EARTH SCIENCES (Geomagnetism)

On their way westward from the Canary Islands to the Caribbean, Christopher Columbus (1451–1506, Italian) and his crew note that their magnetic needle no longer properly points in the direction of the North Star. (This is traditionally the first recognition of the unreliability of the magnetic needle to point North. Decades later the difference between the geographic and magnetic north poles is realized.)

EARTH SCIENCES (Globe)

The first globe of the earth is made by Martin Behaim (1459–1507, German). The indicated Atlantic, roughly the correct size in relationship to Europe and Asia, is scattered with presumed islands and bordered on the west by the east coast of Asia, with the undiscovered Americas and Pacific naturally omitted.

1494

MATHEMATICS (Arithmetic; Algebra)

Luca Pacioli's (c.1445–1517, Italian) *Summa de arithmetica* presents an overview of the mathematics handed down from the Middle Ages. Methods are presented for solving equations of the first and second degree, and eight separate methods are presented of simple multiplication of two integral numbers written in Hindu-Arabic numerals, illustrating the general lack of standardization.

1498

SUPPLEMENTAL (Translations: Aristotle)

Aldus Manutius (1450–1515, Italian) completes publication of a five-volume set of the works of Aristotle.

1500

CHEMISTRY (Distillation)
HEALTH SCIENCES (Herbs)

Publication of Hieronymus Brunschwig's (c.1450–c.1512, German-French) *Liber de arte distillandi de simplicibus* [also referred to as: Little book of distillation], where he describes methods and apparatus for treating plants to obtain their medicinal benefits. The plants are placed in water or alcohol baths, after which the liquids are distilled from them.

EARTH SCIENCES (World Map)

Juan de la Cosa (c.1460–1510, Spanish) presents an elaborate map of the world, including on it the routes of Columbus (for whom he was the pilot of the *Santa Maria* in 1492), Vasco da Gama, and others. The map is among the first to indicate the possibility that Asia is separated from the new lands being discovered on the western shore of the Atlantic.

HEALTH SCIENCES (Cesarean Operation)

Jakob Nufer (Swiss) performs the first recorded cesarean operation on a living woman.

c.1500

MATHEMATICS (Hindu-Arabic Numerals)

Hindu-Arabic numerals have finally superseded Roman numerals for most computational purposes.

SUPPLEMENTAL (General Science)

Among his many scientific speculations, Leonardo da Vinci (1452–1519, Italian) considers the possibilities of human flight, seeks a law for the acceleration of falling bodies (erroneously concluding that the distance fallen is proportional to the time of fall), compares the movement of animal bones to the working of levers, compares various human anatomical and physiological features with corresponding features in other animals, describes how the rings visible in the cross section of a tree can be used to determine the age of the tree, designs a diving suit for the purpose of drilling holes in enemy ships, and speculates on the nature and size of the heavenly bodies and on the origin of rivers, mountains, and the saltiness of ocean water. Regarding meteorology, Leonardo asserts, contrary to accepted contemporary scientific opinion, that air is a mixture or compound rather than an element. He notes that air is partially consumed in respiration and combustion and that if burning cannot take place in a particular "air" then neither can an animal breathe that air. (Three centuries later it is discovered that both respiration and combustion require oxygen.) He rejects the accepted Renaissance view that fossils are "jokes of Nature," asserting instead that they are remains of real animals, and that, for instance, the existence of fish fossils in the mountains of Northern Italy provides evidence that at one point these mountains had been covered by the sea. The notebooks he fills covering such topics—and including many erroneous as well as correct speculations—reflect the wide range of his Renaissance artistic and scientific interests and will, centuries later, serve as an inspiration to future generations; however, they do not have an impact on the science of his day.

1502

EARTH SCIENCES (Descriptive Geography; New World)

Amerigo Vespucci (1451–1512, Italian) concludes from his voyages to the New World that South America is a separate continent, not a part of Asia.

EARTH SCIENCES (Stones)

Leonardus Camillus, in *Speculum lapidum,* presents his views on the formation of stones, asserting that three specific cooperating influences are essential: power from God, influences from the celestial bodies, and material qualities of heat, cold, dryness, and dampness.

PHYSICS (Watch)

Peter Henlein (1480–1542, German) constructs a pocket watch, with steel wheels driven by springs.

1507

EARTH SCIENCES (Descriptive Geography; New World)

Maps by Martin Waldseemüller (1470–c.1522, German) in his *Cosmographiae introductio* depict the New World of North and South America as separate from Asia, these being perhaps the first maps to do so. Waldseemüller suggests the name "America" after Amerigo Vespucci (1451–1512, Italian).

1512

CHEMISTRY (Distillation)

Publication of Hieronymus Brunschwig's (c.1450–c.1512, German-French) *Great Book of Distillation,* an enlarged version of his *Little Book of Distillation* (1500—CHEMISTRY).

1513

EARTH SCIENCES (Gulf Stream)

Juan Ponce de Léon (c.1460–1521, Spanish) provides the first known written description of the North Atlantic's Gulf Stream.

c.1513

BIOLOGY (Anatomy)

Leonardo da Vinci (1452–1519, Italian) creates an extensive set of anatomical drawings, with the intention of collecting them into a book. Through dissections of many animals, repeated drawings of the same anatomical parts from various angles, and the construction of models to simulate various bodily functions, he attains a substantial knowledge of human and other animal anatomy, although much of his anatomical work is done in secret and remains unknown to his contemporaries.

1514

ASTRONOMY (Solar System)

Seeking a theory of the planetary motions which is more pleasing mathematically than the complex Ptolemaic system, Nicolaus Copernicus (1473–1543, Polish) shifts the center of the supposed circular orbits from the earth to a point near the sun, hoping thereby to reduce the number of deferents and epicycles required to match the observations. He writes the first version of his revolutionary *On the Revolutions of the Heavenly Spheres* by May 1514, but delays publication until the year of his death, 1543, in the meantime refining and elaborating his theory.

1515

MATHEMATICS (Algebra; Cubic Equations)

Scipione Ferro (Scipio Del Ferro, 1465–1526, Italian) discovers—but will not divulge—a general method of solving cubic equations of the form $x^3 + ax = b$.

1521

EARTH SCIENCES (Deep Sea Sounding)

Efforts are made at deep sea sounding in the Pacific by crew members during an around-the-world voyage led by Ferdinand Magellan (c.1480–1521, Portuguese). They fail to reach the ocean bottom, using 1,200-foot hand lines and thus falling far short of the approximately 12,000-foot ocean depth.

1522

ASTRONOMY (Theodolite)

An early version of the theodolite is constructed by Martin Waldseemüller (1470–c.1522, German) for the purposes of surveying and astronomical observation.

EARTH SCIENCES (Earth Shape)

The general roundness of the earth is verified through the completion of the first circumnavigation of the globe. Begun in 1519 under Ferdinand Magellan (c.1480–1521, Portuguese) with five ships and 230 sailors, the voyage is completed in September 1522 with the return to Spain of one ship and 18 men under Juan Sebastian de Elcano (d.1526, Spanish), the latter having assumed command after the death of Magellan in the Philippines.

MATHEMATICS (Conics)

Johann Werner (1468–1522, German) publishes the first modern treatise on conic sections.

1524

EARTH SCIENCES (Cartography)

Peter Apian (Peter Bennewitz, Petrus Apianus, 1495–1552, German) creates a map projection appropriate for large areas, with latitude circles appearing as parallel straight lines, the prime meridian as a straight line, and the other meridians of longitude as circles with curvature increasing away from the prime meridian. His *Cosmographia,* published in 1524, includes some of the earliest maps of America and is considered the first textbook on theoretical geography.

1530

BIOLOGY (Herbals)

Otto Brunfels (c.1489–1534, German) publishes volume 1 of his herbal *Herbarum vivae eicones* [Living portraits of plants], with volumes 2 and 3 appearing in 1531 and 1536 respectively. The book describes about 230 species of plants, largely through verbatim quotations from earlier works but containing also 40 species first described by Brunfels. Each chapter is devoted to a single plant, providing descriptions of the root, stem, leaves, fruit, blossoms, and especially the medicinal properties and the methods of preparing and administering extracts for medical purposes. However, the book is noted most not for the text but for the illustrations, drawn by artist Hans Weiditz with a detail far surpassing that in most contemporary and earlier scientific works.

Kuchenſchell. Hacketkraut.

FIGURE 2. *A woodcut illustration of a pasqueflower delicately drawn by artist Hans Weiditz for Otto Brunfels's* Herbarum vivae eicones [*Living portraits of plants*] *(see 1530—*BIOLOGY *entry). (By permission of the Houghton Library, Harvard University.)*

HEALTH SCIENCES (Syphilis)

Girolamo Fracastoro (c.1478–1553, Italian) writes a description of syphilis and coins the name in his treatise *On Syphilis, or the French Disease.* Philippus Aureolus Paracelsus (c.1493–1541, Swiss) writes about syphilis in *Three Chapters on the French Disease.*

MATHEMATICS (Decimals)

Calculations in a work by Christoff Rudolff (1500–1545, Austrian) reflect a computational understanding of decimal fractions.

c.1530

CHEMISTRY (Zinc)

Philippus Aureolus Paracelsus (c.1493–1541, Swiss) describes the properties of many substances; for instance, he characterizes zinc as little known, not malleable, unique in color, capable of being melted, and not easily mixed with other metals. He defines alchemy as the science of transforming raw materials into useful products.

HEALTH SCIENCES (Disease; Iatrochemistry)

Philippus Aureolus Paracelsus (c.1493–1541, Swiss) applies alchemy to medicine and founds the study of iatrochemistry. He views the primary purpose of alchemy to be the development of effective medicines, encouraging the alchemist to use chemical processes such as distillation, extraction, and solution to extract the healing "essence" from plant, animal, or mineral substances. By making medicines from minerals as well as from plants, he deviates from the Greek practice of using only plants and animals. More importantly, he challenges many ancient medical traditions, in particular opposing the Greek theory that disease is determined by an imbalance among the four humors. Paracelsus argues instead that the human body is a chemical system and that disease is determined by an imbalance among three major principles: mercury, sulfur, and salt. Disease hence can be countered by mineral as well as other medicines, and (again in contrast to Greek practices) specific cures can be found for specific diseases, thus encouraging the study of individual illnesses. Paracelsus begins to make such associations as of paralysis with injuries to the head, cretinism with goiter, and silicosis and tuberculosis as serious occupational hazards for miners. He advances the understanding of many individual diseases, including epilepsy, gout, arthritis, ulcers, and cretinism, and he rejects the concept that nervous illnesses are caused by demons, attributing them instead to physical causes.

1531

SUPPLEMENTAL (Scientific Methods: Observation, Experimentation)

In a work anticipating some of the views of Francis Bacon, Juan Luis Vives (1492–1540, Spanish) condemns the dogmatic acceptance of Aristotelian teachings, encouraging instead a return to the observational emphases expressed by Aristotle and to the greater use of experimentation.

1533

MATHEMATICS (Practical Trigonometry)

Reiner Gemma Frisius (1508–1555, Flemish) modernizes surveying by replacing the pacing out of distances with triangulation techniques. After measuring a single base line, other distances are obtained trigonometrically.

1534

MATHEMATICS (Sine Tables)

Peter Apian (Peter Bennewitz, Petrus Apianus, 1495–1552, German) publishes a table of sines for every minute of arc in his *Instrumentum sinuum sive primi mobilis.*

1535

HEALTH SCIENCES (Pharmacopoeias)

Valerius Cordus (1515–1544, German) publishes *Dispensatorium,* one of the first pharmacopoeias, describing drugs, chemicals, and medical preparations.

MATHEMATICS (Cubic Equations)

Niccolò Tartaglia (1499/1500–1557, Italian) discovers a general method for solving $x^3 + ax = b$, independently of Scipione Ferro (1515—MATHEMATICS), and, like Ferro, seeks to keep the method secret.

1537

HEALTH SCIENCES (Dissections)

Andreas Vesalius (1514–1564, Flemish) departs from tradition by performing dissections himself while instructing his medical classes. Traditionally, the dissections were performed by barber-surgeons while the instructor read from the works of earlier writers, such as Galen.

PHYSICS (Mechanics)

Niccolò Tartaglia (1499/1500–1557, Italian), in *Della nova scientia* [The new science (of artillery)], examines the trajectories of moving bodies, countering and advancing beyond the analysis of Aristotle. In his investigations of the range of projectiles, he correctly states that the range is maximized with a projection angle of 45°, although he does not discover the parabolic shape of the orbit, later discovered by Galileo Galilei (1638—PHYSICS).

1538

BIOLOGY (Mind)

Publication of an early Western work on empirical psychology, the *De anima et vita* of Juan Luis Vives (1492–1540, Spanish). Vives deemphasizes philosophical inquiry into the meaning of the mind and emphasizes instead the examination of mental processes.

1539

BIOLOGY (Bloodletting)

Andreas Vesalius (1514–1564, Flemish) supports the revived classical procedure of bloodletting in his *Epistola docens venam axillarem dextri cubiti in dolore laterali secundam.* Specifically, Vesalius rejects the then more commonly used Muslim method of taking blood from a location distant from the injured part, encouraging the classical procedure of taking it from near the difficulties. Furthermore, he introduces a new, more-scientific method of determining the precise location for the bloodletting.

PHYSICS (Mechanics)

Girolamo Cardano (1501–1576, Italian) objects to portions of Niccolò Tartaglia's analysis of moving bodies (1537—PHYSICS), and correctly asserts that a body can move uniformly in one direction while moving nonuniformly in another.

1540

ASTRONOMY (Comets)

Peter Apian (1495–1522, German) publishes *Astronomicon Caesareum,* in which he includes descriptions of five comets, including the comet later known as Halley's comet, and states that the tails of comets are pointed away from the sun.

ASTRONOMY (Solar System)

Georg Joachim Rheticus (1514–1574, German) publishes a summary of Nicolaus Copernicus's heliocentric perceptions of the solar system in *Narratio prima de libris revolutionum* [First account of the books of the revolutions], while continuing his efforts to persuade Copernicus to publish his more detailed work (1543—ASTRONOMY).

CHEMISTRY (Ether)

Valerius Cordus (1515–1544, German) records what is reputed to be the first definite mention of ether and description of its preparation. Others, including Philippus Aureolus Paracelsus (c.1493–1541, Swiss), are believed to have prepared ether earlier although not to have described any recipe for it. Cordus mixes 6 ounces of rectified spirit of wine with 6 ounces of oil of vitriol, then distils the mixture after allowing it to stand for two months. The result separates into a watery part on the bottom and an oily part presumed to be vitriolic, or sulfuric, ether.

CHEMISTRY (Metallurgy)

Vannoccio Biringuccio (1480–c.1539, Italian), in *De la pirotechnia* [On pyrotechnics], covers much of the field of metallurgy as he describes the practical aspects of mining, the smelting of metals, the casting of cannon and bells, and the preparation of chemicals for treating ores. Biringuccio bases the work on his own observations and experience.

1541

EARTH SCIENCES (Globe)

Gerardus Mercator (1512–1594, Flemish) creates a globe of the earth.

1542

BIOLOGY (Herbals)

In an 896-page, well-illustrated herbal, Leonhard Fuchs (1501–1566, German) summarizes the accepted knowledge regarding roughly 400 German and 100 foreign plants. He introduces his European readers to the pumpkin and Indian corn of the Americas.

1543

ASTRONOMY (Solar System)

Nicolaus Copernicus (1473–1543, Polish) in *De revolutionibus orbium coelestium* [On the revolutions of the heavenly spheres], finally publishes a description of his revised system of the universe, with significant differences from the accepted authority of Ptolemy. In the Copernican system the earth moves, revolving yearly around the center of the universe, which is close to but not at the location of the sun, and rotating daily on its axis. The axis gyrates in a manner which accounts for the precession of the equinoxes. Copernicus retains the uniform circular motions of the Greeks, continuing to use epicycles and deferents to describe the planetary motions. While the moon revolves around the earth, the planets all revolve around the sun.

ASTRONOMY (Stars)

The altered conception of the solar system presented in *De revolutionibus orbium coelestium* (1543—ASTRONOMY) requires Copernicus to alter radically the conception of stellar distances as well. Those believing in a stationary earth, and the consequent full revolution of the sphere of stars every 24 hours, perceived the stars as relatively near. Copernicus, however, has to assume the stars are very far away, since otherwise there would be an observable convergence and divergence in star groupings as the earth revolves around the sun in its yearly orbit. Copernicus does retain the concept of a sphere of fixed stars.

BIOLOGY (Anatomy)

Publication of Andreas Vesalius's (1514–1564, Flemish) *De humani corporis fabrica* [On the fabric of the human body], describing his anatomical observations with splendid illustrations by Jan van Calcar (1499–c.1546, Italian), a pupil of Titian. Having dissected human cadavers, a practice forbidden in the time of Galen, Vesalius is able to correct approximately 200 errors in Galen's anatomical teachings; for example, establishing that human thigh bones are straight rather than curved like those of a dog and the human brain does not contain the *rete mirabile* network of arteries found in the brains of some animals. Vesalius shows the location of human blood vessels and the structure of the heart, and suggests that Galen's presumed pores between the left and right ventricles of the heart do not exist. The book, which also includes chapters on the skeleton, muscles, nervous system, and brain, enrages traditionalists.

PHYSICS (Aristotelian Physics)

Peter Ramus (Pierre de la Ramée, 1515–1572, French) attacks Aristotelian physics in his *Animadversions on Aristotle.*

PHYSICS (Gravity)

With the earth no longer envisioned as stationary and at the center of the universe, the Copernican system (1543—ASTRONOMY) requires a conception of gravity revised somewhat from the Aristotelian notion that objects fall to the earth in response to seeking their natural place at the universe's center. Copernicus suggests that gravity is a tendency of matter to aggregate roughly in spherical form; each localized system has its own private gravity, the gravity on earth not influencing or being influenced by that on the other planets. In an explanation reminiscent of many in earlier works he is seeking to overthrow, he asserts that the rapidly moving earth is able to retain its form because rotation is a natural motion for the earth and no natural motion would destroy the moving object.

SUPPLEMENTAL (Translations: Archimedes)

Many of the works of Archimedes become available in Latin translations. These include attempts to analyze physical problems mathematically and to extrapolate from realizable to ideal conditions.

1544

BIOLOGY (Botany)

Valerius Cordus (1515–1544, German) leaves *Historia plantarum,* including descriptions of 500 newly identified plant species, for posthumous publication.

EARTH SCIENCES (Geography)

Sebastian Münster (1489–1552, German) publishes a 6-book compendium of world geography entitled *Cosmographia* [Cosmography].

MATHEMATICS (Logarithms, Irrationals)

Michael Stifel (c.1487–1567, German) presents the current knowledge regarding arithmetic and algebra, plus some original contributions, in his major work, the *Arithmetica integra.* He emphasizes the need to establish general laws, uses letters of the alphabet to designate unknowns, and employs the "+" and "−" signs earlier introduced by Johannes Widman (1489—MATHEMATICS). In chapter 2 he indicates an early step toward logarithmic computation, correlating a geometric series with an arithmetic one. However, he also argues against the use of irrationals.

MATHEMATICS (Quartic Equations)

Ludovico Ferrari (1522–1565, Italian) solves the quartic equation of the form $x^4 + ax^2 + b = cx$, doing so by reducing the equation to a cubic equation in a new variable y, solving the cubic for y, and then substituting to determine the value of x.

1545

HEALTH SCIENCES (Cauterizing Wounds)

Ambroise Paré (1510–1590, French) completes his *Méthode de traicter les plaies* on treating wounds. He strongly discourages the practice of cauterizing gunshot wounds, having accidentally discovered the faster recovery of uncauterized wounds when he ran out of oil part way through treatment of a group of wounded men.

MATHEMATICS (Algebra)

Both the first major treatment in Western civilization of negative numbers and also the first usage of a complex number in a computation appear in the *Ars magna* [The great art] of Girolamo Cardano (1501–1576, Italian). While including and acknowledging the solutions of the cubic equation by Niccolò Tartaglia (1535—MATHEMATICS) and of the quartic by Ludovico Ferrari (1544—MATHEMATICS), Cardano is the first to find all three roots of a cubic and even indicates the suspicion that all cubics have three roots. However, negatives and imaginaries still prove partial stumbling blocks and both are termed "fictitious," although certain correct rules for their usage are stated, among them the fact that the product of two negative numbers is positive. Cardano provides a method of eliminating the second degree term in a cubic. Symbolism remains limited, with, for example, the equation $x^3 + 6x = 20$ being written as "cubus p 6 rebus aequalis 20."

c.1545

MATHEMATICS (Probability Theory)

Girolamo Cardano (1501–1576, Italian) initiates the study of probability theory with his work *Liber de ludo aleae* [The book on games of chance] (not actually published until 1663), a study of various types of gambling, with warnings against cheaters, the influences of the stars, and the supernatural. Among the many stated results, one of the major correct ones is the power law for the probability of n successes in n independent repetitions of an event: if the event itself has probability p, then the probability of n successive successes is p^n.

1546

EARTH SCIENCES (Mineralogy)

After reviewing and rejecting earlier systems of classification, Georgius Agricola (Georg Bauer, 1494–1555, German) in *De natura fossilium* describes and classifies minerals according to the physical properties of color, luster, odor, shape, taste, texture, transparency, and weight. He also defines and explains mineral form, hardness, friability, smoothness, solubility, fusibility, brittleness, cleavage, and combustibility. His mineral classification, recognized by some as the first empirically derived, comprehensive such system, is part of a broader classification including all inanimate subterranean solids, fluids, and vapors.

EARTH SCIENCES (Mountains)

Georgius Agricola, in book 3 of his *De ortu et causis subterraneorum* (1546—EARTH SCIENCES), considers the origin of mountains, hypothesizing that mountains are produced by water erosion, atmospheric winds, subterranean winds, earthquakes, and fire from the earth's interior. The primary mechanism for most mountains is believed to be the erosive action of moving water, with mountains forming along the banks of deepening stream beds.

EARTH SCIENCES (Origin of Minerals and Rocks)

Rejecting earlier theories of an origin through stellar influences or self-reproduction, Georgius Agricola (Georg Bauer, 1494–1555, German) in *De ortu et causis subterraneorum* advances the theory that minerals and rocks derive from a subterranean "lapidifying juice" (or "mineral-bearing solution"), from which the mineral or stony matter is deposited through either heat or cold. This lapidifying juice arises as water moving through the earth's crust becomes impregnated with mineral matter, then it creates mineral veins through deposition in preexistent cracks or fissures. It also transforms uncompacted earth into solid rocks. Agricola also identifies clastic rocks as rocks produced from rock detritus and suggests that the binding together of the rock detritus derives from the deposition of interstitial matter from the lapidifying juices percolating through the detritus.

EARTH SCIENCES (Volcanoes)

Georgius Agricola suggests in his *De ortu et causis subterraneorum* (1546—EARTH SCIENCES) that the subterranean heat apparent in volcanic eruptions is localized under the volcanic centers and derives from combustion of beds of coal, bitumen, or sulfur, ignited by intensely heated vapors.

HEALTH SCIENCES (Germ Theory)

Girolamo Fracastoro's (c.1478–1553, Italian) treatise *De contagione* includes descriptions of how infections and epidemics can arise, including the concept that disease can be transmitted from person to person through imperceptible organisms. Although his arguments are largely without experimental basis, he does anticipate the nineteenth-century germ theory (HEALTH SCIENCES entries for 1840 and 1862) and does suggest the possibility that the germs (or atoms or seeds) can be transmitted through the air as well as by contact.

MATHEMATICS (Cubic and Quartic Equations)

Niccolò Tartaglia (1499/1500–1557, Italian) publishes his solution for the general cubic polynomial equation in *Quesiti et inventioni diverse,* expressing also his anger with Girolamo Cardano for having published his method of solving quartic equations (1545—MATHEMATICS).

1550

BIOLOGY (Evolution)

Publication of Girolamo Cardano's (1501–1576, Italian) *De subtilitate rerum* on natural history. Portions of the work imply a belief in evolutionary changes.

METEOROLOGY (Air Composition)

Girolamo Cardano (1501–1576, Italian) includes in his *De subtilitate rerum* (1550—BIOLOGY) a division of air into two parts, one being destructive of inanimate objects and supportive of animate ones, and the second being destructive of animate objects and supportive of inanimate ones.

c.1550

CHEMISTRY (Metals)

In an effort to distinguish metals from other solids, Girolamo Cardano (1501–1576, Italian) suggests two properties as the defining criteria for a metal: (1) it can be melted; (2) once melted, it will harden upon cooling.

1551

ASTRONOMY (Solar System)

Erasmus Reinhold (1511–1553, German) prepares *Prussian Tables* of the planetary motions based on the new Copernican theory (1543—ASTRONOMY). The accuracy is comparable to that of the *Alfonsine Tables,* constructed in the thirteenth century and based on the Ptolemaic system (1272—ASTRONOMY).

MATHEMATICS (Trigonometric Tables)

Georg Joachim Rheticus (1514–1574, German) presents tables of the six standard trigonometric functions, listing the functional value to seven places for every ten seconds of arc. Rheticus is also the first to relate trigonometric functions to angles and ratios of the sides of a right triangle rather than exclusively to circle arcs. The work is published, after being completed and edited by Valentin Otho, as the two-volume *Opus palatinum de triangulis* [Canon of the doctrine of triangles] in 1596 (see 1596—MATHEMATICS).

1551–1558

BIOLOGY (Animal Classification)

Konrad Gesner (1516–1565, German) publishes the encyclopedic four-volume *Historia animalium,* helping to found modern descriptive zoology, in which he lists and describes each known animal species, the descriptions including physical appearance, emotions, skills, habits, locale, diseases, uses for mankind, and occurrence in literature.

1552

BIOLOGY (Anatomy)

Bartolomeo Eustachi (c.1510–1574, Italian), with the assistance of artist Pier Matteo Pini, prepares a set of 47 distinguished anatomical illustrations depicting in sequence the abdominal structures, thorax, nervous system, vascular system, muscles, and bones. Eight of the illustrations are included in a later volume by Eustachi (1564—BIOLOGY), but the remaining 39 do not appear in published form until 1714 when the full set of 47 plates is published under the title *Tabulae anatomicae Bartholomaei Eustachi quas a tenebris tandem vindicatas.*

1553

BIOLOGY (Lesser Circulation of the Blood)

Believing, counter to Galen, that the heart's septum cannot be penetrated by the blood, Michael Servetus (Miguel Serveto, 1511–1553, Spanish) advances the theory that the blood travels through the lungs from the right to the left chamber of the heart, thereby becoming one of the first proponents of the pulmonary or lesser circulation of the blood. Aeration of the blood takes place in the lungs rather than the left ventricle as Galen had suggested. Servetus announces his new theory in a small segment of his *Christianismi restitutio* [Restoration of Christianity], a book which, due to the theology presented, leads to an investigation by the Inquisition, followed by condemnation, an escape from prison, capture by followers of John Calvin, a trial, and finally, on October 27, 1553, death by burning.

1554

BIOLOGY (Plant Classification)

Publication of Ulisse Aldrovandi's (1522–1605, Italian) *Herbarium.*

HEALTH SCIENCES (Appendicitis, Endocarditis)
BIOLOGY (Heart, Spinal Canal)

Jean François Fernel (1497–1558, French) includes in his text *Medicina* descriptions of appendicitis, endocarditis, peristalsis, the systole and diastole of the heart, and anatomical details of, for instance, the spinal canal.

1555

BIOLOGY (Comparative Anatomy)

In a section of his *L'histoire de la nature des oyseaux* [History of the nature of birds], Pierre Belon (1517–1564, French) compares the bone anatomy of birds with that of humans.

BIOLOGY (Heart)

In the second edition of *De humani corporis fabrica* (first edition 1543), Andreas Vesalius (1514–1564, Flemish) counters Galen's assertion that blood flows from the right to left chamber of the human heart through the septum, arguing that the septum is too thick and muscular. He does not, however, suggest an alternate route through the lungs, as is done by Michael Servetus (1553—BIOLOGY) and Realdo Colombo (1559—BIOLOGY).

1556

BIOLOGY (Human Anatomy)

Juan de Valverde (c.1520–c.1588, Spanish) publishes the anatomical treatise *Historia de la composición del cuerpo humano* [History of the construction of the human body]. Although partly based on Vesalius's work (1543—BIOLOGY), the book includes several corrections and additions to the earlier work, including a description of the pulmonary circulation of the blood based on experiments of Realdo Colombo (BIOLOGY entries for 1553 and 1559). The book is widely read and used.

EARTH SCIENCES (Metallurgy)

Georgius Agricola's (Georg Bauer, 1494–1555, German) *De re metallica* [On the subject of metals] provides an extensive and illustrated analysis of varied aspects of mining operations, including the relevant geology, smelting techniques, construction, and drainage considerations. It becomes a standard text in metallurgy for over a century. Agricola explains the formation of ore veins in rock beds by attributing them to stream deposition.

1557

CHEMISTRY (Platinum)

In what is believed to be the first definite allusion to platinum, Julius Caesar Scaliger (1484–1558, Italian) briefly mentions a refractory noble metal found in Central America.

EARTH SCIENCES (Origin of Mountains)

Gabriele Fallopio (1523–1562, Italian) follows Aristotle in asserting that mountains arise through the condensation of a dry exhalation emerging from the earth's interior.

MATHEMATICS (Symbolism)

Robert Recorde (1510–1558, English), in *Whetstone of Witte,* becomes the first to use the "=" sign, the "+" sign, and the "−" sign in an English mathematics text.

1558

MATHEMATICS (Integration)

Federico Commandino (1509–1575, Italian) translates into Latin mathematical works of Archimedes, these translations helping to stimulate subsequent interest in integration techniques.

1559

BIOLOGY (Human Embryo)

Realdo Colombo (Realdus Columbus, c.1510–1559, Italian) describes the human embryo.

BIOLOGY (Lesser Circulation of the Blood)

Realdo Colombo (c.1510–1559, Italian), independently of Michael Servetus (1553—BIOLOGY), asserts that the blood circulates from the right chamber of the heart through the lungs to the left chamber. Columbus is the first prominent anatomist to advance this viewpoint, doing so in his book *De re anatomica.*

1560

SUPPLEMENTAL (Scientific Societies)

The first major society for investigating physical phenomena, the Academia Secretorum Naturae (or I Segreti), is established at Naples, with Giambattista Della Porta (1535–1615, Italian) as president.

c.1560

BIOLOGY (Fallopian Tubes)

Gabriele Fallopio (1523–1562, Italian) discovers and describes the tubes, later named the Fallopian tubes, connecting the ovaries to the uterus.

1561

EARTH SCIENCES (Mountain Formation)

Through a dialogue between two friends in a rustic setting contemplating the origin of the mountains within their view, Valerius Faventies presents in *De montium origine* ten causes of mountain formation, attributing specific mountains to specific causes. The ten causes are earthquakes, moisture-induced swelling, the uplifting power of air enclosed within the earth, fire, souls of mountains, stars, erosion, wind, the sun, and man.

1563

BIOLOGY (Teeth)

In a general treatise on the teeth, *Libellus de dentibus,* Bartolomeo Eustachi (c.1510–1574, Italian) describes the tooth's hard outer and soft inner structure, the first and second dentitions, and the various nerve and blood supplies.

HEALTH SCIENCES (Healing)

Ambroise Paré (1510–1590, French) discusses in his book *Cinq livres de chirurgie* [Five books on surgery] many of his medical innovations. Over the course of his career as a surgeon, Paré introduces or helps establish several important practices, including the use of artificial limbs, the tying of severed arteries to stop bleeding after amputation, the use of soothing applications rather than boiling oil for gunshot wounds, and the turning of a child in difficult childbirths to encourage a feet-first delivery.

1564

BIOLOGY (Anatomy)

Bartolomeo Eustachi (c.1510–1574, Italian) publishes treatises on the organ of hearing, *De auditus organis,* the kidney, *De renum structura,* the system of veins, *De vena quae azygos graecis dicitur,* and his earlier *Libellus de dentibus* on the teeth (1563—BIOLOGY), collected under the title *Opuscula anatomica.* His descriptions of the Eustachian tube and valve in *De auditus organis* lead subsequently to their being named after him, although actually discovered two millennia earlier by Alcmaeon of Crotona. This treatise also contains descriptions of the cochlea and the ear's chorda tympani, identified by Eustachi as a nerve. The treatise on the venous system contains a description of the thoracic duct, discovered by Eustachi, and *De renum structura* includes a description of the suprarenal gland, also believed to have been discovered by Eustachi.

1565

BIOLOGY (Plant Classification)

Konrad Gesner (1516–1565, German) leaves at his death a classification of plant genera based on the reproductive organs. The work is finally printed in 1751 as part of *Opera botanica,* a two-volume collection of most of Gesner's botanical works.

MATHEMATICS (Centers of Gravity)

Federico Commandino (1509–1575, Italian) applies integration techniques of Archimedes to the determination of centers of gravity.

1565–1586

BIOLOGY (Fundamental Forces)

Bernardino Telesio (1509–1588, Italian) publishes his nine-volume *De natura rerum juxta propria principia* [On the nature of things according to their own principles], in which he seeks to explain all forms of life on the basis of the tensions and interactions between two presumed fundamental forces: the dry-warm force and the moist-cold force, the former arising from the sky and inducing motion and expansion, the latter arising from the earth and inducing calm and contraction. Telesio strongly encourages basing scientific knowledge on experience and experiment rather than on blind acceptance of scholastic Aristotelianism.

1566

SUPPLEMENTAL (Religion and Science: The Creation)

Within his book *De gemmis,* Franciscus Rueus tries to reconcile religious theories of an instantaneous Creation and geological evidence of continued development, claiming that at the Creation some things were created in their final form but others only in their beginnings, along with principles and secondary causes which would allow them to reach their completed form at some later time.

1568

EARTH SCIENCES (Cartography; Mercator Projection)

Gerardus Mercator (1512–1594, Flemish) publishes a description of a projection for world maps later termed the Mercator Projection. It is particularly suited for navigation in that constant compass bearings are plotted as straight lines. Longitude lines are equally spaced and parallel, whereas latitude lines are parallel with the spacing between consecutive ones increasing poleward.

1570

EARTH SCIENCES (Geographical Atlases)

Abraham Ortelius (1527–1598, Flemish) publishes one of the first atlases of maps, the *Theatrum orbis terrarum.* Containing 53 maps, it is widely used for generations.

MATHEMATICS (Geometry; Euclid)

The first English edition of Euclid's (c.330–c.275 B.C., Greek) *Elements* is published, the translation being by Henry Billingsley (d.1606, English). In a preface, John Dee (1527–1608, English) encourages the wide application of mathematics to other fields of study.

1572

ASTRONOMY (Supernova)

Tycho Brahe (1546–1601, Danish) notices a very bright new star—a supernova—in the constellation of Cassiopeia. The new star quickly becomes the next brightest object in the sky after the sun, moon, and Venus, and remains visible to the naked eye throughout 1573 and the start of 1574. Tycho attempts to determine any measurable parallax, and, finding essentially none, concludes that the new star is well outside the earthly sphere. This finding contributes to the eventual abandonment of the traditional Aristotelian concept of the changelessness of the realm of fixed stars.

HEALTH SCIENCES (Skin Diseases)

Geronimo Mercuriali (1530–1606, Italian) describes various skin diseases in *De morbis cutaneis et omnibus corporis humani excrementis tractatus.*

MATHEMATICS (Complex Numbers)

The first consistent theory of imaginary numbers appears in Rafael Bombelli's (1526–1572, Italian) *Algebra.* Bombelli also presents the first explicit use of imaginaries, in solving the cubic equation $x^3 = 15x + 4$, and he employs continued fractions to approximate square roots.

1573

ASTRONOMY (Supernova)

Tycho Brahe (1546–1601, Danish) publishes *De nova stella,* where he records his observations of the supernova first observed during the previous year (1572—ASTRONOMY).

BIOLOGY (Brain Anatomy; Brain Dissection; Pons Varolii)

Anatomical studies of the brain are described in two letters by Costanzo Varolio (1543–1575, Italian) published without his consent under the title *De nervis opticis* [The optic nerve], which includes also a reply to the first letter by Girolamo Mercuriale. Varolio introduces a new procedure for dissecting the brain, first removing the brain from the skull, then dissecting it from the base up. This allows better observation of the various structures of the brain and in particular leads to Varolio's observation and description of the optic nerve and of the pons Varolii group of nerve fibers.

1576

ASTRONOMY (Calendar Reform)

Luigi Lilio (d.1576, Italian) completes his *Compendium novae rationis restituendi kalendarium* [Compendium of the new plan for the restitution of the calendar], which eliminates the problem of the gradual shifting of the date of Easter, as encountered in the currently used Julian calendar, by adjusting so that years divisible by 100 are no longer leap years unless also divisible by 400. With this reform, the calendar should remain accurate to within one day for over 2,400 years, and the vernal equinox should continue to fall approximately on March

21 for over 3,550 years. The suggested revisions are presented to Pope Gregory XIII who then replaces the Julian calendar with the revised Gregorian calendar (1582—ASTRONOMY).

ASTRONOMY (Observational)

Tycho Brahe (1546–1601, Danish) begins assembling instruments for astronomical observation on the island of Hveen, Denmark, given to him by King Frederick II. He devotes much of the next quarter century to measuring and recording stellar and planetary positions (see 1588—ASTRONOMY).

ASTRONOMY (Stellar Distances)

Thomas Digges (c.1546–1595, English) suggests that the stars are not arranged on a sphere but are scattered throughout space, at varying distances from the solar system. He is the first known to assert that the universe is infinite, which he does in *A perfit description of the coelestiall orbes,* in which he also presents a translation of a portion of Copernicus's *De revolutionibus orbium coelestium* (1543—ASTRONOMY) and in general supports the Copernican theory.

BIOLOGY (Botany)

Charles de L'Écluse (Carolus Clusius, 1526–1609, French) publishes the treatise *Rariorum aliquot stirpium per Hispanas observatarum historia* on Spanish and Portuguese flowers.

1577

ASTRONOMY (Comets)

Tycho Brahe (1546–1601, Danish), Michael Mästlin (1550–1631, German) and other astronomers observe and track a comet accurately enough to conclude that: (1) it moves around the sun and is not a terrestrial phenomenon as conventional belief and Aristotelian theory would suggest, and (2) its path cuts across the spheres of the planetary orbits and hence through the supposed solid crystalline shells of Aristotelian cosmology.

1578

ASTRONOMY (Copernican Theory; Comets)

Michael Mästlin (1550–1631, German) announces his acceptance of the Copernican theory of the earth's revolution around the sun (1543—ASTRONOMY), having been converted through analysis of the orbit of the comet of the previous year (1577—ASTRONOMY).

EARTH SCIENCES (Oceanography; Sea Water)

William Bourne (c.1535–1582, English) discusses various aspects of oceanography in the fifth book of his *Treasure for Travellers,* including processes by which the sea shapes and modifies the coastline, the impact of the moon on diurnal tides, some broad features of the world ocean's current system, and the cause of the saltiness of the sea. Regarding the latter, Bourne discounts the belief that the sun induces the saltiness and suggests instead that the salt content is due to dissolved mineral matter. Bourne justifies his rejection of the former view by noting that reports from explorers indicate a fairly uniform saltiness of the sea irrespective of latitude.

1579

HEALTH SCIENCES (Glass Eyes)

Reputedly the first glass eyes are constructed. (The ancient Egyptians had constructed artificial clay eyes.)

MATHEMATICS (Trigonometry Tables)

François Viète (Vieta, 1540–1603, French) extends the trigonometric tables of Georg Joachim Rheticus (1551—MATHEMATICS), presenting the functional values for every second of arc. He obtains the law of tangents and includes that along with other trigonometric formulae and relationships in his *Canon mathematicus.*

1580

EARTH SCIENCES (Springs and Rivers)

Bernard Palissy (c.1514–1589, French) discusses a wide range of geological and chemical subjects in his *Discours admirables de la nature des eaux et fonteines, tant naturelles qu'artificielles, des metaux, des sels & salines, des pierres, des terres, du feu et des emaux* [Admirable discourses on the nature of waters and fountains, natural as well as artificial, on metals, on salt and salt marshes, on rocks, on earths, on fire, and on enamels], a book written as a dialogue between "Practice" and "Theory." Regarding springs and rivers, Palissy concludes from observations (and in contrast to many then-current theories) that these are fed by rainwater. The rainwater in turn derives through evaporation from the ocean and gathers into clouds, from which it falls back to earth as rain at the time of God's choosing.

c.1580

MATHEMATICS (Algebra)

François Viète (Vieta, 1540–1603, French) introduces into mathematics a more systematic use of letters for unknowns.

MATHEMATICS (Prosthaphaeresis)

Paul Wittich (1555–1587, Silesian), Tycho Brahe (1546–1601, Danish), and other astronomers shorten their calculations by a method of "prosthaphaeresis," whereby trigonometric identities are used to replace products and quotients by sums and differences. (The term "prosthaphaeresis" is Greek, meaning "addition and subtraction.") The method is a forerunner to John Napier's invention of logarithms (1614—MATHEMATICS).

1581

EARTH SCIENCES (Geomagnetism)

William Borough (1536–1599, English) writes a treatise on magnetic compass variations, basing the work largely on observations obtained during several voyages.

EARTH SCIENCES (Geomagnetism)
PHYSICS (Magnetism)

After extensive experimentation, Robert Norman (fl. c.1590, English) publishes *The New Attractive,* an exposition of his researches on magnetism. Included are descriptions of his discoveries that a magnetic needle, in addition to pointing north, also dips with respect to the vertical; and that a freely floating magnet only orients itself (in contrast to propelling itself) in the north-south direction.

1582

ASTRONOMY (Calendar Reform)

Pope Gregory XIII (1502–1585, Italian), accepting the basic calendar revisions devised by Luigi Lilio (1576—ASTRONOMY), introduces the Gregorian calendar to replace the long-used Julian calendar. By this time, after sixteen centuries of use, the Julian calendar has resulted in approximately an 11-day error in the timing of Easter. Hence Gregory eliminates 10 days (October 5–14) in 1582 to provide the immediate needed adjustment, and then accepts Lilio's further scheme of eliminating the leap day in all years divisible by 100 but not divisible by 400. In spite of opposition from the mathematician François Viète (1540–1603, French) and the astronomer Michael Mästlin (1550–1631, German), the revised calendar becomes widely adopted.

1583

BIOLOGY (Plant Classification)

A major early system of plant classification is put forward by Andrea Cesalpino (1519–1603, Italian) in the treatise *De plantis* [On plants]. Cesalpino classifies according to two characteristics: the root and the fruit organs. Lacking roots, lichens and mushrooms are placed at the bottom of the plant hierarchy.

PHYSICS (Statics; Fluid Pressure)

Simon Stevin (Stevinus, 1548–1620, Flemish) originates the study of modern statics with his theory of statical equilibrium. His work in hydrostatics establishes that the depth of a liquid in a container determines the pressure it exerts, with the shape of the container and the actual quantity of liquid being irrelevant.

c.1583

PHYSICS (Pendulum)

Galileo Galilei (1564–1642, Italian) discovers that the period of oscillation of a pendulum is independent of the amplitude.

1584

ASTRONOMY (Solar System)

Giordano Bruno (1548–1600, Italian) defends the sun-centered theory of Nicolaus Copernicus (see 1543—ASTRONOMY) in *Cena de le ceneri* and postulates in *De l'infinito, universo e mondi* an infinite universe and innumerable planetary systems with life. His science is tied very closely to his magico-religious views, heliocentricity, for instance, being a sign of the imminent return of the supposed true Egyptian religion. The earth and other worlds are perceived as living beings infused with magical animation and moving in space. Bruno's magico-religious views lead to imprisonment by the Inquisition in 1591, a lengthy trial, condemnation as a heretic, and finally burning at the stake in Rome in 1600. The specific charges against him are not known, though likely include his beliefs in a pantheistic God, in the nonexistence of absolute truth, and in the relativeness of one's perception of the world to one's location in time and space.

1585

MATHEMATICS (Decimals)

In *De Thiende* [The tithe; frequently referred to by the title of its French translation, *La Disme,* also published in 1585], Simon Stevin (1548–1620, Flemish) presents the first systematic account of decimal fractions and strongly advocates their usage. From the modern standpoint, his notation of $24^{0}3^{1}7^{2}5^{3}$ for 24.375 is a regression from the earlier usage of 24|375 by Christoff Rudolff in 1530. (Not until the French Revolution, in the 1790s, does the large-scale use of decimals, as advocated by Stevin, come into being.)

PHYSICS (Mechanics)

In *On Mechanics,* Giovanni Battista Benedetti (1530–1590, Italian) critiques the Aristotelian theory of motion and notes aspects with which he disagrees.

1586

BIOLOGY (Physiognomy)

In *De humana physiognomonia* [On human physiognomy], Giambattista Della Porta (1535–1615, Italian) attempts to establish a direct correspondence between a human being's inner character and his external bodily shape.

PHYSICS (Falling Bodies)

Simon Stevin (1548–1620, Flemish) publishes results of an experiment in which he dropped two lead balls, one ten times heavier than the other, from a height of 30 feet. The heavy and light bodies fell at the same rate, invalidating yet another notion engrained in Aristotelian philosophy, that being that the heavier body would fall faster.

PHYSICS (Hydrostatics)

Galileo Galilei (1564–1642, Italian) writes the unpublished treatise *La bilancetta* on hydrostatics. Following Archimedes, he analyzes the issue of determining the compositional purity of a substance, specifically the relative weights of gold and silver, by comparing specific gravities, and he suggests an improved hydrostatic balance for making such comparisons.

1588

ASTRONOMY (Solar System)

Using instruments an order of magnitude more accurate than those previously built, Tycho Brahe (1546–1601, Danish) catalogs the positions of the bright stars and the paths of the planets but remains unable to measure a nonzero parallax for a star. Convinced that he can detect any angle as large as one minute of arc and unable to accept as realistic the possibility of distances so great that the parallax angle would be less than one minute, he concludes that the earth cannot be moving. He publishes a new scheme of the universe which is qualitatively a compromise between the Ptolemaic and Copernican systems: the sun revolves around the earth, while the other planets revolve around the sun. The accuracies of planetary positions calculated by each of the three systems—those of Ptolemy, Nicolaus Copernicus, and Tycho Brahe—are all comparable. In fact, the scheme of Brahe is geometrically equivalent to that of Copernicus.

BIOLOGY (Ileocecal Valve)

Gaspard Bauhin (1560–1624, Swiss) describes the ileocecal valve, also named Bauhin's valve.

BIOLOGY (Plant Physiognomy)
HEALTH SCIENCES (Medicinal Plants)

Giambattista Della Porta (1535–1615, Italian), in a treatise on plant physiognomy resembling his earlier treatise on human physiognomy (1586—BIOLOGY), attempts to establish a direct correspondence between the medicinal properties of a plant and its external shape, as well as identifying resemblances between different plants and other objects such as heavenly bodies, parts of animals, and minerals. Plants resembling human organs are believed to be potentially useful in healing diseases of those organs.

EARTH SCIENCES (Statistical Geography)

Thomas Harriot's (1560–1621, English) *A Brief and True Report of the New Found Land of Virginia* presents one of the earliest large-scale statistical studies.

1589

PHYSICS (Optics; Eyeglasses)

Giambattista Della Porta (1535–1615, Italian) issues an extended, revised version of a 20-book volume first published in 1558: *Magiae naturalis* [Natural magic] on the secrets of nature. Among the practical, theoretical, and experimental material included are the sections on optics: book 1 details the construction and use of lenses for correcting poor vision, while book 17 discusses refraction. The discussion in the latter chapter is the basis of early seventeenth-century claims that Della Porta had essentially invented the telescope. The work can more realistically be viewed as contributing to the theoretical preparation for the later invention (see ASTRONOMY entries for 1608 and 1609).

1589-1592

PHYSICS (Mechanics)

Galileo Galilei (1564–1642, Italian) initiates his experiments on the motions of bodies, dealing with freely falling bodies, inclined planes, projectiles, and pendulums. In this period he writes the unpublished treatise *De motu* in which he attacks Aristotelian physics, uses the buoyancy principle of Archimedes to develop a theory of falling bodies, and begins to relate vertical fall to descent along inclined planes and circular arcs. At this point he has not yet abandoned the concept of an earth-centered universe, and he has not yet formulated clear conceptions of inertia or acceleration (see PHYSICS entries for 1604, 1609, and 1638 for Galileo's further developments of these topics).

PHYSICS (Projectile Motion)

In dealing with projectile motion, Galileo Galilei (1564–1642, Italian) separates the horizontal and vertical components of motion and is thereby able to determine that the path followed by a projectile is a parabola.

1590

EARTH SCIENCES (Bering Land Bridge)

In *The Natural and Moral History of the Indies,* José de Acosta (1539–1600, Spanish) postulates that the Indians arrived in the New World by migrating from Asia over a bridge of land (the Bering Land Bridge) in high northern latitudes.

1591

MATHEMATICS (Algebra)

In *In artem analyticam isagoge* [Introduction to the analytic art; published in 1591] and *De aequationum recognitione et emendatione* [On the review and correction of equations; written in 1591, published posthumously in 1615], François Viète (Vieta, 1540–1603, French) makes such improvements in algebraic symbolism that he is often referred to as the "father of algebra." The modern equation $3x^3 - 6x^2 = 4x + 5$ is written by Viète as $3Acu - 6Aq$ aequatur $4A + 5$. Viète's work is additionally important to the transition from ancient and medieval mathematics to modern mathematics because of his significant advances in generality. For example, he performs a linear transformation to remove the second degree term in solving quadratic, cubic and quartic equations and recognizes the importance of the generality of the procedure. As noted above (c.1580—MATHEMATICS), he also introduces the more important general procedure of employing letters to stand for numbers. Although he rejects negative numbers, thereby retrogressing in a sense from some earlier mathematical works, he does present a uniform method for numerically solving algebraic equations, solves an equation of degree 45, expresses $\sin n\phi$ as a polynomial in $\sin \phi$ and $\cos \phi$, and systematically applies algebra to trigonometry, obtaining many of the fundamental trigonometric identities.

1592

METEOROLOGY (Thermometer)

Galileo Galilei (1564–1642, Italian) invents the thermometer, or, more precisely, since its measurements are influenced by pressure, the barothermoscope. The instrument consists of a small glass bulb opened to a slender glass tube which extends downward to a vessel of water. The bulb and upper portion of the tube contain air, while the lower portion of the tube contains water. Since the water in the tube rises as the bulb is cooled, the air temperature is reflected by the water level in the tube.

1593

BIOLOGY (Vision)
PHYSICS (Optics)

Giambattista Della Porta (1535–1615, Italian) expands book 17 of his *Magiae naturalis* (1589—PHYSICS) into *De refractione optices,* describing the properties of refracting lenses and specifically describing the binocular vision of human eyes. Della Porta compares the pupil of the eye to the lens of the camera obscura.

MATHEMATICS (π)

François Viète (Vieta, 1540–1603, French) derives a precise analytical expression for the ratio (π) of the circumference to the diameter of a circle, obtaining π as an infinite product through consideration of regular polygons with 2^n sides inscribed in a circle.

MATHEMATICS (Prosthaphaeresis)

Christoph Clavius (1537–1612, Italian) advances current methods of prosthaphaeresis (c.1580—MATHEMATICS) in his *Astrolabium* by essentially proving the formula $2 \sin A \sin B = \cos (A - B) - \cos (A + B)$ and doing so for all three possible cases of $A + B$ equaling 90°, exceeding 90°, and being less than 90°.

1594

EARTH SCIENCES (Geographical Atlases)

Posthumous publication of the atlas *Atlas sive cosmographicae* begun by Gerardus Mercator (1512–1594, Flemish) in 1585 and completed by his son.

1596

ASTRONOMY (Planetary Orbits)

Based on the belief that God designed the universe according to a simple mathematical plan, Johannes Kepler's (1571–1630, German) *Mysterium cosmographicum* [The mystery of the universe] elaborates the precise placement of Plato's five regular solids (cube, tetrahedron, octahedron, dodecahedron, icosahedron) between the spheres of the orbits of the six known planets.

MATHEMATICS (π)

Ludolph van Ceulen (1540–1610, Dutch) calculates the ratio (π) of the circumference to the diameter of a circle to 20 decimal places, using a method of Archimedes.

MATHEMATICS (Trigonometric Tables)

Georg Rheticus's two-volume *Opus palatinum de triangulis* (1551—MATHEMATICS) is published, after being completed and edited by Valentin Otho (c.1550–1605, German). In it are tables of the values of the six standard trigonometric functions for every ten seconds of arc.

1597

CHEMISTRY (General)
HEALTH SCIENCES (Chemical)

Andreas Libau (Libavius, 1540–1616, German) describes a wide range of chemical methods and preparations in the popular chemical textbook *Alchemia.* In addition to other chemical works as well, Libau writes medical texts, emphasizing, like Philippus Aureolus Paracelsus before him (c.1530—HEALTH SCIENCES), the importance of chemistry for medicine, but criticizing Paracelsus's mysticism.

1598

ASTRONOMY (Instruments)

Tycho Brahe (1546–1601, Danish) presents descriptions of his instruments and astronomical discoveries in *Astronomicae instauratae mechanica.* Among the most spectacular instruments depicted are a 19-foot quadrant and a five-foot-diameter celestial globe, both of which were constructed starting in 1569.

1599

EARTH SCIENCES (Cartography)

Edward Wright (1561–1615, English) places the Mercator projection (1568—EARTH SCIENCES) on a mathematical basis and proceeds to construct navigational charts based on it, providing details in his *Certaine Errors in Navigation.*

1600

ASTRONOMY (Mars)

Johannes Kepler (1571–1630, German) becomes Tycho Brahe's (1546–1601, Danish) assistant at Brahe's Prague observatory and begins examining data on the eccentric orbit of Mars.

ASTRONOMY (Solar System)

Seeking a physical explanation for the planetary motions, William Gilbert includes in *De magnete* (1600—EARTH SCIENCES) a magnetically based concept of the solar system, with each of the individual bodies influencing each of the others through magnetic forces. Gilbert agrees with Tycho Brahe that the planets revolve round the sun, while the sun revolves round the earth (1588—ASTRONOMY). However, he agrees with Nicolaus Copernicus that the stars are stationary and that the earth rotates daily on its axis (1543—ASTRONOMY). He suggests that between the individual planetary atmospheres there is likely a vacuum.

EARTH SCIENCES/PHYSICS (Geomagnetism)

William Gilbert's (1544–1603, English) *De magnete* [Concerning the magnet; extended title *De magnete magneticisque corporibus et de magno magnete tellure physiologia nova*] extends the experimental work of Robert Norman (1581—PHYSICS) and postulates that the earth, underneath the external layer of sediment, is a giant magnet and that gravity is simply a realization of that magnetism. Gilbert also shows that the dip of a magnetic needle varies with latitude. However, he erroneously believes that the major compass deviations from true geographic north result from deflections produced by land masses.

SUPPLEMENTAL (Religion and Science)

Giordano Bruno (1548–1600, Italian) is burned at the stake, in Rome, as a heretic (see 1584—ASTRONOMY).

1600–1601

BIOLOGY (Larynx; Ear)

Giulio Casseri (c.1552–1616, Italian) publishes *De vocis auditusque organis historia anatomica,* containing two anatomical treatises, one on the larynx and the other on the ear. Both discuss the anatomical structure of the human organ and compare it to the corresponding organ in other animals, including mammals, birds, amphibians, and insects. The first also discusses phonation, whereas the second also discusses acoustics, the physiology of hearing, and the possible connection between the shape of the earlobe and criminal tendencies.

1602

ASTRONOMY (Star Locations)

Tycho Brahe (1546–1601, Danish) provides locations of 777 stars with far greater accuracy than previously achieved and details his observations of the 1572 supernova in his posthumous *Astronomiae instauratae progymnasmata,* edited by Johannes Kepler (1571–1630, German). The work also contains estimates of the diameters of the sun, the moon, the planets, and the supernova, plus revisions to the theory of solar motions and significant advances in the theory of lunar motions.

1603

ASTRONOMY (Star Identifications)

Johann Bayer (1572–1625, German) attempts in his *Uranometria* to provide unambiguous identifications of each star visible to the naked eye. Using as his major divisions the 48 constellations from Ptolemy's *Syntaxis,* he labels each star in a given constellation by a letter, using the Greek alphabet for up to 24 stars of a constellation and the Latin alphabet for the remaining stars in constellations with more than 24 stars. This significant simplification over the confusing and unsystematic earlier naming of stars provides the basis of a stellar nomenclature which, in spite of flaws, continues in usage in the late twentieth century. Bayer also includes in *Uranometria* stellar maps of 12 newly discovered constellations visible from the Southern Hemisphere and recently identified by a Dutch navigator.

BIOLOGY (Human Veins)

Girolamo Fabrici (Hieronymus Fabricius, c.1533–1619, Italian) describes the structure and placement of valves in the veins hindering the flow of blood outward from the heart. He concludes that the function of the valves is to prevent the blood from accumulating too rapidly in the hands and feet; it remains for his pupil William Harvey (1628—BIOLOGY) to conclude that the blood in the veins is not flowing from but toward the heart. (Fabrici's work is presented in 1603 in the 24-page *De venarum ostiolis* [On the valves in the veins].)

EARTH SCIENCES (Rock Crystal)

Andreas Baccius notes that, in contrast to ice, which floats, rock crystal is heavier than water. He uses this contrast to counter the popular viewpoint that rock crystals are hardened pieces of ice.

HEALTH SCIENCES (Pulsilogium)

Santorio Santorio (Sanctorius, 1561–1636, Italian) describes the pulsilogium for counting pulse beats, invented by Santorio by applying the pendulum to medical practice following discussions with Galileo Galilei on Galileo's pendulum experiments (PHYSICS entries for c.1583, 1589–1592, and 1609).

SUPPLEMENTAL (Scientific Societies)

Founding of the Accademia dei lincei [Academy of the lynx-eyed] at Rome. Among its 32 members are Nicolas Fabri de Peiresc (1580–1637, French), Giambattista Della Porta (1535–1615, Italian), and Galileo Galilei (1564–1642, Italian). The society continues until 1630.

40 LIVRE I. DE LA NATVRE

Portraict de l'amas des os humains, mis en comparaiſon de l'anatomie de ceux des oyſeaux, faiſant que les lettres d'icelle ſe raporteront à ceſte cy, pour faire apparoiſtre combien l'affinité eſt grande des vns aux autres.

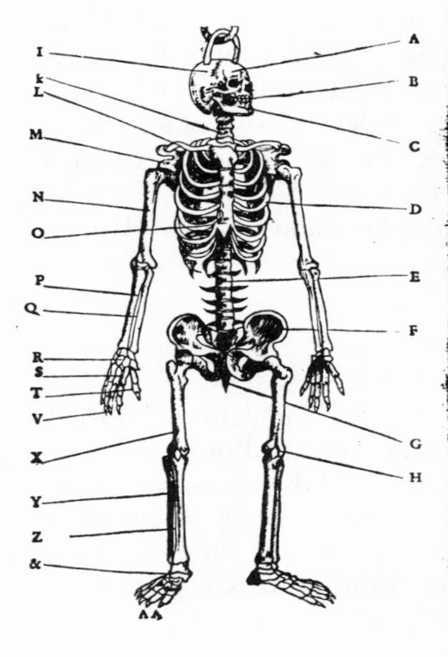

DES OYSEAVX, PAR P. BELON. 41

La comparaiſon du ſuſdit portraict des os humains monſtre combien ceſtuy cy qui eſt d'vn oyſeau, en eſt prochain.

Portraict des os de l'oyſeau.

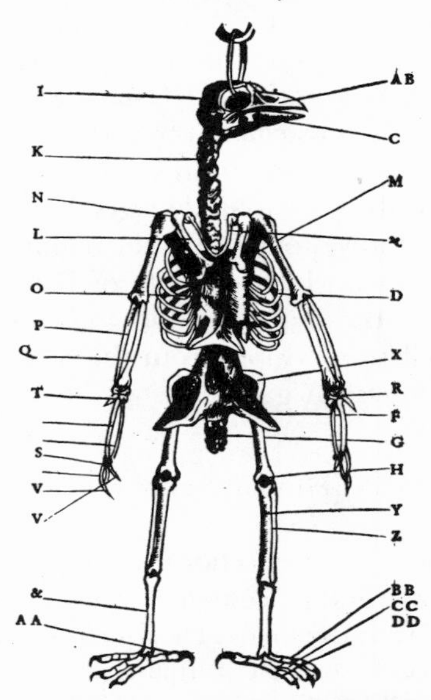

AB Les Oyſeaux n'ont dents ne leures, mais ont le bec tranchant fort ou foible, plus ou moins ſelon l'affaire qu'ils ont eu à mettre en pieces ce dont ils viuent.

M Deux pallerons longs & eſtroicts, vn en chaſcun coſté.

L'os qu'on nommé la Lunette ou Fourchette n'eſt trouvé en aucun autre animal, hors mis en l'oyſeau.

D Six coſtes, attachees au coffre de l'eſtomach par deuāt, & aux ſix vertebres du dos par derriere.

F Les deux os des hanches ſont longs, car il n'y a aucunes vertebres au deſſoubs des coſtes.

G Six oſſelets au cropion.

H La rouelle du genoil.

I Les ſutures du teſt n'apparoiſſent gueres ſinon qu'il ſoit boully.

k Douze vertebres au col, & ſix au dos.

d iii

FIGURE 3. *Coded comparisons between the skeletons of a man and a bird in Pierre Belon's* L'histoire de la nature des oyseaux *[History of the nature of birds] (see 1555—BIOLOGY entry), an early study in comparative anatomy. (By permission of the Houghton Library, Harvard University.)*

1604

ASTRONOMY (Nova)

Johannes Kepler (1571–1630, German) observes and writes an initial description of a new star (nova), further (see 1572—ASTRONOMY) discrediting the Aristotelian notion of the fixed sphere of stars.

BIOLOGY (Fetus; Umbilical Cord)

Girolamo Fabrici (Hieronymus Fabricius, c.1533–1619, Italian) presents in *De formato foetu* a description and discussion of the fetus along the lines of earlier works of Aristotle and Giles of Rome (c.1276—BIOLOGY) but adding also a detailed description of the blood system within the umbilical cord.

CHEMISTRY (Antimony)

Basil Valentine's (Johann Thölde, fl. 1604, German) *Triumph-wagen des antimonii* [Triumphal chariot of antimony] is perhaps the first chemical monograph devoted to an individual metal or metalloid. Detailed descriptions are presented of antimony, the preparation of its compounds, and its value in medicine. (Thölde, listed as the editor of the work, claims that it was written by a Benedictine monk named Basil Valentine in the early sixteenth century, but the work may have been written by Thölde himself, using Basil Valentine as a pseudonym.)

MATHEMATICS (Geometry; Infinity)

Johannes Kepler (1571–1630, German) suggests the interpretation that the opposite ends of a straight line meet at infinity and that parallel lines also meet at infinity. This allows a complete continuity between the points of intersection of a line L and the lines through any point P external to L. It also allows Kepler to describe the ellipse, parabola, hyperbola, and circle as transformable from one to the other through continuous change, with no discontinuous jumps.

PHYSICS (Free Fall)

Galileo Galilei (1564–1642, Italian) states in a letter that a freely falling body will fall such that the distance traveled from the moment of rest is proportional to the square of the time elapsed. His assumptions are flawed at the time, specifically in the belief that velocity is proportional to distance, but he is later able to derive the law from the corrected assumption that velocity is proportional to the time elapsed and from a revised conception of uniform acceleration (1609—PHYSICS).

PHYSICS (Optics)

Johannes Kepler (1571–1630, German) publishes *Astronomiae pars optica* [The optical part of astronomy], leading to his later characterization as "the father of modern optics." Working within the perspectivist tradition of Alhazen, Roger Bacon, Witelo, John Pecham, and others, Kepler accepts that sight is possible due to rays emitted from visible objects, with these rays emitted in all directions from each point on the object's surface. He overcomes a major shortcoming of the earlier theories by establishing a logically acceptable one-to-one correspondence between the points on the observed portion of the object and the points on the image produced on the surface of the eye's retina: instead of rejecting all rays hitting the eye nonperpendicularly, as Alhazen had done, Kepler argues that all the rays incident on the eye from any specific point on the object will arrive at a single point on the retina after refraction in the eye's humors. In consequence, the retina receives an unconfused, although inverted, image of the object. Kepler also presents experimental results, examining, for instance, the range of applicability of Ptolemy's direct proportionality between the angle of incidence and the angle of refraction; and he formulates the principle that the intensity of illumination is inversely proportional to the square of the distance from the illuminating source.

1605

ASTRONOMY (Solar System)

In various letters, Johannes Kepler (1571–1630, German) compares the heavenly motions to the workings of a clock, emphasizing that these motions are caused by a force, presumably at least partly magnetic. He supports the Copernican system (1543—ASTRONOMY) rather than that of Tycho Brahe (1588—ASTRONOMY), stating that as the sun is higher ranking than the earth, its motions are unlikely to be caused by the earth.

SUPPLEMENTAL (Scientific Methods: Experiment)

Francis Bacon's (1561–1626, English) *The Advancement of Learning* encourages widespread usage of experimentation.

1606

ASTRONOMY (Nova)

Johannes Kepler (1571–1630, German) publishes *De stella nova* on the new star first observed in October 1604. In addition to his extensive astronomical observations of the star, Kepler includes commentary on the star's astrological significance, urging his readers to heed the sign and repent of their sins.

1608

ASTRONOMY (Telescope)

Hans Lippershey (1587–1619, Dutch) and Zacharias Jansen (1580–c.1630,

Dutch) construct the first practical telescopes, though they use them only for magnifying distant objects on earth, not for examining the heavens. Lippershey offers his "looker" to the Dutch army for military operations.

SUPPLEMENTAL (Scientific Methods: Applied Mathematics)

Joachim Jungius (1587–1657, German) asserts the importance of giving scientific theories a mathematical foundation.

1609

ASTRONOMY/PHYSICS (Gravitation)

Johannes Kepler (1571–1630, German) theorizes that a force of gravity exists which can act across empty space and which has a magnitude related to the amount of matter in the bodies involved.

ASTRONOMY (Planetary Motions)

Johannes Kepler (1571–1630, German) publishes his major work, *Astronomia nova* [New astronomy]. Contained unobtrusively in this book are Kepler's first two laws of planetary motion: (1) the orbit described by each planet is an ellipse with the sun located at a focus, (2) the radius vector joining a planet to the sun sweeps out equal areas in equal time intervals. By adjusting to elliptical planetary orbits, he creates a system more accurate than those of Ptolemy, Nicolaus Copernicus, or Tycho Brahe.

ASTRONOMY (Solar System)

After determining the elliptical orbits of the planets (above), Johannes Kepler (1571–1630, German) attempts to explain the noncircularity by hypothesizing that the sun is a magnet with only one pole while the planets have both north and south poles and an axis oriented constantly in space. Hence the sun attracts the planet during part of its orbit and repels it during the remainder. (In contrast to Galileo Galilei [1613—PHYSICS], Kepler believes that for a body to continue moving it needs the continuing action of a force upon it.)

ASTRONOMY (Telescope)

Upon hearing of Hans Lippershey's invention of an instrument for magnifying distant objects (1608—ASTRONOMY), Galileo Galilei (1564–1642, Italian) combines lenses to construct an instrument with a nine-power magnification, three times better than the magnification of Lippershey's instrument. Galileo continues to improve the instrument, while retaining the basic structure of using a convex lens for an object glass and a concave lens for an eyepiece, until obtaining a 30-power magnification by the end of the year.

BIOLOGY (Bees)

Charles Butler (1559–1647, English) publishes *The Feminine Monarchy, or the History of Bees.*

BIOLOGY (Microscope)

Hans Lippershey (1587–1619, Dutch) and Zacharias Jansen (1588–c.1630, Dutch) invent the compound microscope, attaching lenses to both ends of a central tube, in an arrangement similar to the recently invented telescope (ASTRONOMY entries for 1608 and 1609).

BIOLOGY (Sense Organs)

Giulio Casseri (c.1552–1616, Italian) completes the anatomical work *Pantaesthеseion* consisting of five books on the organs of hearing, sight, smell, taste, and touch. The work is noted for its 34 illustrative plates as well as its popular writing style.

EARTH SCIENCES (Geomagnetism)

Henry Hudson (d.1611, English), on a voyage to North America, verifies William Gilbert's prediction (1600—EARTH SCIENCES) that in the polar regions a magnetic dipping needle aligns itself nearly vertically.

MATHEMATICS (Ellipses)

In order to establish his result that the planets move in elliptical orbits (1609—ASTRONOMY), Johannes Kepler further develops various aspects of the geometry of the ellipse, including these in his *Astronomia nova.* Among his innovations is the introduction of the term "focus" to the study of conic sections.

PHYSICS (Mechanics; Falling Bodies)

Galileo Galilei (1564–1642, Italian) begins a systematic treatise on motion, including motion along inclined planes, free fall, and the motion of pendulums. His conception of acceleration has advanced from his 1604 work (1604—PHYSICS) and he now assumes a direct proportionality between the velocity of a falling body and time. He is highly innovative in the extent to which mathematics enters his reasoning. Setting a equal to the proportionality constant between velocity v and time t, so that $v = at$, he estimates an average velocity of $at/2$, and a fall distance of $(at/2)t = at^2/2$. He tests the result with a ball rolling down an inclined plane, the incline serving to slow the motion and make it more easily measured. His experiments show decisively that bodies do not fall with velocities proportional to their weights, though they also show that heavier objects do fall very slightly faster than lighter ones.

1610

ASTRONOMY (Earth-Moon Distance)

Galileo Galilei (1564–1642, Italian) estimates the distance from the earth to the moon at 60 times the diameter of the earth.

ASTRONOMY (Observational)

Galileo Galilei (1564–1642, Italian) turns his new magnifying instrument (1609—ASTRONOMY) to the sky, making the first known use of the telescope for astronomical observation. Within months he discovers mountains on the moon, many new fixed stars, and four satellites of Jupiter (also discovered by Simon Marius), and also resolves the Milky Way into thousands of individual stars. He quickly publishes *Sidereus nuncius* [Celestial messenger, or Starry messenger] in March, detailing his observations and various theoretical implications, such as the striking evidence provided by the satellites of Jupiter that the earth is not the center of all motion. Many refuse to accept Galileo's interpretations and reject his supposed observations as resulting either from flaws in the lenses of the telescope, or from intervening clouds between the supposed observed object and the observer. Later in the year Galileo discovers phases of the planet Venus, strongly supporting the sun-centered Copernican universe, and the apparent oval shape of Saturn. The rings around Saturn cannot yet be distinguished as rings—this requires more powerful telescopes—but are referred to by Galileo as appendages and, initially, as perhaps two moons.

ASTRONOMY (Orion Nebula)

Nicolas Fabri de Peiresc (1580–1637, French) discovers the Orion nebula.

c.1610

BIOLOGY (Insect Anatomy)

Galileo Galilei (1564–1642, Italian) uses the microscope to study the anatomy of insects.

1611

ASTRONOMY (Telescope)

Johannes Kepler (1571–1630, German) designs an improved telescope and provides a description of it and an analysis of lenses and refracting telescopes in general in *Dioptrice.*

CHEMISTRY (General)

Jean Beguin (c.1550–c.1620, French) publishes the important early chemical text *Tyrocinium chymicum e naturae fonte et mannali experientia depromptum.*

1611–1613

CHEMISTRY (Preparations)

In *Syntagma,* Andreas Libau (1540–1616, German) describes methods of preparing various chemical compounds, including sulfuric and hydrochloric acids.

1612

ASTRONOMY (Andromeda Nebula)

Simon Marius (1573–1624, German) discovers the Andromeda nebula.

ASTRONOMY (Saturn's Rings)

Galileo Galilei (1564–1642, Italian) records his present inability to see the two appendages previously seen (in 1610; see 1610—ASTRONOMY) on either side of Saturn. Later the appendages are again seen by him and others. (The temporary invisibility is explained in 1655 when Christiaan Huygens interprets the apparent appendages as part of a continuous disk, invisible from earth during periods when the earth is in the plane of the disk.)

ASTRONOMY (Sunspots)

Christoph Scheiner (1573–1650, German) publishes a work on his 1611 discovery of sunspots.

PHYSICS (Hydrostatics)

Galileo Galilei (1564–1642, Italian) develops elementary theorems of hydrostatics in his *Discorso intorno alle cose che stanno in su l'acqua* [Discourse on things that float], using principles from Archimedes.

1613

ASTRONOMY (Sunspots)

Galileo Galilei (1564–1642, Italian) writes *Letters on Sunspots,* claiming to have discovered such spots in 1611, independently of Christoph Scheiner (1612—ASTRONOMY). He objects to Scheiner's interpretation that the spots are actually planets, and seeks instead to demonstrate that they are actually a part of the sun and revolve with it. Under Galileo's interpretation, the sunspots refute the generally held belief that the sun is perfect and unchanging. Galileo also includes in the *Letters* his first strong published statement in support of the Copernican system (1543—ASTRONOMY).

METEOROLOGY (Thermometer)

Francesco Sagredo (Italian) constructs several versions of the primitive thermometer invented by Galileo Galilei (1592—METEOROLOGY) and discovers through investigations with them such circumstances as that winter air can be colder than ice or snow.

PHYSICS (Inertia)

Galileo Galilei includes in his *Letters on Sunspots* (1613—ASTRONOMY) clear notions regarding inertia, stating that a body on a horizontal surface and with no external impediments would be indifferent to motion or rest and would maintain itself in whichever state it was initially placed, whether motion or rest. Furthermore, if the initial state was one of motion, then the body would maintain a constant direction. This inertial concept had developed for Galileo from primitive beginnings in his *De motu* (1589–1592—PHYSICS), where he determined mathematically that any nonzero force should be sufficient to set a body in motion along a horizontal plane. By 1607 he had implied that, if once set in motion with a sudden impulse, an unhindered body on a horizontal surface would retain a constant speed without the necessity of a continuing external force. His conclusions apparently derived from mental extrapolations from increasingly friction-free surfaces. In spite of his innovativeness in thinking along the lines of inertial concepts, he is not consistent in his application of these concepts, so that the further development by René Descartes in the 1630s and 1640s (1644—PHYSICS) is a significant advance.

1614

HEALTH SCIENCES (Quantification; Metabolism)

Santorio Santorio (1561–1636, Italian) publishes a book of aphorisms, *De medicina statica aphorismi,* based on his attempted studies of metabolism, then known as "insensible perspiration." He substantiates his arguments with the results of a long-term experiment in which he himself served as the main participant. During the duration of the experiment, Santorio consumed food and drink only while sitting in an elaborate weighing chair, the purpose being to quantify human weight increases according to food intake.

MATHEMATICS (Logarithms)

In *Mirifici logarithmorum canonis descriptio,* John Napier (1550–1617, Scottish) simplifies the representation of decimal fractions and introduces logarithms, which are accepted almost immediately. In the absence of any exponential notation or concept of bases, it required 20 years for Napier to develop his system, and in the resulting scheme, the logarithm of n becomes what in twentieth-century notation is $10^7\log_e(10^7n^{-1})$. The fundamental idea of relating an arithmetic and a geometric series is physically represented by Napier through the motion of two points on separate straight lines, one point moving with uniform velocity, the other with accelerated velocity. Napier becomes aware of the system of prosthaphaeresis used by many astronomers to speed their calculations (c.1580—MATHEMATICS) while working on the idea of logarithms.

1615

MATHEMATICS (Calculus)

Johannes Kepler (1571–1630, German), in *Stereometria doliorum vinorum* [Measurement of the volume of wine casks], employs primitive integration techniques in attempting to find volumes of bodies with curved surfaces, his researches in this area having been spurred by comparison of the current methods used to find the volume of wine casks with the work of Archimedes on volume measurement. Kepler views solids as composed of infinitesimal pieces and proceeds to determine volumes of various solids of revolution, some not considered by Archimedes.

1616

BIOLOGY (Circulation of the Blood)

William Harvey (1578–1657, English) attributes the internal movement of human blood to a mechanical cause when, in a lecture to the Royal College of Physicians, he compares the heart to a pump and attributes the blood's motion to the heart's muscular contraction.

EARTH SCIENCES (Sea Ice)

William Baffin (1584–1622, English), searching for the Northwest Passage under the command of Robert Bylot, records the existence and recurring nature of the North Water polynya, an open water region in the midst of the pack ice of northern Baffin Bay.

SUPPLEMENTAL (Religion and Science; Astronomy)

Nicolaus Copernicus's *On the Revolutions of the Heavenly Spheres* (1543—ASTRONOMY) is, for the first time, placed on the Index of forbidden books by the Catholic church, which forbids, in particular, the teaching of a sun-centered universe. Galileo Galilei (1564–1642, Italian) is summoned before the Inquisition for teaching the sun-centered theory and for suggesting that it is not the Scriptures but misinterpretations of them which have led to the supposition that the Bible confirms the geocentric theory. Galileo is dismissed with a warning to stop supporting the Copernican viewpoint.

1617

CHEMISTRY (Quantification)

Having carefully weighed the combining substances, Angelo Sala (1576–1637,

Italian) prepares copper vitriol from copper, sulfuric acid, and water, then decomposes the result into its initial components, finding the individual weights to be in the same proportions as initially, a result not necessarily expected from a traditional, Aristotelian context.

1619

ASTRONOMY (Planetary Orbits)

Johannes Kepler's (1571–1630, German) *Harmonice mundi* [The harmony of the world] includes his third law of planetary motion: the square of the period of a planet's revolution is proportional to the cube of its mean distance from the sun.

BIOLOGY (Vision)

In *Oculus,* Christoph Scheiner (1573–1650, German) summarizes his research on vision.

1620

ASTRONOMY (Comets)

Johannes Kepler (1571–1630, German) publishes a treatise on comets. He describes the tendency for a comet's tail to point away from the sun, and he explains this as a result of the sun's emissions and their effect on sweeping material from the head of the comet outward away from the sun.

ASTRONOMY (Philosophical)

In his *Novum organum* (1620—SUPPLEMENTAL) Francis Bacon rejects the Greek belief in the uniform and circular motion of the heavenly bodies but accepts the belief in the earth as the center of the universe.

MATHEMATICS (Slide Rule; Logarithms)

Edmund Gunter (1581–1626, English) invents Gunter's Scale, a forerunner to the slide rule, with logarithms plotted along a scale for the purpose of multiplying and dividing through the addition and subtraction of line segments. He also writes a treatise on logarithms: *Canon triangulorium.*

SUPPLEMENTAL (Scientific Methods: Induction)

Francis Bacon's (1561–1626, English) widely read *Novum organum* (or, more completely, *Instauratio magna: Novum organum scientiarum*) encourages observation and experimentation, stressing the inductive method of science rather than the a priori method emphasized in medieval scholasticism. His emphasis is on the straightforward collection of facts and the attempt to construct a systematic method of deriving useful conclusions from those facts. He lays little emphasis on creativity or mathematics and in fact strongly opposes Galileo Galilei's attempts to isolate specific measurable aspects of phenomena and construct a mathematical theory around those aspects. Bacon encourages instead a more holistic and qualitative approach, with the central aim being to benefit mankind. He desires a union of the scholarly and craft traditions and feels that such inventions as printing, gunpowder and the compass indicate the superiority of modern knowledge over that of the ancient Greeks.

1621

ASTRONOMY (General)

Johannes Kepler's (1571–1630, German) *Epitome astronomiae Copernicanae* presents a systematic treatment of the field of astronomy, perhaps the fullest such treatment since the *Almagest* of Ptolemy.

ASTRONOMY (Gravitation)

In *Epitome* (1621—ASTRONOMY), Johannes Kepler, reiterating ideas which he had expressed earlier, hypothesizes that force is needed to sustain motion and that hence some force must be acting on the planets. This force, he speculates, originates from the sun, decreases with distance from the sun, can act over a vacuum, and may be magnetic. In contrast to many scientists of the time, Kepler believes much of space to be a vacuum.

BIOLOGY (Embryology)

Girolamo Fabrici (Hieronymus Fabricius, c.1533–1619, Italian) presents detailed sequential illustrations of the developing chick embryo in his *On the Formation of Eggs and Chickens.*

MATHEMATICS (Number Theory)

Claude-Gaspar Bachet de Méziriac (1581–1638, French) publishes Diophantus's *Arithmetica,* including the Greek plus a Latin translation. Although translations had appeared in the sixteenth century, without significant impact, this 1621 edition is read by Pierre de Fermat, with notable consequences for the development of number theory (see MATHEMATICS entries for 1630–1665 and 1670).

PHYSICS (Snell's Law)

Willebrord Snell (Snel, Snellius, 1580–1626, Dutch) discovers Snell's Law of Refraction for light, stating that the sine of the angle of incidence is proportional to the sine of the angle of refraction for any given interface.

c.1621

MATHEMATICS (Slide Rule)

William Oughtred (1575–1660, English) invents the slide rule, through proper positioning of two of Gunter's Scales (1620—MATHEMATICS) and allowance for relative motion.

1622

PHYSICS (Magnetism; Declination)

Edmund Gunter (1581–1626, English) records that a magnetic needle's declination varies temporally at one spot as well as varying spatially.

1623

ASTRONOMY (Comets)
SUPPLEMENTAL (Scientific Methods: Applied Mathematics)

Galileo Galilei (1564–1642, Italian) analyzes three comets that were visible in 1618 in *Il saggiatore* [The assayer] and uses the occasion to elaborate a more general scientific approach to investigating nature. Following the fourth-century B.C. Greek Democritus, he distinguishes measurable, or primary, properties of matter, such as length, mass, shape, and velocity, from nonmeasurable or secondary properties, such as color, taste, odor, and sound (nonmeasurable at that time), and declares that to understand nature we must use the language of mathematics. He confines his studies to the primary qualities, reasoning that these are the ones amenable to mathematical description.

BIOLOGY (Classification; Binomial Names)

Gaspard Bauhin (1560–1624, Swiss), in *Pinax theatri botanici,* introduces binomial names and classifies roughly 6,000 plants. The names include one term for plant genus, one for species.

HEALTH SCIENCES (Tropical Medicine; Scurvy; Yellow Fever)

Aleixo de Abreu (1568–1630, Portuguese) describes various tropical diseases and medical issues in his *Tratado de las siete enfermedades.* Among his notable descriptions are those of scurvy, yellow fever, the Brazilian flea *Tunga pentrans,* and the Guinea worm *Dracunculus medinensis.* His description of yellow fever, which he termed "enfermedad del gusano," is reputedly the first account of the disease and includes mention of the associated headache, fever, thigh pain, vomiting, ulcers, and sudden death.

MATHEMATICS (Calculators)

Wilhelm Schickard (1592–1635, German) constructs a wooden calculating machine that can add and subtract fully automatically and can multiply and divide partially automatically. A year later a copy under construction for Johannes Kepler is destroyed in a fire.

SUPPLEMENTAL (Scientific Methods: Empiricism)

The science undertaken by the utopian society of Tommaso Camparella's (1568–1639, Italian) *Civitas solis* [City of the sun] is dominated by close observation and experimentation.

1624

CHEMISTRY (Gases)

Johann Baptista van Helmont (1579–1644, Belgian) coins the term "gas" to describe any compressible fluid.

MATHEMATICS (Logarithms)

Both Johannes Kepler (1571–1630, German) and Henry Briggs (1561–1630, English) publish tables of logarithms, which by this time are being used extensively. Briggs simplifies John Napier's logarithms (1614—MATHEMATICS) by converting to base 10 calculations and lists the common (or base 10) logarithms—to 14 places—for 30,000 natural numbers (from 1 to 20,000 and from 90,000 to 100,000) in his *Arithmetica logarithmica.*

SUPPLEMENTAL (Dogmatism)

Pierre Gassendi (1592–1655, French), in *Exercitationes paradoxicae adversus Aristoteleos,* strongly rejects the dogmatic teaching of Aristotelian science and philosophy as unquestioned truth.

1625

BIOLOGY (Microscopic Observations)

Publication of drawings by Francesco Stelluti (1577–1640, Italian) from microscopic observations of animate and inanimate objects marks the first publication of results of scientific observations made through a microscope.

1626

BIOLOGY (Anatomy)

Posthumous publication of two anatomical works noted for their impressive illustrations: *Tabulae anatomicae* by Giulio Casseri (c.1552–1616, Italian) and *De humani corporis fabrica* by Adriaan van den Spiegel (1578–1625, Italian) with illustrations by Casseri.

CHEMISTRY (Atomic Theory)

Pierre Gassendi (1592–1655, French) proclaims his adherence to the philosophy of Epicurus and, in particular, to his belief in the atomic structure of matter.

HEALTH SCIENCES (Temperature)

Santorio Santorio (1561–1636, Italian) initiates the practice of measuring human temperatures with a thermoscope.

1627

ASTRONOMY (Planetary Orbits; Star Catalogs)

Johannes Kepler's (1571–1630, German) *Rudolphine Tables* tabulate the motions of the planets based on the observations of Tycho Brahe and Kepler's geometry of the planetary orbits. The tables prove measurably more accurate than the Prussian Tables based on the Copernican system (1551—ASTRONOMY) or the Alfonsine Tables based on the Ptolemaic system (1272—ASTRONOMY). The tables also present positions of 1,000 fixed stars.

MATHEMATICS (Logarithms)

Ezechiel de Decker (fl. c.1630, Dutch) publishes a complete table of the logarithms of all natural numbers to 100,000, having filled in the 20,000–90,000 gap in the table of Henry Briggs (1624—MATHEMATICS).

SUPPLEMENTAL (Scientific Methods: Experiment)

Francis Bacon (1561–1626, English), in the incomplete, posthumously published *The New Atlantis,* describes a sojourn at an imaginary island with a research institute named Solomon's House, using the fictional tale to elaborate his concepts of proper scientific method. Observation and experiment are emphasized; however, it is one group of individuals doing the experiments, another group recording and classifying, another concluding from the experiments, and another determining which new experiment should be performed. Bacon justifiably criticizes Aristotle's tendency to deduce without empirical basis, but he perhaps shifts too radically in the opposite direction, not acknowledging the importance of conjecture or the necessity for selecting which facts to record. *The New Atlantis* is remembered also for the mention of capabilities realized centuries later through the inventions of robots, telephones, tape recorders, and the electric motor.

1628

BIOLOGY (Circulation of the Blood)

William Harvey (1578–1657, English), in *Exercitatio anatomica de motu cordis et sanguinis in animalibus* [An anatomical disquisition on the motion of the heart and blood in animals], aggressively presents his theory of the circulation of blood in humans and other animals. Basing his conclusions on numerous dissections carried out over the past decade on 40 animal species, Harvey describes the heart as a pump that sends blood into the arteries, from which it returns in the veins via presumed anastomoses between the arteries and veins. He proceeds to examine how much blood the heart pumps out with each beat and determines that the weight of blood pumped in two hours far exceeds the weight of a human adult, substantiating the thesis that the blood returns to the heart. Harvey traces most of the circulation but is unable to see the capillaries connecting the veins to the arteries (see 1660—BIOLOGY), and hence hypothesizes unseen anastomoses. A major incentive for the theory derived from his recognition that the valves discovered by Girolamo Fabrici (1603—BIOLOGY) regulate the direction, not just the volume, of blood flow. Although the theory counters the ancient authority of Galen and does not contain the classical emphasis on purpose, Harvey does discuss how the circular nature of the blood's motion is analogous to such other circular motions—deemed natural by Aristotle—as those of the heavenly bodies and the evaporation/condensation cycle occurring on earth. Harvey's work is considered by many to be the first major application of mathematics and experiment to biology. Although unable to carry out vivisections on humans, he experiments with several other animal species having four-chambered hearts and extrapolates the results to human beings.

1629

MATHEMATICS (Algebra)

In *L'Invention nouvelle en l'algèbre* [New invention in algebra], Albert Girard (1595–1632, Dutch) accepts negative numbers as roots of equations and correctly conjectures, though with limited basis, that if an equation of degree n has r real roots, where $r < n$, then it will also have $n - r$ imaginary roots.

MATHEMATICS (Analytic Geometry)

Pierre de Fermat (1601–1665, French) begins development of the study of analytic geometry, including the determination of the general equation of a straight line and the equations of a circle, ellipse, parabola and hyperbola. In some respects his geometry is more complete and systematic than the later work (1637—MATHEMATICS) of René Descartes; however, Fermat does not publish his results, and hence his work in analytic geometry does not pass into the mainstream of mathematical development, as Descartes's will. A treatise by Fermat on analytic geometry is finally published posthumously in 1679. (Neither Fermat nor Descartes allows negative numbers.)

1630

ASTRONOMY (Telescopes)

Christoph Scheiner (1573–1650, German) constructs the first telescope to use more than one convex lens, following a design given by Johannes Kepler (1611—ASTRONOMY).

CHEMISTRY (Calcination)

Jean Rey (c.1582–c.1645, French) asserts that the cause of the well-known weight increase of a metal when calcinated is that absorption of air takes place during the process.

MATHEMATICS (Circular Slide Rule)

Richard Delamain (died c.1645, English) writes the 32-page pamphlet *Grammelogia, or the Mathematicall Ring,* in which he describes a circular slide rule, leading to a priority dispute with William Oughtred, who earlier invented the rectilinear slide rule (c.1621—MATHEMATICS) but did not publish a description of the circular rule until later (1632—MATHEMATICS).

1630–1665

MATHEMATICS (Number Theory)

Pierre de Fermat (1601–1665, French) initiates the study of modern arithmetic, or number theory, stating many conjectured theorems and introducing proof by infinite descent (see MATHEMATICS entries for 1637, 1640, 1659, and 1670).

1631

ASTRONOMY (Transit of Mercury)

Pierre Gassendi (1592–1655, French) observes a transit of Mercury across the disk of the sun, describing the event the following year in his *Mercurius in sole visus.*

MATHEMATICS (Arithmetic; Algebra)

William Oughtred (1575–1660, English) writes a short (100 pages), compact treatise on arithmetic and algebra, *Clavis mathematicae,* summarizing the current status of these two fields and employing an unusual amount of mathematical symbolism.

1632

ASTRONOMY (Solar System)

Written in the form of a dialogue among a supporter of the Copernican system, a supporter of the Ptolemaic system, and a supposedly neutral third party, Galileo Galilei's (1564–1642, Italian) *Dialogo sopra i due massimi sistemi del mundo, tolemaico e copernico* [Dialogue concerning the two chief world systems—Ptolemaic and Copernican] states arguments for and against both theories. Galileo thereby avoids having explicitly to commit himself, although he is clearly presenting his own views through the discussant Salviati, who attacks Aristotelian physics and supports a new physics conforming to the Copernican universe. Among the arguments advanced for the sun's being at the center of the planetary orbits are the following: the distance from any other planet to the earth varies over time; the outer planets are closer to the earth when in opposition to the sun and further when in conjunction; Venus and Mercury are never far from the sun. Galileo attacks the Aristotelian doctrine of the essential differences between the earth and the celestial bodies, entering as evidence recent astronomical discoveries. He retains the ancient notion of circular planetary and lunar motions, in contrast to Johannes Kepler's ellipses (1609—ASTRONOMY). (On a philosophical level, Galileo criticizes the ancient notion that changelessness is a sign of perfection.)

MATHEMATICS (Circular Slide Rule)

William Oughtred (1575–1660, English), inventor of the rectilinear slide rule (c.1621—MATHEMATICS), describes the circular slide rule in his *Circles of Proportion and the Horizontal Instrument.*

1633

SUPPLEMENTAL (Religion and Science: Copernican Theory)

The Roman Inquisition, its members angered by the 1632 publication of *Dialogue Concerning the Two Chief World Systems,* forces Galileo Galilei (1564–1642, Italian) publicly to recant the Copernican theory that the earth revolves around the sun and sentences him to house arrest.

1634

SUPPLEMENTAL (Science Fiction)

One of the first science fiction tales is published, a story by Johannes Kepler (1571–1630, German) presenting a supposed dream about life on the moon.

1635

EARTH SCIENCES (Longitude; Geomagnetism)

The hope of determining longitude with a compass is weakened when Henry Gellibrand (1597–1636, English) discovers that the magnetic deflection from true north varies temporally as well as spatially.

MATHEMATICS (Integration)

Bonaventura Cavalieri (1598–1647, Italian) expands upon Johannes Kepler's earlier work on infinitesimals (1615—MATHEMATICS), advancing the emerging "calculus" of integration in his *Geometria indivisibilibus continuorum.*

MATHEMATICS (Spherical Trigonometry)

In *Trigonometria plana et sphaerica* [Plane and spherical trigonometry], Bonaventura Cavalieri (1598–1647, Italian) proves that for any triangle on a sphere the sum of the three angles exceeds 180° but is no greater than 450°.

1635–1648

SUPPLEMENTAL (Scientific Conferences)

A series of gatherings arranged by Marin Mersenne (1588–1648, French) brings together such French scientists as Pierre Gassendi (1592–1655), Girard Desargues (1591–1661), Gilles de Roberval (1602–1675), René Descartes (1596–1650), Blaise Pascal (1623–1662), and Étienne Pascal (1588–1651). Mersenne serves as an active intermediary between scientists, encouraging communication and the rapid transmission of scientific results.

1636

PHYSICS (Acoustics)

Marin Mersenne (1588–1648, French) publishes his *Harmonie universelle* [Universal harmony] on the study of music and its physical basis. He includes descriptions of experiments with various instruments and empirically derived rules on, for instance, the contrast in tones produced with strings of different physical characteristics.

1636–1639

MATHEMATICS (Projective Geometry)

Girard Desargues (1591–1661, French) and Blaise Pascal (1623–1662, French) develop the beginnings of synthetic projective geometry.

1637

[As the following entries reflect, René Descartes's *Discours de la méthode,* a work published in 1637, had relevance to many different subject areas. The main entry for the work can be found under 1637—SUPPLEMENTAL.]

ASTRONOMY (Lunar Librations)

Galileo Galilei (1564–1642, Italian) discovers the diurnal and monthly librations of the moon, exposing slightly different portions of its face to the earth.

MATHEMATICS (Analytic Geometry)

René Descartes (1596–1650, French) makes major advances toward the development of analytic geometry in *La Géométrie,* one of three appendices to *Discours de la méthode* (1637—SUPPLEMENTAL). In the appendix, Descartes reformulates geometry analytically by developing conceptually the one-to-one correspondence between curves in the plane and equations in x and y, including the key fact that solving two equations simultaneously yields the points of intersection of the corresponding curves. By thus combining algebra and geometry he produces what he views as a more systematic attack on geometrical problems than is presented in traditional Euclidean geometry. Descartes also discusses properties of curved lines and classifies algebraic curves by their degrees. Although indicating that the analysis can be extended to three variables in three dimensions, he does not do so himself.

MATHEMATICS (Number Theory; Fermat's Last Theorem)

Pierre de Fermat (1601–1665, French) asserts his famous, unproved "Last Theorem," that no natural numbers x, y, z, n with $n > 2$ satisfy $x^n + y^n = z^n$. This is posthumously published in 1670.

MATHEMATICS (Symbolism)

René Descartes (1596–1650, French) revises exponential notation—though only for integral exponents—almost to its twentieth-century form, replacing François Viète's *N, Q, C, QQ, QC,* . . . by x, xx, x^3, x^4, x^6, ... In the previous few years others had approached this notation, with James Hume employing $5a^{iv}$ in 1636, Pierre Hérigone $5a4$ in 1634, and finally Descartes $5a^4$ in 1637. Descartes was likely the first, date unknown, to allow powers higher than the third, that is, powers without the clear geometrical interpretations—lengths, areas, volumes—available for the first three.

MATHEMATICS (Theory of Equations; Descartes's Rule of Signs)

In addition to founding analytic geometry, René Descartes's *La Géométrie* (1637—MATHEMATICS) contains advances in the theory of equations notable especially for their generality. Descartes shows that the number of roots of a polynomial equation can be no more than the degree, and that if a is a root of $f(x) = 0$ then $x - a$ is a factor of $f(x)$ (though without the $f(x)$ notation). He proceeds to assert the rule of signs: $f(x) = 0$ can have no more positive (or "true") roots than the number of changes of sign in $f(x)$ and no more negative (or "false") roots than the number of pairs of successive positive or successive negative signs.

METEOROLOGY (Atmospheric Phenomena; Clouds)

In *Les Météores,* a treatise on meteorology appearing as an appendix to *Discours de la méthode* (1637—SUPPLEMENTAL), René Descartes (1596–1650, French) centers his description of atmospheric phenomena around cloud formation and dissipation, describing also winds, precipitation, thunder, lightning, and the occurrence and positioning of primary and secondary rainbows. Clouds are perceived as consisting of water droplets or small ice particles formed by the coalescence of vapors given off by terrestrial objects. When the droplets or ice pieces become too large, they fall as rain if warm enough, otherwise as snow or hail, the latter resulting after an ice piece has melted and then refrozen. Thunder results from cloud collisions, and lightning from inflammable material between clouds.

PHYSICS (Optics; Speed of Light)

In *La Dioptrique,* one of three appendices to *Discours de la méthode* (1637—SUPPLEMENTAL), René Descartes (1596–1650, French) demonstrates the principles enunciated in the *Discours* by applying them to the study of optics, merging results from physics and geometry, and showing connections between theory and practical application. The appendix is divided into two parts, the first being a description of the organs used in seeing and the second a description of methods by which sight might be improved, including an analysis of improved lenses for the telescope and microscope. Descartes derives Snell's Law of Refraction (1621—PHYSICS) from more basic principles and relatedly shows that if a ball travelling in a straight line path slows in moving from one medium to another, then its linear path is rotated toward the boundary, whereas if the ball accelerates, then the path is rotated away from the boundary. Since the path of a ray of light is rotated away from the boundary when passing from air to water, Descartes reasons that the speed of light must be greater in a dense medium than in a less dense medium. Although conflicting with indications elsewhere by Descartes that the propagation of light is instantaneous, this conclusion that light moves faster through a dense medium is used by Descartes in *La Dioptrique* to support his conception of light as consisting of small particles in rapid motion in the "subtle matter" filling the pores of other bodies. The harder, firmer boundaries containing the subtle matter in dense media allow the quicker passage of light just as a hard surface allows a ball bounced on it to retain more of its motion than does a soft surface.

SUPPLEMENTAL (Scientific Methods: Deduction)

René Descartes (1596–1650, French), in *Discours de la méthode pour bien conduire le raison et chercher la vérité dans les sciences* [Discourse on the method of rightly conducting the reason and seeking truth in the sciences, or, briefly, Discourse on method], analyzes the mathematical-deductive method and outlines his views of the physical world. Descartes encourages derivation of results in all fields through step by step deduction from self-evident truths, emphasizing such deductive methodology over induction from observations. He seeks the establishment of other branches of knowledge on a basis similar to that of Euclidean geometry: deductive, mathematical, rational.

1638

ASTRONOMY (Moon; Extraterrestrial Life)

Making use of telescopic observations and the yet to be fully accepted Copernican astronomy (1543—ASTRONOMY), John Wilkins (1614–1672, English) in *Discovery of a New World* attempts to prove that the moon is populated by rational beings.

MATHEMATICS (Calculus)

Pierre de Fermat (1601–1665, French) takes major strides toward differential calculus, investigating tangents, maxima, and minima by essentially the procedures used today, though without modern terminology or notation. He bases his method of finding tangents on the principle that the tangent is the limit of secants; and he determines the x values of maxima and minima for a simple algebraic function by solving the equation that essentially sets the derivative of the function equal to zero.

MATHEMATICS (Infinite Classes)

In *Mathematical Discourses and Demonstrations on Two New Sciences* (1638—PHYSICS), Galileo Galilei calls attention to the equivalence of two specific infinite classes one of which contains the other, by having discourser Salviatus describe, without the terminology, the one-to-one correspondence between the set of natural numbers (1, 2, 3, . . .) and the set of square natural numbers (1, 4, 9, . . .). This points out a fundamental difference between infinite and finite classes, as no finite class has the same number of elements as a proper subset of itself.

METEOROLOGY (Air Density)

Galileo Galilei (1564–1642, Italian) publishes a description of an experiment to determine the density of air. After pumping excess air into a vessel, he compares the weight of the excess air with its volume as determined by allowing it to escape through a tube and displace water.

PHYSICS (Mechanics)

Galileo Galilei's (1564–1642, Italian) major work on mechanics—*Discorsi e dimostrazioni matematiche intorno à due nuove scienze* [Mathematical discourses and demonstrations concerning two new sciences, or, Dialogues concerning two new sciences]—is published, summarizing his many theories and experiments developed over the past half century on the resistance of bodies and the theory of motion. Among the topics discussed are the concepts of acceleration, friction and inertia and experiments with falling weights, projectiles, and inclined planes. He defines uniform motion, then uniform acceleration, shows that free fall conforms to uniform acceleration, determines that a body falling through air will eventually reach a constant terminal velocity, due to air resistance, and establishes that the path of a projectile is a parabola. The latter results from combining a uniform horizontal motion with an accelerated vertical motion, Galileo having perceived the advantage of theoretically separating the horizontal and vertical components. Although he strongly encourages observation and experimentation, it is probable that many of the experiments discussed were thought experiments not actually carried out physically by Galileo. In contrast to Aristotle, Galileo does not concern himself with attempted explanations of observed motions but instead tries to describe the motions mathematically. He specifically states that for his purposes—the intent being the mathematical description of position as a function of time—the cause of the body's acceleration or lack of acceleration is irrelevant. This neglect of causes is disputed by contemporary scientists and philosophers.

1639

ASTRONOMY (Transit of Venus)

Jeremiah Horrocks (1618–1641, English) makes the first recorded observations of a transit of Venus across the projected sphere of the sun. Horrocks predicts the transit, then observes it on a white screen set up behind a telescope to intercept the focused solar rays.

MATHEMATICS (Projective Geometry)

The architect Girard Desargues (1591–1661, French) develops synthetic projective geometry in his *Brouillon projet d'une atteinte aux évenements des rencontres d'un cone avec un plan* [Proposed draft of an attempt to deal with the events of the meeting of a cone with a plane]. Desargues discusses cross ratio, projective properties of quadrilaterals, Johannes Kepler's principle of continuity (1604—MATHEMATICS), the concept of an asymptote as a tangent at infinity, and other topics of projective geometry. He allows straight lines to extend to infinity in both directions and suggests that a straight line can be viewed as a circle with infinite radius.

MATHEMATICS (Topology)

The origins of topology can be traced to an unpublished manuscript by René Descartes (1596–1650, French) on simple polyhedrons. He shows in essence that for any closed convex polyhedron the sum of the number of vertices and the number of faces is two more than the number of edges.

METEOROLOGY (Rain Gauge)

Benedetto Castelli (1578–1643, Italian) describes in a letter to Galileo Galilei the use of a cylindrical glass container to measure rainfall. This is the first known record of a rain gauge in Western civilization.

1639–1640

MATHEMATICS (Conics)

Blaise Pascal (1623–1662, French) writes his “Essai pour les coniques et génération des sections coniques” [Essay on conics and the generation of conic sections].

1640

ASTRONOMY (Copernicus and Religion)

In *A Discourse Concerning a New Planet,* John Wilkins (1614–1672, English) defends Copernican astronomy and attempts to reconcile it with the Bible.

BIOLOGY (Plant Classification)

John Parkinson (1567–1650, English) describes and classifies nearly 3,800 plants in *Theatrum botanicum.*

EARTH SCIENCES (Tides; Earth’s Rotation)

Jeremiah Horrocks (c.1617–1641, English) begins an observational study of the tides at Toxteth, England, in the hopes of gathering from it convincing evidence of the rotation of the earth. The study is never completed, due to Horrock’s premature death in January 1641.

MATHEMATICS (Geometry; Pascal’s Theorem)

Blaise Pascal (1623–1662, French) publishes the one-page *Essay pour les coniques* in which he presents the theorem that if a hexagon is inscribed in a conic, and P and Q are the points of intersection of two pairs of opposite sides of the hexagon, then either the point of intersection of the third pair of opposite sides is collinear with P and Q, or those two opposite sides are both parallel to PQ. Called *mysterium hexagrammicum* by Pascal, this result is later known as Pascal’s Theorem.

MATHEMATICS (Number Theory)

Pierre de Fermat (1601–1665, French) states in letters several new theorems in number theory. For instance, in a letter to Bernard Frénicle de Bessy he states Fermat’s “Little Theorem” that for any prime p and any whole number n, p divides $(n^p - n)$. This is later proved by Leonhard Euler (1736—MATHEMATICS). Also in 1640, in a letter to Marin Mersenne, Fermat states that a prime of the form $4n + 1$ can be expressed in exactly one way as the sum of two squares (for example, $13 = 4 + 9$) and hence is the hypotenuse of exactly one right triangle with sides of integral length. Furthermore, he generalizes to state that $(4n + 1)^m$ is the hypotenuse of exactly m right triangles with sides of integral length.

MATHEMATICS (Number Theory; Fermat Numbers)

Pierre de Fermat (1601–1665, French) speculates that all numbers of the form $F_t = 2^{2^t} + 1$ (termed Fermat numbers) are prime. (Leonhard Euler later shows the conjecture to be inaccurate [1732—MATHEMATICS].)

PHYSICS (Minimum Principles)

In his *Discourse Concerning a New Planet* (1640—ASTRONOMY), John Wilkins expresses the belief that nature economizes, using the shortest and easiest means available to accomplish its ends. This is a forerunner to the "minimum" principles later to become so important to physics (see PHYSICS entries for 1657, 1682, 1744).

c.1640

CHEMISTRY (Atomic Theory)

A biography of Epicurus by Pierre Gassendi (1592–1655, French) aids in the revival of the atomic theory of the Greeks.

1641

ASTRONOMY (Telescope)

William Gascoigne (c.1612–1644, English) makes two important advances in the telescope: introducing cross hairs to permit accurate alignment and inventing a micrometer for angular measurements. Although he has a working model, his developments remain unknown and hence do not pass into the mainstream of research until they are reinvented by Adrien Auzout and Jean Picard in the 1660s (1666—ASTRONOMY).

PHYSICS (Mechanics)

Pierre Gassendi (1592–1655, French) confirms some of Galileo Galilei's earlier speculations by dropping a weight from a ship both while at rest and while moving and by recording and analyzing the results.

1642

MATHEMATICS (Calculators)

Blaise Pascal (1623–1662, French) constructs a machine capable of adding and subtracting.

1643

ASTRONOMY (Gravitation)

Gilles de Roberval (1602–1675, French) asserts that gravitation is inherent to matter throughout the universe and that the counterbalancing force allowing bodies to remain separated is the resistance of the intervening ether.

METEOROLOGY (Air Pressure; Barometer)

Evangelista Torricelli (1608–1647, Italian) invents the mercury barometer for measuring atmospheric pressure. He finds that upon upending a six-foot tube of mercury into a dish of mercury, the mercury level in the tube falls to approximately 30 inches, and concludes that the 30 inches of mercury exactly balances the weight of the atmosphere on the mercury in the dish. Torricelli correctly predicts that on a mountain summit the height of the column of mercury will be reduced, as the pressure of the air will be decreased by the pressure of the air intermediate between the mountain's foot and summit. The experiments with mercury are precipitated by Torricelli's desire to understand why lift pumps are unable to raise water more than 34 feet; mercury is selected rather than water because of its higher density.

PHYSICS (Vacuum)

Evangelista Torricelli's experiments with the upended tube of mercury (1643—METEOROLOGY) provide the first strong refutation of the generally held view of the impossibility of a vacuum. Torricelli reasons that a vacuum exists in the tube in the space above the 30-inch mercury column. He then confirms that the mercury is being maintained at the 30-inch level by the pressure of the atmosphere, and that the size of the vacuum is irrelevant, by upending a second tube with more volume at the top and finding that the mercury again falls to the 30-inch level.

1644

ASTRONOMY (Cosmology; Vortex Theory)

In *Principia philosophiae* (1644—SUPPLEMENTAL), René Descartes presents hypotheses regarding both the origin and present state of the universe. Vortices are central throughout; for example, spherical bodies such as the sun and earth are perceived as originally formed through the whirling within the primal mass. The solar system is a vast vortex, with the sun at the center and with layers rotating at varying speeds; whereas the earth is seen in this scheme as having evolved from a star to a planet when its speed slowed and it became enveloped by a layer of the sun's vortex system. Each heavenly body contains its own vortex system, though, as in the case of the earth, this system might be revolving within the larger system of another body. Terrestrial gravity is explained as the suction effect of the earth's vortex, and planetary orbits are deemed circular, not elliptical [thus regressing from the work of Johannes Kepler (1609—ASTRONOMY) on this issue]. Descartes rejects the possibility of a void and insists that throughout the universe an object can move only by displacing other objects, visible or otherwise.

EARTH SCIENCES (Composition of the Earth)

René Descartes includes in part 4 of his *Principia philosophiae* (1644—SUPPLEMENTAL) his perception of the composition of the earth, the materials of which he believes to be the same as those composing the heavens. The perceived center of the earth consists of an incandescent, self-luminous nucleus; the middle region consists of an opaque solid; and the outer crustal region has solidified to its present condition as the earth has cooled. Under the influence of solar heat and light, the crust ruptured at various periods early in the earth's history, with portions of the projecting pieces rising above the ocean and forming dry land.

PHYSICS (Conservation of Momentum)

In his *Principia philosophiae* (1644—SUPPLEMENTAL), René Descartes deduces the principle of the conservation of momentum by arguing that all motion—defined as mass times velocity—was given by God at the Creation.

PHYSICS (Inertia)

René Descartes (1596–1650, French) includes in his *Principia philosophiae* (1644—SUPPLEMENTAL) a principle of inertia in which natural motion involves uniform speed in a straight line. His presentation is a decided advance over the less-consistent statements of Galileo Galilei (1613—PHYSICS). Descartes's principle is essentially equivalent to Isaac Newton's first law of motion (1687—ASTRONOMY/PHYSICS).

SUPPLEMENTAL (Scientific Methods: Deduction)

René Descartes (1596–1650, French) publishes *Principia philosophiae* [Principles of philosophy], an elaboration of the second part of his 1637 *Discourse on Method.* As in the earlier work, the scientific method advocated centers on pure thought heavily infused with mathematical reasoning. Descartes differs radically from Francis Bacon (1620—SUPPLEMENTAL), encouraging deduction from general principles rather than Bacon's induction from experiment. Descartes lays more emphasis on mathematics than did Bacon, and, in sharp contrast to Bacon, supports focusing on individual measurable aspects of phenomena, in the manner of Galileo Galilei. However, he criticizes Galileo for not listing fundamental principles on which his results are based.

1645

PHYSICS (Vacuum Pump)

Otto von Guericke (1602–1686, German) constructs an air or vacuum pump, after a sequence of attempts starting in 1635. The pump leads to much experimentation over the succeeding decades, with researchers moving ever closer to a true vacuum and testing how the rarefied space affects heat, sound, light, and, later, electricity and magnetism. The air pump and barometer (1643—METEOROLOGY) excite much interest and wonder as well as experimentation.

c.1645

METEOROLOGY (Air Density)

Otto von Guericke (1602–1686, German) uses his air pump (1645—PHYSICS) to determine air density, by weighing a metal sphere before and after pumping the air out, subtracting, and dividing by volume.

1646

HEALTH SCIENCES (Plague)

Isbrand van Diemerbroeck (1609–1674, Dutch) describes the plague in *De peste* and recommends using burning sulfur to fumigate the rooms of the sick.

METEOROLOGY (Air Pressure)
PHYSICS (Vacuum)

Upon hearing of Evangelista Torricelli's experiments with the mercury barometer (1643—METEOROLOGY), Blaise Pascal (1623–1662, French) repeats the experiments to examine specifically whether a vacuum exists above the mercury column. He also experiments with a 40-foot tube, filled alternately with water and with red wine. The upended tube yields a 34-foot column of water and a 34.6 foot column of wine, both columns being considerably longer than the earlier 30-inch column of mercury. Since the relative heights correspond inversely with the relative densities of mercury, water and wine, Pascal concludes that a precise force is at work and that indeed a void is created in the space above the liquid.

c.1646

EARTH SCIENCES (Subterranean Rivers)

John Greaves (1602–1652, English) concludes from observations of fairly continuous (surface level) inflow of water to the Mediterranean through the Strait of Gibraltar and no apparent outflow there or elsewhere, that there must be subterranean tunnels through which water flows from the Mediterranean to other seas.

1647

ASTRONOMY (Moon)

Johannes Hevelius (1611–1687, Polish) names many of the mountains, craters, and other topographic features of the moon and presents detailed lunar maps in his work *Selenographia.*

ASTRONOMY (Solar System)

Robert Boyle (1627–1691, Irish-English) expresses the general lack of agreement on the proper description of the solar system when he writes in a letter that he is no longer strongly inclined toward the Copernican view and instead feels that it is impossible at the moment to decide among the various rival theories, including the Copernican (1543—ASTRONOMY), the Ptolemaic (second century), and that of Tycho Brahe (1588—ASTRONOMY).

MATHEMATICS (Calculus)

In *Exercitationes geometricae sex* [Six geometrical exercises], Bonaventura Cavalieri (1598–1647, Italian) calculates areas by dividing regions into smaller and smaller parts and summing. The work is an extension of Archimedes's method of exhaustion and employs Johannes Kepler's concepts of infinitely small quantities (1615—MATHEMATICS). Although vague and criticized at the time, Cavalieri's methods are essentially those of the definite integral, and lead toward the development of the calculus later in the century.

PHYSICS (Vacuum)

Blaise Pascal (1623–1662, French) publishes an account of his experiments (1646—PHYSICS) on the vacuum above a column of liquid in *Experiences nouvelles touchant le vide.*

1648

BIOLOGY (Digestion)

In his *Ortus medicinae* (1648—HEALTH SCIENCES), Johann Baptista van Helmont (1579–1644, Belgian) introduces the concept of digestion as being a series of six fermentations transforming the ingested food into flesh. He, like Paracelsus (c.1530—HEALTH SCIENCES), emphasizes the chemical basis of physiological processes.

BIOLOGY (Spontaneous Generation)

Johann Baptista van Helmont indicates in his *Ortus medicinae* (1648—HEALTH SCIENCES) that mice can arise either through spontaneous generation or through the coupling of male and female parents. Either way, the resulting mice will have similar sexual characteristics.

CHEMISTRY (Elements; Gases)

Johann Baptista van Helmont's *Ortus medicinae* (1648—HEALTH SCIENCES) includes a discussion of chemical concepts and experiments. Rejecting the four-element quartet of Aristotelian theory, Helmont considers water to be the primary matter and the basis of all substances. Neither fire nor earth is accepted as an element, although air is. As evidence for his belief that earth is obtainable from water, Helmont offers the results of an experiment in which he watered a tree for five years, with the net 164 lb. weight gain of the tree being attributed to the water inputs. Helmont is one of the first to distinguish other gases from air, isolating several such gases, including oxides of carbon, nitrogen, and sulfur. The vital force of a substance is believed to be contained within gases within the substance.

HEALTH SCIENCES (Disease)

Manuscripts left by Johann Baptista van Helmont (1579–1644, Belgian) are collected under the title *Ortus medicinae* [On the development of medicine] and published posthumously. Van Helmont regards each human organ (and other objects) as possessing an individual "archeus" or vital principle actively directing and governing it. He rejects the traditional view of disease as reflecting a humoral imbalance, suggesting instead that diseases are living beings generally activated by a pathogenic agent from outside the body. Disease occurs upon unfavorable interactions between the archeus of the pathogenic agent and the human archei. In contrast to the Greek concept of a diseased state, van Helmont regards separate diseases—as well as their corresponding pathogenic agents—as distinct and hence to be treated by distinctive remedies.

MATHEMATICS (Projective Geometry)

Abraham Bosse (1602–1676, French) publishes the theorem developed by his friend Girard Desargues (1591–1661, French) stating that the three points of intersection of pairs of corresponding sides of two triangles lie along the same line if and only if the three lines joining pairs of corresponding vertices of the two triangles intersect at one point. Later known as Desargues's Theorem, this result is a fundamental theorem of projective geometry.

METEOROLOGY (Air Pressure)

Blaise Pascal (1623–1662, French) instructs his brother-in-law to measure the height of the column of mercury supported at the base and at the summit of Mount Puy-de-Dôme. A careful experiment is performed, with witnesses and a control tube of mercury left at the base of the mountain and observed periodically to ensure that the height of the mercury at the base is not fluctuating as the second tube is carried to the mountain summit. As predicted, the level of mercury supported in the column is less at the summit, verifying the notion that the column is supported by the pressure of the air and that air pressure reduces with altitude. Pascal describes the results in "Récit de la grande expérience de l'équilibre des liqueurs" and suggests use of the barometer to determine altitudes. He further uses the measurements to estimate the total weight of the atmosphere, calculating a value of 8.3×10^{18} pounds.

SUPPLEMENTAL (Scientific Methods: Intuition)

Johann Baptista van Helmont, in his *Ortus medicinae* (1648—HEALTH SCIENCES), encourages insight as the road to knowledge and rejects Aristotelian logic and deduction.

c.1648

PHYSICS (Fluid Pressure; Pascal's Principle)

Blaise Pascal (1623–1662, French) performs extensive experimentation on fluid pressure, examining the pressure exerted by one fluid on another placed in it in an expandable bag. He states Pascal's Principle that the pressure exerted on an enclosed fluid is transmitted without reduction throughout the fluid. He relates air pressure to liquid pressure.

1650

EARTH SCIENCES (Geography)
METEOROLOGY (Winds)

Bernhard Varen (Bernhardus Varenius, 1622–1650, German) publishes what is to become a standard geographical text for over a century, *Geographia generalis* [General geography]. Varen emphasizes the importance of not simply describing specifics but of relating them to general concepts. Geography is viewed as encompassing all aspects of the surface of the earth, including its geologic and oceanographic features, climate, plant and animal life. Varen advances toward an explanation of the global wind system by suggesting that the air in the equatorial regions is thinned by the sun's heat and in response the cold, heavier air of the polar regions flows equatorward.

HEALTH SCIENCES (Rickets)

In *De rachitide,* Francis Glisson (c.1597–1677, English) describes the symptoms of rickets, including the enlarged heads, the swollen wrists and ankles, the crooked growth of the shank bone, the narrowing of the breast, the decreasing ability to walk or stand, and the general weakening and enervation.

1650–1654

EARTH SCIENCES (Date of Creation)

In his two-volume *Annales veteris et novi testamenti,* Archbishop James Ussher (1581–1656, Irish) dates the creation of earth and man at 4004 B.C., a date later incorporated into various editions of the King James version of the Bible.

1651

ASTRONOMY (Moon)

Giambattista Riccioli (1598–1671, Italian) includes in his *Almagestum novum*

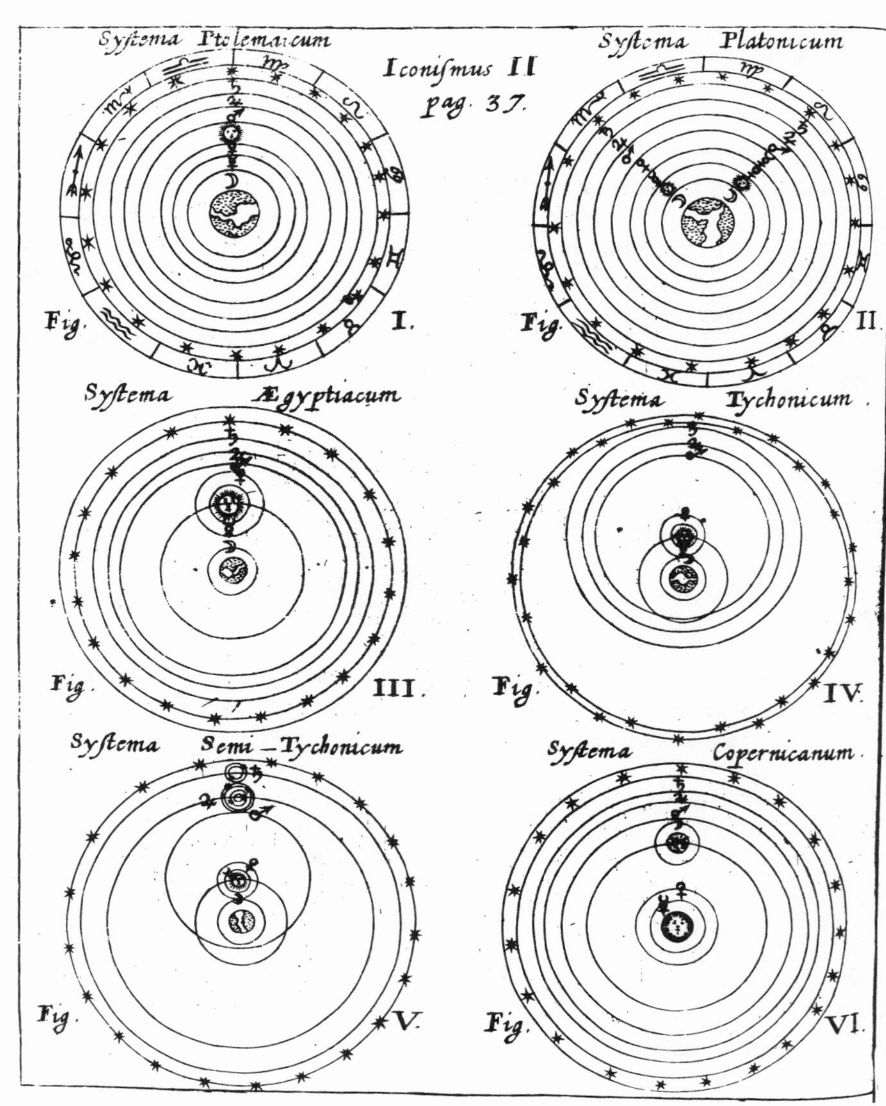

FIGURE 4. *Schematic representations of six major schemes of the universe in vogue in the mid-seventeenth century: the Ptolemaic and Platonic systems with the circular orbits of the planets, sun, and moon all centered on the earth; the Egyptian, Tychonic (see 1588—*ASTRONOMY *entry), and semi-Tychonic (or Ricciolian) (see 1651—*ASTRONOMY *entry) systems with the sun, moon, and sphere of fixed stars revolving around the earth but one or more planets revolving around the sun; and the Copernican system (see 1543—*ASTRONOMY *entry), with the moon revolving around the earth but the earth and all other planets revolving around the sun. (By permission of the Houghton Library, Harvard University.)*

(1651—ASTRONOMY) a new map showing and naming the visible features on the earth side of the moon. Riccioli is aided in the work by Francesco Grimaldi (1618–1663, Italian), who contributes measurements of the heights of lunar mountains and introduces the practice of naming lunar features after noted scientists. Many of the names used in the 1651 map are still used.

ASTRONOMY (Solar System)

Giambattista Riccioli (1598–1671, Italian) publishes his *Almagestum novum* [New almagest], in which he presents a geocentric system of the universe modified from that of Tycho Brahe (1588—ASTRONOMY) and argues its superiority to the Copernican sun-centered universe (1543—ASTRONOMY). In Riccioli's system, Jupiter and Saturn as well as the sun revolve around the earth, while Mercury, Venus, and Mars revolve around the sun.

BIOLOGY (Animal Generation)

William Harvey (1578–1657, English) publishes *Exercitationes de generatione animalium* [Anatomical exercises on the generation of animals], correcting what he denounces as patently erroneous conceptions of animal generation by earlier physicians. He, like Aristotle, rejects the preformationist belief that the embryo begins as a miniature version of the later organism, favoring instead an epigenetic theory, with the growth of the embryo involving a successive emergence and development of new structures. His "ex ovo omnia" [All creatures come from an egg] dictum presented on the title page reflects his opposition to a currently popular viewpoint that the primary generative agent derives from the male parent rather than the female. An original aspect of the work is Harvey's concentration on oviparous generation, including his own anatomical observations and particularly his examination of a hen's egg at various stages of incubation, plus his use of oviparous generation to interpret viviparous generation. He also discusses the embryology of deer and questions whether either menstrual blood or semen has any influence on the formation of the fetus.

BIOLOGY (Circular Motion)

Convinced of the superior nature of circular motion and desirous to place terrestrial beings on a par with the heavenly bodies, William Harvey includes in his *Exercitationes de generatione animalium* (1651—BIOLOGY) descriptions of explicit examples of presumed circularity in terrestrial life, such as the circulation of the blood and the general notion that the succession of individual members of a species is a type of circular motion.

MATHEMATICS (Pre-Calculus)

Andreas Tacquet (1612–1660, Belgian) publishes *Cylindricorum et annularium* [On cylinders and rings], in which he further develops the method of exhaustion of Archimedes for finding areas and volumes.

1652

SUPPLEMENTAL (Religion and Science)

John Cotton (1584–1652, American) declares that it is man's Christian duty to seek an understanding of all God's creations, thus encouraging his Puritan followers toward scientific study.

1654

EARTH SCIENCES (Date of Creation)

John Lightfoot (English) takes the biblically oriented calculations of James Ussher on the date of Creation (1650–1654—EARTH SCIENCES) a step further, declaring that the world was created at 9:00 A.M. on October 26, 4004 B.C.

MATHEMATICS (Arithmetical Triangle)

Blaise Pascal (1623–1662, French) writes, although does not publish, his "Traité du triangle arithmétique" and associated treatises, where he introduces and analyzes the arithmetical triangle, each number of which is calculated from the initial top row and left column of 1's by adding the numbers in the positions to the left and top of the given position. Pascal demonstrates various uses of the arithmetical triangle, including its application in determining the coefficients in the expansion of $(x + y)^n$. (The "Traité du triangle arithmétique" is published posthumously in 1665.)

MATHEMATICS (Probability Theory)

In response to a request by Chevalier de Méré (1610–1685, French) to determine the equable distribution of the pot in an unfinished dice game in which each player has an equal chance of making a point and the winner is the first player to score *n* points, Blaise Pascal (1623–1662, French) consults with Pierre de Fermat (1601–1665, French) and the two together, though with separate methods, inaugurate the mathematical theory of probability.

METEOROLOGY (Air Pressure)
PHYSICS (Vacuum)

Otto von Guericke (1602–1686, German) experiments with two evacuated brass hemispheres to demonstrate the effects of atmospheric pressure and the creation of a near vacuum. This includes a demonstration before the Imperial Diet in which the hemispheres could not be separated by two opposing teams of eight horses each.

1655

ASTRONOMY (Saturn; Titan)

Christiaan Huygens (1629–1695, Dutch) discovers Titan, a satellite of Saturn, and suggests that the appendages of Saturn seen by Galileo Galilei in 1610 (later recognized as the rings of Saturn) are in actuality the edges of a flat disk surrounding the planet. He explains Galileo's 1612 failure to relocate the appendages (1612—ASTRONOMY) by presuming that the earth was at the time of the failure aligned in the plane of the disk. Titan is the first solar-system satellite to be discovered after the moon and the four satellites of Jupiter discovered by Galileo in 1610. It revolves around Saturn every two weeks.

HEALTH SCIENCES (Surgery)

Johann Schultes (1595–1645, German) describes and illustrates surgical procedures and tools in the posthumously published *Armamentarium chirurgicum* [The hardware of the surgeon]. Among the procedures included is one for amputating the female breast.

MATHEMATICS (Integration; Symbolism)

In *Arithmetica infinitorum* [Arithmetic of infinitesimals], John Wallis (1616–1703, English) reports his investigations of the quadrature of curves and introduces the concept of limit, negative and fractional exponents (using x^{-n} for $1/x^n$ and $x^{(1/n)}$ for the nth root of x), and the symbol ∞ for infinity. Although advanced in terms of accepting negative numbers, Wallis considers negatives to be both greater than infinity and less than 0, arguing that as n decreases to 0, $1/n$ increases to infinity and hence as n decreases beyond 0 to negative values, $1/n$ must be greater still. Wallis further advances the emerging "calculus" of integration by replacing the largely geometrical arguments of Bonaventura Cavalieri (1635—MATHEMATICS) by a more arithmetical treatment.

1656

BIOLOGY (Wharton's Duct)

Thomas Wharton (1614–1673, English) describes the duct of the submaxillary salivary gland, later termed "Wharton's duct."

MATHEMATICS (Infinity)

Andreas Tacquet (1612–1660, Belgian) asserts that if x is less than 1 and n is infinite, then $ax^n = 0$.

PHYSICS (Mechanics)

Christiaan Huygens (1629–1695, Dutch) suggests that in elastic collisions between two bodies of masses m_1 and m_2 and velocities v_1 and v_2, the quantity $m_1v_1^2 + m_2v_2^2$ is conserved. Earlier, statements by Galileo Galilei in *Two New Sciences* (1638—PHYSICS) also implied a conservation of mv^2, though for motions of falling bodies. Later, Isaac Newton instead insists on the conservation of mv (momentum).

PHYSICS (Pendulum Clocks)
MATHEMATICS (Dynamics)

Partly in response to the need of astronomers for more accurate time measurement, Christiaan Huygens (1629–1695, Dutch) applies the law of pendulums to time-keeping and constructs the first pendulum clock. To do so, he examines mathematically the small oscillations of a compound pendulum, providing perhaps the first mathematical treatment of a dynamical problem concerned with something other than individual particles.

1657

MATHEMATICS (Algebra)

Jan Hudde (1628–1704, Dutch) allows letter coefficients to stand for negative as well as positive numbers, an innovation soon adopted by others.

MATHEMATICS (Probability Theory; Mathematical Expectation)

Christiaan Huygens (1629–1695, Dutch), aware of the ongoing work of Blaise Pascal and Pierre de Fermat (1654—MATHEMATICS), publishes a treatise on probability, entitled *De ratiociniis in ludo aleae* [On reasoning in games of dice], which introduces the concept of "mathematical expectation."

PHYSICS (Gravitation)

Robert Hooke (1635–1703, English) improves the air pump (1645—PHYSICS) and uses the improved version to test Galileo Galilei's hypothesis that in a vacuum all objects will fall at the same rate. Hooke confirms the hypothesis with a feather and a coin.

PHYSICS (Optics; Least Time)

Pierre de Fermat (1601–1665, French) formulates the principle—later labelled Fermat's Principle—of least time in the study of optics, that is, that light in proceeding from point to point travels along the route which minimizes the time required. From it Fermat derives Snell's Law (1621—PHYSICS) of the refraction of light.

SUPPLEMENTAL (Scientific Societies)
METEOROLOGY (Observing Network)

The Accademia del Cimento is founded, with the encouragement and support of Ferdinand II, grand duke of Tuscany (1610–1670, Italian). Among the important activities the society initiates during its ten years of existence is the establishment of a network of meteorological observers, all having standard forms on which to record temperature, pressure, humidity, the state of the sky, and wind direction. The network begins with observers in several Italian cities, then expands to attain a fuller European representation, adding observers in Paris, Innsbruck, Warsaw, and Osnabruck. Ferdinand II provides many of these observers with thermometers, barometers, hygrometers, and other instruments. He himself designs a primitive condensation hygrometer and considerably improves upon current thermometers by constructing one consisting of bulbs at both ends of a tube, sealed hermetically and partially filled with alcohol. His is perhaps the first thermometer yielding temperature measurements independent of atmospheric pressure.

1658

BIOLOGY (Red Blood Cells)

Jan Swammerdam (1637–1680, Dutch), conducting microscopic examinations of aspects of human anatomy, becomes reputedly the first to observe and record red blood cells.

MATHEMATICS (Calculus)

Several treatises by Blaise Pascal (1623–1662, French) on the cycloid, circle, parabola, and spirals establish his work as a forerunner to both differential and integral calculus. In particular, his *Traité des sinus du quart de cercle* and his *Traité générale de la roulette* [General treatise on the cycloid] influence Gottfried Wilhelm Leibniz in his development of the concept of differentials (1673–1676—MATHEMATICS).

MATHEMATICS (Continued Fractions)

William Brouncker (1620–1684, English) represents the quantity $4/\pi$ as an infinite continued fraction.

MATHEMATICS (Polynomial Equations; Calculus)

Jan Hudde (1628–1704, Dutch) develops various rules regarding polynomials and polynomial equations, including that if the polynomial $g(x) = a_o x^n + a_1 x^{n-1} + \ldots + a_{n-1}x + a_n$ has a relative maximum or minimum at $x = a$, then a is a root of the equation $na_o x^n + (n-1)\, a_1 x^{n-1} + \ldots + 2a_{n-2}x^2 + a_{n-1}x = 0$. Later, with the development of differential calculus, this is translated to $a \cdot g'(a) = 0$ if

$g(a)$ is a relative maximum or minimum (g' = derivative). In this work, Hudde generalizes the earlier methods of Pierre de Fermat for finding maxima and minima for a simple algebraic function (1638—MATHEMATICS).

PHYSICS (Balance Spring)

Robert Hooke (1635–1702, English) devises the balance spring for use in watches.

1659

ASTRONOMY (Saturn)

Christiaan Huygens (1629–1695, Dutch) describes his earlier hypotheses and discoveries concerning the rings and a satellite of Saturn (1655—ASTRONOMY) in *Systema saturnium*.

HEALTH SCIENCES (Typhoid Fever)

Thomas Willis (1621–1675, English) describes typhoid fever.

MATHEMATICS (Calculus)

Under the pseudonym A. Dettonville, Blaise Pascal (1623–1662, French) sets forth methods of solving problems through use of infinitesimals and applies them to determining centers of gravity, quadrature of surfaces, cubature of volumes, and rectification of curves. Like his work of the previous year (1658—MATHEMATICS), his four *Lettres de A. Dettonville* help establish Pascal as having taken major steps toward the development of the calculus.

MATHEMATICS (Infinite Descent)

Pierre de Fermat (1601–1665, French) introduces proof by infinite descent, sketching the method in a letter to Pierre de Carcavi (c.1600–1684, French) and stating that he had used the method to prove his earlier conjecture that a prime of the form $4n + 1$ can be expressed in exactly one way as a sum of two squares (1640—MATHEMATICS). Such a proof proceeds by showing that if any such prime exists without the stated property then a smaller such prime without the stated property also exists. Hence, since the smallest such prime, $5 = 4 + 1$, does have the property, all such primes must have it.

PHYSICS (Mechanics; Centripetal Force)

Continuing his experimentation with pendulums, Christiaan Huygens (1629–1695, Dutch) discovers that a centripetal force is required in order to maintain a body in circular motion and derives a mathematical expression to govern such a force. (Believing in the vortex theory of René Descartes [see 1644—ASTRONOMY], he does not apply the centripetal force concept to the heavens and thereby move a step closer to the idea that gravitation perhaps provides the required force for maintaining the planetary motions.)

1660

BIOLOGY (Blind Spot in the Eye)

Edme Mariotte (1620–1684, French) discovers the blind spot at the back of the eye.

BIOLOGY (Capillaries)

Marcello Malpighi (1628–1694, Italian) observes through a microscope the capillaries in a frog's lungs. These passages allowing the blood to pass to the veins from the arteries are the final links needed to complete William Harvey's classic description of the circulation of the blood (1628—BIOLOGY).

BIOLOGY (Classification)

Marcello Malpighi (1628–1694, Italian) attempts to classify all living creatures on one scale, based on an inverse relationship between the relative size of the respiratory system and the level of the being. Plants are at the bottom of the scale, humans at the top.

METEOROLOGY (Air)

Robert Boyle (1627–1691, Irish-English) describes experiments on atmospheric phenomena, including researches on near-vacuum states obtained with an air pump, in his *New Experiments Physico-mechanical, Touching the Spring and Weight of the Air and Its Effects.*

PHYSICS (Electricity)

Otto von Guericke (1602–1686, German) produces static electricity by rubbing a ball of sulfur. The ease of obtaining electricity encourages others, both amateur and professional, to examine electrical phenomena.

SUPPLEMENTAL (Royal Society)

The Royal Society of London formally receives its charter from Charles II.

1661

ASTRONOMY (Mercury; Star Catalogs)

Johannes Hevelius (1611–1687, Polish) observes and records a transit of Mercury across the solar disk and finishes cataloging 1,500 stars from naked eye observations aided by alidades but not telescopes.

BIOLOGY (Duct of Steno)

During investigations of the glandular system, Nicolaus Steno (Niels Stensen, 1638–1686, Danish) discovers the duct of Steno, the excretory duct of the parotid gland.

CHEMISTRY (Elements)

In *The Sceptical Chymist,* Robert Boyle (1627–1691, Irish-English) disputes Aristotelian and Paracelsan ideas on chemical composition via a dialogue between a sceptical chemist Carneades and supporters of the traditional views. Boyle reiterates earlier definitions of a chemical element as a substance which cannot be constructed from or broken into other substances but which, with the other elements, combines to make up all substances. He then argues that such elements do not exist. The four Greek "elements" of air, earth, fire and water are specifically rejected as chemical elements; and the fundamental nature of the Paracelsan triumvirate of sulfur, salt, and mercury (c.1530—HEALTH SCIENCES) is also rejected. Boyle defines "chemical reaction," distinguishes chemical compounds from mixtures, and supports an early form of the atomic theory, called corpuscular philosophy. (Although Boyle argues against the existence of elements, his definition of "element" becomes the most referred-to sentence of the book.)

EARTH SCIENCES (Observational Oceanography)

The Royal Society of London (1660—SUPPLEMENTAL) prepares a list of recommended oceanographic observations to be made on a 1661–1662 naval expedition to the Mediterranean commanded by Edward Montagu (Earl of Sandwich, 1625–1672, English). Entitled *Propositions of Some Experiments to be Made by the Earl of Sandwich in his Present Voyage,* the list includes the following variables: sea depth, to be determined by sounding with a wooden float tied to a lead weight; the three-dimensional distribution of salinity, to be determined by collecting water samples, then evaporating them and weighing the remaining salt; sea water pressure; water motions in the Strait of Gibraltar; and luminescence. The *Propositions* are regarded as the first planned schedule of strictly oceanographic scientific research, but it is not known whether Montagu actually made the recommended observations.

METEOROLOGY (Air Pollution)

John Evelyn (1620–1706, English) addresses the issue of air pollution in the pamphlet *Fumifugium, or The Inconvenience of the Air and Smoke of London Dissipated,* written to King Charles II. Evelyn proclaims the unhealthfulness and general unpleasantness of the polluted air, mentioning particularly the damage caused to the lungs. He suggests specific steps for improvements, such as moving the factories out of the city and planting trees and shrubbery. His recommendations are not acted upon, and increased use of coal further worsens the air quality.

PHYSICS (Manometer)

Christiaan Huygens (1629–1695, Dutch) invents the manometer to measure fluid pressure.

1662

BIOLOGY (Glands)

Nicolaus Steno (1638–1686, Danish) determines that material necessary for later glandular secretion is brought to the gland by arterial blood.

CHEMISTRY (Boyle's Law)

Robert Boyle (1627–1691, Irish-English) asserts Boyle's Law: for an "ideal" gas under constant temperature, the pressure and volume are roughly inversely proportional, at least over a limited pressure range. This assertion is based partly on experimentation with Boyle's Engine, an air pump devised by Boyle after hearing about Otto von Guericke's invention (1645—PHYSICS). The Boyle Engine becomes very popular among both professionals and amateurs.

CHEMISTRY (Gases)

To explain his empirically derived Boyle's Law (1662—CHEMISTRY), Robert Boyle (1627–1691, Irish-English) theorizes that a gas is composed of minute particles which are either spheres moving randomly or stationary springs.

MATHEMATICS (Statistics; Mortality Rates)

John Graunt (1620–1674, English) publishes *Natural and Political Observations made upon the Bills of Mortality,* a work contributing to the later labeling of Graunt as the "father of vital statistics."

1662–1666

PHYSICS (Gravity)

In a test of the hypothesis that gravity might, like magnetism, be a force whose magnitude varies according to the distance separating the bodies, Robert Hooke (1635–1702, English) weighs the same bodies at the earth's surface, in deep mines, and above the surface. The results prove inconclusive.

1663

ASTRONOMY (Reflecting Telescope)

James Gregory (1638–1675, Scottish) publishes *Optica promota,* in which he includes a theoretical description of an unconstructed reflecting telescope.

BIOLOGY (Capillaries)

By injecting coloring into the blood stream, Robert Boyle (1627–1691, Irish-English) is able to trace the blood as it passes through the capillaries from the arteries to the veins.

EARTH SCIENCES (North Atlantic Circulation; Tides)

In his *De motu marium et ventorum liber* [A treatise concerning the motion of the seas and winds], Isaac Vossius (1618–1689) correctly (on the large-scale) suggests that the ocean circulation in the North Atlantic is basically clockwise while incorrectly attributing tides and currents to a rise in sea level in the equatorial region produced by volume expansion due to solar heating.

METEOROLOGY (Observational Network)

Robert Hooke (1635–1702, English) encourages the systematic recording of standardized weather observations, including the wind direction and strength, moisture, temperature, pressure, the existence and movement of clouds, and the existence and form of precipitation. Hooke constructs a chart to be used as a standard form and suggests filling out the form at least once a day plus at the occurrence of any significant weather change.

PHYSICS (Fluid Pressure)

In the posthumously published *Traité de l'equilibre des liqueurs*, Blaise Pascal (1623–1662, French) presents the postulate that in a fluid at rest the pressure is transmitted equally in all directions.

1664

ASTRONOMY (Gravitation)

In discussing the orbit of a comet appearing in 1664, Robert Hooke (1635–1702, English) disagrees with Christopher Wren's (1632–1723, English) estimate that the comet moved in a straight line, claiming instead that the path is curved in the vicinity of the sun and that perhaps the curvature results from the sun's gravitational force.

ASTRONOMY (Jupiter's Red Spot)

Robert Hooke (1635–1702, English) reports observations of a large oval shape on Jupiter—the Great Red Spot.

BIOLOGY (Heart)

In his *De musculis & glandulis observationum specimen,* Nicolaus Steno (1638–1686, Danish) classifies the heart as a muscle, due to its tendons, flesh, and nerves, and states that the movements of the heart are produced by the muscle fibers.

BIOLOGY (Mechanical View)

Posthumous publication of René Descartes's (1596–1650, French) *Traité de l'homme et de la formation de foetus,* in which he describes animals as purely mechanical beings, a point of view he was reluctant to publish in his lifetime for fear of widespread disapproval.

BIOLOGY (Nervous System)

Thomas Willis (1621–1675, English) publishes *Cerebri anatome* on the structure of the nervous system.

1664–1666

ASTRONOMY/PHYSICS (Gravitation)

Isaac Newton (1642–1727, English) begins grappling with the concepts which later lead to his development of the theory of universal gravitation (1687—ASTRONOMY). Contemplating the possible connection between the fall of a body on earth and the motions of the moon, he performs initial calculations to determine whether the force of gravity could account for the moon's motions, and then, further, whether the sun's gravitation could account for Johannes Kepler's laws of planetary motion (ASTRONOMY entries for 1609, 1619). After deriving from Kepler's third law (1619—ASTRONOMY) and the law of centrifugal force the result that the force which maintains the elliptical orbits of the planets must vary reciprocally to the square of the orbit's radius, he proceeds to examine the moon's orbit in this context. He calculates a value of approximately $g/4000$ for the acceleration of the moon toward the earth, where g is the acceleration of a freely falling body near the earth's surface. In order to establish that the gravitational force of the earth could indeed be extending to the moon and could be the force maintaining its orbit, the 4,000 denominator should be the square of the ratio of the earth-moon distance and the earth's radius. This is confirmed roughly with the accepted 60 earth radii distance between the earth and the moon, but it is confirmed much more closely two decades later when Newton inserts a revised value for the earth's radius (1687—ASTRONOMY). The insights obtained by Newton in the 1664–1666 period are major, but require considerable further development before attaining the finished form of Newton's *Principia* (1687—ASTRONOMY). For instance, Newton's development of the calculus over the intervening years allows him by 1687 to justify the assumption that the earth and moon can be treated as point masses located at their respective centers.

MATHEMATICS (Calculus)

Isaac Newton (1642–1727, English) begins development of his method of fluxions, leading to the calculus. He develops the concept of a curve as the tracing of the motion of a point, defining the "fluxion" as the quotient of the "infinitely short" path traced divided by the "infinitely short" time in which the tracing is done. (Newton does not publish any work on fluxions or gravitation for many years, but begins developing many of the important concepts in the period 1664–1666, during much of the last half of which he is freed from academic obligations as Cambridge University is closed due to the plague and Newton retreats to the family home of Woolsthorpe.)

PHYSICS (Angular Momentum; Inertia)

In his early work on mechanics, Isaac Newton (1642–1727, English) arrives at a principle of the conservation of angular momentum and begins unsteady progress toward a concept of inertia, a concept later formalized in Newton's first law of motion (1687—ASTRONOMY).

1665

BIOLOGY (Microscopic Observations)

Robert Hooke (1635–1702, English), in *Micrographia,* describes and presents engravings of plant tissues and other living bodies viewed through a microscope as well as inanimate objects such as a razor edge. He introduces the term "cell" to describe the honeycomb arrangement observed in cork.

CHEMISTRY (Combustion)
METEOROLOGY (Air Composition)

In *Micrographia* (1665—BIOLOGY), Robert Hooke presents a thorough description of a theory of combustion, including a development of the thesis that air contains two parts, the major one in terms of volume or weight being inert and the minor one being essential to combustion and existing in the solid state in saltpeter. These two constituents are later identified as nitrogen and oxygen respectively.

EARTH SCIENCES (Earth)

Athanasius Kircher (1602–1680, German) completes his *Mundus subterraneus* [The subterranean world], soon to become a standard geological text. The earth is viewed as containing a central fire and numerous fire-carrying channels which sometimes reach the surface, producing volcanoes, and sometimes approach or reach underground water chambers, in which case they can result in hot springs, fountains, or rivers. Kircher describes a hydrologic cycle in which water runs in rivers from mountain peaks to oceans, then passes through holes in the ocean bottom and from there through subterranean passages back to the mountain peaks. He explains the upward movement by analogy with mechanical devices allowing an upward flow and speculates that the increased water pressure during high tide could be the acting mechanism. His description elsewhere in the book of weathering as due to a chemical process proved more sound.

HEALTH SCIENCES (Blood Transfusion)

Richard Lower (1631–1691, English) is the first to perform a direct blood transfusion from one animal to another.

MATHEMATICS (Probability Theory)

Blaise Pascal's (1623–1662, French) "Traité du triangle arithmétique" (1654—MATHEMATICS) is published posthumously.

METEOROLOGY (Hygrometry; Atmospheric Pressure; Thermometry)

Robert Hooke includes in his *Micrographia* (1665—BIOLOGY) various meteorological topics, including measurement of atmospheric humidity, pressure, and temperature. He describes a hygrometer he invented based on the relationship between humidity and the curling of a wild oat's beard. He discusses the variation of atmospheric pressure with altitude, using Boyle's Law (1662—CHEMISTRY) to determine the heights of the layers if a column of air were divided into one thousand layers, each containing the same weight of air. Concerned with the need for standardizing temperature measurements, he examines the readings from four-foot tubes containing spirit of wine, adjusting the liquid amount until the liquid is near the top of the tube at the peak of summer and near the bottom at the peak of winter, then graduating the stem, with zero set at the level of the liquid when the bottom bulb is immersed in freezing distilled water.

METEOROLOGY (Wheel Barometer)

Robert Hooke (1635–1702, English) invents the "wheel barometer," with a circular scale and a pulley system connecting the level of the expanding and contracting liquid (mercury) to the index indicator for the scale.

PHYSICS (Light)

Francesco Grimaldi (1618–1663, Italian) presents one of the first major accounts of the phenomena of interference and diffraction in his posthumous work *Physico-mathesis de lumine coloribus et iride*. He conceptualizes light as a fluid with rapid, wavelike motion and offers as evidence for the nonlinear motion the existence of blurred shadow edges. Grimaldi suggests that light has different frequencies corresponding to the different colors seen at the edges of shadows.

PHYSICS (Light; Wave Theory)

In *Micrographia* (1665—BIOLOGY), Robert Hooke presents the first substantial opposing theory to the Pythagorean concept of light as a stream of particles. He hypothesizes that light is a vibration transmitted through a medium and enters as evidence the failure of shadows to have sharp edges. He discusses extensive experimentation on light, including results concerning the phenomenon of refracted light and the nature of the composition of light. In addition to introducing a concept close to that of interference, he speculates that color depends on the form of the light.

SUPPLEMENTAL (Scientific Journals)

Publication of the first issue of the *Philosophical Transactions of the Royal Society of London,* the oldest scientific journal to survive into the twentieth century.

1665–1666

PHYSICS (Composition of Light)

Isaac Newton (1642–1727, English) analyzes the nature of white light and the ability of a prism to separate sunlight into the colors of the rainbow.

1666

ASTRONOMY (Planetary Orbits)

Giovanni Alfonso Borelli (1608–1679, Italian) revives Johannes Kepler's theories (1609—ASTRONOMY) on the planetary orbits. As Kepler had, he envisions the planetary motions as caused by rotating rays of force radiating from the sun. However, unlike Kepler, he believes natural motion to be in a straight line rather than a circle, hence requiring a form of gravitational force to maintain the planets in orbit. Borelli suggests that the elliptical orbit results from a balance between an attractive force of gravity between the sun and the planet and an opposing centrifugal force; but he is unable to calculate the magnitude of the forces required to maintain the observed orbits.

ASTRONOMY (Telescope)

Adrien Auzout (1622–1691, French) and Jean Picard (1620–1682, French) independently follow William Gascoigne (1641—ASTRONOMY) in introducing cross hairs to the telescope to allow precision pointing. Of the two parallel hairs, one is fixed to a mobile chassis. Auzout and Picard use this new micrometer to measure diameters of the moon, sun, and planets.

CHEMISTRY (Atomic Theory)

Robert Boyle (1627–1691, Irish-English) supports a form of atomic theory (or corpuscular philosophy) in his *Origin of Forms and Qualities.* He views the world as composed of small, solid, indivisible particles or atoms, with physical properties determined by size, shape, and motion. He seeks, through these particles, mechanical explanations for all physical properties and chemical reactions.

MATHEMATICS (Logic)

Gottfried Wilhelm Leibniz (1646–1716, German) initiates the study of symbolic logic with his essay "De arte combinatoria." Although the essay itself does not explicitly contain any symbolic logic, because of its status as the first publication calling for a future "calculus of reasoning" it is traditionally viewed as the historic beginning of symbolic and mathematical logic. Leibniz then develops symbolic logic over the next two and a half decades, including in his researches the concepts of null class, class inclusion, logical addition and multiplication, similarity, and implication.

METEOROLOGY (Water Vapor)

Urbano d'Aviso (Italian) proposes that water vapor consists of fire-filled bubbles of water which ascend to the level at which the specific gravity of the air is reduced to their own specific gravity. The explanation gains wide acceptance for about a century.

SUPPLEMENTAL (French Academy of Sciences)

The French Academy of Sciences holds its first official meeting, in the library of Louis XIV. Very different from the Royal Society of London, the Academy of Sciences has about 20 members, all salaried through the king and assigned problems by the royal ministers. (This structure changes after the French Revolution in the 1790s.) Projects include the writing of natural histories of plants and animals, the cataloging of inventions, the recording of meteorological observations of pressure, temperature, and rainfall, and the building of an observatory (1667–1672). Astrology is forbidden.

1667

ASTRONOMY (Telescope)

Adrien Auzout (1622–1691, French) further improves the micrometer devised by himself and Jean Picard (1666—ASTRONOMY) by suggesting a revised placement of the cross hairs to allow a surer line of sight.

ASTRONOMY (Variable Stars)

In an effort to explain the changes in brightness of the star Mira, Ismael Boulliau (1605–1694, French) speculates that the star rotates and has substantial dark spots which then periodically reduce its brightness as seen from earth.

BIOLOGY (Respiration)

Jan Swammerdam (1637–1680, Dutch) performs a series of experiments on animal respiration, describing them in his *Physico-Anatomical Medical Treatise on Respiration and the Functions of the Lungs*. By compressing and expanding an air bellows attached to the windpipe and lungs of various animals, he is able to examine the effects of inflating and deflating the lungs.

MATHEMATICS (Calculus)

James Gregory (1638–1675, Scottish) publishes *Vera circuli et hyperbolae quadratura* on the quadrature of ellipses and hyperbolas. Extending the applications of the methods of Archimedes, he finds the areas of these conics by determining identically converging sequences of inscribed and circumscribed areas.

METEOROLOGY (Anemometer)

Robert Hooke (1635–1702, English) improves upon the pressure-plate anemometer of Leone Alberti (c.1450—METEOROLOGY), making it simpler and more practical. Consisting of a flat plate swinging along a graduated quarter circle, the Hooke version rapidly gains acceptance.

METEOROLOGY (Instrumentation)

Among the earliest published descriptions of the barometer, thermometer, hydrometer, and other instruments are those given in the *Report of Experiments* of the Florentine Accademia del Cimento [Academy of experiments], a group formed in 1657 and disbanded in 1667.

1668

ASTRONOMY (Jupiter's Satellites; Navigation)

Gian Domenico Cassini (1625–1712, Italian-French) prepares tables of the positions of Jupiter's satellites, tables soon used by ship navigators as aids in determining longitude.

ASTRONOMY (Reflecting Telescopes)

After reading James Gregory's (1638–1675, Scottish) description of the principle of reflecting telescopes (1663—ASTRONOMY) and being impressed with the possibility of eliminating the troublesome problem of chromatic aberration, Isaac Newton (1642–1727, English) improves upon Gregory's suggested arrangement of mirrors and constructs the first reflecting telescope. Newton uses a concave mirror to concentrate the light and a small planar mirror to reflect the rays to an eyepiece at the side of the telescope tube. Earlier telescopes were refracting instead, with the light being focused by a lens rather than a mirror and with the eyepiece at the end rather than the side of the telescope tube.

BIOLOGY (Blood Circulation)

Examining both the tail of a tadpole and the foot of a frog through a microscope, Antony van Leeuwenhoek (1632–1723, Dutch) observes the blood circulate through the capillaries, establishing definitively the route of the blood through the body. These are among the first of van Leeuwenhoek's microscopic observations of hundreds of animate and inanimate objects over the next half century. These observations are made mostly with single lens microscopes having magnifications of approximately 300 and resolutions of about one thousandth of a millimeter.

MATHEMATICS (Area Under a Curve; Mercator's Series)

By using an infinite series to determine the area of a hyperbola in *Logarithmotechnia,* Nicolaus Mercator (c.1619–1687, Flemish) becomes apparently the first to publish a calculation of the area under a curve using the new analytical geometry. He computes the area under $y = 1/(1 + x)$ to be $(x/1) - (x^2/2) + (x^3/3) - (x^4/4) + \ldots$, by first performing the division and then integrating term by term. As this area was alternatively known to be $ln(1 + x)$ through the work of Gregory of St. Vincent (1584–1667, French), Mercator thereby obtains an infinite series representation of $ln(1 + x)$. The series is known as Mercator's Series. William Brouncker (1620–1684, English) also obtains infinite series representations of logarithms.

MATHEMATICS (Gregory's Series)

James Gregory (1638–1675, Scottish) publishes *Geometriae pars universalis* [The universal part of geometry] and *Exercitationes geometricae* [Geometrical exercises], presenting, with a largely geometrical framework, results concerning integration and infinite series. Included is Gregory's series: $x - (x^3/3) + (x^5/5) - (x^7/7) + \ldots$, determined as the area under the curve $y = 1/(1 + x^2)$ from $x = 0$ to $x = x$. The first work furthermore indicates an understanding of the inverse nature of differential and integral calculus, although expressed geometrically, leading to the fundamental theorem of calculus of Isaac Newton and Gottfried Wilhelm Leibniz.

1669

BIOLOGY (Blood)

Richard Lower (1631–1691, English) establishes that the color change of blood from dark to bright red as it passes through the lungs results from its absorption of a portion of the air, and he reports this work within a broader treatise on the heart, *Tractatus de corde.*

BIOLOGY (Insects)

Marcello Malpighi (1628–1694, Italian) publishes a monograph on the anatomy of the silkworm *Bombyx,* based on microscopic observation. Malpighi shows the insect to have no lungs but instead a system of tracheae distributing the air which enters through holes in the side of the body.

CHEMISTRY (Combustion)

Johann Joachim Becher (1635–1682, German) explains combustion as the escape of *terra pinguis,* the oily earth in his three-part composition of solids (1669—CHEMISTRY). The concept of *terra pinguis* is further developed in the first third of the eighteenth century by Georg Stahl, who renames it "phlogiston."

CHEMISTRY (Composition of Substances)

Johann Joachim Becher (1635–1682, German) modifies the iatrochemical view of substances as containing sulfur for inflammability, mercury for fluidity, and salt for fixity, referring instead to the following three constituents of solids: a fixed earth (found in all solids), an oily earth (found in combustible solids), and a fluid earth.

CHEMISTRY/EARTH SCIENCES (Crystals)

Nicolaus Steno includes in his *Prodromus* (1669—EARTH SCIENCES) plots and descriptions of the characteristic forms of crystals and a set of propositions on the growth of crystals once hardening first begins. A crystal is said to grow through the addition of crystalline matter to the external planes and predominantly to the apex rather than the intermediate planes. The enveloping fluid gradually spreads the crystalline matter across the relevant plane, so that the resulting surface is generally smoother when the hardening of the crystalline matter was slower.

EARTH SCIENCES (Fossils)

Nicolaus Steno includes in his *Prodromus* (1669—EARTH SCIENCES) his conclusion from examining the fossils of northern Italy that these fossils are the remains of once-living plants and animals and that they, along with the strata containing them, can reveal information on the earth's history. He thus explicitly rejects the conception of fossils as the never-living remains of the false starts of Creation.

EARTH SCIENCES (Mountains)

Nicolaus Steno (1638–1686, Danish) concludes from field studies in Tuscany that the primary cause of mountain formation rests in the shifting of earth strata, a shifting that occurs sometimes through violent upthrusting and sometimes through slippage or downfall. He explicitly rejects the belief that mountains can grow from the expansion of isolated rocks. Steno also classifies mountains into three groups—volcanic mountains, erosive mountains, and block or fault mountains, the first two categories having been described before and the third deriving from his Tuscany field studies.

EARTH SCIENCES (Stratigraphy)

Nicolaus Steno (1638–1686, Danish) completes *Nicolai Stenonis de solido intra solidum naturaliter contento dissertationis prodromus* [Forerunner of Nicolaus Steno's dissertation concerning a solid body enclosed by a process of nature within a solid], a geological treatise preliminary to a planned but never completed expanded Dissertation. Steno examines a wide range of geological topics proceeding from the proposition that a solid within, whether rock, bone, teeth, shells, or crystals, must have hardened before the surrounding solid (see separate 1669—EARTH SCIENCES entries on fossils and crystals). He identifies the concept of rock strata, suggests that the observed strata resulted from sediment deposits from turbid water, and presents several rules for strata formation, such as that as one stratum is being formed the one below must already be solid. Also, each stratum forms with an upper surface parallel to the horizon, so that any inclined strata have been altered after formation, major mechanisms for that alteration being running water and subterranean fire. Similarly, each stratum originally is bounded laterally by a solid body or extends around the globe. Recognizing that the contents of the stratum reveal information on the period in which the stratum was deposited, Steno presents from the stratigraphical evidence a geological history of Tuscany divided into six major periods.

MATHEMATICS (Calculus)

Isaac Newton (1642–1727, English) writes "De analysi per aequationes numero terminorum infinitas" [On analysis by equations with an infinite number of terms], an essay unifying the beginnings of his new calculus of fluxions and using infinite series in the calculation of integrals, integrating term by term such fractions as $1/(1 + x^2) = 1 - x^2 + x^4 - x^6 + x^8 - \ldots$ (for small x). Although Newton presents the separate series $1/(1 + x^2) = (1/x^2) - (1/x^4) + (1/x^6) - (1/x^8) + \ldots$ for large x, there are no clear ideas of convergence stated. Newton determines maxima and minima of functions, draws tangents, and finds quadratures and lengths of curves. The essay is sent to Isaac Barrow and John Collins, but is not published until 1711.

METEOROLOGY (Atmospheric Pressure)

In *A Continuation of New Experiments Touching the Spring and Weight of Air,* Robert Boyle (1627–1691, Irish-English) describes several meteorological experiments, including one in which he uses a long, narrow tin tube to determine how high water can be raised by a suction pump. Obtaining a value of 33.5 feet, he compares it against the level of quicksilver (mercury) in a baroscope and confirms that the ratio of the two heights is approximately the ratio of the densities of the two materials, consistent with the explanation that the pressure of the atmosphere is holding up the liquids in their respective tubes.

PHYSICS (Optics; Double Refraction)

Erasmus Bartholin (1625–1692, Danish) discovers double refraction and reports his discovery in *Experimentis crystalli islandici disdiaclastici.* Having noted that images seen through a crystal of Icelandic feldspar (calcite) appear double, Bartholin attributes the effect to refraction at two different angles of an entering ray of light. With proper rotation of the crystal he finds he can keep one image stationary and the other revolving. The existence of double refraction poses difficulties for contemporary theories of light, and in fact is not explained until the early 1800s.

PHYSICS (Gravitation)

Arguing by analogy, Christiaan Huygens (1629–1695, Dutch) demonstrates the centerward movement undergone by pebbles in a whirlpool of water, particularly as the water motion slows, and uses it to support René Descartes's theory of vortices (1644—ASTRONOMY) as the explanation of why objects fall to the earth. Gravity is thus attributed to the vortex motions rather than to a linear attractive force.

1670

BIOLOGY (Nasal Secretions)

In an appendix to the second edition of *Tractatus de corde* (1669—BIOLOGY), Richard Lower (1631–1691, English) refutes the theory that the pituitary gland is the source of nasal secretions.

HEALTH SCIENCES (Diabetes)

Thomas Willis (1621–1675, English) describes the symptoms of diabetes.

MATHEMATICS (Binomial Series)

James Gregory (1638–1675, Scottish), in his work with infinite processes, arrives at the binomial series.

MATHEMATICS (Calculus)

Isaac Barrow (1630–1677, English) publishes his *Lectiones geometricae* [Geometrical lectures], in which he describes tangent problems with methods essentially equivalent to those in differential calculus and area problems with methods similar to those in integral calculus. He furthermore seems to understand that tangent and area problems are inverse to one another, but, in contrast to the contemporaneous and upcoming work of Isaac Newton and Gottfried Wilhelm Leibniz (MATHEMATICS entries for 1669, 1671, 1673–1676, 1677), he fails to make effective use of this relationship. He views both time and geometrical lines as composed of indivisibles, and geometrical magnitudes as generated by flowing points. The work on tangents contains the "characteristic" or "differential" triangle.

MATHEMATICS (Number Theory)

Pierre de Fermat's (1601–1665, French) marginal notes in his copy of Bachet de Méziriac's 1621 edition of the *Arithmetica* of Diophantus are published posthumously by his son. These marginal notes include many conjectured theorems in number theory, including Fermat's famous Last Theorem (1637—MATHEMATICS).

1671

ASTRONOMY (Saturn; Iapetus)

Gian Domenico Cassini (1625–1712, Italian-French) discovers Iapetus, the second satellite of Saturn to be discovered, the first being Titan (1655—ASTRONOMY).

ASTRONOMY (Telescopes)

Isaac Newton (1642–1727, English) sends a version of his reflecting telescope (1668—ASTRONOMY) to the Royal Society, along with a description entitled "An Account of a New Catadioptrical Telescope." The magnifying power claimed is roughly 38.

CHEMISTRY (Hydrogen)

Robert Boyle (1627–1691, Irish-English) dissolves iron in hydrochloric or sulfuric acid and produces hydrogen, although not recognizing it as such (1766—CHEMISTRY).

EARTH SCIENCES (Fossils)

Martin Lister (1638–1712, English) claims in "Fossil Shells in Several Places of England" that the English fossils are not the remains or impressions of living creatures and that there was no petrifying of shells, the rocks instead having been always as at present (*lapides sui generis*). The consistent texture throughout the individual rocks is presented as evidence.

MATHEMATICS (Calculus; Polar Coordinates)

In a second unpublished paper on the calculus, "Methodus fluxionum et serierum infinitarum" [The method of fluxions and infinite series], Isaac Newton (1642–1727, English) expands his "De analysi" (1669—MATHEMATICS) and revises his treatment by conceiving variables as undergoing continuous change rather than changing exclusively in discrete amounts. He also introduces several new coordinate systems, including polar coordinates, with points located by reference to a fixed point and a fixed line through that point, and bipolar coordinates, with a point located by reference to two fixed points. The work remains largely unknown until it is posthumously published in 1736.

MATHEMATICS (Differential Equations)

Isaac Newton (1642–1727, English) separates ordinary differential equations of the first order into three types.

MATHEMATICS (Taylor Series)

James Gregory (1638–1675, Scottish) arrives at the Taylor series expansion of a function, later rediscovered and published by Brook Taylor (1715–1717—MATHEMATICS).

1672

ASTRONOMY (Distance to Mars)

Jean Richer (1630–1696, French) leads an expedition to measure the altitude of Mars from Cayenne, French Guiana. Comparison of the observations made at Cayenne with those made at Paris yield, through triangulation with the earth's diameter as a base line, an approximate value of 60 million kilometers for the minimum distance between earth and Mars. This provides a scale for the solar system (Kepler's laws dealt only in relative distances), with a computed astronomical unit of 150 million kilometers, a significant improvement over the earlier, lower estimates of this earth-sun distance.

ASTRONOMY (Saturn; Rhea)

Gian Domenico Cassini (1625–1712, Italian-French) discovers Rhea, a satellite of Saturn.

PHYSICS (Generators)

The first description of an electrical generator is given in a one paragraph segment of Otto von Guericke's (1602–1686, German) *Experimenta nova Magdeburgica de vacuo spatio* [The new Magdeburg experiments]. The static generator is a sulfur globe built by Guericke which when spun attracts objects and, after rubbing, glows in the dark. Guericke interprets the effects as magnetic rather than electric.

PHYSICS (Gravitation; Pendulums)

A pendulum clock taken by Jean Richer (1630–1696, French) from Paris to Cayenne, French Guiana for astronomical observations (1672—ASTRONOMY) is found to lose 2.5 minutes a day at the new location. Since previous work by Christiaan Huygens (1629–1695, Dutch) showed that the period of a given pendulum varies in inverse proportion to the square root of g, it is correctly concluded that the magnitude of the force of gravity differs from one location to another. This draws attention to the distinction between mass and weight and also initiates the use of pendulums to measure gravity.

PHYSICS (Vacuum)

In *Experimenta nova Magdeburgica de vacuo spatio* [The new Magdeburg experiments], Otto von Guericke (1602–1686, German) records his initial difficulty and eventual success in creating a vacuum. Having found wood too porous, he succeeds with a large copper sphere, creating the vacuum within the sphere by first filling it with water and then pumping the water out.

PHYSICS (White Light)

Isaac Newton (1642–1727, English) describes to the Royal Society his studies on the spectral dispersion of light rays, formalized in the paper "New Theory about Light and Color." After separating white light into the colors of the spectrum by passing it through a glass prism, he isolates the colors and passes them individually through a second prism, showing that they are not thereby further resolved. He asserts that white light is composed of an infinite number of colors, each of which is distinguished by its refrangibility and none of which can be transformed into any other. The latter is a difficult conclusion for many contemporary scientists to accept, and the work encounters some strong criticism.

1673

BIOLOGY (Embryology)

Marcello Malpighi (1628–1694, Italian) describes the development of the chick embryo from the early hours of incubation in the treatise *On the Formation of the Chick in the Egg*, publishing also the separate treatise *On the Incubation of Eggs*.

BIOLOGY (Red Blood Corpuscles)

Through microscopic observations, Antony van Leeuwenhoek (1632–1723, Dutch) discovers and describes the red blood corpuscles in the human blood stream.

EARTH SCIENCES (Sea Water)

Robert Boyle (1627–1691, Irish-English) rejects the concept that the sea is salty only at the surface in his *Observations and Experiments About the Saltness of the Sea.*

MATHEMATICS (Calculators)

Gottfried Wilhelm Leibniz (1646–1716, German) invents a calculator capable of multiplication and division as well as addition and subtraction. Multiplication is performed automatically through repeated additions.

MATHEMATICS (Geometry)

In *Horologium oscillatorium* (1673—PHYSICS), Christiaan Huygens (1629–1695, Dutch) defines and analyzes involutes and evolutes of a plane curve and proves that a cycloid symmetric about a vertical axis with its vertex at the bottom is a tautochrone, meaning that the time an object takes to descend solely under the force of gravity from rest at any point along the curve to the vertex is independent of the starting point.

MATHEMATICS (Series Convergence)

Gottfried Wilhelm Leibniz (1646–1716, German) distinguishes series which converge from those which diverge.

PHYSICS (Centripetal Force)

Christiaan Huygens (1629–1695, Dutch) publishes his major work: *Horologium oscillatorium* [The oscillation of pendula], concerned largely with time measurement. As an outgrowth of his experimentation with pendulums and with circular motion, Huygens is able to derive the law of centripetal force, that a body moving in uniform circular motion has an acceleration $a = v^2/r$ directed toward the center of the circle, where v is the speed of the body and r is the radius of the circle.

1673–1676

MATHEMATICS (Calculus)

Gottfried Wilhelm Leibniz (1646–1716, German) develops the methods and notation of differential and integral calculus. He acknowledges Blaise Pascal's *Traité des sinus* (1658—MATHEMATICS) as the principal earlier work influencing him.

1674

ASTRONOMY (Gravitation)

Robert Hooke (1635–1702, English) writes an "Attempt to Prove the Motion of the Earth" in which he includes a paragraph postulating that all celestial bodies attract all other celestial bodies and that planetary motions result as such attractive forces deflect the planets from what would otherwise be straight line paths (tangential to the actual orbits). The paragraph suggests a step toward the theory of gravitation developed a decade later by Isaac Newton (1687—ASTRONOMY).

CHEMISTRY (Oxygen)
METEOROLOGY (Air)

In *Tractatus quinque medico-physici* [Five medico-physical treatises], John Mayow

(1641–1679, English) publishes conclusions from experiments which lead him to recognize the existence of two constituents in the air and to explain combustion by asserting that it is only the "Spiritus nitro-aereus" portion (later identified as oxygen) that is consumed in either combustion or animal respiration. The experiments consist of burning substances in jars inverted over water. As the burning occurs, the water level rises, filling the space formerly taken by the "Spiritus nitro-aereus" portion of the air. When this portion is fully depleted, the remaining air can no longer support either animal life or combustion. Mayow found this "Spiritus" to exist in the acid part of saltpeter.

EARTH SCIENCES (Rivers)

Pierre Perrault (1611–1680, French) presents quantitative evidence for the previously little-accepted claim that rain water is sufficient to furnish the water outflow of rivers, doing so in the anonymous *De l'origine des fontaines*. Using the upper portion of the Seine River, with measurements for 1668–1670, he calculates that the annual outflow through the Seine canal at Aynay le Duc is only about one sixth the rainfall into the appropriate drainage area. He cautiously extrapolates to speculate on the possibility that indeed all the world's rivers may be supplied by rain and snow water.

1675

ASTRONOMY (Observatories)

The Greenwich Observatory is founded in Great Britain, with a major purpose being the more accurate determination of the positions of the stars and moon. This has practical importance in its relevance to calculating longitudes from locations without fixed terrestrial reference points, for example, from ships in the open ocean.

ASTRONOMY (Saturn's Rings: Cassini Division)

Gian Domenico Cassini (1625–1712, Italian-French) discovers the Cassini division, a dark circular band dividing the ring system of Saturn. This discovery damages the hypothesis of Christiaan Huygens that the ring system is a single flat disk surrounding the planet (1655—ASTRONOMY).

CHEMISTRY (Atomic Theory)

Cours de chymie, a textbook by Nicolas Lémery (1645–1715, French), increases public interest in the subject of chemistry. Lémery includes some atomic theory, emphasizing the shape of the individual atoms and the influence of this shape on their physical and chemical properties. For instance, acids derive their taste from sharp spikes, and alkalies are able to neutralize acids due to a porous nature which allows the spikes of the acids to penetrate.

PHYSICS (Light)

In a memoir to the Royal Society, Isaac Newton (1642–1727, English) discusses three hypotheses on the nature of light, those being that light consists of corpuscles (or particles), that it consists of waves in the ether, and that it consists of a wave/corpuscle combination. Newton shows that, by assuming different wavelengths for different colors, the wave theory can explain the ability of a prism to split light into the colors of the rainbow.

PHYSICS (Speed of Light)
ASTRONOMY (Revolution of the Earth)

Olaus Römer (1644–1710, Danish) makes the first quantitative measurement of the speed of light, having hypothesized the finite velocity from observations of the satellites of Jupiter. The timing of the eclipses of the satellites varies systematically with the relative positions of the earth, sun, and Jupiter; and Römer notices that these variations can be accounted for by the varying distance of travel of the light rays. His calculations both require and, for the remaining doubters, help substantiate the Copernican theory of the earth's revolution around the sun (1543—ASTRONOMY). The measurements indicate that light takes 11 minutes to cross the diameter of the earth's orbit.

1675–1679

BIOLOGY (Plant Anatomy)

Publication of Marcello Malpighi's (1628–1694, Italian) *Anatome plantarum.*

1676

BIOLOGY (Protozoa)

Antony van Leeuwenhoek (1632–1723, Dutch) announces in a letter to the Royal Society his discovery of living creatures (protozoa) in rain water and describes their shape and motions. He also examines pepper water and estimates the number of living creatures in a single drop to be one million.

CHEMISTRY (Gas Laws)

In *De la nature de l'air,* Edme Mariotte (1620–1684, French) states Boyle's Law (1662—CHEMISTRY; also known as Mariotte's Law) and emphasizes the necessity for temperature to remain constant to insure the law's validity.

HEALTH SCIENCES (Measles)

Thomas Sydenham (1624–1689, English) provides a detailed description of measles in *Observationes medicae circa morborum acutorum historiam et curationem.*

189 DE MOTIB. STELLÆ MARTIS

CAP. LIX.

PROTHEOREMATA.

I.

SI intra circulum deſcribatur ellipſis, tangens verticibus circulum, in punctis oppoſitis; & per centrum & puncta contactuum ducatur diameter; deinde a punctis aliis circumferentiæ circuli ducantur per pendiculares in hanc diametrum: eæ omnes a circumferentia ellipſeos ſecabuntur in eandem proportionem.

Ex l. 1. Apollonii Conicorum pag. XXI. *demonſtrat* COMMANDINUS *in commentario ſuper* V. *Sphæroideon* ARCHIMEDIS.

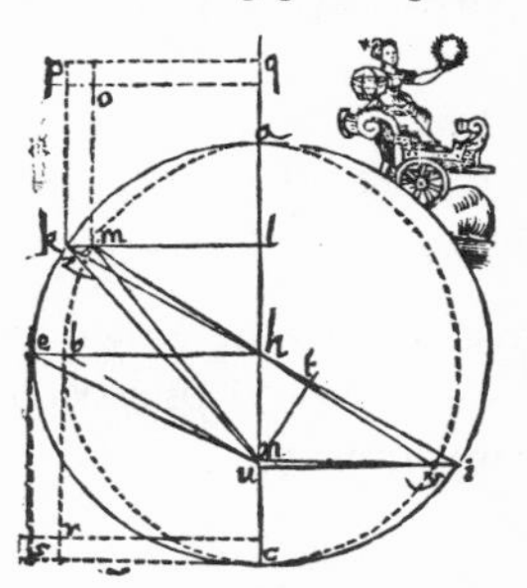

Sit enim circulus AEC. *in eo ellipſis* ABC *tangens circulum in* AC. *& ducatur diameter per* A. C. *puncta contactuum, & per* H *centrum.* *Deinde ex punctis circumferentiæ* K. E. *deſcendant perpendiculares* KL, EH, *ſecta in* M. B. *a circumferentia ellipſeos. Erit ut* BH *ad* HE, *ſic* ML *ad* LK. *& ſic omnes alia perpendiculares.*

II.

Area ellipſis ſic inſcriptæ circulo, ad aream circuli, habet proportionem eandem, quam dictæ lineæ.

Vt enim BH *ad* HE, *ſic area ellipſeos* ABC *ad aream circuli* AEC. *Eſt quinta Sphæroideon* ARCHIMEDIS.

III.

Si a certo puncto diametri educantur lineæ in

FIGURE 5. *A page from Johannes Kepler's* Astronomia nova [*New astronomy*] *(see 1609—*MATHEMATICS *and 1609—*ASTRONOMY *entries), where he demonstrates various properties of the ellipse prior to showing the elliptical shape of the planetary orbits around the sun. (By permission of the Houghton Library, Harvard University.)*

MATHEMATICS (Binomial Theorem)

The first formal statement of the Binomial Theorem is given by Isaac Newton (1642–1727, English) in a letter to Henry Oldenburg, secretary of the Royal Society.

MATHEMATICS (Calculus)

Isaac Newton (1642–1727, English) continues toward a sounder calculus in his unpublished paper "De quadratura curvarum" [Quadrature of curves], eliminating infinitesimals and flowing quantities. (This essay is finally published in 1704, as an appendix to Newton's *Opticks* [1704—PHYSICS].)

1676–1678

ASTRONOMY (Southern Hemisphere Stars)

Edmond Halley (1656–1743, English) journeys to St. Helena to make the first systematic study of the stars of the Southern Hemisphere. The observations lead to a catalog of the celestial coordinates of 341 stars (1678—ASTRONOMY).

1677

ASTRONOMY (Mercury)

Edmond Halley (1656–1743, English) records observations of a complete transit of Mercury across the solar disk.

BIOLOGY (Protozoa)

Robert Hooke (1635–1702, English) demonstrates to the Royal Society confirmation of Antony van Leeuwenhoek's (1676—BIOLOGY) microscopic discoveries concerning the existence of microorganisms in water.

CHEMISTRY (Phosphorus)

The alchemist and physician Hennig Brand (c.1630–c.1692, German) discovers elemental phosphorus and obtains it by distilling urine. Its ability to shine in the dark arouses much excitement. ("Balduin's phosphorus," a phosphorescent calcium nitrate, had been known since 1630.)

EARTH SCIENCES (Fossils)

Robert Plot (1640–1696, English) includes in his *Natural History of Oxford-Shire* a discussion of the origin of fossils concluding that most likely they are not the remains of formerly living beings but instead were probably formed by some "virtue" latent within the earth.

HEALTH SCIENCES (Sperm)

Observing sperm in the semen of a man suffering from gonorrhea, Johan Ham erroneously concludes that the sperm are a consequence of the disease.

MATHEMATICS (Calculus)

Gottfried Wilhelm Leibniz (1646–1716, German) announces, in a letter to Henry Oldenburg, his discovery of the differential calculus, including a clear statement of the methods and notation.

1677–1704

METEOROLOGY (Rainfall)

Richard Towneley (1629–1707, English) records detailed rainfall measurements in England.

1678

ASTRONOMY (Southern Hemisphere Stars)

Edmond Halley (1656–1743, English) publishes the first catalog containing telescopically determined celestial coordinates of Southern Hemisphere stars. The catalog lists the 341 stars observed by Halley from St. Helena between 1676 and 1678.

EARTH SCIENCES (Shells; Fossils)

Martin Lister (1638–1712, English) publishes *Historiae animalium angliae tres tractatus,* an illustrated history and description of shells. Although he includes a chapter on English fossil shells, he continues to maintain (1671—EARTH SCIENCES) that the fossils did not derive from living creatures.

PHYSICS (Hooke's Law)

Robert Hooke (1635–1702, English) presents Hooke's Law for elastic bodies, that the deformation is proportional to the applied stress ($F = -kx$), in the pamphlet *De potentia restitutiva; or Of Spring, Explaining the Power of Springing Bodies.* The law is an approximate empirical relation actually discovered by Hooke in the early 1660s but not published until 1678. If, for any particular body, the deformation exceeds the elastic limit, then the original shape is not regained.

PHYSICS (Polarization)

Christiaan Huygens (1629–1695, Dutch) discovers that an unpolarized light ray will be divided into two perpendicular polarized rays by a crystal of calcite (or Iceland spar).

PHYSICS (Wave Theory of Light; Ether)

Christiaan Huygens (1629–1695, Dutch) advances beyond Francesco Grimaldi's initial conceptions of a wavelike nature of light (1665—PHYSICS), postulating a stationary "ether" through which light propagates by longitudinal waves. The luminiferous ether postulated by Huygens consists of hard, oscillating, elastic particles existent throughout space, which allow light to be transmitted through the ether in a manner similar to the transmission of sound through air.

1679

ASTRONOMY/PHYSICS (Gravitation)

By 1679, Robert Hooke (1635–1702, English), Christopher Wren (1632–1723, English), and Edmond Halley (1656–1743, English) have all deduced from Kepler's third law (1619—ASTRONOMY) and Huygen's law of centripetal force (1673—PHYSICS) that the force exerted by a gravitational source is inversely proportional to the square of the distance from that source. Hooke in 1679 writes Isaac Newton asking if the centripetal force law and this inverse square law together imply elliptical orbits for the planets, but Newton does not reply. That they do imply elliptical orbits is proved by Newton in his major work of 1687.

MATHEMATICS (Analysis)

Gottfried Wilhelm Leibniz's (1646–1716, German) *Characteristica geometrica* marks the beginnings of analysis situs.

MATHEMATICS (Analytic Geometry)

Pierre de Fermat's (1601–1665, French) *Ad locus planos et solidos isagoge* [Introduction to plane and solid loci] is published posthumously, providing a short exposition of his analytic geometry (1629—MATHEMATICS). He presents an exhaustive analysis of linear and quadratic equations in two unknowns, plotting the straight line solutions to linear equations, and the hyperbolic, elliptic, and parabolic solutions to various quadratic equations. Although using two perpendicular axes, he restricts his plots to the first quadrant (with nonnegative coordinates). In his equations, he follows the notational conventions of François Viète (1591—MATHEMATICS).

1680

BIOLOGY (Fermentation)

Johann Joachim Becher (1635–1682, German) shows that sugar is essential for fermentation.

BIOLOGY (Insects)

John Banister (1650–1692, American) classifies 52 species of American insects.

BIOLOGY (Mechanical Analogy)

In the posthumous publication *On the Motions of Animals,* Giovanni Alfonso Borelli (1608–1679, Italian) presents a mechanical conception of animal movements. Walking, running, flying, swimming, and lifting weights are all analyzed in the context of a mechanical analogy, supported by diagrams and equations. The arm, for instance, is described as a second order lever, with the fulcrum at the elbow.

BIOLOGY (Yeast)

Antony van Leeuwenhoek (1632–1723, Dutch) observes yeast under a microscope and describes its structure as composed of small spherical granules.

PHYSICS (Diffraction)

Christiaan Huygens (1629–1695, Dutch) utilizes the emerging wave theory (PHYSICS entries for 1665 and 1678) to explain the diffraction of light.

1681

EARTH SCIENCES (Earth History; Noah's Flood)

Thomas Burnet (1635–1715, English) publishes the popular *Telluris theoria sacra* [Sacred theory of the earth], speculating on the origin of the earth, its original structure, and the past and future general changes in that structure. The account ascribes a pivotal position to Noah's Flood, before which the earth's axis is assumed to have been perpendicular to the plane of the ecliptic, with a perpetual spring at the latitude of England, and the ocean was assumed to be constrained within the earth. Solar heat caused the earth's crust to crack, upon God's decree, and the constrained waters to emerge, creating the Great Flood and changing what had been a smooth surface into a very rough one, marked particularly by continent/ocean and mountain/valley contrasts.

1682

BIOLOGY (Insects)

Publication of Jan Swammerdam's (1637–1680, Dutch) *Histoire générale des insectes* [General history of insects].

BIOLOGY (Plant Anatomy; Sexual Reproduction)

Publication of Nehemiah Grew's (1641–1712, English) *The Anatomy of Plants,* an outgrowth of his numerous microscopic observations. Grew discovers sexual reproduction in plants and identifies the stamen and pistil as the male and female organs respectively, as well as presenting detailed drawings of plant anatomy.

PHYSICS (Optics)

Gottfried Wilhelm Leibniz (1646–1716, German) enunciates the principle that light travels along the path offering the least resistance.

1683

BIOLOGY (Bacteria)

Antony van Leeuwenhoek (1632–1723, Dutch) discovers bacteria in the human mouth.

EARTH SCIENCES (Geologic Mapping)

Martin Lister (1638–1712, English) proposes mapping soils, rocks, and minerals in his "Ingenious Proposal for a New Sort of Maps of Country, together with tables of sands and clays." It remains until 1743 before such a geological map is produced, at which time Christopher Packe maps the region of East Kent, England (1743—EARTH SCIENCES).

EARTH SCIENCES (Ocean Tides)

Isaac Newton (1642–1727, English) presents a mathematical theory of ocean tides, based on the gravitational attraction of the sun and moon.

MATHEMATICS (Algebra)

Ehrenfried Tschirnhaus (1651–1708, German) advances the methodology of solving polynomial equations algebraically when he applies a rational substitution to remove terms from a higher-degree equation, generalizing the earlier removal by Girolamo Cardano (1545—MATHEMATICS), François Viète (1591—MATHEMATICS), and others, of the second degree term from cubic and quartic equations.

1684

ASTRONOMY (Gravitation)

Robert Hooke (1635–1702, English) claims to have a proof that the hypothesis of a force of gravity varying with the inverse square of the distance will produce elliptical planetary orbits. Christopher Wren (1632–1723, English) informally offers a prize for such a proof, but Hooke declines to submit his proof for inspection.

ASTRONOMY (Saturn; Dione, Tethys)

Gian Domenico Cassini (1625–1712, Italian-French) discovers Dione and Tethys, two satellites of Saturn.

BIOLOGY (Spontaneous Generation)

Francesco Redi (1626–1697/1698, Italian) refutes the widely held belief in the spontaneous generation of small, living creatures, presenting evidence in particular against the spontaneous creation of maggots in rotting meat. Redi compares a piece of meat left unprotected with one protected by gauze, and finds that the maggots arise only in the uncovered meat. He hypothesizes that the maggots can arise only where flies are able to deposit eggs.

EARTH SCIENCES (Earth's Radius)
ASTRONOMY (Earth-Sun Distance)

Posthumous publication of Jean Picard's (1620–1682, French) calculations and improved values for the radius of the earth and for the earth-sun distance.

MATHEMATICS (Calculus)

Gottfried Wilhelm Leibniz (1646–1716, German) publishes his first formal paper on calculus, "Nova methodus pro maximis et minimis, itemque tangentibus, quae nec fractas nec irrationales quantitates moratur, et singulare pro illi calculi genus" [A new method for maxima and minima as well as tangents, which is impeded neither by fractional nor by irrational quantities, and a remarkable type of calculus for this], after discovering the fundamental theorem of the calculus, that is, that the process of integration by summing infinitesimals to find areas is the inverse of the process of differentiation. His notation, including the symbols dx and dy, becomes widely adopted. Among the results presented in the paper are the following: $dy = 0$ for maxima and minima; $d^2y = 0$ for points of inflection; and such rules as that for the differential of a product of functions, $d(uv) = udv + vdu$. (This paper is regarded by many as the first published paper on calculus, although several earlier works had employed integration or differentiation techniques.)

METEOROLOGY (Trade Winds)

Martin Lister (1638–1712, English) suggests that the trade winds are the "breath" of the sargasso weed, an explanation in harmony with Aristotle's conception of winds as caused by exhalations or emanations.

1685

ASTRONOMY (Gravitation)

Responding to a request by Edmond Halley (1656–1743, English), Isaac Newton (1642–1727, English) provides a derivation of elliptical planetary orbits assuming the centripetal force and inverse square laws (the derivation sought by Christopher Wren in 1684). The proof impresses Halley, who then encourages Newton to elaborate the details and publish, leading ultimately to *Principia mathematica* (1687—ASTRONOMY).

ASTRONOMY (Gravitation)
MATHEMATICS (Calculus)

Through his newly developed calculus, Isaac Newton (1642–1727, English) is able to prove that for the mathematics of gravitation a spherical body can be treated as a point mass concentrated at the sphere's center. This provides a foundation for a portion of his initial 1664–1666 concepts of solar system dynamics, and leads toward the finalization of his theory of universal gravitation in the *Principia* (1687—ASTRONOMY).

MATHEMATICS (Complex Numbers)

John Wallis (1616–1703, English), in *Treatise of Algebra, Both Historical and Practical,* presents the first graphical representation of complex numbers $a + bi$, locating on an x-axis a point at distance a from the origin and then drawing a perpendicular to this axis with length equal to the coefficient of the imaginary i. The one step which remains before arriving at the twentieth-century notation is the construction of an actual y-axis for the axis of imaginaries. This latter step is finally accomplished by Caspar Wessel over a century later (1797—MATHEMATICS). (Wallis's treatise is also noted for its sections on the history of mathematics.)

MATHEMATICS (Geometry)

In his *Sectiones conicae,* Philippe de la Hire (1640–1718, French) presents arguments for the power and potentials of synthetic projective geometry. Despite the efforts of La Hire and, earlier, Girard Desargues (1639—MATHEMATICS), the mathematical community in general is not convinced that the synthetic methods can match in power the analytic methods of René Descartes (1637—MATHEMATICS).

METEOROLOGY (Three-Liquid Barometer)

Robert Hooke (1635–1702, English) constructs a three-liquid barometer, with the pressure of the atmosphere determined from the level of the boundary between the two top liquids.

1686

MATHEMATICS (Calculus)

Gottfried Wilhelm Leibniz (1646–1716, German) publishes, within a book review, rules for integral calculus, thereby complementing his earlier paper presenting rules for differentiation (1684—MATHEMATICS). He here introduces the symbol $\int$ for integration.

METEOROLOGY (Winds)

In "An Historical Account of the Trade Winds and Monsoons, observable in the Seas between and near the Tropicks; with an Attempt to assign the Physical Cause of the Said Winds," Edmond Halley (1656–1743, English) publishes the first meteorological chart, a world map indicating the prevailing winds over the tropical oceans, and offers an attempted explanation of the trade winds and monsoons. He explains the equatorward flow of the trades as resulting from the combination of (1) the rising of air near the equator due to the intensity of solar heating, and (2) the resulting surface inward flow of air toward the updraft region. His explanation of the trade winds's east-to-west flow, based on the east-to-west apparent movement of the sun, is less soundly based, with an improved explanation based on the rotations of earth and atmosphere awaiting the later work of George Hadley (1735—METEOROLOGY). Halley explains the Asian monsoon basically as a giant sea breeze.

1686–1704

BIOLOGY (Classification)

In the three-volume *Historia plantarum,* John Ray (1627–1705, English) presents an extensive botanical classification based on a scheme of Aristotle but incorporating many of the new plant forms discovered on the sixteenth and seventeenth century voyages of discovery.

1687

ASTRONOMY (Gravitation)
PHYSICS (Laws of Motion)

Publication of Isaac Newton's (1642–1727, English) *Principia mathematica philosophiae naturalis* [Mathematical principles of natural philosophy], in which he fully develops his theory of universal gravitation (see ASTRONOMY entries for 1664–1666 and 1685). Early in the work, Newton introduces the concept and definition of "mass" as well as his three laws of motion: (1) a body will remain at rest or in a state of uniform motion in a straight line unless acted upon by an external force. This law is generally known as the law of inertia; (2) the force exerted upon a body is proportional to the change of motion induced. Often expressed $\mathbf{F} = m\mathbf{a}$ [or, more accurately, $\mathbf{F} = d(m\mathbf{v})/dt$, $\mathbf{a}$ symbolizing acceleration, $\mathbf{v}$ velocity, and m mass] and considered the fundamental equation of classical mechanics, this second "law" can be viewed instead as a definition of force. With $\mathbf{F} = O$, the law becomes a statement of the conservation of momentum for a system with no external forces acting upon it; (3) any action is countered by an equal and opposite reaction. Proceeding from these three laws of motion, Newton unifies terrestrial and celestial mechanics by formulating a law of gravitation to govern both the fall of bodies on earth and celestial motions. That law, later expressed notationally as $F = Gm_1m_2/r^2$, asserts the universal existence of an attractive gravitational force between any two point masses, with the force being directly proportional to the product of the masses, m_1 and m_2, and inversely proportional to the square of the intervening distance r. With this Newton is able to *derive* the empirically obtained laws of Johannes Kepler (ASTRONOMY entries for 1609 and 1619), making the appropriate theoretical adjustment in the third law from (in more recent notation)

$$p_1^2/p_2^2 = D_1^3/D_2^3 \quad \text{to} \quad p_1^2(m_s + m_1)/p_2^2(m_s + m_2) = D_1^3/D_2^3,$$

where m_1, m_2, m_s are the masses of the two planets and the sun respectively, p_1, p_2 are the periods, and D_1, D_2 are the diameters of the planetary orbits. Jean Picard's (1684—EARTH SCIENCES) revised value for the earth's radius enables Newton to improve upon his earlier results (1664–1666—ASTRONOMY) thereby showing the earth's gravity to have the proper magnitude to provide the force for maintaining the moon's observed orbit. Other results included in *Principia mathematica* are the following: (1) tides result from the gravitational effects of the sun and moon; (2) the precession of the equinoxes results from the non-spherical shape of the earth and the pull of the sun and moon; and (3) each planet should be flattened at the poles (a phenomenon observable at the time in the case of Jupiter). The *Principia* is both acclaimed and criticized, among the criticizers being Christiaan Huygens and Gottfried Wilhelm Leibniz, who claim that the work reinstates the erroneous belief that a force can act across empty space.

ASTRONOMY (Star Atlas)

Johannes Hevelius (1611–1687, Polish) maps the stars along with pictorial representations of the constellation stories in his sky atlas *Firmamentum sobrescianum*.

CHEMISTRY (Gas Laws)

Isaac Newton (1642–1727, English) shows in *Principia* (1687—ASTRONOMY) that the assumption that a gas is composed of particles which repel each other with a force inversely proportional to the intervening distance leads mathematically to Boyle's Law (1662—CHEMISTRY) that the pressure and volume of a gas are inversely proportional.

EARTH SCIENCES (Earth's Shape)
ASTRONOMY (Rotation of Earth)

In his *Principia* (1687—ASTRONOMY), Isaac Newton (1642–1727, English) interprets the major geophysical result of the Richer expedition to Cayenne, French Guiana (1672—PHYSICS)—that the gravity at Cayenne is less than that at Paris—as evidence that the earth is shaped as an oblate spheroid flattened at the poles. He suggests that the shape resulted from the earth's rotation, and hence provides an indirect confirmation that the earth rotates.

MATHEMATICS (Applied)

Isaac Newton's (1642–1727, English) *Principia* (1687—ASTRONOMY) is often considered the origin of modern applied mathematics. In deriving aspects of the planetary geometry from more basic hypotheses (1687—ASTRONOMY), Newton departs radically from the earlier purely descriptive discussions of planetary orbits. To a large although not complete extent, the mathematics is framed in the language of classical Euclidean geometry rather than Newton's more appropriate but new and little known calculus.

METEOROLOGY (Hygrometer)

Guillaume Amontons (1663–1705, French) develops an improved hygrometer, for measuring atmospheric humidity.

1689

SUPPLEMENTAL (Science and Politics: Newton)

Isaac Newton (1642–1727, English) is elected to the House of Commons.

1690

ASTRONOMY (Extraterrestrial Life)

Christiaan Huygens (1629–1695, Dutch) publishes his *Cosmotheoros,* asserting that there exists life on other planets and that among this life are beings comparable to mankind. His argument incorporates the nonprivileged position of the earth in the now widely accepted Copernican system (1543—ASTRONOMY) and the belief that God's wisdom is most manifest in the creation of life, which is needed for the contemplation of the magnificence of the Creation.

BIOLOGY (Blank Mind)

In his *Essay Concerning Human Understanding,* John Locke (1632–1704, English) concludes that at birth the mind is essentially a blank sheet and that all knowledge gained derives from experience. The mind through reflection organizes and perhaps expands the ideas formed from experience.

MATHEMATICS (Calculus; Tautochrone)

Jakob Bernoulli (1654–1705, Swiss) uses the technique of separation of variables—earlier discovered by Gottfried Wilhelm Leibniz—to solve for the equation of the tautochrone, the tautochrone being the locus of points in a vertical plane such that the time of descent under gravity alone to the lowest point from any other point is the same.

PHYSICS (Wave Theory of Light)

Christiaan Huygen's (1629–1695, Dutch) *Traité de la lumière* [Treatise on light; written in 1678, published in 1690] provides an explicit development of the wave theory of light, including elaboration of how the wave theory can account for reflection, refraction, and rectilinear propagation. Space is assumed to be permeated by hard, elastic ether particles, and light is assumed to be a nonmaterial pulse traveling through the ether in the manner that sound travels through air, with pulses of compression and rarefaction. The nonmaterial nature of light is concluded from the observed crossing of different light rays without major hindrance.

1691

BIOLOGY (Classification)

John Ray (1627–1705, English) presents a major classification of organic life in *The Wisdom of God Manifested in the Works of Creation,* bringing some unity to the vast array of new plants and animals being discovered during the voyages of discovery around the world over the previous few centuries.

MATHEMATICS (Polar Coordinates)

Jakob Bernoulli (1654–1705, Swiss) invents and publishes a paper on polar coordinates, with points located by reference to a fixed point and a line through that point. Although Isaac Newton had earlier also devised such a coordinate system (1671—MATHEMATICS), his work was not known, so that the credit for the discovery generally goes to Bernoulli.

MATHEMATICS (Rolle's Theorem)

Michel Rolle (1652–1719, French) presents as a sideline in a book on geometry and algebra the result later termed Rolle's Theorem. Using modern terminology, this theorem states that if $f(x)$ is differentiable from a to b, $f(a) = 0$, and $f(b) = 0$, then $f'(x) = 0$ has a root between a and b.

1692

BIOLOGY (Bacteria)

Antony van Leeuwenhoek (1632–1723, Dutch) describes the living creatures (bacteria) lodged between the teeth in the human mouth.

MATHEMATICS (Differential Equations)

Gottfried Wilhelm Leibniz (1646–1716, German) reduces homogeneous linear differential equations of the first order to quadratures.

1693

MATHEMATICS (Calculus)

The first published account of Isaac Newton's (1642–1727, English) method of fluxions (his version of the calculus) appears as a brief section entitled "Quadrature of Curves" in volume 2 of John Wallis's (1616–1703, English) *Opera mathematica,* this volume being an expanded Latin translation of Wallis's *Algebra* (1685—MATHEMATICS). Volume 1 of Wallis's *Opera mathematica* also appears in 1693, whereas volume 3 appears in 1699.

MATHEMATICS (Statistics, Mortality Tables)

Edmond Halley (1656–1743, English) constructs mortality tables for Breslau, presenting death rates as a function of age.

c.1693

EARTH SCIENCES (Earth History)

Gottfried Wilhelm Leibniz (1646–1716, German) speculates, from consideration of exposed rocks, that the earth was once smooth, incandescent, and molten. Upon sufficient cooling, a crust formed and much of the atmospheric water vapor condensed to water, thereafter forming the world's oceans. These waters became salty as they dissolved soluble ingredients of the crust. The crustal face of the earth has had many major changes, resulting both from internal heat and from water. Leibniz publishes a short statement of his views in 1693, but the fuller statement appearing in his *Protogaea* is not published until 1749, long after his death.

1694

BIOLOGY (Classification; Plant Sexuality)

Rudolph Jakob Camerarius (1665–1721, German) elaborates the plant classification scheme of Andrea Cesalpino (1583—BIOLOGY), incorporating the nature and number of stamens and pistils. Following careful observation and experiment, Camerarius develops the clearest picture to date of sexual reproduction in plants, presenting his descriptions in *De sexu plantarum epistola*.

EARTH SCIENCES (Noah's Flood)

Edmond Halley (1656–1743, English) submits a paper to the Royal Society suggesting that Noah's Flood occurred during an unusually close passage of a large comet, the nearness to the earth producing the crucial tidal wave.

MATHEMATICS (Calculus)

Although accepting the results being obtained with the emerging calculus, Bernard Nieuwentijt (1654–1718, Dutch) criticizes the methods being employed, in particular objecting to the vagueness of Isaac Newton's "evanescent quantities" and what he sees as a failure on Gottfried Wilhelm Leibniz's part to provide a clear definition of higher-order differentials.

MATHEMATICS (Lemniscate)

Jakob Bernoulli (1654–1705, Swiss) introduces the lemniscate, a symmetric self-intersecting curve resembling a figure eight and defined by the condition that the product of the distances of any point on the curve from two fixed points is $(d/2)^2$, where d is the distance between the fixed points.

PHYSICS (Temperature Scales)

Anticipating aspects of the Celsius temperature scale (1742—PHYSICS), Carlo Renaldini (1615–1698, Italian) recommends universal acceptance of the melting point of ice and the boiling point of water as two fixed points on all temperature scales. The recommendation is not accepted, thereby prolonging the widespread problems resulting from the absence of an accepted standard in temperature measurements. Among the reasons for the lack of acceptance is the uncertainty regarding whether the melting and boiling points are indeed invariant.

SUPPLEMENTAL (History of Science)

Publication of William Wotton's (1666–1727, English) *Reflections upon Ancient and Modern Learning,* one of the earliest works on the history of science.

1695

EARTH SCIENCES (Earth History; Stratigraphy; Fossils)

John Woodward (1665–1728, English) publishes *An Essay Towards a Natural History of the Earth,* in which he supplements his own extensive observations of earth composition, strata, and fossils by responses received to questionnaires he distributed worldwide. He concludes that throughout the world stone is layered in strata and enclosed within it are shells or fossils. He believes the fossils to be the remains of once-living plants and animals, a viewpoint which by this time has both vocal opponents and vocal proponents. Like others, he believes the earth to have been filled with water until the crust broke and the water poured out during Noah's Flood. Fossils are among the relics of the Flood, left behind as the waters retreated.

HEALTH SCIENCES (Salt Wells)

Nehemiah Grew (1641–1712, English) analyzes the health value of the salt wells at Epsom, Surrey, believed since 1618 to have a healing effect.

MATHEMATICS (Calculus; Symbolic Methods)

Gottfried Wilhelm Leibniz (1646–1716, German) notes that the operator d used in differentiation satisfies certain common algebraic properties, among these being the addition of integral exponents when multiplying two quantities with the same base.

SUPPLEMENTAL (Encyclopedias)

Publication of the French encyclopedia *Historical and Critical Dictionary.*

1696

EARTH SCIENCES (Earth History)

William Whiston (1667–1752, English) proposes in his *New Theory of the Earth* that the earth was stationary until the Fall of Man, when it began to rotate on its axis, and that Noah's Flood began on November 18, 2349 B.C., as a comet passed over the equator, causing extensive rains, and the waters from the earth's interior broke out and inundated the land. The sediments of the Flood created the now-visible stratified layers of the earth's crust.

HEALTH SCIENCES (Obstetric Forceps)

Hugh Chamberlen (English), in the preface to a translation of a work on obstetrics, refers to the obstetric forceps earlier invented by Peter Chamberlen (1572–1626, English) and used by the Chamberlen family for generations in their midwifery practice to ease difficult deliveries. Chamberlen mentions how much safer, faster, and less painful deliveries are when the forceps are used, although still, in the tradition of the Chamberlen family, refuses to reveal the details that would allow others to follow their methods.

MATHEMATICS (Integral Calculus)

Jakob Bernoulli (1654–1705, Swiss), who had in 1690 introduced the term "integral," and Gottfried Wilhelm Leibniz (1646–1716, German) agree to replace Leibniz's earlier "calculus summatorius" by "calculus integralis" for the inverse of the "calculus differentialis."

MATHEMATICS (L'Hospital's Rule)

Guillaume de L'Hospital (1661–1704, French) publishes *Analyse des infiniment petits,* probably the first textbook on differential calculus and one which continues to be influential for most of the eighteenth century. It contains L'Hospital's Rule that if two functions $f(x)$ and $g(x)$ each have a limit of 0 as $x \rightarrow a$, then the limit of the ratio of the two functions as $x \rightarrow a$ equals the limit of the ratio of the two derivatives, provided that the latter limit exists. In spite of its name, the rule was actually obtained by Johann Bernoulli (1667–1748, Swiss) in 1694, but L'Hospital and Bernoulli had an agreement whereby Bernoulli sent L'Hospital his mathematical discoveries, to be used as L'Hospital chose, in return for a regular salary. Much of the rest of the book is also based on the work of Bernoulli.

MATHEMATICS (Series)

Difficulties with power series representations are indicated when Jakob Bernoulli (1654–1705, Swiss) notes that at $x = -1$, $1/(1-x) = 1/2$ and yet the series for $1/(1-x)$ is $1 + x + x^2 + \ldots$, which at $x = -1$ yields $1 - 1 + 1 - 1 + \ldots$.

1697

CHEMISTRY (Nomenclature)

In a revised edition of his *Cours de chymie* (first edition published in 1675), Nicolas Lémery (1645–1715, French) takes care to increase the clarity of chemical terminology, reducing in particular the use of allegory and mysticism. Although objecting to the common association of planets and metals, he does retain some astrologically based names for some metals.

HEALTH SCIENCES (Bathing)

John Floyer (1649–1734, English) describes his research on bathing in "An Enquiry into the Right Use of Baths."

MATHEMATICS (Calculus of Variations)

A major advance toward a theory of the calculus of variations is made when Johann Bernoulli (1667–1748, Swiss) determines the curve down which an object would descend in the least time from a fixed point to a fixed vertical line, under the restriction that the only force acting on the object is the earth's gravity. By containing a variable end point, the problem is a conceptual step beyond earlier problems with both end points fixed. Johann and his brother Jakob Bernoulli (1654–1705, Swiss) also solve the corresponding problem with two fixed end points, finding that the brachystochrone, that is, the curve allowing the most rapid descent, is a cycloid.

1698

MATHEMATICS (Symbolism)

Gottfried Wilhelm Leibniz (1646–1716, German) introduces the dot as a symbol for multiplication, being unsatisfied with the standard $\times$ due to its confusion with the letter x.

PHYSICS (Steam Pump)

Thomas Savery (c.1650–1715, English) constructs a steam pump to raise water in mines. The force generated by steam pressure replaces that of teams of horses commonly used with other pumps. This is the first instance of a commercially successful steam engine.

1698–1700

EARTH SCIENCES (Latitude; Longitude; Magnetic Variations)

Under the leadership of Edmond Halley (1656–1743, English), the ship *Paramour Pink* undertakes the first specifically scientific voyage. The object is to record magnetic compass variations and to determine accurate latitudes and longitudes for various ports bordering the South Atlantic. Halley's resulting plots of the magnetic data (1701—EARTH SCIENCES) are thought to be the first isarithmic maps, showing lines of equal magnetic variation.

1699

CHEMISTRY (Gas Laws)

Guillaume Amontons (1663–1705, French) notes the uniform relationship between the volume and temperature of air.

EARTH SCIENCES (Ocean Currents)
METEOROLOGY (Atmospheric General Circulation)

William Dampier (1652–1715, English), in a *Discourse of Winds, Breezes, Storms, Tides and Currents,* describes maritime meteorological and oceanographic conditions and distinguishes the variable nature of the midlatitude westerlies and the steady nature of the low-latitude trade winds. He suggests that major ocean currents might be caused by winds, thereby moving beyond the earlier recognition that winds can cause temporary wave motions in the ocean. Previously currents had often been explained as resulting from ocean height differences produced by evaporation and rain, hence from the hydrological cycle rather than from the winds.

1700

CHEMISTRY (Neutralization)

Wilhelm Homberg (1652–1715, German) experiments with several acids, measuring the amount of base necessary to neutralize each.

CHEMISTRY (Phlogiston Theory)

Georg Ernest Stahl (1660–1734, German) initiates the phlogiston theory of combustion (later replaced by the oxygen theory of Antoine Lavoisier), asserting that any object that is capable of burning contains a substance "phlogiston," which it releases to the air during the burning process. An object stops burning when its full content of phlogiston is expelled. Stahl further asserts that the rusting of metals is similar to but simply slower than burning, not giving off phlogiston rapidly enough for it to become visible as a flame. In some respects Stahl's "phlogiston" replaces the "oily earth" of Johann Joachim Becher (1669—CHEMISTRY).

SUPPLEMENTAL (Berlin Academy)

Founding of the Berlin Academy of Sciences, with Gottfried Wilhelm Leibniz (1646–1716, German) as president.

c.1700

MATHEMATICS (Number Systems)

Gottfried Wilhelm Leibniz (1646–1716, German) develops the first generalized exposition of positional number systems, including the base 2, binary system.

1701

EARTH SCIENCES (Magnetic Variations)

Following the 1698–1700 voyage of the *Paramour Pink,* Edmond Halley (1656–1743, English) publishes the first charts of the Atlantic and Pacific showing contours of magnetic variation.

MATHEMATICS (Isoperimetric Figures; Calculus of Variations)

Jakob Bernoulli (1654–1705, Swiss) presents a solution to an isoperimetrical problem which he had openly challenged his brother Johann Bernoulli (1667–1748, Swiss) to solve in 1697. The problem is to obtain the parametric representation of the curve which has a given length L and maximizes a particular areal integral. The work done over the years by Jakob and Johann Bernoulli on this and other isoperimetrical problems contributes to the emergence of a calculus of variations.

PHYSICS (Acoustics)

Joseph Sauveur (1653–1716, French) introduces the term "acoustics" for the study of sound and examines the relations of the tones of the musical scale in the memoir "Système général des intervalles des sons, et son applications à tous les systèmes et à tous les instrumens de musique" [General system of sound intervals and its applications to all musical systems and instruments]. He shows that a string can vibrate at integral multiples of a fundamental frequency simultaneously with vibrating at the fundamental frequency itself, calling the tones produced by the vibrations at the multiples the "harmonics" of the fundamental tone.

PHYSICS (Newton's Law of Cooling)

Following thermometric experiments on the temperature changes as a hot iron cools while warming surrounding objects, Isaac Newton (1642–1727, English) suggests that the rate of cooling of a solid is proportional to its temperature. Later termed "Newton's Law of Cooling," Newton uses it to calculate temperatures beyond the range of the scale on his thermometer, which consists of a three-foot tube, a two-inch diameter bulb, linseed oil for the measuring liquid, and a scale determined by fixing points for the level of the liquid when measuring the temperature of melting snow and the level of the liquid when measuring the temperature of the human body.

1702

CHEMISTRY (Alkali)

Georg Ernest Stahl (1660–1734, German) draws the distinction between natural and artificial alkali, soda and potash.

CHEMISTRY (Boric Acid)

Wilhelm Homberg (1652–1715, German) makes the first preparation of boric acid ("sedative salt").

METEOROLOGY (Aneroid Barometer)

Gottfried Wilhelm Leibniz (1646–1716, German) outlines the basic principles of the yet-to-be-constructed aneroid barometer, describing in a letter a pocket-sized instrument measuring air pressure through the amount of compression undergone by a small metallic bellows.

METEOROLOGY (Thermometer)

Guillaume Amontons (1663–1705, French) invents the air pressure thermometer.

1703

MATHEMATICS (Convergence)

The current difficulties resulting from the absence of rigorous notions of series' convergence are exemplified in Guido Grandi's (1671–1742, Italian) *Quadratura circuli et hyperbolae* [Quadrature of circles and hyperbolas], where he examines, for example, $1/(1 + x) = 1 - x + x^2 - x^3 + x^4 - \dots$ at $x = 1$, obtaining $1/2 = (1 - 1) + (1 - 1) + (1 - 1) + \dots = 0$ and concluding that something (and hence everything) can be created from nothing.

1704

CHEMISTRY (Atomic Theory)

In his *Opticks* (1704—PHYSICS), Isaac Newton (1642–1727, English) proposes—similarly to the Greek atomists before him—that all matter is composed of small, hard, impenetrable particles that cannot be broken into smaller pieces.

MATHEMATICS (Calculus)

In an appendix to *Opticks* (1704—PHYSICS), Isaac Newton publishes a description of his method of fluxions (calculus), under the title "Tractatus de quadratura curvarum" [A treatise on the quadrature of curves] (see 1676—MATHEMATICS).

MATHEMATICS (Geometry; Cubic Curves)

In the treatise *Enumeratio linearum tertii ordinis* [Enumeration of curves of the third degree] appended to his *Opticks* (1704—PHYSICS), Isaac Newton distinguishes 72 types of cubic curves in the plane and plots an example of each. (Six additional types are determined later.) He defines a transcendental curve as one intersecting a straight line at an infinity of points; and he mentions various properties of cubics such as the impossibility of having greater than three asymptotes.

METEOROLOGY (Barometer)

Guillaume Amontons (1663–1705, French) determines that heat, in addition to pressure, affects barometers.

PHYSICS (Gravitation; Ether)

In an appendix to the *Opticks* (1704—PHYSICS), Isaac Newton proposes an explanation of gravitation, based on the existence of an all-pervasive ether consisting of very fine particles. An object falls toward the earth because the particles within it are repelled by the ether particles, which are less dense as the earth is approached.

PHYSICS (Nature of Light)

Isaac Newton (1642–1727, English) publishes detailed discussions of his many experiments on optical phenomena in *Opticks: A Treatise of the Reflections, Refractions, Inflections and Colours of Light.* Newton discusses why different objects appear different colors, how the placement of a prism with respect to the incident light affects the emerging light, how different surfaces affect the respective percentages of reflection and refraction, and many other optical issues. The work basically supports the theory that light is composed of particles which move in straight lines, at finite velocity, although Newton does include elements of the wave theory of light as well, in particular postulating that the light particles create vibrations within the all-pervasive ether.

SUPPLEMENTAL (Encyclopedias)

John Harris (1667–1719, English) completes the first edition of the general scientific encyclopedia *Lexicon Technicum. Or, an Universal Dictionary of Arts and Sciences.* The work is more technical and less critical and theoretical than the recently published *Historical and Critical Dictionary* (1695—SUPPLEMENTAL). Harris makes use of the works and advice of key contemporary scientists, including Isaac Newton, John Ray, Edmond Halley, Nehemiah Grew, and John Woodward. In volume 2, published in 1710, he includes an English translation ("Quadrature of Curves") of Newton's "Tractatus de quadratura curvarum" (1704—MATHEMATICS).

1705

ASTRONOMY (Comets; Halley's Comet)

Edmond Halley (1656–1743, English) publishes in *A Synopsis of the Astronomy of Comets* an analysis of all available comet observations. Halley assumes that the orbits of comets are determined by the same law of gravitation as are the orbits of planets (that law having been formulated by Isaac Newton, 1687—ASTRONOMY), and suggests that the bright comets observed in 1531, 1607, and 1682 are the same comet, reappearing with a 76 year periodicity. He correctly predicts a reappearance in 1759, and the comet is subsequently named Halley's Comet.

EARTH SCIENCES (Earthquakes; Fossils)

Robert Hooke's (1635–1702, English) "Lectures and Discourses of Earthquakes and Subterraneous Eruptions, explicating the Causes of the Rugged and Uneven Face of the Earth; and what Reasons may be given for the frequent finding of Shells and other Sea and Land Petrified Substances scattered over the whole Terrestrial Superficies" are published posthumously along with other lectures left in manuscript form at his death. In these lectures, the first of which was probably given orally in 1668, Hooke advances the theory that earthquakes have had major consequences in raising and lowering portions of the surface of the earth, and that the earth has, due in large part to earthquakes, undergone substantial changes since its creation. Hooke also deals with fossils, arguing, in agreement with scattered individuals before him, that fossils are the remains of or impressions left by once-living organisms, further asserting that fossils can reveal much about the history of the earth and should be examined from that perspective. They suggest, for instance, that much of the earth was once submerged under water, and this for a much longer period than Noah's Flood.

MATHEMATICS (Series)

Gottfried Wilhelm Leibniz (1646–1716, German) notes a sufficient condition for the convergence of an alternating series.

PHYSICS (Electricity)

Francis Hauksbee (c.1666–1713, English) observes that when a glass vessel containing mercury is shaken light is produced, an effect that he attributes to friction.

c.1705

BIOLOGY (Blood Pressure)

Stephen Hales (1677–1761, English) initiates experiments on methods of determining blood pressure in man and other animals. The results are published decades later in Hales's *Haemastaticks* (1733—BIOLOGY).

1706

MATHEMATICS (π)

William Jones (1675–1749, English) introduces the Greek letter π to use as a symbol for the ratio of the circumference to the diameter of a circle. It gains popularity once Leonhard Euler adopts it in 1739.

PHYSICS (Electricity)

Francis Hauksbee (c.1666–1713, English) constructs a machine for generating electricity. The machine consists of a glass globe mounted in a manner to allow spinning, with a valve in the globe to allow extraction and reintroduction of air. The arrangement becomes a centerpiece of many popular scientific lectures across Europe over the succeeding decades.

1707

BIOLOGY (Soul)

In *Theoria medica vera,* Georg Ernest Stahl (1660–1734, German) theorizes that the structure as well as the function of the body is controlled by the soul.

HEALTH SCIENCES (Pulse Rate)

John Floyer (1649–1734, English) initiates the counting of pulse beats, describing the procedure in *The Physician's Pulse Watch.*

MATHEMATICS (Algebra; Newton's Identities)

Isaac Newton (1642–1727, English) publishes *Arithmetica universalis,* written in 1673–1683 in conjunction with his lectures on algebra during that period. Included are "Newton's identities" providing expressions for the sums of the ith powers of the roots of any polynomial equation, for any integer i, plus a rule providing an upper bound for the positive roots of a polynomial, and a generalization, to imaginary roots, of René Descartes's Rule of Signs (1637—MATHEMATICS).

1709

BIOLOGY (Descriptive)

In *A New Voyage to Carolina,* the explorer John Lawson (d.1711, English) describes the flora and fauna of the Carolinas.

EARTH SCIENCES (Fossils)

Johann Scheuchzer (1672–1733, Swiss) publishes his *Complaints and Claims of the Fishes,* in which fossil fish are made to argue against the interpretation that they were never living, organic beings. They also claim to have witnessed Noah's Flood, thereby explaining their existence at altitudes currently well above sea level.

SUPPLEMENTAL (Philosophy of Science)

In a criticism centered on Isaac Newton's *Principia* (1687—ASTRONOMY), George Berkeley (1685–1753, Irish) objects to attempts at reducing phenomena to general rules, criticizing them as "grammatical" studies of nature, where the meaning is secondary to the grammar employed.

1711

MATHEMATICS (Calculus)

Isaac Newton's (1642–1727, English) *De analysi per aequationes numero terminorum infinitas* (1669—MATHEMATICS) is finally published.

1712

ASTRONOMY (Star Catalogs)

Volume 1 of John Flamsteed's (1646–1719, English) *Historia coelestis,* the first of the star catalogs from Greenwich Observatory, appears through an unauthorized printing arranged by Edmond Halley (1656–1743, English) and Isaac Newton (1642–1727, English), who are anxious for certain portions of the data and claim that Flamsteed, as Astronomer Royal, is excessively slow in making his observations available in published form. Flamsteed succeeds in getting the edition suppressed, and the official, more carefully edited three-volume edition appears posthumously (1725—ASTRONOMY).

MATHEMATICS (Newton-Leibniz Dispute)

A committee appointed by the Royal Society issues a report entitled *Commercium epistolicum* concluding that Isaac Newton had developed the calculus before Gottfried Wilhelm Leibniz and containing implications that Leibniz might have plagiarized. This is in the midst of continuing heated arguments, largely divided along national lines, over which of the two men should be credited with the invention.

1713

MATHEMATICS (Difference Equations)

In the posthumous work *Ars conjectandi* (1713—MATHEMATICS), Jakob Bernoulli defines Bernoulli numbers and Bernoulli polynomials.

MATHEMATICS (Probability and Statistics)

Posthumous publication of Jakob Bernoulli's (1654–1705, Swiss) *Ars conjectandi,* in which he develops the science of probability, proves the binomial theorem for positive integral exponents, and states Bernoulli's Theorem that any desired degree of statistical accuracy can be obtained by sufficiently increasing the number of observations.

1714

MATHEMATICS (Cotes Formula)

Roger Cotes (1682–1716, English) establishes the key formula: $ix = \log_e(\cos x + i \sin x)$, where $i = \sqrt{-1}$ and e is the base of the natural logarithms.

PHYSICS (Temperature Scales)

Daniel Fahrenheit (1686–1736, German) devises the Fahrenheit temperature scale, tied to two points, 0°F for the temperature of a mixture of ice, water, and sal-ammoniac, and 96°F for the body temperature of an average healthy human, found by placing the thermometer in the mouth or armpit. Fahrenheit believed the first to be the coldest artificially produced temperature. On this scale, water freezes at 32°F and boils at 212°F. Fahrenheit shows that the boiling point varies somewhat with atmospheric pressure.

1715

ASTRONOMY (Solar Eclipse)

Due to the efforts of Edmond Halley (1656–1743, English), for the first time a total solar eclipse is carefully timed and recorded by a large number of observers throughout Europe. Halley not only organized the observations of the eclipse, but also accurately predicted and mapped the expected path of the moon's shadow across England before the April 22 event, in part to forestall superstitious fears.

EARTH SCIENCES (Ocean Salinity)

Edmond Halley (1656–1743, English) writes a treatise on the salinity of the oceans and major seas.

HEALTH SCIENCES (Smallpox Inoculation)

Giacomo Pylarini (1659–1718, Italian) describes his research into smallpox inoculation in the work *Nova et tuta variolas excitandi per transplantationem methodus; nuper inventa et in usum tracta.* Included is a description of his 1701 inoculation of three children with smallpox virus in Constantinople, Turkey. Contemporaneously, Emmanuel Timoni (fl. 1714) writes "An account, or history, of the procuring of the smallpox by incision or inoculation, as it has for some time been practised at Constantinople," a letter written to John Woodward and published in the *Philosophical Transactions* in 1714–1716.

1715–1717

MATHEMATICS (Taylor and Maclaurin Series)

Brook Taylor (1685–1731, English) originates the study of the calculus of finite differences in his *Methodus incrementorum directa et inversa.* The book includes the first published statements both of the Taylor series expansion of the function $f(x)$ about the point a, discovered by Taylor a few years earlier, and of the specific subcase with $a = 0$, later termed the Maclaurin series (1742—MATHEMATICS).

1716

ASTRONOMY (Earth-Sun Distance)

Edmond Halley (1656–1743, English) recommends calculation of the sun-earth distance through analysis of transits of Venus across the solar disk. The method is put into practice at the next transit of Venus, but this is not until after mid-century (1761—ASTRONOMY).

EARTH SCIENCES (Fossils)

Johann Scheuchzer (1672–1733, Swiss) publishes a catalog of his sizable fossil collection.

EARTH SCIENCES (Glaciology)

Publication of Jean-Jacques D'Ortous de Mairan's (1678–1771, French) *Dissertation sur la glace,* on ice and freezing. The work also contains sections on optics, particularly refraction and color.

MATHEMATICS (Elliptic Integrals)

Giulio Fagnano (1682–1766, Italian) helps inaugurate the study of elliptic integrals when he shows the difference of any two elliptic arcs to be algebraic.

PHYSICS (Ether)

John Theophilus Desaguliers (1683–1744, English) performs an experiment before the Royal Society which is interpreted as providing evidence for a mechanical ether. The experiment, apparently done at the request of Royal Society president Isaac Newton, involves warming two thermometers in separate glasses, one with its air removed. The thermometer in the vacuum warms almost as much, although not as rapidly, as the other thermometer. Desaguliers's written version of the experiment—"An account of an experiment to prove an interspersed vacuum"—is published in 1717.

1716–1718

EARTH SCIENCES (Geology)

Johann Scheuchzer (1672–1733, Swiss) describes his geological studies of the Alps in *Helvetiae stoicheiographia.*

1717

HEALTH SCIENCES (Smallpox Inoculation)

Mary Montagu (1689–1762, English) reports on a smallpox inoculation procedure of ingrafting common in Turkey: smallpox matter is inserted into the vein by needle, after which the wound is bound and the patient typically remains healthy for eight days, then suffers a fever for two to three days, then emerges healthy and protected against the disease (see also 1715—HEALTH SCIENCES). A second method for inoculating against smallpox is imported to England from China at about the same time. This latter method consists of soaking a thread in the fluid of a smallpox pustule and then drawing it through a scratch in the patient's arm.

PHYSICS (Polarization)

Isaac Newton (1642–1727, English) includes his discovery of the polarization of light in the second edition of his *Opticks* (first edition 1704—PHYSICS).

1718

ASTRONOMY (Proper Motions)

Edmond Halley (1656–1743, English) publishes data on the current celestial coordinates of Sirius, Aldebaran, and Arcturus, three of the brightest stars listed in Ptolemy's *Almagest,* and shows that the changes in their positions, after adjusting for precession, are too great to be attributed to observational errors. Hence he determines the "proper motions" of the stars examined, and, more fundamentally, shatters the belief in a realm of fixed stars.

CHEMISTRY (Affinity)

Étienne Geoffroy (1672–1731, French) asserts that if substances *A* and *B* are united but *C* is brought into contact with them and *C* has more affinity to unite with *A* than *B* has, then *A* and *B* will break apart and *A* and *C* will unite. Based on this concept, Geoffroy presents a 16-column table listing, for each of 16 substances, a set of other substances which can unite with it, in order of apparent affinity. This, presented in a work entitled *Table des différents rapports observés entre différentes substances,* is the first of many "Tables of Affinity." Geoffroy lists the substances according to their alchemical symbols, helping to revive use of such symbolism.

HEALTH SCIENCES (Mastectomy)

Zabdiel Boylston (1679–1766, American) performs a mastectomy.

MATHEMATICS (Probability)

Abraham de Moivre (1667–1754, French-English) publishes a systematic treatise on probability entitled *Doctrine of Chances.*

1719

EARTH SCIENCES (Paleogeology)

Emanuel Swedenborg (1688–1772, Sweden) asserts from empirical evidence that Scandinavia was once covered by an ocean, presenting his thesis in *Om watnens högd och förra werldens starcka ebb och flod* [On the level of the seas and the great tides in former times]. The evidence discussed includes sedimentary deposits, gravel ridges, and the fact that the land is rising along the Baltic coast.

EARTH SCIENCES (Stratigraphy)

John Strachey (1671–1743, English) identifies and sketches the existence of strata in the earth's crust. He presents observations on the sequences of strata in coal fields and surrounding hills in southwest England.

1720

HEALTH SCIENCES (Scurvy)

J. G. H. Kramer (Austrian) notes that fresh vegetables have a curative effect on soldiers stricken with scurvy.

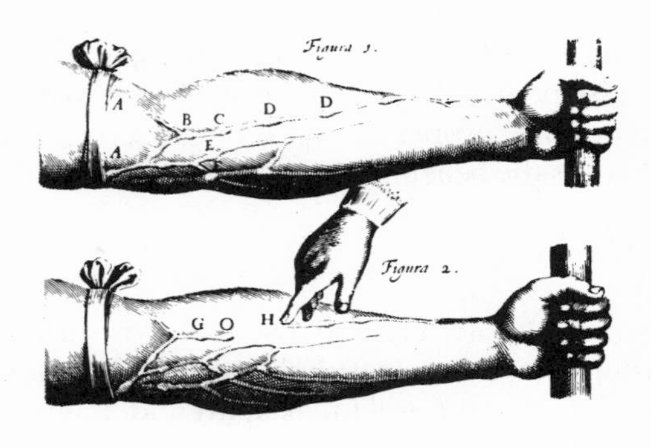

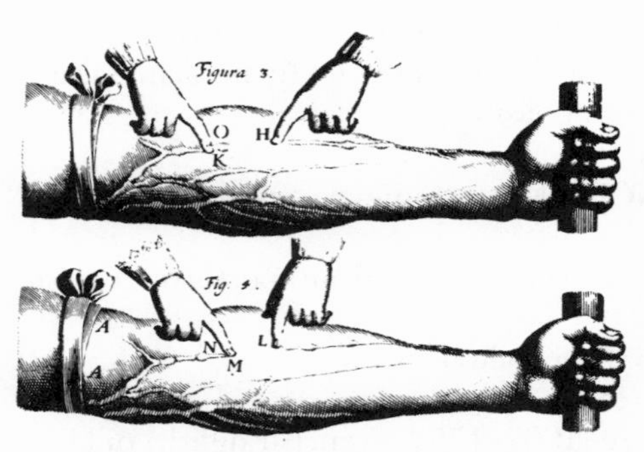

FIGURE 6. *An illustration from William Harvey's* De motu cordis *[On the motion of the heart] (see 1628—*BIOLOGY *entry), demonstrating the circulation of the blood through the simple experiment of applying pressure to a vein in an arm lightly bound above the elbow. (By permission of the Countway Medical Library.)*

MATHEMATICS (Geometry)

Colin Maclaurin (1698–1746, Scottish) publishes *Geometria organica* on the general properties of planar curves. He determines an expression for the maximum number of double points on an irreducible curve of degree n, generalizes from individual cases to conjecture that there are mn points of intersection of two curves of degrees m and n, presents various properties of conics and curves inscribed in conics, and notes that $n(n + 3)/2$ points do not always uniquely determine a curve of the nth order. Furthermore, he uses cinematic methods to describe curves of different degrees and shows that the cubic and quartic can be described by rotating about their vertices two angles of fixed size.

MATHEMATICS (Partial Differentiation)

Nikolaus Bernoulli (1687–1759, Swiss) asserts that when the partial derivative of a function $f(x,y)$ of two variables is taken with respect to one variable and the result is then differentiated with respect to the other variable, the net result is independent of the order of differentiation. This is later proved by Leonhard Euler (1755—MATHEMATICS).

1721

EARTH SCIENCES (Springs)

Antonio Vallisneri (1661–1730, Italian) denounces the accepted belief that springs arise through the recycling of sea water through subterranean passages.

HEALTH SCIENCES (Smallpox Inoculations)

Violent controversy flares in Massachusetts as Zabdiel Boylston (1679–1766, American) attempts to arrest an epidemic of smallpox and provide long-term protection by inoculating the populace against the disease. The inoculations had been suggested by Cotton Mather (1663–1728, American) after hearing from a slave that the procedure had been carried out successfully in Africa. The report later published by Boylston and Mather is noted for its early use of statistical techniques to determine the value of a medical procedure. They judge the procedure a success since far fewer deaths occurred from artificially induced infection due to the inoculation than the expected number of smallpox deaths in the same group if allowed normal exposure to the disease.

1723

METEOROLOGY (Observational Network)

James Jurin (English), secretary of the Royal Society, recommends annual submission by all weather observers of their meteorological observations. Jurin emphasizes standardization, recommending specific thermometer and barometer models, conditions for where to place the instruments, and frequency and type of measurements to be recorded. He distributes a standard form encouraging at least daily measurements of pressure, temperature, wind force and direction, rain, snow, and cloud cover, with greater frequency suggested during storms. Monthly and yearly averages are to be calculated at each station, with a summary of all measurements published each year in the Royal Society's *Transactions*. The system works successfully for a few years, with participation from observers in Sweden, Finland, North America, Italy, and India as well as England.

1724

HEALTH SCIENCES (Geriatrics)

John Floyer (1649–1734, English) writes the first book on geriatrics: *Medicina gerocomica.*

1725

ASTRONOMY (Star Catalogs)

Posthumous publication of the three-volume *Historia coelestis Britannica* of John Flamsteed (1646–1719, English) is completed under his former assistants Joseph Crosthwait and Abraham Sharp. The work includes the vast majority of Flamsteed's observations over four decades of work as Royal Astronomer.

EARTH SCIENCES (Oceanography)

Luigi Marsili (1658–1730, Italian) publishes a monograph on oceanography in which he discusses bottom topography, ocean plants and animals, and ocean circulation. He reports tidal measurements and water temperature measurements to 120 fathom depths.

1726

EARTH SCIENCES (Noah's Flood; Fossils)

Johann Scheuchzer (1672–1733, Swiss) interprets remains found in the quarries near Oeningen, Germany, as fossilized skeletal elements of sinners who perished in the biblical Noah's Flood, classifying the finds as *Homo diluvii testis* [The man who witnessed the flood]. A century later Georges Cuvier, in 1825, reports that he has reexamined what is considered to be the best of Scheuchzer's specimens and has determined the skeleton to be from a salamander rather than a human.

HEALTH SCIENCES (Emphysema)

John Floyer (1649–1734, English) provides one of the first medical descriptions of the effects of emphysema on the lungs.

1727

BIOLOGY (Plant Physiology)

In *Vegetable Staticks,* Stephen Hales (1677–1761, English) reports on years of experiments regarding the flow of fluids in plants and plant respiration. Hales quantifies his explanations with data on root pressure, leaf growth, and other variables. After attaching glass tubes to cut branches, he measures the level to which sap rises under varying plant conditions, weather conditions, and time of day. The book, which also describes experiments determining the generation or depletion of air by various substances, contributes to the later recognition of Hales as one of the founders of the study of plant physiology.

1727–1728

MATHEMATICS (*e*)

Leonhard Euler (1707–1783, Swiss) suggests symbolizing the base of the natural logarithms by the letter *e*. He uses the symbol 16 times in a manuscript entitled "Meditatio in experimenta explosione tormentorum nuper instituta" [Meditation upon experiments made recently on the firing of canon], believed to have been written in 1727 or 1728 although not actually published until 1862.

1727–1747

ASTRONOMY (Star Catalogs)

James Bradley (1693–1762, English) catalogs the positions of 3,000 stars to an accuracy of a few seconds of arc.

1728

ASTRONOMY (Stellar Aberration)

James Bradley (1693–1762, English), in attempting to measure the distance to stars by determining the stellar parallax, is unable to find a non-zero parallax but does discover stellar aberration—the angular displacement of the apparent direction of starlight due to the earth's motion—and correctly attributes it to the combined effect of the finite velocity of light and the earth's orbital velocity. He publishes a description of his discovery in 1729 as "An Account of a New Discovered Motion of the Fixed Stars."

HEALTH SCIENCES (Dentistry)

Pierre Fauchard (1678–1761, French) publishes a comprehensive account of the field of dentistry, the two-volume *Le chirurgien dentiste, ou traité des dents* [The surgeon dentist, or treatise of the teeth], which leads later to his being regarded as the "Father of Dentistry." He includes detailed discussions of the treatment of caries, the making and use of removable dentures, and a variety of dental instruments. After removal of the carious material, with pain reduced through application of oil of cinnamon, the cavity is filled in with small pieces of thin foil of tin, lead, or gold.

MATHEMATICS (Fourier Series)

Daniel Bernoulli (1700–1782, Swiss) authors the first publication containing the Fourier series' equations, obtaining them as the solution to the one-dimensional wave equation for the motion of a string confined by fixed end points and vibrating in a plane.

METEOROLOGY (Observational Network)

Isaac Greenwood (1702–1745, American) recommends that the Royal Society expand its meteorological observations from the plans laid down by James Jurin (1723—METEOROLOGY) by additionally compiling all ship weather observations. Greenwood encourages compilation and analysis of past as well as current observations in search of an understanding of typical large-scale patterns and the typical timing, location, and movement of major storm systems such as hurricanes. Work proceeds enthusiastically at first, although continues for only a few years.

SUPPLEMENTAL (Religion and Science: Creation Date)

Isaac Newton's (1642–1727, English) *Chronology* is published posthumously. The work is the result of much effort on Newton's part to determine a chronology of biblical and other events that satisfies an assumed 4004 B.C. date of Creation. In order to shorten world history to such an extent, Newton resorts to such measures as assuming several different rulers are really the same person.

1728–1729

EARTH SCIENCES (Fossils)

John Woodward (1665–1728, English) uses the knowledge gained from years of collecting fossils to produce the two-volume *Attempt Towards a Natural History of the Fossils of England, or a Catalogue of English Fossils.*

1729

MATHEMATICS (Eulerian Integrals)

Leonhard Euler (1707–1783, Swiss) introduces the functions later to be known as Eulerian integrals of the first and second kind, or beta and gamma functions.

PHYSICS (Electricity)

Intrigued by Francis Hauksbee's primitive electrical generator (1706—PHYSICS), Stephen Gray (1666–1736, English) finds that by rubbing a corked glass tube he can cause an 800-foot thread attached to the cork to attract a feather at the other end, thereby demonstrating that the attractive force is carried at least that distance. He also demonstrates that metals can conduct current and concludes that electricity is a fluid.

1730

MATHEMATICS (De Moivre's Theorem)

Abraham de Moivre (1667–1754, French-English) derives De Moivre's Theorem, $\cos nx + i \sin nx = (\cos x + i \sin x)^n$, for positive integral n, a ready consequence of an earlier formula of Roger Cotes (1714—MATHEMATICS).

MATHEMATICS ($n!$)

James Stirling (1692–1770, Scottish) publishes *Methodus differentialis,* in which he presents both an approximation, for large n, to $n!$ and an infinite series representation for $\log n!$. The former, equal to $\sqrt{2 \pi n}\,(n/e)^n$, has since been termed Stirling's approximation, and the latter has been termed the Stirling series.

PHYSICS (Réaumur Temperature Scale)

René de Réaumur (1683–1757, French) constructs an alcohol thermometer, using a temperature scale which is later named after him. The scale is based on 0° for the freezing point of water and 80° for the boiling point of water, so that 0°R = 0°C and 80°R = 100°C, where C refers to the Centigrade scale, constructed later (1742 and 1743—PHYSICS), and R to the Réaumur scale.

1731

HEALTH SCIENCES (Dieting)

John Arbuthnot (1667–1735, Scottish) advocates dieting in his "Essay Concerning the Nature of Ailments."

MATHEMATICS (Solid Analytic Geometry)

While still a teenager, Alexis Claude Clairaut (1713–1765, French) publishes *Recherches sur les courbes à double courbure* on solid analytic geometry, a pioneering study of the differential geometry of space curves.

1732

MATHEMATICS (Number Theory)

Leonhard Euler (1707–1783, Swiss) disproves Pierre de Fermat's conjecture that all numbers of the form

$$2^{2^t} + 1$$

are prime (1640—MATHEMATICS) by showing that

$$2^{2^5} + 1 = 6{,}700{,}417 \times 641.$$

PHYSICS (Electricity)

Stephen Gray (1666–1736, English) distinguishes conductors and insulators and announces his discovery of electrical induction.

1733

ASTRONOMY (Telescope; Achromatic Lens)

Chester Hall (1703–1771, English) constructs an achromatic telescope by inserting a two-lens objective, with the glass of one lens slightly denser than that of the other.

BIOLOGY (Blood Flow and Pressure)

Stephen Hales (1677–1761, English) relates his experiments on the flow and force of naturally circulating blood in *Haemastaticks: Or an Account of Some Hydraulic and Hydrostatical Experiments Made on the Blood and Blood-Vessels of Animals.* Complementing his experiments on fluid flow in plants (1727—BIOLOGY), these experiments with a variety of live animals, including frogs, horses, dogs, sheep, oxen, doe, and humans, lead him to the discovery of fundamental characteristics of blood circulation. For instance, he demonstrates the lesser resistance and greater velocity of blood in the pulmonary versus systematic circulation, measures a speed of approximately one and a half feet per second for the blood in the aorta, and determines the blood pressure in different animals under a variety of conditions. As he measures the height that blood rises in a glass tube attached to ligated arteries of different animals, he finds in some instances that the blood rises to a height exceeding eight feet.

BIOLOGY (Climatic Impact)

In *An Essay Concerning the Effects of Air on Human Bodies,* John Arbuthnot (1667–1735, Scottish) discusses the impact of the air on the fibers of plants and animals, including humans. He includes physicochemical explanations and experimental verification, and extends his discussion to the large-scale picture, including the influence of air temperature and moisture on molding characteristic qualities of various nationalities.

MATHEMATICS (Non-Euclidean Geometry)

The Jesuit priest Girolamo Saccheri (1667–1733, Italian) completes *Euclides ab omni naevo vindicatus* [Euclid vindicated from all fault], a work designed to demonstrate the necessity of Euclid's Parallel Postulate (the fifth postulate of Euclidean geometry, stating that through a given point not on a given line there is exactly one line parallel to the given line in the plane determined by the line and point) by assuming alternatives and showing their absurdity. In order to do this, Saccheri develops the consequences of retaining the first four postulates while denying the fifth and thereby inadvertently lays the foundations of non-Euclidean geometry. He actually develops the first two non-Euclidean geometries, although they are not recognized by him as such, and indeed another 90 years pass before it is realized that the Parallel Postulate is not a necessary consequence of the other four.

MATHEMATICS (Statistics; Normal Distribution)

The work of Abraham de Moivre (1667–1754, French-English) on aspects of the binomial expansion leads him to the discovery of the normal distribution function as an approximation to the binomial law. He presents the function in a seven-page unpublished paper entitled "Approximatio ad summam terminorum binomii $\overline{a+b}\backslash^{n}$ in seriem expansi"[A method of approximating the sum of the terms of the binomial $\overline{a+b}\backslash^{n}$ expanded into a series].

METEOROLOGY (Aurora Borealis)

In *Physical and Historical Treatise on the Aurora Borealis,* Jean-Jacques D'Ortous de Mairan (1678–1771, French) disputes the explanation that the northern lights are caused by volcanic exhalations and suggests instead that the cause derives from the sun's atmosphere.

1734

ASTRONOMY (Cosmology)

Emanuel Swedenborg (1688–1772, Sweden) publishes *Principia rerum naturalium,* in which he seeks a comprehensive world system based on mathematical and mechanical principles. He believes the world to have developed ultimately from a mathematical point and the solar planetary system to have developed from the sudden explosion of an expanding shell of material from the sun concentrated along the solar equatorial plane.

EARTH SCIENCES (Mountains)

Marie-Pompée Colonne (Italian) contends that mountains grow similarly to plants, explicitly presenting the mountain/plant analogy in volume 1 of his *Histoire naturelle de l'univers.*

MATHEMATICS (Calculus)

George Berkeley (1685–1753, Irish) publishes *The Analyst; Or a Discourse Addressed to an Infidel Mathematician* (the infidel being Edmond Halley), criticizing the lack of rigor in the calculus works of Isaac Newton, Gottfried Wilhelm Leibniz, and others, and suggesting that any infidel (specifically Edmond Halley) who criticizes points of religious faith on the basis of lack of rigor should consider whether the lack of rigor there is any more than the lack of rigor in the study of calculus. Berkeley refers to fluxions as "ghosts of departed quantities . . . he who can digest a second or third fluxion . . . need not, methinks, be squeamish about any point in Divinity."

PHYSICS (Electricity; Two-Fluid Theory)

Charles François de Cisternai DuFay (1698–1739, French) distinguishes positive and negative electricity, terming them "vitreous" and "resinous," based on their respective properties of being obtainable by rubbing glass and by rubbing resin. (The modern names are introduced in 1750 by Benjamin Franklin.) DuFay regards the two electricities as distinct and separable, and formulates a two-fluid theory of electricity based upon them.

SUPPLEMENTAL (Spread of Newtonian Science)

François Marie Arouet de Voltaire's (1694–1778, French) *Lettres sur les Anglais* [Letters on the English], including four chapters discussing Newtonian natural philosophy, helps spread Newtonian science through the European intellectual community.

1734–1742

BIOLOGY (Insects)

Publication of the six-volume *History of the Insects* by René de Réaumur (1683–1757, French), who elsewhere shows that corals are animals rather than plants.

1735

BIOLOGY (Plant Classification)

Carolus Linnaeus (1707–1778, Swedish) presents his first plant classification in *Systema naturae* [System of nature], classifying plants according to the number of stamens and pistils. Linnaeus devotes much of his life to discovering and classifying the plants of the world, though he also classifies minerals, diseases, and animals.

MATHEMATICS (Calculus)

Benjamin Robins (1707–1751, English) undertakes a revision of Isaac Newton's calculus of fluxions in which he eliminates all reference to quantities becoming "infinitely small."

MATHEMATICS (Topology; Koenigsberg Bridge Problem)

Leonhard Euler (1707–1783, Swiss) solves the Koenigsberg bridge problem, proving that it is not possible to cross the seven bridges in Koenigsberg each exactly once in one continuous walk. More generally, in the process of solving this problem he proves that a network can be traversed without overlap if and only if the number of its odd vertices is either zero or two.

METEOROLOGY (Trade Winds; Hadley Cell)

George Hadley (1685–1768, English), using the concept of the conservation of angular momentum, revises Edmond Halley's earlier explanation of the trade winds (1686—METEOROLOGY), doing so in his paper "Concerning the cause of the general trade winds." According to Hadley, solar heating causes air near the equator to rise, the risen air moves poleward, cools, sinks, and returns equatorward, producing a cell circulation in each hemisphere. However, the low-latitude surface air moving toward the equator has a smaller linear component of west-to-east motion due solely to the earth's rotation about its axis than does the equator itself. This results, from the perspective of the earth-bound observer, in an easterly component to the low-latitude trade winds, thus accounting for northeasterly trades in the Northern Hemisphere and southeasterly trades in the Southern Hemisphere. Hadley similarly explains the southwesterlies in the northern mid-latitudes and the northwesterlies in the southern mid-latitudes, the earth's rotation inducing a west-to-east component in the equator-to-pole flow.

1735–1744

EARTH SCIENCES (Geoid)

Charles de la Condamine (1701–1774, French), Louis Godin (1704–1760, French), and Pierre Bouguer (1698–1758, French) participate (although not very compatibly) in an expedition to the high Andes of Ecuador and Peru to determine the length of a degree of latitude (see 1744—EARTH SCIENCES). On the return trip, Condamine explores the Amazon River.

1736

CHEMISTRY (Alkali)

Henri-Louis Duhamel du Monceau (1700–1782, French) demonstrates that common salt, Glauber's salt (sodium sulfate), and borax all contain the mineral alkali (soda).

CHEMISTRY (Salts)

Using taste as one criterion, Henri-Louis Duhamel du Monceau (1700–1782, French) distinguishes sodium salts from potassium salts, establishing that they have different bases.

HEALTH SCIENCES (Scarlet Fever)

William Douglass (c.1691–1752, American) publishes an account and clinical diagnosis of scarlet fever.

MATHEMATICS (Calculus)

Thomas Bayes's (1702–1761, English) *Introduction to the Doctrine of Fluxions* defends Isaac Newton's calculus (MATHEMATICS entries for 1664–1666, 1669, 1671) against the attacks by George Berkeley on its logical foundations (1734—MATHEMATICS).

MATHEMATICS (Calculus of Variations)

Leonhard Euler (1707–1783, Swiss) presents necessary conditions for insuring that the solution to a differential equation yields the minimizing curve.

MATHEMATICS (Number Theory)

Leonhard Euler (1707–1783, Swiss) publishes a proof of Pierre de Fermat's Little Theorem that for any prime p and any whole number n, p divides $n^p - n$ (1640—MATHEMATICS), doing so through induction on n.

PHYSICS (Mechanics)

In *Mechanica, sive motus scientia analytice exposita,* Leonhard Euler (1707–1783, Swiss) develops the study of the kinematics and dynamics of a point mass through analytic methods, replacing the predominantly synthetic methods of Isaac Newton's *Principia* (1687—ASTRONOMY). Volume 1 deals with the free motion of a point mass, volume 2 with constrained motion, the latter requiring solution of problems in the differential geometry of surfaces.

1736–1737

EARTH SCIENCES (Geoid)

Pierre de Maupertuis (1698–1759, French) leads an expedition to Lapland in Sweden to measure a degree of latitude and thereby, with results from a similar expedition to Peru, to determine whether the earth is elongated at the poles, as believed by Cartesians, or flattened at the poles, as indicated by Isaac Newton (1687—EARTH SCIENCES). Results (1744—EARTH SCIENCES) confirm the Newtonian theory and establish Maupertuis's fame.

1737

BIOLOGY (Descriptive Botany)

Carolus Linnaeus (1707–1778, Swedish) publishes *Flora Lapponica* [The flora of Lapland], describing his research in Lapland on the local flora and on the customs of the Lapp people.

BIOLOGY (Insects)

Posthumous publication of Jan Swammerdam's (1637–1680, Dutch) *Biblia naturae,* presenting conclusions from his microscopic dissections of such insects as the bee, the mayfly, the head louse, and the rhinoceros beetle.

BIOLOGY (Plant Classification)

In *Genera plantarum,* Carolus Linnaeus (1707–1778, Swedish) explains his plant-classification scheme, based predominantly on the number of stamens, which determines the class, and the number of pistils, which determines the order. His is an extension of the scheme of Andrea Cesalpino (1583—BIOLOGY). He uses Gaspard Bauhin's binomial nomenclature (1623—BIOLOGY) for genera and species and classifies 18,000 species of plants, recording the discontinuous nature of the gradation of species.

EARTH SCIENCES (Isobaths)

Philippe Buache (1700–1773, French) maps the English Channel using isobaths, or lines of equal depth.

MATHEMATICS (Continued Fractions)

Leonhard Euler (1707–1783, Swiss) provides the first systematic discussion of continued fractions.

MATHEMATICS (Zeta Function)

Leonhard Euler (1707–1783, Swiss) examines the sum $\Sigma\ (1/n^s)$ for $n = 1$ through $n = \infty$, later termed the zeta function $\zeta(s)$, and determines that it equals the product of all $[1/(1-p^{-s})]$ for integer primes p.

1737–1738

CHEMISTRY (Cobalt)

The element cobalt is discovered and isolated by Georg Brandt (1694–1768, Swedish).

1738

CHEMISTRY (Kinetic Theory; Gas Laws)

Daniel Bernoulli (1700–1782, Swiss) initiates the mathematical study of the kinetic theory of gases in *Hydrodynamica* (1738—PHYSICS) and analytically deduces Boyle's Law that the volume and pressure of a gas are inversely related, a law originally obtained empirically (1662—CHEMISTRY). To establish the analytical derivation, Bernoulli follows Robert Hooke in visualizing the pressure of a gas as resulting from huge numbers of impacts on the walls of the container by hard, fast-moving gas particles. Bernoulli's explanation, based on random motions of the gas particles, is more modern than an earlier attempt by Isaac Newton to explain Boyle's Law by assuming relatively motionless particles which repel each other with a force inversely proportional to the distance between them.

PHYSICS (Fluid Dynamics)
EARTH SCIENCES (Oceanic Fluid Flow)

Publication of Daniel Bernoulli's (1700–1782, Swiss) *Hydrodynamica,* one of the major works initiating the mathematical study of fluid flow. Bernoulli presents the following equation for steady, nonviscous, incompressible flow: $p + \rho v^2/2 + \rho gy = A$, where p symbolizes pressure, ρ density, v velocity, g the acceleration of gravity, y height, and A a constant. He also examines the equilibrium oscillation of an inertialess ocean, and explicitly states that the flow equations are appropriate not only for the more common applications of fluid dynamics but also for the flow of blood in veins and arteries. Bernoulli, like Galileo Galilei in 1638 and Christiaan Huygens (1656—PHYSICS), assumes conservation of mv^2 rather than conservation of momentum mv, m and v symbolizing a body's mass and velocity respectively.

PHYSICS (Heat)

Leonhard Euler (1707–1783, Swiss) receives a prize for an *Essay on Heat,* in which heat is viewed as an oscillation of particles.

PHYSICS (Optics; Perception)

In addition to describing the theories and discoveries of Isaac Newton, Voltaire (1694–1778, French), in his *Elements of Newton's Philosophy,* discusses the religious and other implications of Newtonian science, and also presents new ideas in the field of optics, most particularly regarding the psychology of perception.

1739

BIOLOGY (Combination of Ideas)

In *Treatise of Human Nature,* David Hume (1711–1776, Scottish) postulates that when ideas are combined this results from one of three possibilities: they resemble each other, they are close in time or space, there exists a cause-effect relation. Ideas attract similarly to mechanical attraction. Hume declares that we know only our sensations and have no assurance that objects outside ourselves exist.

BIOLOGY (Scale of Life)

Abraham Trembley (1710–1784, Swiss) discovers a fresh-water polyp, the *Hydra,* which is interpreted by some as the missing link joining the vegetable and animal kingdoms.

MATHEMATICS (Differential Equations)

Leonhard Euler (1707–1783, Swiss) creates the method of variation of parameters for solving differential equations.

1740

BIOLOGY (Reproduction)

Charles Bonnet (1720–1793, Swiss) discovers that fertilization is not a prerequisite for the production of offspring by female aphids.

BIOLOGY (Spontaneous Generation)

John Needham (1713–1781, English) becomes a foremost proponent of the theory of spontaneous generation after experiments with a sealed tube of boiled beef broth lead him to conclude that living organisms have arisen in the tube without animate progenitors.

EARTH SCIENCES (Earth History; Mountains)

Lazzaro Moro (1687–1764, Italian) provides a history of the early earth in *Dei crostacei e degli altri corpi marini che si truovano sui monti*. At first covered entirely by fresh water, the earth's surface is disrupted on the third day by subterranean fires which lead to the rising of dry land and mountains above the water. Eruptions of some of the mountains lead to the discharge of salts to the sea, accounting for the conversion from fresh to salt water. More eruptions lead to the laying of strata on the sea floor and the rising of more dry land. Moro contends that the primary cause of origin of all mountains is the central fire within the earth. Basing his position on explicit accounts of the emergence of islands and hills through volcanic activity, he extrapolates to all islands and all mountains, reasoning that Newtonian principles that like causes lead to like effects support such extrapolations. He also discusses the origin and development of fossils.

MATHEMATICS (Exponents)

Leonhard Euler (1707–1783, Swiss) expands the repertoire of possible exponential values to imaginary exponents.

1740s

CHEMISTRY (Catalysts)

Joshua Ward (1685–1761, English), a manufacturer of sulfuric acid (for use in bleaching), finds that adding potassium nitrate (saltpeter) to burning sulfur greatly speeds the conversion to sulfur trioxide, an intermediate product needed in the manufacture of sulfuric acid. The amount of saltpeter is not depleted during the process, a welcome though unexplained result at the time (see 1806—CHEMISTRY). This is among the earliest recorded uses of a catalyst to speed a chemical reaction.

1742

CHEMISTRY (Zinc)

Zinc is distilled from the allcy calamine by Anton von Svab.

MATHEMATICS (Maclaurin Series)

Colin Maclaurin (1698–1746, Scottish), in *Treatise of Fluxions,* specifies the $a = 0$ subcase of the Taylor series expansion of the function $f(x)$ about the point a (1715–1717—MATHEMATICS), obtaining the infinite series later termed the Maclaurin series. The book, written in part as a defense of Isaac Newton against the attacks of George Berkeley (1734—MATHEMATICS), also contains Maclaurin's studies on ellipsoids of revolution.

MATHEMATICS (Number Theory; Goldbach Conjecture)

In a letter to Leonhard Euler, Christian Goldbach (1690–1764, Prussian) conjectures that every even integer greater than 2 is a sum of two primes.

PHYSICS (Celsius Temperature Scale)

Anders Celsius (1701–1744, Swedish) proposes a temperature scale based on 100° for the freezing point of water and 0° for the boiling point of water. This scale is later inverted, with 0° for freezing and 100° for boiling (1743—PHYSICS), and termed the Celsius (or centigrade) scale.

1743

EARTH SCIENCES (Geological Mapping)

Christopher Packe (1686–1749, English) produces "A New Philosophico-chorographical Chart of East Kent," mapping the geology of East Kent, England.

EARTH SCIENCES (Shape of Earth)
ASTRONOMY (Stars)

Alexis Claude Clairaut (1713–1765, French) publishes *Théorie de la figure de la terre,* a theoretical study on the equilibrium of fluids and the shape of the earth and stars. He examines the gravity of points on the surface of rotating ellipsoids and specifically obtains a formula, later termed the Clairaut formula, for the earth's gravity as a function of latitude. For stars, he assumes an inhomogeneous fluid and examines a variety of gravitational laws. (Clairaut had participated in the Maupertuis expedition to Lapland to determine the shape of the earth [1736–1737—EARTH SCIENCES].)

PHYSICS (Celsius Temperature Scale)

Jean Pierre Christin (French) reverses the temperature scale proposed by Anders Celsius (1742—PHYSICS), so that 0° is set at the freezing point of water and 100° at the boiling point.

PHYSICS (D'Alembert's Principle)

Jean le Rond d'Alembert (1717–1783, French) states d'Alembert's Principle that an equilibrium exists among the actions and reactions internal to any closed system of rigid moving bodies, doing so in *Traité de dynamique,* a treatise on mechanics.

1744

ASTRONOMY (Distances to Stars)

Jean-Philippe Loÿs de Cheseaux (1718–1751, Swiss) calculates an estimate of the distance to first magnitude stars. Being based on the erroneous assumption that all stars have luminosity equal to that of the sun, the answer of roughly 3.7 light years is wrong; however, it is an advance simply to make an estimate at all of absolute stellar distances. This work is presented in the appendix "Sur la force de la lumière et sa propagation dans l'ether, et sur la distance des étoiles fixes" to Loÿs de Cheseaux's *Traité de la comète qui a paru en décembre 1743* [Treatise on the comet which appeared in December 1743].

ASTRONOMY (Night Sky)

Jean-Philippe Loÿs de Cheseaux (1718–1751, Swiss) asks the fundamental question of why, with a presumed infinity of stars, the sky remains basically dark at night. His answer, presented in the same appendix as his stellar distance calculation (1744—ASTRONOMY), is that part of the emitted starlight is absorbed within the all-pervading ether (Heinrich Olbers repeats the question and the explanation in 1826, and the problem becomes known as Olbers's Paradox [1826—ASTRONOMY]).

EARTH SCIENCES (Geoid)

Results of expeditions to Peru (1735–1744—EARTH SCIENCES) and Lapland (1736–1737—EARTH SCIENCES) establish that the earth is flattened at the poles. The flattening, predicted theoretically by Isaac Newton (1687—EARTH SCIENCES), is confirmed by the expeditions' measurements of 110,600 m for the length of a degree of latitude in Peru and 111,900 m for the length in Lapland.

MATHEMATICS (Algebra)

Leonhard Euler (1707–1783, Swiss) states, but does not prove, the law of quadratic reciprocity.

MATHEMATICS (Calculus of Variations)

In *Methodus inveniendi lineas curvas maximi minimive proprietate gaudentes,* Leonhard Euler (1707–1783, Swiss) expands and revises his earlier methods for determining minimizing curves from differential equations (1736—MATHEMATICS), and more systematically develops his studies in the calculus of variations. Earlier work, such as the proof by the Bernoulli's that the catenary is the curve of fixed length between two fixed points which has the lowest center of gravity, had suggested to Euler the likely importance of maxima and minima problems to the study of natural phenomena. He distinguishes between absolute and relative extrema and shows how problems of relative extrema can be reduced to those of absolute extrema. He applies the work to the study of physics in two appendices (see 1744—PHYSICS).

MATHEMATICS (Transcendental Numbers)

Leonhard Euler (1707–1783, Swiss) distinguishes transcendental numbers from algebraic numbers, the latter being solutions to polynomial equations in integral coefficients.

PHYSICS (Elasticity)

In an appendix to his *Methodus* (1744—MATHEMATICS), Leonhard Euler applies the calculus of variations to the study of elasticity, examining the bending and vibrations of elastic bands and plates. Included is the Euler buckling formula on the strength of columns.

PHYSICS (Fluid Dynamics)

Publication of Jean le Rond d'Alembert's (1717–1783, French) *Traité de l'équilibre et du mouvement des fluides.*

PHYSICS (Least Action)

In the paper "Accord of different laws of nature which hitherto had appeared incompatible," Pierre de Maupertuis (1698–1759, French) states his principle of least action and applies it to optics and mechanics. Maupertuis defines "action" as the product of mass m, velocity v, and distance s, Maupertuis's principle stating that a "natural" motion of a particle from one point to another will tend to minimize this product. Although generally referred to with Maupertuis's name, Leonhard Euler in an appendix to his *Methodus* (1744—MATHEMATICS) also provides a formulation, that a point under the influence of a central force will move along a path which minimizes the integral $\int mvds$, and lays the foundation for future mathematical studies.

1745

BIOLOGY (Blood)

Vincenzo Menghini (1704–1759, Italian) accidentally discovers that red blood corpuscles contain iron and subsequently attributes the red coloring to the presence of the iron. The discovery occurs as Menghini analyzes a control case in an experiment with dogs devised to demonstrate that, if eaten, iron will emerge in the blood stream.

HEALTH SCIENCES (Lead Poisoning; Decalcification)

Thomas Cadwalader (1708–1799, American) publishes the first account of lead poisoning (although not recognizing it as such), describing in *An Essay on the West-India Dry-Gripes: with the Method of Preventing and Curing that Cruel Distemper. To which is added, an Extraordinary Case* a case brought on by drinking rum that had been distilled through lead pipes. Appended at the end of the book is an essay on "An Extraordinary Case in Physick," describing massive decalcification of the skeletal structures of a middle-aged diabetes patient who in four years lost 17 inches in height.

HEALTH SCIENCES (Paralysis; Brain)

O. Dalin records the case of a patient paralyzed on the right side after a serious illness, who, although no longer able to speak, is able to sing specific hymns learned before the illness. (Over a century later, this and similar such cases are viewed as evidence that singing, in contrast to speech, is controlled by the brain's right hemisphere rather than the left.)

PHYSICS (Leyden Jar)

Ewald von Kleist (c.1700–1748, German) and Petrus van Musschenbroek (1692–1761, Dutch) independently invent the Leyden jar to store electricity from a Hauksbee-type machine (1706—PHYSICS). The glass jar, partially filled with water and containing a nail projecting from its cork stopper, is a simple electrical capacitor.

1746

CHEMISTRY (Nomenclature)

Considerable revision in chemical terminology is effected with a new edition of the *London Pharmacopoeia,* the revision being accomplished over eight years by a committee of the Royal College of Physicians.

CHEMISTRY (Zinc)

Andreas Marggraf (1709–1782, German) isolates zinc.

EARTH SCIENCES (Geological Mapping)

Jean-Étienne Guettard (1715–1786, French) maps the minerals and rocks of France and southern England and determines a general banded distribution centered approximately at Paris. The central sandy band is surrounded by a marly band which in turn is surrounded by a metalliferous band containing mines of various different minerals. The truncation of the bands at the English Channel leads Guettard to conjecture their continuation in England, a conjecture that he verifies through study of English geological writings. This map and another one of coarser scale for western Europe, both presenting surface distributions of rocks and minerals, are included in Guettard's *Mémoire et carte minéralogique*.

MATHEMATICS (Complex Numbers)

Leonhard Euler (1707–1783, Swiss) establishes that the complex power of a complex number can be real by providing the example $i^i = e^{-\pi/2}$.

PHYSICS (Electricity)

Jean Antoine Nollet (1700–1770, French) describes and illustrates numerous electrical experiments in his *Essai sur l'électricité des corps* [Essay on the electricity of bodies].

PHYSICS (Electricity; One-Fluid Theory)

William Watson (1715–1787, English) presents the hypothesis that electricity is a fluid existent in all matter in its natural state. This fluid can be transferred from one body to another and will tend to equalize itself by flowing from a body with a greater to a body with a lesser amount, always conserving the total amount. This hypothesis, brought to greater prominence with the supporting work of Benjamin Franklin over the next several years, is termed the "one-fluid theory" of electricity.

1747

CHEMISTRY (Beet Sugar)

Andreas Marggraf (1709–1782, German) discovers beet sugar.

MATHEMATICS (Partial Differential Equations)

Jean le Rond d'Alembert (1717–1783, French) publishes an analysis of vibrating strings, a pioneering study in partial differential equations, as he is led to the equation $\partial^2 u/\partial t^2 = \partial^2 u/\partial x^2$, which he proceeds to solve.

METEOROLOGY (Winds)

Publication of Jean le Rond d'Alembert's (1717–1783, French) *On the General Theory of the Winds.* Rejecting the conception of Edmond Halley (1686—METEOROLOGY) and George Hadley (1735—METEOROLOGY) that the general circulation of the atmosphere is significantly controlled by the distribution of solar heating, d'Alembert develops a mathematical theory based on Isaac Newton's law of gravitation (1687—ASTRONOMY) and explaining the winds by means of the gravitational forces from the sun and moon. Although the theory is later roundly rejected, d'Alembert's work encourages a more scientific study of meteorology.

PHYSICS (Electricity)

Members of the Royal Society conduct experiments on the velocity of electricity.

PHYSICS (Projectiles)

Benjamin Robins (1707–1751, English) investigates the physics of a spinning projectile, including the effects of air resistance.

1748

ASTRONOMY (Earth's Motion)

James Bradley (1693–1762, English) announces discovery of the nutation of the earth's axis (the oscillation resulting from the gravitational attraction of the moon). This rotational wobble is superimposed on the shifts due to the precession of the axis.

BIOLOGY (Agriculture)

Jared Eliot (1685–1763, American) applies scientific procedures and theory to agriculture in his *Essay on Field Husbandry.*

BIOLOGY (Evolution)

In his *Telliamed* (1748—EARTH SCIENCES), Benoît de Maillet theorizes that all land animals, including humans, have evolved from sea animals, changing over time as a result of changing environments and habits.

BIOLOGY (Man as Machine)

In *L'Homme machine* [Man a machine], Julien Offray de LaMettrie (1709–1751, French) develops his theory that all animals, including humans, are machines.

CHEMISTRY (Matter)
PHYSICS (Forces)

In *An Attempt to Demonstrate That All the Phenomena in Nature May be Explained by Two Simple Active Principles, Attraction and Repulsion,* Gowin Knight (1713–1772, English) dismisses the concept of the reality of matter and suggests instead that the forces of attraction and repulsion account for all observed phenomena. The work is believed probably to have influenced Humphry Davy, Michael Faraday, and others in the nineteenth century.

CHEMISTRY (Osmosis)

Jean Antoine Nollet (1700–1770, French) discovers osmosis while experimenting with water diffusing into a sugar solution from which it is separated by an animal membrane.

CHEMISTRY (Platinum)

In the notes of a trip to South America, Antonio de Ulloa (1716–1795, Spanish) refers briefly but definitively to platinum, a new element found in several gold mines in the sands of Rio Pinto, Columbia, in 1736. De Ulloa remarks on the extreme resistance of the metal, the probable great difficulty in extracting it, and the nuisance it causes for those attempting to mine the contiguous gold.

EARTH SCIENCES (Earth History)

Posthumous publication of Benoît de Maillet's (1656–1738, French) *Telliamed,* in which he discusses the past and future history of the earth through conversations between an Indian philosopher and a French missionary. De Maillet contends that the earth at one point was completely surrounded by water, that the topography of the earth has been shaped by this water as it has retreated over time, and that the water amount will continue to diminish by evaporation until the planet is entirely desiccated. Afterward, volcanoes will extinguish all life and cause the burning of the globe and its transformation to a sun. From limited observations, de Maillet estimates that sea level lowers by 3 to 4 inches per 100 years.

EARTH SCIENCES (Volcanism)

Benoît de Maillet asserts in his posthumous *Telliamed* (1748—EARTH SCIENCES) that volcanoes result from the combustion of the oils and fats of animals entombed in the mountain sediments.

HEALTH SCIENCES (Diphtheria; Scarlet Fever)

John Fothergill (1712–1780, English) provides a careful clinical description of recent sore throat outbreaks in London in his *An Account of the Sore Throat Attended with Ulcers: A Disease Which Hath of Late Years Appeared in This City, and the Parts Adjacent.* Later readings of Fothergill's account have led to various conclusions regarding the precise disease being described, some possibilities being that the outbreaks were of diphtheria, diphtheritic pharyngitis, malignant scarlatina, or a mixture of scarlet fever and diphtheria.

MATHEMATICS (Algebra)

Leonhard Euler (1707–1783, Swiss) publishes the two-volume *Introductio in analysin infinitorum* [Introduction to infinitesimal analysis], in which he defines "function" and argues the centrality of the function concept, and treats functions defined both implicitly and by parametric representation. Euler permits imaginary values for the independent variable of a function, generalizes De Moivre's Theorem (1730—MATHEMATICS) beyond integral n to any real or complex n, presents $e^{ix} = \cos x + i \sin x$ and the Euler identities providing expressions for $\sin x$ and $\cos x$ as functions of e^{-ix} and e^{ix} (these had been known before, but not extensively used), treats trigonometric functions analytically as numbers or ratios, discusses the zeta function $\zeta(s)$ (see 1737—MATHEMATICS) and its applications to prime number theory, introduces several new notational conventions, provides a systematic development of polar coordinates (see MATHEMATICS entries for 1671 and 1691), including transformation equations from rectangular to polar coordinates, provides a systematic study of curves and surfaces, not restricted to conic sections, and extols the advantages of algebraic over geometric methods of analysis. Volume 1 is devoted to developing the theory of elementary functions through algebra and infinite processes (infinite products, infinite continued fractions, infinite series), and volume 2 is devoted to analytic geometry and particularly to second-order and third-order curves. An appendix provides a textbook of solid analytic geometry, with substantial original contributions to the study of surfaces, both algebraic and transcendental.

MATHEMATICS (Calculus)

Maria Agnesi (1718–1799, Italian) publishes *Istituzioni analitiche* [Analytical institutions], a widely used, comprehensive two-volume text on algebra and analysis covering a range from elementary algebra and coordinate geometry to differential and integral calculus, infinite series, and elementary differential equations. Emphasizing developments within the past century, she combines the methods and results of numerous mathematicians, including the fluxions of Isaac Newton (1664–1666—MATHEMATICS), the differentials of Gottfried Wilhelm Leibniz (1673–1676—MATHEMATICS), and some of her own methods and generalizations.

MATHEMATICS (Periodicity)

Leonhard Euler (1707–1783, Swiss) initiates the formal mathematics of periodicity through his work on circular functions.

PHYSICS (Electroscope)

Jean Antoine Nollet (1700–1770, French) invents the electroscope.

1749

BIOLOGY (Classification; Evolution)

Georges Buffon (1707–1788, French) sets forth his general views on species classification in the first volume of his *Histoire naturelle* (1749–1788—EARTH SCIENCES; see also 1749—EARTH SCIENCES). Buffon objects to the so-called "artificial" classifications of Andrea Cesalpino (1583—BIOLOGY) and Carolus Linnaeus (1735—BIOLOGY), stating that in nature the chain of life has small gradations from one type to another and that the discontinuous categories are all artificially constructed by mankind. Buffon suggests that all organic species may have descended from a small number of primordial types; this is an evolution predominantly from more perfect to less perfect forms.

BIOLOGY (Competition)

Carolus Linnaeus (1707–1778, Swedish) publishes *Oeconomia naturae,* in which he describes the economy of nature in limiting unnecessary competition between species by having each species have a specified place in the scheme of nature, including allotted geographical regions and allotted placements in the food chain. The proper numerical proportions between species are viewed as maintained by the individual reproductive rates and the predation characteristics of the interacting species.

CHEMISTRY (Nomenclature)

In his *Élémens de chymie théorique,* Pierre Macquer (1718–1784, French) calls for a more systematic chemical terminology, criticizing such usages as the term "regenerated tartar" naming a nontartar salt.

EARTH SCIENCES (History of the Earth)

Georges Buffon (1707–1788, French) presents his theory of the earth (*Théorie de la terre*) in the first volume of his *Histoire naturelle* (1749–1788—EARTH SCIENCES), including the earth's origin, along with the other planets and satellites of the solar system, from a collision between the sun and a comet. He sketches the history of the earth based on his interpretations of the fossil and stratigraphic evidence. The land he believes to have been submerged under the sea for an extended period, with dry land and mountains appearing after some of the waters retreated into the earth's interior. In his scheme, large earthquakes arise from gaseous explosions and volcanoes from the combustion of sulfur and bitumen.

HEALTH SCIENCES (Neurology; Psychology)

In *Observations on Man,* David Hartley (1705–1757, English) attempts to show the mechanism by which ideas are generated from sense impressions. The sense organs receive stimuli which create vibrations in the nervous system. The vibrations reach the brain and ideas are formed. Through this and other works, Hartley founds associative psychology.

MATHEMATICS (Statistics)

The first usage of the term "statistik" occurs in a work by Gottfried Achenwall (1719–1772, German), with the term signifying a comprehensive description of various features of a state—social, political, and economic.

1749–1788

BIOLOGY/EARTH SCIENCES (Natural History)

Georges Buffon (1707–1788, French) publishes the popular 36-volume *Histoire naturelle* [Natural history], extended to 44 volumes by his followers between 1788 and 1804.

1750

ASTRONOMY (Cosmology)

Pierre de Maupertuis (1698–1759, French) describes the universe from a solidly mechanistic framework in *Essai de cosmologie.*

ASTRONOMY (Milky Way)

Thomas Wright (1711–1786, English) presents two possible models of the Milky Way galaxy in his volume *An Original Theory, or New Hypothesis of the Universe.* He agrees with the conclusion of Democritus and of Galileo that the "Milky Way" visible in the sky is actually unresolved starlight, and then suggests the two alternatives that the earth is in a thin spherical shell of stars or that the Milky Way is a ring of stars revolving around a large body at the center. He also suggests that what appear to be nebulae are actually galaxies, the solar system comprising only a small portion of one of the universe's countless galactic structures.

ASTRONOMY (Moon)

Tobias Mayer (1723–1762, German) maps major mountainous features of the moon.

ASTRONOMY (Parallax)

Nicolas de LaCaille (1713–1762, French) heads a scientific expedition to the Cape of Good Hope in order to make measurements appropriate for determining the parallaxes of the sun and moon (see 1751–1753—ASTRONOMY).

CHEMISTRY(Phlogiston)

Mikhail Lomonosov (1711–1765, Russian) publishes a strong criticism of the phlogiston theory, although he himself does not fully abandon the concepts (see 1751—CHEMISTRY).

PHYSICS (Electricity)
METEOROLOGY (Lightning)

Benjamin Franklin (1706–1790, American) identifies lightning with the electric spark, invents the lightning rod to pass potentially dangerous lightning bolts to the ground, and, while retaining the one-fluid theory of electricity (1746—PHYSICS), asserts that an object can be electrified either positively, representing an excess of the electrical fluid, or negatively, representing a deficiency of the electrical fluid. In this year Franklin authors both "Experiments and observations on electricity" and "Opinions and conjectures concerning the properties and effects of the electrical matter."

c.1750

MATHEMATICS (Number Theory)

Leonhard Euler (1707–1783, Swiss) proves that all even perfect numbers are of the form $2^{n-1}(2^n - 1)$, with $2^n - 1$ a prime.

MATHEMATICS (Symbolism)

Leonhard Euler (1707–1783, Swiss) introduces the functional notation $f(x)$, the use of Σ for summation, i for imaginaries, e for the base of the natural logarithms, and π for the ratio of the circumference to the diameter of a circle.

1751

BIOLOGY (Origin of Species)

Pierre de Maupertuis (1698–1759, French) revives the theory of the Atomists that species derived ultimately from chance combinations of small units (or atoms).

CHEMISTRY (Metallic Luster)

Mikhail Lomonosov (1711–1765, Russian) publishes a paper, "On the Luster of Metals," in which he attempts to prove that it is the phlogiston content in metals which determines their luster and ductility.

CHEMISTRY (Nickel)

Nickel is isolated by Axel Cronstedt (1722–1765, Swedish).

CHEMISTRY (Nomenclature)

Pierre Macquer (1718–1784, French) continues to criticize chemical terminology (see 1749—CHEMISTRY) in his *Élémens de chymie pratique.* Among the terms criticized are "sel febrifuge de Sylvius" for a substance which cannot be relied upon to cure fevers, and "mercury precipitate" for a substance which can be obtained without precipitation.

EARTH SCIENCES (Deep Ocean)

Henry Ellis (1721–1806, English) takes ocean measurements at depths considerably deeper than any before him. Situated at 25°13′N, 25°12′W, he measures temperatures at depths ranging to 5346 feet, finding surface temperatures of 84°F and deep water temperatures of 53°F.

HEALTH SCIENCES (Childbed Fever)

John Burton (1710–1771, English) becomes the first to suggest in writing that puerperal (or childbed) fever comes to the patient from the outside, perhaps through the carelessness of the medical attendants.

METEOROLOGY (Dew Point; Relative Humidity)

Charles LeRoy (1726–1779, French) discovers the principle of the dew point temperature, finding experimentally that cooling of damp air sealed in a glass container results in dew formation, with the dew disappearing upon heating and reappearing upon subsequent recooling to the temperature of previous dew formation. Having thus demonstrated that warm air can hold more water vapor than cool air, LeRoy suggests that humidity be recorded as "relative humidity," comparing the actual amount of water vapor in the air with the maximum possible at the given temperature.

METEOROLOGY (Precipitation)

Charles LeRoy (1726–1779, French) conjectures that atmospheric moisture arrives in the air by dissolving into it and leaves the air by precipitating out. Although the process indeed differs from chemical precipitation, the term "precipitation" becomes widely used for rain and snowfall.

1751–1753

ASTRONOMY (Moon)

Observations by Nicholas de LaCaille (1713–1762, French) at the Cape of Good Hope and by Joseph de Lalande (1732–1807, French) in Berlin allow computation of the value of the moon's parallax. The result indicates a moon-to-earth distance of about 60 times the earth's radius and provides a more exact value of the estimates known since antiquity.

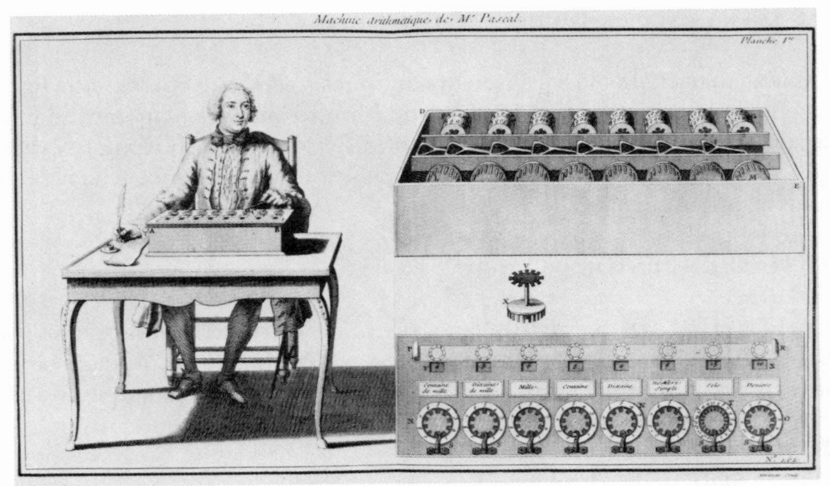

FIGURE 7. *The calculating machine of Blaise Pascal (see 1642—*MATHEMATICS *entry), depicted in use and with details of the top cover (bottom right) and of the interior toothed wheels for performing additions and subtractions (above right). (By permission of the Houghton Library, Harvard University.)*

1751–1772

SUPPLEMENTAL (Encyclopedias)

Publication of the highly successful but also controversial 28-volume *Encyclopédie; ou, dictionnaire raisonné des sciences, des arts, et des métiers* [Encyclopedia, or critical dictionary of the sciences, arts, and trades] by the French philosophes. Consisting of 17 volumes of text and 11 volumes of illustrations, the *Encyclopédie* describes the processes and principles of mid-eighteenth century art, industry, and science. Denis Diderot (1713–1784, French) as main editor for the arts and trades and Jean le Rond d'Alembert (1717–1783, French) as main editor for mathematics and the physical sciences, prepare many of the articles and illustrations themselves, as well as securing contributions from other noted French intellects, among them Voltaire and Rousseau. The introduction, or "Discours préliminaire" [Preliminary discourse], written by d'Alembert, states the aims of the project and a general, empirically based philosophy of science. Diderot, who compiles the 11 volumes of illustrations, largely depicting the machines and workings of individual trades, regards a central purpose of the *Encyclopédie* to be to dignify common pursuits and to lead to the improvement of those pursuits by spreading knowledge of scientific advances. Although highly successful, the project encounters major difficulties, being viewed with suspicion by many tradesmen reluctant to abandon traditional policies of maintaining trade secrets, and being severely attacked from various religious factions, due to its failure to accord a prominent place to religious truths. The attacks lead d'Alembert to resign in 1758, leaving a disappointed Diderot to complete the many remaining volumes largely by himself.

1752

BIOLOGY (Irritability; Sensibility)

Albrecht von Haller (1708–1777, Swiss) publishes *A Dissertation on the Sensible and Irritable Parts of Animals,* in which he experimentally defines and distinguishes irritability and sensibility, the former being viewed as a muscular contraction following stimulation and the latter as a property of the nervous system rather than the muscles. Sensibility is demonstrated in man by a conscious notice of a stimulus and in other animals by a sign of unrest.

CHEMISTRY (Platinum)

Henric Scheffer (1710–1759, Swedish) shows platinum to be a distinct metal and publishes a detailed description of it, including its properties of being less pliable than gold and almost infusible by itself, although capable of being fused with the aid of arsenic.

EARTH SCIENCES (Earth's Wobble)

Leonhard Euler (1707–1783, Swiss) calculates, based on the laws of inertia, that the earth should have a 304-day periodic wobble. Others proceed to seek observational evidence of such a wobble, although these efforts are unsuccessful. Over a century later Seth Chandler finally discovers a wobble, but the period is 428 days rather than 304 days (1891—EARTH SCIENCES).

EARTH SCIENCES (Extinct Volcanoes)

Jean-Étienne Guettard (1715–1786, French) reads to the French Academy of Sciences a "Mémoire sur quelques montagnes de la France qui ont été des volcans" [Memoir on certain mountains in France that have once been volcanoes] (not published until 1756), asserting the existence of extinct volcanoes in central France, the whole region around Auvergne having been volcanic. Guettard arrived at this surprising conclusion after tracing black stone building material to its mountain origin and then traveling throughout areas of central France, noting pumice deposits and supposed remains of lava streams and old volcanic cones.

EARTH SCIENCES (Paleogeography)

Nicholas Desmarest (1725–1815, French) concludes that a land bridge must at one time have connected England and France, later to be broken by the North Sea currents. He bases his argument on the correspondence between the opposite cliffs of the two countries, the form of the intervening shallow strait, and the former existence in England of wild animals which could not have swum across the sea and would not have been transported by man. An essay on the topic wins for Desmarest first prize in a contest devoted specifically to the issue of the possibility of a land connection, reflecting a widespread interest in the question.

MATHEMATICS (Polyhedra)

Leonhard Euler (1707–1783, Swiss) proves that for any closed convex polyhedron the number of vertices plus the number of faces is two more than the number of edges: $N_0 + N_2 = N_1 + 2$ (a result known earlier to René Descartes [1639—MATHEMATICS] but commonly accredited to Euler). Euler further classifies polyhedra.

METEOROLOGY (Lightning)
PHYSICS (Electricity)

Experimenting with a kite, Benjamin Franklin (1706–1790, American) confirms his conjecture that lightning is an electrical discharge.

1752–1754

PHYSICS (Efficiency)

In an early application of science to engineering, John Smeaton (1724–1792, English) varies component parts of model windmills and waterwheels to determine the effects on efficiency.

1753

BIOLOGY (Nomenclature; Genera; Species)

Carolus Linnaeus (1707–1778, Swedish) elaborates a systematic botanical binomial nomenclature, naming plants in terms of genera and species in his two-volume *Species plantarum.*

EARTH SCIENCES (Longitude)
ASTRONOMY (Navigational)

Tobias Mayer (1723–1762, German) publishes tables on the positions of the moon which allow accurate enough determinations of longitude for him to collect the first awarding of a prize offered since 1714 (and sought by many) for a successful method of calculating longitude at sea. With the tables in hand, a seaman can obtain his longitude through a method suggested by Peter Apian in the sixteenth century, but these determinations are only accurate to about 20 miles and they require hours of calculation (see 1765—PHYSICS).

EARTH SCIENCES (Metals)

Johann Lehmann (1719–1767, German) expresses in his *Abhandlung von den Metall-Müttern und der Erzeugung der Metalle* the concept that mineral veins are merely the offshoots of a great mineral trunk within the earth. He also supports the concept that certain sites are favorable for the development of metals and metallic ores. These sites, or "Metalmütters" ("mothers of metals" or "wombs" or "matrices") are found within the earth's crust, and not also, as others had earlier suggested, within the air or the bodies of plants and animals.

HEALTH SCIENCES (Scurvy)

James Lind (1716–1794, Scottish) publishes *A Treatise on the Scurvy,* describing his firsthand studies on the disease while a ship surgeon in 1746. He asserts the value of fresh fruits and vegetables for preventing and treating the disease and specifically relates the rapid recovery brought about by feeding oranges and lemons to sufferers.

METEOROLOGY (Atmospheric Electricity)
PHYSICS (Electricity)

Giambatista Beccaria (1716–1781, Italian) publishes *Dell elettricita* supporting the electrical theories (1750—PHYSICS) of Benjamin Franklin, and conducts experiments on atmospheric electricity.

1753–1767

BIOLOGY (Quadrupeds)

As part of his *Histoire naturelle* (1749–1788—BIOLOGY), Georges Buffon (1707–1788, French) publishes 12 volumes on quadrupeds.

1754

ASTRONOMY (Precession)

Jean le Rond d'Alembert (1717–1783, French) explains mathematically the precession of the equinoxes and shows that the 26,000 year precession cycle and the countering slow rotation of the earth's elliptical orbit result in a 22,000 year cycle of the equinoxes along the earth's orbit.

BIOLOGY (Animal Creation)

In a revival of a theory of Empedocles of Acragas, Denis Diderot (1713–1784, French) suggests that, at some point in the distant past, various animal organs existed and lived independently and that eventually sets of rather random combinations of these organs joined together, forming not only those creatures we recognize today but also many creatures which did not survive and which today would be viewed as strange and grotesque.

BIOLOGY (Sense Perception)

In his *Treatise on Sensations,* Étienne Condillac (1714–1780, French) analyzes each sense separately by postulating in turn five men each with only one of the five senses. He contends that ideas generated by any of the senses are identical in kind, although not necessarily in degree, to those generated by any of the others or any combination thereof. When impressions are derived from different senses, they combine through a process analogous to a mechanical mixing.

CHEMISTRY (Carbon Dioxide)

Joseph Black (1728–1799, Scottish), in his dissertation *De humore acido a cibis orto, et magnesia alba* [The acid humor arising from food, and magnesia alba], shows magnesia alba to be distinct from lime and clearly characterizes the gas carbon dioxide, then termed "fixed air" and obtained by heating magnesia alba. For centuries lime from a kiln was known to retain the volume but not the weight of the original limestone. Black demonstrates that this results from the loss of fixed air. He shows fixed air to have different chemical properties from those of ordinary air, thereby rebutting a common misconception that all "airs" are the same.

MATHEMATICS (Limit; Derivative)

Jean le Rond d'Alembert (1717–1783, French), in the article "Différentiel" for volume 4 of the *Encyclopédie* (1751–1772—SUPPLEMENTAL), replaces Isaac Newton's conception of "fluxion" (1664–1666—MATHEMATICS) by a conception in which the derivative becomes the limit of the ratio of finite differences: $dy/dx = \lim(\Delta y/\Delta x)$, the limit being taken as Δx approaches 0. D'Alembert furthermore claims that the tangent is the limit of the secant to a curve as the two points of intersection of the secant with the curve approach each other. These ideas are not widely accepted at the time, although they gain acceptance through the work of Augustin Louis Cauchy in the nineteenth century.

1755

ASTRONOMY (Nebulae)

Nicholas de LaCaille (1713–1762, French) publishes the first extensive list of nebulous astronomical objects.

ASTRONOMY (Star Systems; Stellar Evolution)

Immanuel Kant's (1724–1804, German) *Universal Natural History and Theory of the Heavens* is published anonymously. The work was triggered by an article on Thomas Wright's *Original Theory* (1750—ASTRONOMY). Kant conjectures that the solar system is part of a huge system of stars constituting a galaxy, infuses physics into a disk model of that galaxy, and elaborates Wright's hypothesis that the "nebulous" stars are also galaxies, or island universes containing numerous individual stars. Kant bases his conclusions on a Principle of Uniformity that each large segment of the universe is like each of the others. He suggests that all the star systems might compose a vaster system revolving around a common center. He also suggests an evolution of the universe, beginning with a chaos of particles possessing both attractive and repulsive properties.

The attractive properties led to the conglomeration of the particles into the various heavenly bodies, whereas the repulsive properties led to the various motions of the solar, sidereal and cosmic systems.

MATHEMATICS (Differential Calculus)

Leonhard Euler (1707–1783, Swiss) attempts to place differential calculus on a more rigorous base in his *Institutiones calculi differentialis* (for integral calculus, see 1768–1770—MATHEMATICS). He rejects geometry as the proper foundation, emphasizing instead algebraic functions and arithmetic. He begins with the calculus of finite differences, then treats differential calculus as a limiting case, arising when differences approach 0. In an effort to address earlier criticisms of the calculus, largely aimed at the concept of infinitesimals, he sets all infinitesimals precisely at 0, then elaborates a calculus of "zeros of different order," $(dx)^2$, for instance, being a zero of higher order than dx. He also develops methods of differentiation through substitution of variables, and proves Nikolaus Bernoulli's theorem on the irrelevance of the order of partial differentiation (1720—MATHEMATICS). The book leads to later, sounder developments of the numerical foundations of calculus in the nineteenth century. Euler, in spite of his advances, erroneously believes that any series with terms of continually decreasing magnitude will converge.

METEOROLOGY (Fluid Flow)
PHYSICS (Fluid Dynamics)

Leonhard Euler (1707–1783, Swiss) derives the equations of motion of nonviscous, compressible fluids.

METEOROLOGY (Wet Bulb Thermometer)

William Cullen (1710–1790, Scottish) explains that the energy used for evaporation will cause a thermometer with a wet bulb to register a lower temperature than a thermometer with a dry bulb.

1756

BIOLOGY (Blood Circulation)

Albrecht von Haller (1708–1777, Swiss) publishes *Two Memoirs on the Motions of the Blood,* describing details of blood motion determined through injection of colored fluids into blood vessels.

CHEMISTRY (Carbon Dioxide)

Joseph Black (1728–1799, Scottish) publishes chemical results from his doctoral dissertation (1754—CHEMISTRY) and subsequent clarifying studies under the title *Experiments upon Magnesia Alba, Quick-Lime, and some other Alcaline Substances.* This work, through its description of "fixing" the gas carbon dioxide (then termed "fixed air"), that is, combining it chemically with a solid to produce a new substance, contributes to eliminating the misconception that gases cannot participate in chemical reactions.

EARTH SCIENCES (Earthquakes)

Nicholas Desmarest (1725–1815, French) publishes *Conjectures physico-méchaniques sur la propagation des secousses dans les tremblements de terre.*

EARTH SCIENCES (Mountains; Stratigraphy)

Johann Lehmann (1719–1767, German) determines from field work in northern Europe that the earth's crust consists of a structured sequence of layers or strata. He presents in *Versuch einer Geschichte von Flötz-Gebürgen* stratigraphic sequences from the areas he researched, plus a classification of mountains into three groups: Primary Mountains formed at the Creation, Secondary Mountains composed of well-defined beds deposited during the Great Flood, and a less numerous group formed later by agencies such as volcanoes and lesser floods.

HEALTH SCIENCES (Dentistry; False Teeth)

Philipp Pfaff (1716–1780, Prussian), dentist to Frederick the Great, publishes an important treatise on dentistry, *Abhandlung von den Zähnen,* in which he describes the casting of models for making false teeth.

1757

ASTRONOMY (Telescopes; Achromatic Lens)

John Dolland (1706–1761, English) invents and patents his achromatic lens, combining a lens of crown glass with one of flint glass, for astronomical telescopes.

ASTRONOMY (Venus)

Alexis Claude Clairaut (1713–1765, French) calculates the mass of Venus.

EARTH SCIENCES (Fossils)

Jean-Étienne Guettard (1715–1786, French) describes trilobites and other organic remains found in the Silurian slates of Angers. (Guettard's memoir, read to the French Academy of Sciences in 1757, is published in 1762.)

1758

ASTRONOMY (Comets)

Alexis Claude Clairaut (1713–1765, French) predicts that Halley's Comet (last seen in 1682) will return in April, 1759, plus or minus 30 days. The comet returns in March.

BIOLOGY (Classification)

In the tenth edition of his *Systema naturae* (first edition, 1735), Carolus Linnaeus (1707–1778, Swedish) fully explains his system of binomial nomenclature, and extends his plant classification scheme (1753—BIOLOGY) based on genera and species to the animal kingdom.

CHEMISTRY (Flame Test)

Andreas Marggraf (1709–1782, German) records that sodium salts impart a yellow color to a flame whereas potassium salts impart a lavender color, thereby establishing the flame test as a tool in chemical analysis.

CHEMISTRY (Matter; Atoms)

Rudjer Bošković (1711–1787, Yugoslavian-Italian) substitutes the forces of attraction and repulsion for the reality of matter, going a step further than Gowin Knight (1748—CHEMISTRY) by combining the two forces into a single point atom. Chemical elements are considered to result from combinations of point atoms and chemical compounds from combinations of chemical elements. The atoms of Bošković are indivisible, homogeneous, and identical to geometric points except that they possess inertia and mutually interact. The work later influences both Humphry Davy and Michael Faraday, with Faraday in particular (1844—CHEMISTRY) stating a preference for the atoms of Bošković over the atomic theory of John Dalton (1803—CHEMISTRY). Bošković's ideas are elaborated in his *Philosophiae naturalis theoria redacta ad unicam legem virium in natura existentium.*

MATHEMATICS (Calculus)

John Landen (1719–1790, English) attempts to provide firmer foundations for the calculus in his *Discourse Concerning the Residual Analysis,* the derivative being determined by dividing the difference $f(x) - f(y)$ by the "residual" $x - y$ and then setting $y = x$. He advocates a purely algebraical method of solving problems in the calculus, eliminating the need for considering infinitesimals or the motion of points and lines as in Isaac Newton's method of fluxions (1664–1666—MATHEMATICS).

1759

ASTRONOMY/PHYSICS (Gravitation)

Gabrielle-Émile Châtelet (1706–1749, French) translates Isaac Newton's *Principia* (1687—ASTRONOMY) into French.

BIOLOGY (Embryology)

In *Theoria generationis,* a comparative study of the development of plants and animals, Caspar Wolff (1734–1794, German) refutes the established preformation theory of reproduction in which the female egg is viewed as containing in miniature all future generations. He advocates instead a theory of epigenesis, whereby an embryonic organism passes through a succession of changes, with new organs appearing which were not preformed.

EARTH SCIENCES (Earth's Crust)

Giovanni Arduino (1714–1795, Italian) divides the earth's crust into four classifications: primary mountains generally composed of rocks with metallic ores; secondary mountains composed of marble and limestones; low mountains and hills composed of such substances as gravel, sand, and clay; and alluvial materials washed down the mountains by streams and rivers.

PHYSICS (Electricity)

Theoretical and experimental work with the air capacitor invented by himself and Johan Wilcke (1732–1796, Swedish) leads Franz Theodosius (1724–1802, German) to reject the current mechanical theories of electricity. In *Tentamen theoriae electricitatis et magnetismi* he offers an alternative theory of electrostatics along the lines of the gravitational theory of Isaac Newton (1687—ASTRONOMY).

1759–1792

MATHEMATICS (Difference Equations)

In a sequence of publications, in 1759, 1775, and 1792, Joseph Lagrange (1736–1813, French) develops a theory of recurring series with a level of mathematical rigor uncommon for the eighteenth century.

1760

CHEMISTRY (Specific Heats)

Joseph Black (1728–1799, Scottish) discovers that the capacity for absorbing heat varies with substance, leading to the determination of specific heats. By reexamining and reinterpreting earlier experiments of Daniel Fahrenheit and others, he is able to reject conclusively the formerly accepted viewpoint that the amount of heat required to raise the temperature of a body by a given amount is proportional to the amount of matter and independent of the type of matter. The earlier experiments showed, for example, that when equal amounts of quicksilver (mercury) and water are heated simultaneously by the same fire, the temperature of the quicksilver rises almost twice as fast as the water's temperature. Black's conclusion is that the fire communicates equal quantities of heat to the two substances but the quicksilver has less capacity for absorbing that heat. He is thereby able to explain also why when equal quantities of quicksilver at 150° and water at 100° are mixed, the equilibrium temperature is less than the average of 125°, being instead 120°. Similarly, if equal quantities of quicksilver at 100° and water at 150° are mixed, the equilibrium temperature is 130°, again confirming that the heat being transferred between the two substances has a greater effect on the temperature of the quicksilver than on the temperature of the water, due to quicksilver's lesser capacity to hold the heat.

EARTH SCIENCES (Glacial Sliding)

Gottlieb Gruner (1717–1778, Swiss) suggests in his three-volume *Die Eisgebirge des Schweizerlandes* that glacial movement is a sliding process.

EARTH SCIENCES (Seismology)

John Michell (1724–1793, English) hypothesizes in his "Essay on the Causes and Phenomena of Earthquakes" that earthquakes result from a sudden access of water to subterranean fires and a buildup of gas pressure due to an underground boiling from volcanic heat, thereby explaining the frequency of occurrence of earthquakes near volcanoes. He also suggests that quakes can begin under the ocean and explains the destructive Lisbon earthquake of 1755 on that basis, its origin being located under the Atlantic. Believing that earthquakes generate waves within the earth, Michell further suggests that the location of the center of the quake can be determined from the differing times the quake is felt at various locations. This is finally accomplished by John Milne in the late 1800s. Michell's speculation that the wave motion is due to the underground propagation of vapor has been rejected.

MATHEMATICS (Calculus of Variations)

Joseph Lagrange (1736–1813, French) publishes his "Essai d'une nouvelle méthode pour déterminer les maxima et les minima des formules intégrales indéfinies" [Essay on a method for determining the maxima and minima of indefinite integral formulas], presenting his treatment, developed over the past decade, of the calculus of variations through purely analytical methods rather than the earlier geometric-analytic methods.

PHYSICS (Hydrodynamics)

Preceding Laplace, Joseph Lagrange (1736–1813, French) makes use of Laplace's Equation to solve problems in hydrodynamics.

1760–1762

BIOLOGY (Comparative Anatomy)

Peter Camper (1722–1789, Dutch) publishes the two-volume *Demonstrationum anatomico-pathologicarum* on the comparative anatomy of different human races.

1761

ASTRONOMY (Sun; Venus)

Several expeditions, including one at the Cape of Good Hope headed by Charles Mason (1728–1786, English) and Jeremiah Dixon (1733–1799, English), one at Newfoundland headed by John Winthrop (1714–1779, American), and one at St. Helena headed by Nevil Maskelyne (1732–1811, English), observe and record the June 6 solar transit of Venus. The major objective of this international effort is to determine more accurately the sun's parallax and hence the earth-sun distance.

ASTRONOMY (Venus)

Mikhail Lomonosov (1711–1765, Russian) infers that Venus has an atmosphere.

HEALTH SCIENCES (Pathological Anatomy)

Giovanni Battista Morgagni (1682–1771, Italian) initiates the study of pathological anatomy with his work *On the Causes of Diseases.*

HEALTH SCIENCES (Percussion)

Leopold Auenbrugger (1722–1809, Austrian) develops the method of percussion, tapping portions of the body and diagnosing chest diseases based on the resulting sounds. The method is ignored for about 40 years, then is revived by Jean Nicolas Corvisart (1755–1821, French) and widely adopted.

MATHEMATICS (Analysis)

Leonhard Euler (1707–1783, Swiss) proves the addition theorem for elliptic integrals.

MATHEMATICS (Number Theory)

Leonhard Euler (1707–1783, Swiss) proves that all primes of the form $6n + 1$ can be expressed as $x^2 + 3y^2$ for integral x, y.

PHYSICS (Fluid Dynamics)
EARTH SCIENCES (Fluid Flow)

In *Principia motus fluidorum,* Leonhard Euler (1707–1783, Swiss) presents two general equations for incompressible flow and develops the procedure of describing fluid flow by determining at fixed points the time variation of the density and velocity of the fluid (Eulerian flow).

1761–1768

BIOLOGY (Evolution)

Publication of a five-volume work by Jean Baptiste Robinet (1735–1820, French) in which he asserts that the Creator created a complete linear scale of all organic beings, vegetable and animal, but that placement along this scale does not remain static with time, as all species have an internal energy through which they can move upward toward the peak, which is mankind. Robinet further believes that all matter possesses life, and that inorganic matter can evolve, through various combinations, into a living being.

1762

CHEMISTRY (Iron)

Cast iron is converted into malleable iron at Carron Ironworks, in Stirlingshire, Scotland.

CHEMISTRY (Latent Heats)

Joseph Black (1728–1799, Scottish) discovers that for a solid to melt or a liquid to vaporize requires the absorption of heat even though the temperature remains constant; that is, he develops the concept of latent heat, and indeed introduces the term, arguing that so far as temperature is concerned, the heat is concealed, or "latent," within the body.

EARTH SCIENCES (Stratigraphy)

Extrapolating from his field studies in his native region of Thuringia, George Füchsel (1722–1773, German) publishes *Historia terrae et maris, ex historia Thuringiae per montium descriptionem erecta* [A history of the earth and sea, based on a history of the mountains of Thuringia]. Like Johann Lehmann (1756—EARTH SCIENCES), Füchsel determines that the rocks of the earth's crust are layered in strata and that the series of strata reveal processes and changes undergone in the history of the earth. He presents the succession of strata in Thuringia and notes that the individual strata have distinctive fossil remains.

MATHEMATICS (Calculus of Variations)

Joseph Lagrange (1736–1813, French) unifies and further develops the calculus of variations, introducing new analytic methods and the variational operator δ.

1763

BIOLOGY (Fertilization)

Joseph Kölreuter (1733–1806, German) analyzes the mechanism of plant fertilization through pollen transport by animals.

HEALTH SCIENCES (Disease Classification)

Carolus Linnaeus (1707–1778, Swedish) classifies diseases in *Genera morborum.*

MATHEMATICS (Non-Euclidean Geometry)

Georg Klügel (1739–1812, German) suggests that perhaps Euclid's Parallel Postulate of a unique line through a given point parallel to a given line cannot be proved from the other Euclidean postulates, and declares that the geometric results of Girolamo Saccheri with alternative hypotheses of no parallel lines and many parallel lines (1733—MATHEMATICS) are merely strange, not internally contradictory.

MATHEMATICS (Statistics)

Thomas Bayes (1702–1761, English), in his posthumous "Essay Towards Solving a Problem in the Doctrine of Chances," becomes perhaps the first to use probability theory in an inductive sense, arguing from the sample to the full population.

1764

PHYSICS (Condenser)

James Watt (1736–1819, Scottish) invents the condenser, a major step toward invention of the steam engine.

1765

CHEMISTRY (Food Preservation)

Lazzaro Spallanzani (1729–1799, Italian) suggests preserving the quality of food by sealing it in airtight containers.

EARTH SCIENCES (Basalt; Volcanism)

Nicholas Desmarest (1725–1815, French) establishes the volcanic origin of basalt columns, showing that they are the remains of once continuous sheets of lava. Struck by the regularity of the columnar structure and observant of the occasional association of such structure with the remains of former volcanoes, he convinces himself that the prismatic basalts in Auvergne, central France, were part of an old lava stream, then that the Irish basalts in the Giant's Causeway must be volcanic in origin also in view of the close similarity of the substances. This leads to the broader conclusion and prediction that wherever the polygonal basalt columns are found there must also exist formerly active volcanoes. Desmarest announces his theory to the French Academy of Sciences in 1765, although does not publish until 1774.

EARTH SCIENCES (Fossils)

Jean-Étienne Guettard (1715–1786, French) attempts in the paper "On the Accidents that have befallen Fossil Shells compared with those which are found to happen to Shells now living in the Sea" to prove to the remaining doubters that fossils are the remains of once-living plants and animals, doing so by noting the similarities between the beds of fossil shells and the bed of the sea floor. He thus explains fossil beds, laid down in the distant past, on the basis of processes still occurring. Such explanations become more frequent in succeeding decades.

HEALTH SCIENCES (Diphtheria)

Francis Home (1719–1813, Scottish) includes a clinical description of diphtheria in his monograph *An Enquiry Into the Nature, Cause, and Cure of the Croup.*

PHYSICS (Chronometers)
EARTH SCIENCES (Longitude)

Having worked on improving chronometers since 1728, John Harrison (1693–1776, English) finally wins a long-standing prize from the Board of Longitude for a method of accurate determination of longitude from ships. Although it is large (66 pounds) and therefore cumbersome, tests of Harrison's chronometer on ocean voyages in 1761 and 1764 established the instrument as accurate to within 0.1 seconds per day, leading to longitude determinations accurate to within about 1.3 minutes of arc, well within the 30 minute of arc accuracy required for the prize. Harrison's advances, along with contemporaneous ones by Pierre Le Roy (1717–1785, French), greatly reduce the importance of the moon for navigation and therefore also reduce the importance of the three-body problem in applied mathematics. (Harrison actually receives only half the prize's monetary award in 1765, then is subjected to several new requirements by the Board of Longitude, such as constructing two more instruments, before receiving the remainder of the award money in 1773, after obtaining the support of George III, among others.)

PHYSICS (Mechanics)

In *Theoria motus corporum solidorum seu rigidorum,* Leonhard Euler (1707–1783, Swiss) develops the mechanics of solid bodies similarly to his earlier development of the mechanics of the point mass (1736—PHYSICS). He gives particular attention to rotational motion, after establishing that the instantaneous motion of a body can be decomposed to a rotation and a rectilinear translation.

1766

ASTRONOMY (Bode's Law)

Johann Titius (1729–1796, German) constructs the empirical relationship later known as Bode's Law (see 1772—ASTRONOMY).

CHEMISTRY (Hydrogen)

Henry Cavendish (1731–1810, English), in the paper "On Factitious Airs," publishes a study about "inflammable air," or, in later terminology, hydrogen. Although Robert Boyle and others probably had earlier examined the gas somewhat, and although Cavendish at first considers it perhaps to be phlogiston, Cavendish is generally considered the discoverer of the element hydrogen because of his careful description and his work establishing it as a unique air. Cavendish obtains hydrogen through the action of dilute acids on zinc, iron, and tin, and terms it "inflammable air" because of the ease with which it burns. He isolates this air—the term "gas" was not yet current—as well as others by means of the pneumatic trough, which he also further develops.

CHEMISTRY (Nomenclature)

In the two-volume *Dictionnaire de chymie* [Dictionary of chemistry], Pierre Macquer (1718–1784, French) continues to identify and revise inconsistencies and other flaws in the current chemical nomenclature (see also 1751—CHEMISTRY), for instance, objecting to the use of separate names for an individual salt depending upon point of origin.

MATHEMATICS (Non-Euclidean Geometry)

Johann Lambert (1728–1777, German-French) notes that the surface of a sphere satisfies the postulate that if two perpendiculars *AD* and *BC* of equal length are drawn at the ends of a line segment *AB* and if *D* and *C* are joined, then the angles *ADC* and *BCD* are equal but obtuse rather than right angles as in Euclidean geometry. This is equivalent to one of the two alternative hypotheses to Euclid's Parallel Postulate—equivalent itself to having angles *ADC* and *BCD* both right angles—which Girolamo Saccheri examined in an attempt to prove them invalid (1733—MATHEMATICS). Lambert becomes convinced that Euclid's Parallel Postulate cannot be proved from the other Euclidean postulates and that it is possible to build a logically consistent system satisfying the other postulates but explicitly rejecting the Parallel Postulate.

1767

ASTRONOMY (Ephemeris)

Publication of the first astronomical ephemeris, *The British Nautical Almanac and Astronomical Ephemeris for the Meridian of the Royal Observatory at Greenwich,* including tabulations of the moon's positions at midnight and noon.

BIOLOGY (Spontaneous Generation)

Lazzaro Spallanzani (1729–1799, Italian) helps disprove John Needham's assertions (see 1740—BIOLOGY) on the spontaneous generation of microorganisms by repeating Needham's boiled-broth-in-a-sealed-tube experiments with the additional restriction of withdrawing the air from the flasks. Since no microorganisms arise when the air is removed, Spallanzani concludes that the microorganisms appearing in Needham's experiments did not arise spontaneously but derived from germs transported in the air.

MATHEMATICS (Theory of Equations)

Joseph Lagrange (1736–1813, French) examines the roots of algebraic equations in the memoir "Sur la résolution des équations numériques" [On the solution of numerical equations] and provides methods of separating the real and imaginary roots and of approximating the real roots with continued fractions.

1768–1770

MATHEMATICS (Calculus)

Leonhard Euler (1707–1783, Swiss) publishes his three-volume *Institutiones calculi integralis,* in which he presents methods of definite and indefinite integration, having invented many of the methods himself, such as the use of an "Euler substitution" for rationalizing particular irrational differentials. His treatment is near exhaustive for integrals expressible as elementary functions. He also develops the theory of ordinary and partial differential equations and presents many properties of the beta and gamma function Eulerian integrals introduced by Euler earlier (1729—MATHEMATICS).

1768–1771

BIOLOGY (Observational Botany)

Joseph Banks (1743–1820, English) and several assistants collect specimens of 3,607 species of plants, approximately 1,400 of them being previously unidentified, during an around-the-world voyage of H.M.S. *Endeavour* captained by James Cook (1728–1779, English).

1768–1774

SUPPLEMENTAL (Scientific Expeditions)

Pierre Simon Pallas (1741–1811, Prussian) leads a large-scale scientific expedition throughout the Russian Empire, gathering information on the geography, meteorology, geology, plant and animal life, and social customs, and taking detailed observations of the 1769 transit of Venus. Government-supported, the expedition was organized by the St. Petersburg Academy of Sciences after being commissioned by the Empress Catherine II. Pallas publishes results of the expedition in three quarto volumes (1772–1776).

1769

ASTRONOMY (Venus)

The June 3 transit of Venus across the solar disk is recorded from several locations, including Tahiti by observers on the 1768–1771 *Endeavour* voyage led by Captain James Cook (1728–1779, English) and Cavan, Ireland, by Charles Mason (1728–1786, English) and colleagues (though the view from Cavan is partially obscured by cloud cover). Analysis of such transits provides data for estimating the sun-earth distance, accounting for the interest and expense provided for the transit expeditions.

CHEMISTRY (Phosphorus)

Carl Scheele (1742–1786, Swedish) and Johan Gahn (1745–1818, Swedish) find phosphorus in bones and isolate it.

CHEMISTRY (Quantitative Methods)

Quantitative studies of Antoine Lavoisier (1743–1794, French) throughout his scientific career contribute significantly toward making the study of chemistry—formerly almost exclusively qualitative—into a quantitative science. Among Lavoisier's early successes is his determination in 1769 that the sediment found in flasks after the boiling of water derives from the flask itself rather than from a transmutation of water into earth. This he is able to determine through carefully weighing the materials before and after a 101-day period of boiling in a closed flask.

HEALTH SCIENCES (Excisions)

The first excision of a hip is carried out by Charles White (1728–1813, English).

1769–1770

MATHEMATICS (Diophantine Analysis)

Joseph Lagrange (1736–1813, French) derives a method for obtaining all integral solutions to the equation $x^2 - Ay^2 = B$, where A and B are given, arbitrary integers. This marks a turning point in the study of Diophantine analysis, as earlier researchers had been satisfied with limited sets of solutions to equations. The method, later to prove fundamental in the development of the Gaussian theory of binary quadratic forms, allows Lagrange to obtain the first determination of the units of an algebraic number field other than the rationals.

1769–1772

PHYSICS (Steam Engine)

John Smeaton (1724–1792, English) improves the Newcomen steam engine, invented in 1705, by adjustments decided upon through experimental determination of the length of the cylinder and the rate of piston rotation yielding maximum efficiency.

1770

BIOLOGY (Catastrophic Evolution)

In his *Philosophical Palingenesis, or Ideas on the Past and Future States of Living Beings,* Charles Bonnet (1720–1793, Swiss) adheres to the view that miniature forms of all future generations are contained within the females of each species. However, he also postulates that at very long intervals major catastrophes occur, the most recent being the biblical Flood, during which times the miniatures within the females can evolve upward. He points to fossils as evidence, and predicts that with the next catastrophe stones will gain organic structure, plants will become self-moving, animals will be able to reason, and humans will become angels.

CHEMISTRY (Combustion)

On observing a burning candle in a closed container, Joseph Black (1728–1799, Scottish) notes the eventual extinguishing of the flame and questions why, with both air and candle remaining, the flame should go out (1775—CHEMISTRY).

EARTH SCIENCES (Basalt)

In "On the Basalt of the Ancients and the Moderns," Jean-Étienne Guettard (1715–1786, French) concludes that basalt forms by crystallization in an aqueous fluid, a conclusion reversed by him at the end of the decade when he aligns himself instead with those asserting a volcanic origin (1779—EARTH SCIENCES).

EARTH SCIENCES (Gulf Stream)

Benjamin Franklin (1706–1790, American) publishes the first scientific chart of the North Atlantic's Gulf Stream, a map subsequently widely used by navigators and highly regarded by scientists. The Gulf Stream appears on the map as a river of warm water originating near the Florida Strait, moving northeastward off the North American coast, then eastward well beyond Newfoundland. Franklin hypothesizes that the trade winds cause the Gulf Stream, by driving warm waters into the Gulf of Mexico, from which they exit by way of the Florida Strait and proceed to form the Gulf Stream.

MATHEMATICS (Algebra)

Leonhard Euler (1707–1783, Swiss) publishes the algebra textbook *Vollständige Anleitung zur Algebra,* in which he develops the theories of polynomial and indeterminate equations. He includes a proof, based on infinite descent, of the $n = 3$ case of Fermat's Last Theorem (1637—MATHEMATICS).

MATHEMATICS (Group Theory; Lagrange's Theorem)

The notion of "group," though not the term, emerges with the work of Joseph Lagrange (1736–1813, French) and Alexandre Vandermonde (1735–1796, French) on permutation groups, groups of one-to-one transformations on a set of elements. Later the term "group" is formally defined as a set of elements and an associated operation such that the set is closed under the operation, the associative law is satisfied, there exists an identity element, and each element has an inverse. Lagrange in particular moves toward the later theory of groups through his attempts to solve the general quintic equation (1770—MATHEMATICS). Lagrange includes a proof, essentially, that the number of elements in a group is divisible by the number of elements in any of its subgroups (Lagrange's Theorem).

MATHEMATICS (Number Theory)

Joseph Lagrange (1736–1813, French) proves that every positive integer is a sum of four squares, a proof sought by many.

MATHEMATICS (Number Theory; Waring's Problem)

Edward Waring (c.1736–1798, English) asserts that not only is each positive integer a sum of four squares but it is also a sum of nine cubes and of 19 fourth powers (see 1909—MATHEMATICS).

MATHEMATICS (π)

Johann Lambert (1728–1777, German-French) shows that both π and π^2 are irrational.

MATHEMATICS (Theory of Equations)

In his "Réflexions sur la résolution algébrique des équations," Joseph Lagrange (1736–1813, French) unifies the solution by radicals of quadratic, cubic, and quartic equations, having examined for this purpose the sixteenth-century solutions of Niccolò Tartaglia, Ludovico Ferrari and others. He then proceeds to the problem of solving the general quintic equation by examining the applicability of the methods used in the solutions of the lower order equations. In the process he obtains hints that the higher order equations might not be solvable by radicals, that is, that the answers might not be obtainable as functions of the coefficients of the equation with no operations on these coefficients other than addition, subtraction, multiplication, division, and the extraction of roots.

c.1770

CHEMISTRY (Fixed Air)

Joseph Priestley (1733–1804, English) experiments with the layer of "fixed air" (carbon dioxide) above the beer in vats of a brewery: mice placed in the air die, candles go out, and water poured from one glass to another becomes bubble-infested. The distinctive taste of the resulting soda water gains wide popularity and establishes Priestley's fame.

1770s

CHEMISTRY (Gases)

Joseph Priestley (1733–1804, English) discovers or co-discovers several gases, including ammonia, nitrous and nitric oxide, nitrogen dioxide, sulfur dioxide, oxygen (see 1774—CHEMISTRY), nitrogen (see 1772—CHEMISTRY), and carbon monoxide, and uses the pneumatic trough to isolate them.

1771

ASTRONOMY (Nebulae and Clusters)

Charles Messier (1730–1817, French) publishes a list of 45 nebulae and star clusters, later expanded to the major Messier catalog of 1784, with 103 nebulae and clusters (1784—ASTRONOMY).

BIOLOGY (Blood Coagulation and Color)

William Hewson (1739–1774, English) publishes an *Experimental Inquiry into the Properties of the Blood,* in which, among other things, he establishes the essential features of blood coagulation, describing his discovery that coagulable lymph (or fibrinogen) is required for blood to clot. Hewson also asserts that the blood which in the arteries is red is dark or brackish in the veins and postulates that the color change occurs as the blood passes through the lungs and there comes in contact with air.

CHEMISTRY (Combustion and Respiration)
BIOLOGY (Respiration)

Having determined that candles burn remarkably well in air long sustaining plant life, Joseph Priestley (1733–1804, English) discovers that the air affected by burning candles or breathing mice can be entirely restored by vegetation. (The air affected was believed by Priestley to have acquired phlogiston from the burning; it is recognized today as instead having been depleted of oxygen.)

CHEMISTRY (Hydrofluoric Acid; Oxygen)

Carl Scheele (1742–1786, Swedish) describes hydrofluoric acid and also prepares oxygen, but, due to a six-year delay in publication, the credit for the discovery of oxygen generally goes to Joseph Priestley (1774—CHEMISTRY).

HEALTH SCIENCES (Teeth)

John Hunter (1728–1793, English-Scottish) publishes a treatise on *The Natural History of the Human Teeth.*

MATHEMATICS (Algebra)

Alexandre Vandermonde (1735–1796, French) shows, in the work "Sur la résolution des équations algébriques," that for any prime $p < 13$, the equation $x^p - 1 = 0$ is solvable by radicals, meaning solvable through the five operations of addition, subtraction, multiplication, division, and extraction of nth roots. He then asserts, though does not prove, that the equation is solvable by radicals for any prime p.

METEOROLOGY (Observational Network; Resultant Winds)

Johann Lambert (1728–1777, German-French) proposes a worldwide network for meteorological observations and suggests determining not just wind frequencies from different directions but a resultant mean wind direction. Although it is a forerunner of more sophisticated resultant wind calculations, Lambert's resultant wind has limited usefulness due to its being a function solely of wind frequency, with no weighting according to wind magnitude.

1772

ASTRONOMY (Bode's Law)

Johann Bode (1747–1826, German) publishes Bode's Law, an empirical relationship showing that the relative distances, in astronomical units, of the six known planets from the sun roughly satisfy the sequence formed by adding 4 to the members of the simpler sequence 0, 3, 6, 12, (24, no corresponding planet, see 1801—ASTRONOMY), 48, 96 and then dividing by 10. (The distance from the sun to Uranus, discovered in 1781, fits the sequence, as do the distances from the sun to the four asteroids discovered from 1801 to 1807; however, the distance from the sun to Neptune, discovered in 1846, does not.)

ASTRONOMY (Meteorites)

A memorandum signed by several French scientists including Antoine Lavoisier (1743–1794, French) concludes that stories of stones falling from the sky are all somehow in error or mere fabrications, since the phenomenon is considered impossible.

CHEMISTRY (Combustion)

Antoine Lavoisier (1743–1794, French) extensively reviews the work done before him on combustion, repeats several of the experiments with a new emphasis on quantification, and determines that significant amounts of air are *absorbed* during combustion. This is in direct contrast to the phlogiston theory that phlogiston is given off during combustion. Lavoisier does not publish immediately but, to establish priority, deposits with the French Academy a sealed note regarding experiments on the burning of phosphorus and sulfur. He contends that these and likely all other substances gaining weight during combustion or calcination do so by absorbing air. To determine the amount of air involved, he performs the experiment in a container with a single outlet to an expandable bladder of air. As predicted, the weight increase of the phosphorus or sulfur is balanced by the weight decrease of the air.

CHEMISTRY (Combustion; Phlogiston)

Daniel Rutherford (1749–1819, Scottish) extends the recent experiments of Joseph Black on candles in closed containers (1770—CHEMISTRY). Rutherford finds that when a newly burning candle is placed in the air of a container in which a first candle burned until a natural extinguishing of its flame, the flame of the new candle is extinguished immediately. Similarly, a mouse placed in the affected air quickly dies. From the context of the phlogiston theory, both these phenomena are explicable by assuming the air to have become fully saturated with phlogiston.

CHEMISTRY (Nitrogen or Phlogisticated Air)

Nitrogen—or "phlogisticated" air—is discovered independently by Daniel Rutherford (1749–1819, Scottish), Carl Scheele (1742–1786, Swedish), Joseph Priestley (1733–1804, English), and Henry Cavendish (1731–1810, English). Rutherford, who obtains predominantly nitrogen (or "mephitic air" in his terminology) by subtracting from air those portions removable by respiration or combustion, proceeds to a more careful study of the properties of the gas.

EARTH SCIENCES/CHEMISTRY (Crystallography)

Publication of Jean Romé de l'Isle's (1736–1790, French) *Essai de cristallogra-*

phie, in which he describes the process of crystallization, the geometric regularity of the resulting forms, and the relevance to understanding natural history.

MATHEMATICS (Probability Theory)

Emphasizing the uncertain nature of many aspects of life—for instance, the inability of a judge to know for certain the guilt or innocence of an accused—Voltaire (1694–1778, French) in his *Essay on Probabilities Applied to the Law* encourages the serious study and development of probability theory.

MATHEMATICS (Three-Body Problem)
ASTRONOMY (Celestial Mechanics)

Joseph Lagrange (1736–1813, French) publishes his *Essai sur le problème des trois corps* on the three-body problem of the possible orbits of three bodies, each under the gravitational influence of the other two, providing exact solutions for certain special cases of the problem. One solution has the three bodies orbiting in similar ellipses with a common focus at the center of mass of the three-body system. Another solution has them rotating as a unit at the vertices of an equilateral triangle, the rotation occurring about the center of mass. A third solution has them rotating as a unit at fixed positions on a rotating straight line.

PHYSICS (Optics; Color)

Joseph Priestley's (1733–1804, English) *History and Present State of Discoveries Relating to Vision, Light and Colours,* covering the century 1672–1772, is reflective of the increasing interest in color and related phenomena.

1773

BIOLOGY (Blood)

Hilaire-Marin Rouelle (1718–1779, French) finds sodium carbonate, potassium chloride, and common salt in the blood of humans and other animals.

HEALTH SCIENCES (Hygiene)

Charles White's (1728–1813, English) *A Treatise on the Management of Pregnant and Lying-in Women* encourages cleanliness, fresh air, and hygiene, a major innovation for the hospitals of the time and one not adopted by many.

MATHEMATICS (Number Theory)

Joseph Lagrange (1736–1813, French) introduces general methods into the theory of binary quadratic forms, originating basic techniques for later researches and obtaining immediately numerous individual theorems labored over by his predecessors.

METEOROLOGY (Hygrometer)

Jean Andre De Luc (1727–1817, Swiss) constructs and describes a hygrometer, measuring the atmospheric humidity by the level of mercury in a thermometer attached to an ivory cylinder filled with mercury and closed at the opposite end.

PHYSICS (Gravitational Potential)

Joseph Lagrange (1736–1813, French) originates the idea of scalar gravitational potential.

1774

BIOLOGY (Blood)

William Hewson (1739–1774, English) postulates that the thymus and lymph glands create white blood corpuscles and that these become red during passage through the spleen.

BIOLOGY (Pregnant Uterus)

In a sequence of 34 copper plates in *Anatomia uteri humani gravidi tabulis illustrata* [The anatomy of the human gravid uterus exhibited in figures], William Hunter (1718–1783, Scottish-English) depicts the uterus during the second half of pregnancy.

CHEMISTRY (Chlorine)

Carl Scheele (1742–1786, Swedish) discovers the gas chlorine, preparing it by combining manganese dioxide and hydrochloric acid. (It is not until 1810 that chlorine is recognized as an element rather than a compound.)

CHEMISTRY (Combustion)

Publication of Antoine Lavoisier's (1743–1794, French) *Opuscules physiques et chimiques,* which includes a 15-chapter historical introduction followed by a description of experiments. In research similar to his work of 1772, Lavoisier repeats an experiment by Robert Boyle which determined that tin and lead increase in weight after heating in sealed containers. Lavoisier weighs the air in the container as well as the metal, and finds that the weight increase of the metal derives from absorption of a portion of the air. The total weight of the system remains unchanged. Lavoisier observes that during combustion no more than about one fifth of the air is ever absorbed; but he remains unable to explain this until 1775 (see 1775—CHEMISTRY).

CHEMISTRY (Manganese)

Carl Scheele (1742–1786, Swedish) publishes the treatise "Concerning manganese and its properties" and his disciple Johan Gahn (1745–1818, Swedish) becomes the first to isolate the metal. Scheele and others had believed in the existence of metallic manganese but had failed in attempts at isolating it. Gahn's success comes through intense heating of a mixture of pulverized pyrolusite and oil in a crucible lined with wet charcoal dust.

CHEMISTRY (Oxygen or Dephlogisticated Air)

Joseph Priestley (1733–1804, English) isolates, and in a sense discovers, oxygen by heating mercuric oxide and thereby decomposing it to mercury vapor and oxygen. The oxygen is not recognized as such by Priestley but is interpreted as "dephlogisticated air." It supports combustion remarkably well. In 1775 Priestley places a mouse in the dephlogisticated air and finds that it can survive at least twice as long as a mouse placed in an equal amount of ordinary air. In a 1774 letter to Antoine Lavoisier, Carl Scheele (1742–1786, Swedish) describes the chemical and physiological properties of the new air and details a procedure for producing it.

EARTH SCIENCES (Basalt)

Nicholas Desmarest (1725–1815, French) publishes his conclusion that basalt has a volcanic origin (1765—EARTH SCIENCES), and further suggests that basalt is produced from the melting of granite, a suggestion not recognized as false until a better understanding of chemical composition is developed in the nineteenth century.

EARTH SCIENCES (Earth's Mass)

Nevil Maskelyne (1732–1811, English) computes the mass of the earth by comparing the deviations of a plumb line from the vertical on both sides of a granite mountain, the mass of which he determines from volume and density estimates. His calculations yield an estimate of the earth's average density at 4.5 times the density of water (the twentieth-century estimate is approximately 5.5 times the density of water).

EARTH SCIENCES (Mineral Classification)

Abraham Werner (1749–1817, German) publishes the first edition of *Von den ausserlichen Kennzeichen der Fossilien* on mineral classification. Mentioning the eventual desirability but current impracticality of classifying minerals according to chemical composition, he bases his classification on external characteristics. These include form, color, hardness, cleavage, and luster.

HEALTH SCIENCES (Hypnotism)

Franz Mesmer (1734–1815, Austrian) uses hypnotism therapeutically, leading to the coining of the terms "mesmerize" and "mesmerism." His use of hypnotism developed partly from his earlier interest in animal magnetism.

MATHEMATICS (Binomial Theorem)

Leonhard Euler (1707–1783, Swiss) completes a partial proof of the binomial theorem for any rational exponent.

MATHEMATICS (Differential Equations)

Joseph Lagrange (1736–1813, French) generalizes the method of variation of parameters (1739—MATHEMATICS) for solving differential equations.

METEOROLOGY (Humidity)

Johann Lambert (1728–1777, German) discusses the measurement of atmospheric humidity in his "Suite de l'essai d'hygrometrie," where he also introduces the term "hygrometer." In perhaps the first graphical presentation of meteorological results, Lambert graphs measured humidities as a function of time, and notes a decided annual cycle in humidity at the German cities from which he has measurements. He finds the highest humidities in June and the lowest in December and January, and notes that the temperature curve tends to lag the humidity curve by one and a half months.

1774–1777

CHEMISTRY (Oxygen; Air)

Joseph Priestley (1733–1804, English) describes his experiments with and results from various gases in his three-volume *Experiments and Observations on Different Kinds of Air.* Primary among these are his experiments on the "dephlogisticated" air (later termed "oxygen") first isolated by Priestley in 1774 (1774—CHEMISTRY) and particularly those experiments establishing that mice live longer and candles burn brighter in this than in normal air.

1775

BIOLOGY (Insects)

Johan Fabricius (1745–1808, Danish) classifies insects according to mouth structure (in *Systema entomologiae*).

CHEMISTRY (Acids)

Joseph Priestley (1733–1804, English) discovers gaseous hydrogen chloride and sulfur dioxide gas, which, when dissolved in water, become hydrochloric acid and sulfurous acid respectively.

CHEMISTRY (Affinity)

Torbern Bergman (1735–1784, Swedish) prepares a table of affinities (see

1718—CHEMISTRY) for 59 substances, doing so both for reactions attained through fusion and for those attained through solution.

CHEMISTRY (Oxygen; Combustion)

In the paper "Mémoire sur la nature du principe que se combin avec les métaux pendant leur calcination, et qui en augmente le poids" [On the nature of the principle that combines with metals in calcination and increases their weight], Antoine Lavoisier (1743–1794, French) presents his first major (although still preliminary) statement disavowing the phlogiston theory and presenting a fundamental revision in the theory of combustion. This revision, developed over the following decade, bases analysis of combustion on the absorption of oxygen from the air rather than on the release of phlogiston from the burning substance. In 1774 Lavoisier had been visited by Joseph Priestley, who spoke of his "dephlogisticated air" (see 1774—CHEMISTRY) in which candles burn more readily than in ordinary air. Lavoisier quickly determines that it is this dephlogisticated air, or oxygen, which is the portion of the air absorbed during burning (see CHEMISTRY entries for 1772 and 1774). He also decides (erroneously) that all acids are formed by combination of oxygen and nonmetallic substances. The classic 1775 paper is published in the rapid-publication journal *Observations sur la physique,* while awaiting more formal publication in the Academy of Sciences' *Mémoires.* The final publication occurs three years later, with several important revisions (1778—CHEMISTRY).

EARTH SCIENCES (Volcanism)

Nicholas Desmarest (1725–1815, French) presents to the Paris Academy of Sciences an essay "On the Determination of Three Epochs of Nature from the Products of Volcanoes, and on the Use that may be made of these Epochs in the Study of Volcanoes," in which he groups volcanic remains into three categories reflective of the geologic sequence of gradual degradation. Remains of the most recent eruptions are easily recognizable, as they include distinct cones and continuous lava streams, whereas the remains of the older eruptions are often obscured by the disappearance of the cones, the river-induced excavation of valleys across the former sheets of lava, and the deposition of sediment over the volcanic remains. (The essay is not published until 1806.)

HEALTH SCIENCES (Dropsy)

William Withering (1741–1799, English) relates dropsy to cardiac disease, and introduces the use of digitalis as a diuretic in dropsy. His continued use of the plant-derived drug over the next decade leads to a monograph on its uses and dangers (1785—HEALTH SCIENCES).

PHYSICS (Fluid Flow)

Leonhard Euler (1707–1783, Swiss), using Isaac Newton's second law of motion (1687—ASTRONOMY), derives the general equations of motion for a nonviscid fluid and the equation of continuity.

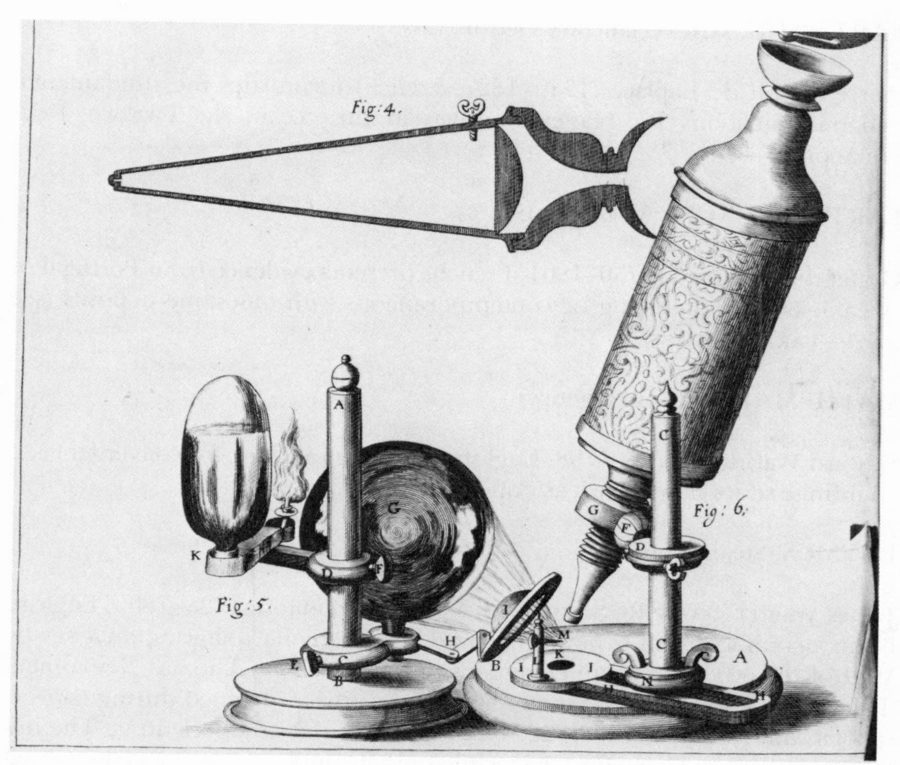

FIGURE 8. *Robert Hooke's microscope, used for examining a wide variety of animate and inanimate objects, as depicted in his* Micrographia *(see 1665—BIOLOGY entry). Focusing is done at the threaded collar at G, while light from the lamp at the left is passed through a spherical condensing system in order to concentrate it on the specimen being examined. A cross section of the main portion of the instrument appears at the top of the illustration. (By permission of the Houghton Library, Harvard University.)*

1776

BIOLOGY (Comparative Human Anatomy)

Johann Blumenbach (1752–1840, German) divides mankind into five races—American Indian, Caucasian, Ethiopian, Malayan, and Mongolian—in his *On the Natural Varieties of Mankind.* He also separates the Primate order of Carolus Linnaeus into two separate orders, the Bimana for humans and the Quadrumana for apes, and suggests that the original human race was the Caucasian, which then "degenerated" into the other varieties as people moved into new environmental conditions.

BIOLOGY (Plant Classification)

William Withering (1741–1799, English) uses the system of Carolus Linnaeus (1735—BIOLOGY) to classify the vegetables of Great Britain.

EARTH SCIENCES (Planetary Ocean Waves)

Pierre Simon de Laplace (1749–1827, French) formulates the fundamental dynamic equations for planetary waves in the ocean, the Laplace Tidal Equations.

EARTH SCIENCES (Volcanoes)

Gratet de Dolomieu (1750–1801, French) presents evidence from Portugal of volcanic activity predating or contemporaneous with limestone deposits (see 1784—EARTH SCIENCES).

MATHEMATICS (Convergence)

Edward Waring (c.1736–1798, English) states the ratio test for convergence of an infinite series now known as Cauchy's Ratio Test.

PHYSICS (Steam Engine)

James Watt (1736–1819, Scottish) and Mathew Boulton (1728–1809, English) begin manufacture of a new, improved steam engine following analyses by Watt of the efficiency of an engine invented in 1705 by Thomas Newcomen. These analyses are undertaken in light of the advice obtained during conversations with Joseph Black (1728–1799, Scottish) and other scientists. The improved engine had been delayed technologically until a method was developed for accurately shaping cylinders. Such a method appeared in 1774 with the precision cannon borer of John Wilkinson (1728–1808, English).

1777

CHEMISTRY (Oxygen; Nitrogen)
METEOROLOGY (Air)

Carl Scheele (1742–1786, Swedish) publishes *Chemische Abhandlung von der Luft und dem Feuer,* in which he describes experiments on air and oxygen. Scheele determines that air contains at least two separate gases, one being "fire air" (or oxygen) and constituting roughly one fourth the total air volume, and the other being "foul air" (or nitrogen), constituting roughly three fourths the total air volume. Scheele shows that fire air supports combustion and that foul air is apparently inert. He retains the phlogiston theory, explaining that the fire air supports combustion because it absorbs into itself the phlogiston given off during burning. When this air becomes saturated with phlogiston, the burning can no longer continue. (Within a year, Antoine Lavoisier will advocate a very different theory, based on air's becoming depleted of oxygen rather than saturated with phlogiston [1778—CHEMISTRY].)

CHEMISTRY (Reaction Rates)

Carl Wenzel (1740–1793, German) determines the rates of various chemical reactions, concluding that reaction rates are approximately proportional to substance concentration and that the extent to which a metal is soluble in a given acid is proportional to the concentration of the acid. Such concern for quantification is common in the work of Wenzel, who also carefully weighs numerous compounds, although he fails to accomplish his main aim in the study, which was to measure chemical affinity.

EARTH SCIENCES (Mountains)

Pierre Simon Pallas (1741–1811, Prussian) determines from analysis of his field observations in the Russian Empire (1768–1774—SUPPLEMENTAL) and the studies of others in Europe and the Andes a pattern in the composition and distribution of mountains, which he details in his *Observations sur la formation des montagnes* [Observations on the formation of mountains]. The highest mountains are composed of granite, with a quartz base and no organic remains. These are accompanied by a band of schistose mountains which also contain no organic remains, and which together with the granite mountains constitute the Primitive Mountains. These in turn are flanked or even covered by limestone bands, constituting the so-called "Secondary Mountains," beyond which the limestone beds flatten into great plains with abundant organic remains of marine origin. Younger still are the "Tertiary Mountains" resting on the limestone beds and composed largely of sandstone and marls. The only large group of Tertiary Mountains known at the time are those examined by Pallas to the west of the Urals.

EARTH SCIENCES (Organic Remains)

Pierre Simon Pallas includes in his *Observations sur la formation des montagnes* (1777—EARTH SCIENCES) discussion of a vast number of organic remains found by himself and others throughout Siberia, particularly the bones and tusks of elephants, rhinoceroses, and buffaloes. He speculates that the rapid destruction of these animals derived from a massive flooding of Siberia at the time of the birth of various islands of the Indian and Pacific Oceans.

HEALTH SCIENCES (Color Blindness)

In a letter to chemist Joseph Priestley, Joseph Huddart (1741–1811, English) describes the condition of color blindness. This has subsequently been acclaimed as the first reliable account of the condition.

MATHEMATICS (Buffon Needle Problem)

Georges Buffon (1707–1788, French) initiates a branch of probability concerning problems with geometrical aspects, doing so through a problem proposed and solved within an "Essai d'arithmétique morale" included in a supplement to his *Histoire naturelle* (1749–1788—BIOLOGY). The problem is to determine the probability that a needle of length L will fall upon a line when thrown on a large plane area with parallel straight lines spaced equidistantly at a distance $d > L$. The theoretical solution that the probability equals $2L/\pi d$ allows experimental estimation of π through numerous tossings of the needle, with the ratio of "successful" tosses (with the needle falling on a line) to the total tosses set equal to $2L/\pi d$. (Pierre Simon de Laplace later extends the problem to consider a plane area crisscrossed with two mutually perpendicular sets of parallel lines, so that it is sometimes referred to as the Buffon-Laplace needle problem although more frequently it is termed the Buffon needle problem.)

METEOROLOGY (Temperature Scales)

The continued lack of standardization in temperature scales for meteorological thermometers is reflected in a list drawn up by Johann Lambert (1728–1777, German-French) of 19 temperature scales in current usage.

PHYSICS (Torsion; Torsion Balance)

Charles Augustin de Coulomb (1736–1806, French) devises the torsion balance for measuring torsional elasticity, devises the theory of torsion in strands of hair and thin silk, and shows that within a specified angular limit torsional oscillation can be closely approximated by simple harmonic motion with the torsional force proportional to the angle of twist. These and other results are presented in the prize-winning memoir "Recherches sur la meilleure manière de fabriquer les aiguilles aimantées." (John Michell [1724–1793, English] may have independently invented the torsion balance, although it is not mentioned in any of his known publications and it appears that he did not actually construct a model until 1793.)

1778

BIOLOGY (Descriptive Botany)

Jean Baptiste Lamarck (1744–1829, French) publishes the three-volume *Flore française* describing and classifying the wild plants of France.

CHEMISTRY (Boric Acid)

Hubert Höfer (German) discovers boric acid in a hot spring in Tuscany.

CHEMISTRY (Calcination; Combustion)

Antoine Lavoisier (1743–1794, French) expands his theory of calcination and combustion from 1775, publishing the revised version of his 1775 paper in the Academy of Sciences' *Mémoires*. He now speaks of the new air (oxygen) not as pure atmospheric air but as "the purest part of the air" and shows that when combined with carbon it produces the "fixed air" of Joseph Black (1754—CHEMISTRY), later recognized as carbon dioxide. Lavoisier determines that only the oxygen in the air supports combustion and respiration and that absorption of the air's oxygen results in the weight increase of a metal during calcination. Lavoisier heats mercury to obtain mercury calx, absorbing all the available oxygen; the remaining air is unable to support either combustion or respiration. Upon heating the calx at higher temperatures, the gas given off (again oxygen) supports combustion remarkably well and when mixed with the remaining gas from the first experiment, produces the original quantity of apparently normal air. Further demolishing the countering theory that combustion involves a release of phlogiston from the burning substance, Lavoisier claims that upon heating mercury in oxygen, the entire volume of the oxygen gas can be absorbed. Still, Lavoisier's conclusion that during combustion the burning substance combines with oxygen in the air is not widely accepted until 1783, when he presents even more convincing evidence (1783—CHEMISTRY).

EARTH SCIENCES (Earth History)

Georges Buffon (1707–1788, French) publishes *Époques de la nature,* volume 34 of his *Histoire naturelle* (1749–1788—BIOLOGY), in which he revises and expands upon his earlier *Théorie de la terre* (1749—EARTH SCIENCES). He considers the six "days" of Creation mentioned in Genesis to be six extended epochs, and describes major events in the Earth's history during each of these plus one additional epoch, the latter being characterized by the influence of humans. Through experiments with cast-iron globes, Buffon makes pioneering attempts at calculating the age of the earth and opposes the supposed 6,000 year age asserted by many theologians. Buffon estimates that the earth is now 75,000 years old and will exist for another 93,000 years, during which period it will continue to cool until attaining a temperature below that of ice. In the first Epoch the still-molten earth assumed its oblate spheroid form, began cooling, and developed a crust. In the second Epoch the molten globe consolidated, forming the interior rock. In the third Epoch water condensed and covered the continents, then escaped in considerable amounts into the earth's interior during the fourth Epoch, through cracks in the surface. This produced extensive tracks of dry land above sea level, these being greatly affected by the former water, with, for instance, valleys being scoured out by the retreating waters. Volcanoes also emerged during the fourth Epoch, first requiring (for Buffon's scheme) sufficient accumulation of combustible materials. The fifth Epoch saw the emergence of terrestrial animals, and the sixth various major topographic shifts such as the continental separation of the Americas from Europe.

HEALTH SCIENCES (Teeth)

John Hunter (1728–1793, Scottish-English) classifies teeth into molars, bicuspids, cuspids, and incisors and publishes *A Practical Treatise on the Diseases of the Teeth.* Hunter establishes that much tooth decay begins on the surface rather than in the interior of the tooth and is particularly likely at locations where food particles lodge.

c.1778

BIOLOGY (Animal Heat)

Antoine Lavoisier (1743–1794, French) determines that animal heat derives from "respirable air" (oxygen), although he believes that this occurs as the respirable air combines with carbon.

1779

BIOLOGY (Animal Heat)
CHEMISTRY (Specific Heats)

In *Experiments and Observations on Animal Heat and the Inflammation of Combustible Bodies,* Adair Crawford (1748–1795, English) discusses heat generation by animals and develops methods of measuring specific heats.

BIOLOGY (Fertilization; Semen)

Lazzaro Spallanzani (1729–1799, Italian) establishes the importance of semen for fertilization.

BIOLOGY (Photosynthesis; Carbon Cycle)

In his paper "Experiments Upon Vegetables, Discovering Their Great Power of Purifying the Common Air in the Sunshine and of Injuring It in the Shade and at Night," Jan Ingenhousz (1730–1799, Dutch) shows that, in the presence of sunlight, the green portions of plants give off oxygen, whereas, in the absence of light, the roots, flowers, and fruits give off carbon dioxide. In the process, Ingenhousz also establishes that plants obtain carbon from the atmosphere, not, as might be supposed, from the soil.

CHEMISTRY (Nomenclature)

Influenced by Carolus Linnaeus's improvements in biological nomenclature (BIOLOGY entries for 1753 and 1758), Torbern Bergman (1735–1784, Swedish) recommends a binomial naming of salts, the first word of the name deriving from the alkali, earth or metal, and the second word an adjective deriving from the acid.

CHEMISTRY (Sassolite)

Paolo Mascagni (1755–1815, Italian) discovers sassolite (solid boric acid) in Montecerboli and Castelnuovo.

EARTH SCIENCES (Basalt)

Jean-Étienne Guettard (1715–1786, French) reverses himself (from 1770—

EARTH SCIENCES) on the origin of columnar basalt, concluding in *Mémoires sur la minéralogie du Dauphiné* that the field evidence is overwhelming for a volcanic origin.

EARTH SCIENCES (Erratic Boulders)

In volume 1 of his *Voyages dans les Alpes* (1779–1796—EARTH SCIENCES), Horace de Saussure (1740–1799, Swiss) rejects the hypothesis that the large erratic boulders found at various spots in the vicinity of the Alps were projected through the air to their present locations, offering as countering evidence the absence of impact craters. Instead he hypothesizes that the boulders were transported in the distant past by powerful currents during the general uplift forming the Alps and were deposited as the current velocities decreased. (It is later decided that the boulders were probably transported by glaciers; see 1837—EARTH SCIENCES.)

EARTH SCIENCES (Glacial Movement; Moraines)

In volume 1 of his *Voyages dans les Alpes* (1779–1796—EARTH SCIENCES), Horace de Saussure (1740–1799, Swiss) advocates the position that glacial movement, at least in the Alps, is a sliding process, the glaciers being divided from the underlying ground by a layer of melt water. He also discusses the size, shape, formation, and distribution of glacial moraines in the Alps and uses them to determine the history of the glaciers.

EARTH SCIENCES/CHEMISTRY (Marble)

Giovanni Arduino (1714–1795, Italian) postulates that marble originates from limestone through volcanically induced grinding, calcination, and repetrification.

1779–1796

EARTH SCIENCES (Mountain Climates; Experimental Geology)

Horace de Saussure (1740–1799, Swiss) publishes his four-volume *Voyages dans les Alpes* [Voyages in the Alps] detailing his geological, meteorological, and botanical studies in the Alps and other European mountains. Among the topics discussed in volume 1, published in 1779, is a series of experiments to determine possible origins of various rocks. De Saussure in particular attempts to create basalt by the fusion of various granites, although is unsuccessful.

1780

BIOLOGY (Animal Electricity)

Luigi Galvani (1737–1798, Italian), while examining the nerves and muscles of a dead frog, observes the frog's legs contract when touched with a metal scalpel. This precipitates further experiments by Galvani over the next 11 years in which he concludes that an animal's nervous response has an electrical nature. Galvani relates this "animal electricity" observed in frogs to the electricity allowing electric eels to startle intruders. (Alessandro Volta later objects to Galvani's interpretation that the electricity derives from the frog.)

CHEMISTRY (Nomenclature)

Étienne Condillac (1714–1780, French) advocates clear, logical language in his *Logique*, a work later to influence Antoine Lavoisier and others in devising an improved system of chemical nomenclature (1787—CHEMISTRY).

METEOROLOGY (Observational Network)

Karl Theodor of Bavaria (German) founds the meteorological society Societas Meteorologica Palatine, which soon establishes a network of 57 observing stations in the Northern Hemisphere, sending each a set of instructions and recording forms, plus, to those who need them, a barometer, thermometer, hygrometer, rain gauge, wind vane, and electrometer, all with standardized scales. Data are recorded at 7 A.M., 2 P.M., and 9 P.M. each day, initiating what could be a tremendous, long-term data record, but the society disbands in 1795.

1781

ASTRONOMY (Binary Stars)

William Herschel (1738–1822, German-English) carries out his second detailed review of the distribution of stars and in so doing notes 269 pairs of double—or binary—stars revolving around each other. The existence of these binary pairs suggests that gravity is not confined to the solar system.

ASTRONOMY (Uranus)

While engaged in his systematic review of the heavens (1781—ASTRONOMY), William Herschel discovers the planet Uranus, at first misinterpreting it to be a comet. The object had been noted before and classified as a star; Herschel recognizes it as within the solar system due to its disk shape and its movement relative to the background stars. Uranus is the first planet to be discovered after the five known to the ancients and the recognition by Nicolaus Copernicus of the earth as a sixth planet. The calculation of 19.19 astronomical units for the radius of the Uranus orbit corresponds well with the 19.6 figure in the sequence of Bode's Law (1772—ASTRONOMY).

CHEMISTRY (Molybdenum)

Peter Hjelm (1746–1813, Swedish) reduces molybdic acid to the metal molybdenum, using carbon as the reducing agent.

CHEMISTRY (Phlogiston)

Richard Watson (1737–1816, English) advocates that phlogiston, like magnetism, gravity, and electricity, is not an object to be held but a power visible only through its effects. (The argument appears in Watson's *Chemical Essays.*)

CHEMISTRY (Water)

Joseph Priestley (1733–1804, English) passes an electric spark through a mixture of Henry Cavendish's inflammable air (hydrogen, see 1766—CHEMISTRY) and his own dephlogisticated air (oxygen, see 1774—CHEMISTRY), and notes the resulting formation of dew. Cavendish (1731–1810, English) repeats the experiment and finds that the oxygen and hydrogen combine in a ratio of 1:2.02 by volume. Instead of concluding that water is a combination of hydrogen and oxygen (as Antoine Lavoisier concludes in 1783), Cavendish interprets the results as suggesting that inflammable air is a combination of phlogiston and water. His paper "Experiments on the Composition of Water" is not published until 1784, by which time Lavoisier has offered his alternative explanation (1783—CHEMISTRY).

MATHEMATICS (Curvature)

After systematic investigations, Gaspard Monge (1746–1818, French) creates a general theory of curvature.

SUPPLEMENTAL (Philosophy: Knowledge; Euclidean Geometry)

Publication of Immanuel Kant's (1724–1804, German) *Kritik der reinen Vernunft* [Critique of pure reason], in which he challenges the Lockean view that all knowledge is derived from the senses. According to Kant, experience cannot establish general proofs, which derive instead from the structure of our minds. The mind is not passive but rather an active molder and coordinator of experience which furthermore determines aspects of our perceptions. Kant lists 12 a priori concepts, including the principle of causality. He considers Euclidean geometry and Newtonian science to be necessary consequences of our a priori perceptions of space and time; he includes among these a priori truths the troublesome Euclidean Parallel Postulate.

1781–1786

BIOLOGY (Birds)
EARTH SCIENCES (Minerals)

As part of his *Histoire naturelle* (1749–1788—BIOLOGY), Georges Buffon (1707–1788, French) publishes ten volumes on birds and minerals.

1782

ASTRONOMY (Binary Stars)

John Goodricke (1764–1786, English) determines the period of Algol's (or beta Persei's) variation (from a star of magnitude 2.3 to one of magnitude 3.5) to be two days and 21 hours, and correctly hypothesizes that Algol is part of a binary pair.

CHEMISTRY (Mineral Classification)

In *Sciagraphia regni mineralis,* Torbern Bergman (1735–1784, Swedish) classifies minerals into four major groups: salts, earths, metals, and inflammable bodies, and then subclassifies each of those, exercising care in trying to provide a more consistent nomenclature than the current standards.

CHEMISTRY (Nomenclature)

Louis Guyton de Morveau (1737–1816, French) publishes an important paper on the need for better systematizing chemical nomenclature, his "Mémoire sur les dénominations chimiques." A forerunner to later work by himself, Antoine Lavoisier, and others (1787—CHEMISTRY), the paper advocates acceptance of Torbern Bergman's binomial terminology for salts (CHEMISTRY entries for 1779 and 1782), elimination of names that have separate meanings in nonchemical language, elimination of names memorializing individuals or groups, use of classical language, and in general a more rational method for naming substances.

CHEMISTRY (Platinum)

Following years of effort, Carl von Sickingen succeeds in devising a method for rendering platinum malleable and publishes his results. The basic procedure consists of the following steps: alloy the platinum with silver and gold, dissolve the alloy, precipitate and ignite ammonium chloroplatinate, and hammer the resulting fine platinum particles.

1783

ASTRONOMY (Nebulae)

Caroline Herschel (1750–1848, English) discovers three nebulae while working as an assistant to her brother William Herschel. One of the first female astronomers of note in modern Europe, she later discovers eight comets, over the period 1786–1797, and publishes two important works (see ASTRONOMY entries for 1798 and 1828).

ASTRONOMY (Sun's Motion)

William Herschel (1738–1822, German-English) discovers that the sun—and along with it the rest of the solar system—is moving through space relative to the other stars. After calculating proper motions for 13 stars, he determines that these motions can be partially explained by assuming that the sun is moving toward a point in the constellation of Hercules.

CHEMISTRY (Heat; Calorimeter)

Antoine Lavoisier (1743–1794, French) and Pierre Simon de Laplace (1749–1827, French) collaborate in a sequence of experiments on heat and write the pamphlet "Mémoire sur la chaleur" (later published in the *Mémoires* of the Academy of Sciences in 1784). Modeled in part on the 1779 work of Adair Crawford, the experiments provide additional evidence for Lavoisier's developing theory of combustion (1783—CHEMISTRY). During the experiments Laplace invents the ice calorimeter, with which heat amounts are determined by measuring the weight of ice melted by the heat.

CHEMISTRY (Tellurium)

Franz Müller (Baron de Reichenstein, 1740–1825, Austrian) discovers tellurium.

CHEMISTRY (Tungsten)

Juan José d'Elhuyar (1754–1796, Spanish) and Fausto d'Elhuyar (1755–1833, Spanish) discover tungsten.

CHEMISTRY (Water; Combustion; Phlogiston)

Antoine Lavoisier (1743–1794, French) hears of and repeats the experiments of Joseph Priestley and Henry Cavendish (1781—CHEMISTRY) in which dew is formed from inflammable air and oxygen. He is the first to offer the interpretation that water is a compound formed from the elements hydrogen and oxygen and renames the "inflammable air" of Cavendish, "hydrogen" or "water producer" (Greek). Lavoisier's interpretation destroys another of the ancient Greek concepts, since water, having been produced from other elements, is presumably not itself an element (although, over a century later, the discovery of radioactive elements refutes such reasoning). The additional support this lends to Lavoisier's earlier theories on combustion (see CHEMISTRY entries for 1775 and 1778) leads to a rapidly growing acceptance of Lavoisier's theory that combustion and calcination involve a chemical combination with oxygen. Lavoisier further weakens the alternative theory that these processes involve an escape of phlogiston when he calculates the various weight gains and losses upon reducing a calx to metal, showing that all weights are accounted for without inserting any phlogiston transfer. The phlogiston theory quickly loses much of its support, though as late as 1790 Lavoisier is still spending considerable time developing arguments against it.

METEOROLOGY (Hot Air Balloon)

Étienne de Montgolfier (1745–1799, French) and Joseph de Montgolfier (1740–1810, French) construct the first hot air balloon and fly over Paris in the first manned flight.

METEOROLOGY (Hydrogen Gas Balloon)

Jacques Charles (1746–1823, French) makes the first ascent in a hydrogen gas balloon, attaining an altitude of 3.2 km.

METEOROLOGY (Hygrometry)

Horace de Saussure (1740–1799, Swiss) discusses the principles of hygrometry in his *Essais sur l'hygrometrie*. Having found the length of individual strands of human hair to vary by 2.4% between dry and saturated atmospheric conditions, he uses this observation to construct a hair hygrometer.

1784

ASTRONOMY (Black Holes)

Reasoning that if light is a particle, as suggested by Isaac Newton, then the gravity of a sufficiently massive body would reduce the outward speed of light and could thereby prevent its escape, John Michell (1724–1793, English) originates the concept of a black hole. Michell calculates that a star with the same density as the sun but with a radius 500 times larger would have a gravitational field strong enough to ensure an escape velocity exceeding the speed of light. Such a star would be invisible to the rest of the universe.

ASTRONOMY (Nebulae and Clusters; Messier Catalog)

Charles Messier (1730–1817, French) and Pierre Méchain (1744–1804, French) publish a catalog of 103 nebulous objects and clusters, compiled as a sideline during 30 years of searching for comets. (Many of the objects are subsequently determined to be galaxies rather than nebulae. The Messier catalog listings become standard designations for the brighter nebulae and clusters; for example, the Andromeda Galaxy is widely known as M31 and the Orion Nebula as M42.)

BIOLOGY (Intermaxillary Bone)

Johann Wolfgang von Goethe (1749–1832, German) discovers the intermaxillary bone.

CHEMISTRY (Nomenclature; Mineral Classification)
EARTH SCIENCES (Minerals)

Torbern Bergman (1735–1784, Swedish) continues in *Meditationes de systemate fossilium naturali* his concern with mineral classification and nomenclature (prominent for example in his 1782 work). Influenced by Carolus Linnaeus's reforms in biological nomenclature (BIOLOGY entries for 1753 and 1758), Bergman sets out general guidelines, including that the major classes of minerals should be given one-word names, that Latin should be used, and that the names of metals should consistently end in "um."

EARTH SCIENCES (Climatology; Volcanic Impacts)

Benjamin Franklin (1706–1790, American) relates the severe Northern Hemisphere winter of 1783–1784 to the powerful volcanic eruption occurring in Iceland in the summer of 1783, suggesting that solar insolation is reduced after a volcanic eruption due to the ash and other particles inserted into the atmosphere.

EARTH SCIENCES (Mineralogy)

Richard Kirwan (1733–1812, Irish) publishes *Elements of Mineralogy*. Kirwan introduces use of chemical composition as a means for classifying minerals.

EARTH SCIENCES (Volcanoes)

Gratet de Dolomieu (1750–1801, French) supplements earlier evidence from Portugal (1776—EARTH SCIENCES) with evidence from Sicily on the interstratification of numerous layers of volcanic material with stratified marine limestone deposits, doing so in the paper "Sur les volcans éteints du Val di Noto en Sicile" [On the extinct volcanoes of the Val di Noto in Sicily]. He concludes that the volcanic material, consisting both of fragmental detritus and sheets of basalt, was discharged over the sea floor contemporaneously with the accumulation of the limestones.

HEALTH SCIENCES (Milk Leg)

Charles White (1728–1813, English) provides the first clinical description of phlegmasia alba dolens, or milk leg.

MATHEMATICS (Legendre Polynomials)

The memoir "Recherches sur la figure des planètes" on celestial mechanics by Adrien Legendre (1752–1833, French) contains the first published appearance of Legendre polynomials, rational integral functions of the cosine of the angle made by two radii vectors, functions which emerged as the coefficients in a series obtained by Legendre while solving a problem in potential theory.

PHYSICS (Mechanics)

George Atwood (1745–1807, English) calculates with improved accuracy the acceleration of a body in free fall.

PHYSICS (Torsion)

Charles Augustin de Coulomb (1736–1806, French) attempts in the memoir "Recherches théoriques et expérimentales sur la force de torsion et sur l'élasticité des fils de métal" [Theoretical and experimental researches on the force of torsion and the elasticity of metal wires] to develop the laws of torsion and present possible applications. He presents the fundamental equation that the torque M in a thin cylinder with diameter D and length L is $M = \mu BD^4/L$, where μ is a constant rigidity coefficient and B is the angle of twist. This corrects a similar equation given by Coulomb in an earlier memoir (1777—PHYSICS), where the dependence on D was to the third rather than the fourth power.

1785

ASTRONOMY (Cosmology; Solar System Formation)

Georges Buffon (1707–1788, French) presents a tidal theory of the formation of the solar system, hypothesizing the collision with the sun of a gigantic comet, causing huge solar tides, with massive amounts of material swept away from the sun and eventually forming the planets.

ASTRONOMY (Nebulae)

In describing his observations to the Royal Society, William Herschel (1738–1822, German-English) asserts that so-called "nebulae," when viewed through increasingly powerful telescopes, become discernible as clusters of stars, while at the same time, new "nebulae" come into view, again presumably discernible as individual stars with greater viewing power. Herschel postulates that most stars have planetary systems and that most, but not all, nebulae are sidereal systems like the Milky Way, the other nebulae being truly gaseous and within the Milky Way. Like Immanuel Kant, he divides the nebulae into four classes and orders the four in an evolutionary sequence: the irregular, diffuse nebulae evolve into star clusters, which later become disc shaped nebulae, and eventually, due to gravitation, globular nebulae.

ASTRONOMY (Saturn's Rings)

Pierre Simon de Laplace (1749–1827, French) concludes from calculations on the likely gravitational tides from the planet Saturn, that the apparent ring around Saturn cannot be a solid ring or even two solid rings divided by the Cassini Division, as these could not continue to rotate unshattered around Saturn. Instead he suggests multiple narrow, although still solid rings, replacing the disk hypothesis of Christiaan Huygens (1655—ASTRONOMY) by a ring system of numerous narrow ringlets.

ASTRONOMY (Shape of Galaxy; Stellar Distances)

In "On the Construction of the Heavens," William Herschel (1738–1822, German-English) presents the first systematic attempt to determine the shape of the Milky Way Galaxy. By supplementing extensive star counts with the assumption that all stars have roughly the same intrinsic brightness, he can estimate relative distances to the solar system. He finds an unsymmetrical distribution of stars, and creates a three-dimensional ellipsoidal model of the galaxy, with the sun near the center. Herschel estimates the number of Milky Way stars at 100 million (versus twentieth-century estimates of 100 billion). He makes no estimates of absolute distances.

CHEMISTRY (Ammonia)

Claude Berthollet (1748–1822, French) determines that ammonia consists of nitrogen and hydrogen.

CHEMISTRY (Argon)
METEOROLOGY (Air)

While studying air, known to contain nitrogen (or phlogisticated air) and oxygen (or dephlogisticated air), Henry Cavendish (1731–1810, English) passes electric sparks through oxygen, air, and an alkali, forming nitrogen oxide. Being unable to oxidize completely the "nitrogen," Cavendish correctly hypothesizes that the small, nonconsumable portion is an unknown gas even more inert than nitrogen itself and accounting for less than 1/120 of the atmosphere. (Argon is finally discovered and named in 1894.)

CHEMISTRY (Carbon)

Tobias Lovits (Tobias Lowitz, 1757–1804, German) discovers carbon's property of changing color.

CHEMISTRY (Steel)

Rudolf Raspe (1737–1794, German) demonstrates that tungsten can be used to harden steel.

EARTH SCIENCES (Geology; Uniformitarianism)

James Hutton (1726–1797, Scottish) introduces the theory of uniformitarianism into geology with his memoir entitled "Theory of the Earth; or an Investigation of the Laws Observable in the Composition, Dissolution and Restoration of Land Upon the Globe," explaining the geological evidence of the earth's past history on the basis of processes still observable in the present such as the weathering of rocks, the excavation of valleys by rivers, the transport of sediment from land to sea by rivers and streams, and the build up of land by volcanoes. He sees no geological evidence of a beginning or reason to expect an end, but instead a recurring cycle consisting of land erosion, deposition in the oceans, consolidation at the ocean floor, and uplift. The work is expanded over the next decade into Hutton's famed *Theory of the Earth* (1795—EARTH SCIENCES).

HEALTH SCIENCES (Aneurysms)

John Hunter (1728–1793, Scottish-English) develops a procedure for ligating aneurysms.

HEALTH SCIENCES (Foxglove; Digitalis)

William Withering (1741–1799, English) publishes *An Account of the Foxglove, and Some of Its Medical Uses,* describing the medical value of the glycoside-yielding plant foxglove and the drug digitalis derived from it, plus the serious dangers of overdosages. Withering details numerous cases since his introduction of the use of digitalis in the treatment of dropsy (1775—HEALTH SCIENCES), describes the preparation of the drug and the appropriate dosages to administer, and cautions other physicians that digitalis intoxication can lead to vomiting, confused vision, convulsions, and even death.

MATHEMATICS (Analysis)
ASTRONOMY (Dynamical)

Laplace coefficients and the potential function are introduced in the work *Théorie des attractions des spheroids et de la figure des planètes* by Pierre Simon de Laplace (1749–1827, French).

1786

CHEMISTRY (Phlogiston)

Antoine Lavoisier (1743–1794, French) attacks the phlogiston theory in the memoir "Réflexions sur le phlogistique." Supporting his statements with results on heat by Adair Crawford (1779—BIOLOGY) and by Lavoisier and Pierre Simon de Laplace (1783—CHEMISTRY), he argues that phlogiston is not given off during the process of combustion.

EARTH SCIENCES (Graded Valleys)

Louis Comte du Buat (1734–1809, French) examines river flow mathematically, determining the slopes needed to maintain an equilibrium between the water velocity and the amount of alluvium being transported.

EARTH SCIENCES (Valleys)

Horace de Saussure (1740–1799, Swiss) establishes in the second volume of his

Voyages dans les Alpes (1779–1796—EARTH SCIENCES) that valleys appear to have been excavated by rivers, rain, and melted snow.

MATHEMATICS (Algebra; Quintic Equations)

Erland Bring (1736–1798, Swedish) reduces the general quintic equation to $x^5 + ax + b = 0$ by use of a Tschirnhaus substitution (1683—MATHEMATICS), taking an important step toward the eventual determination of transcendental solutions to the quintic.

MATHEMATICS (Calculus of Variations; Legendre Conditions)

Working by analogy with the usage of the second derivative in the differential calculus, Adrien Legendre (1752–1833, French) examines the second variation of an integral with the object of determining criteria by which to distinguish whether a solution in the calculus of variations is a maximum or a minimum. His development is incomplete, though leads to further developments by Carl Jacobi (1837—MATHEMATICS) and the naming of "Legendre conditions" for distinguishing maxima and minima in the calculus of variations.

MATHEMATICS (Non-Euclidean Geometry)

Posthumous publication of Johann Lambert's (1728–1777, German-French) *Theory of Parallel Lines,* presenting his work on Euclid's Parallel Postulate and possible alternatives (see 1766—MATHEMATICS).

PHYSICS (Philosophy; Newton's Laws of Motion)

Immanuel Kant (1724–1804, German) publishes his *Metaphysische anfangsgründe der naturwissenschaft* [Metaphysical foundations of natural sciences]. Kant purports to have derived the Newtonian laws of motion from reason alone, and declares that the existent world would not be understandable under any set of assumptions contradictory to those laws. He also suggests that major forces of nature are manifestations of a single force and can be converted from one to another, a doctrine which gains support among many significant scientists over the next century.

1787

ASTRONOMY (Planetary Motions)

Pierre Simon de Laplace (1749–1827, French) publishes his "Mémoire sur les inégalités séculaires des planètes et des satellites" where he largely resolves several of the remaining discrepancies between the observed positions of planets and satellites and the positions expected from Newtonian gravitational theory. Particularly noteworthy are his determinations of (1) the interdependence of the apparent deceleration of Saturn's mean motion and the apparent acceleration of Jupiter's mean motion and (2) the theory behind the intricate relations among the motions of the three inner satellites of Jupiter.

BIOLOGY (Lymphatic Vessels)

Paolo Mascagni (1755–1815, Italian) publishes his *History and Iconography of the Lymphatic Vessels of the Human Body,* in which he depicts in great detail the system of lymphatic vessels, many of which he earlier discovered.

CHEMISTRY (Charles' Law)

Jacques Charles (1746–1823, French) determines the amount of volume expansion of a gas when heated under constant pressure, and thereby discovers the basis of what is later known as Charles' Law: the volume of a gas at constant pressure is directly proportional to its absolute temperature. The relationship is determined empirically for air and extrapolated to all gases, though it later proves valid only for "ideal" gases. Since the law is independently discovered by Joseph Gay-Lussac (1802—CHEMISTRY), it is sometimes known as the Charles-Gay-Lussac Law.

CHEMISTRY (Nomenclature)

In *Méthode de nomenclature chimique,* Louis Guyton de Morveau (1737–1816, French), Antoine Lavoisier (1743–1794, French), Claude Berthollet (1748–1822, French), and Antoine de Fourcroy (1755–1809, French) undertake a major revision of chemical nomenclature. In addition to discussing the need for better systematization, difficulties in the old nomenclature, and advantages of their recommended revisions, they include a full-scale dictionary, both for obtaining the old names from the new and vice versa. The book ends with a statement on it by the French Academy of Sciences. Although the statement is only lukewarm, the chemical community gradually accepts the new nomenclature over the next two decades.

CHEMISTRY (Symbolism)

In two short memoirs included in *Méthode de nomenclature chimique* (1787—CHEMISTRY), Jean Hassenfratz (1755–1827, French) and Pierre Adet (1763–1834, French) advocate a geometric symbolism to accompany the revised chemical nomenclature of Guyton de Morveau et al. Metals are symbolized by a circle, alkalies by a triangle with a horizontal base, and earths by a triangle with a vertex at the base. For compound substances, the symbols of the constituent substances are combined.

EARTH SCIENCES (Geology)

Abraham Werner (1749–1817, German) publishes the 28-page *Kurze Klassifikation und Beschreibung der Verschiedenen Gebirgsarten* sketching his perception of the structure of the earth's crust and the nature and origin of the layers composing it. He believes the earth formerly to have been submerged entirely under a global ocean, from which were deposited by chemical precipitation the Primitive Rocks, the oldest of which was granite. Afterward, as the ocean level lowered, various Floetz or Stratified Rocks, including limestone, sandstone, coal, and salt, were deposited by chemical and mechanical means. Later came the volcanic and pseudovolcanic rocks (the latter being produced by subterranean fire but not by an actual volcanic eruption) and finally the alluvial deposits of gravel, clays, and sands. (Werner later modified the sequence; see c.1796—EARTH SCIENCES.) Werner specifically points out that basalt, as well as all other Primitive and Floetz rocks, is of aqueous rather than volcanic origin, a viewpoint accepted by his immediate followers but later rejected by most geologists. The importance of the ocean in Werner's system leads to the labeling of his followers as "Neptunists," in contrast to the "Vulcanists," who emphasize instead volcanoes and the earth's internal heat.

EARTH SCIENCES (Glacial History; Moraines)

Bernard Friedrich Kuhn (1762–1825, Swiss) concludes from a study of moraines beyond the present terminus of the Grindelwald glacier that the glacier had been more extensive at some time in the past. Similar theses on past glacial extents are reiterated by others over the next half century, but do not receive widespread attention until Louis Agassiz becomes converted, then postulates the past existence of a continental ice cap, and aggressively advocates the position (1837—EARTH SCIENCES).

EARTH SCIENCES (Ocean Circulation)

Richard Kirwan mentions in *An Estimate of the Temperature of Different Latitudes* (1787—METEOROLOGY) the possibility of a continual current from high to low latitudes, produced by the density contrast created through the cooling in the high latitudes (creating denser waters) and the warming in the low latitudes.

METEOROLOGY (Temperature)

Richard Kirwan (1733–1812, Irish) publishes *An Estimate of the Temperature of Different Latitudes.* Among the major factors he indicates as affecting atmospheric temperatures are the elevation of the land and the distance to the ocean.

PHYSICS (Acoustics)

Ernst Chladni (1756–1827, German) studies patterns formed by sand placed on vibrating plates, termed "Chladni figures" or "acoustic figures."

1788

ASTRONOMY (Moon)

Pierre Simon de Laplace (1749–1827, French) solves the nearly century old puzzle of the secular acceleration of the moon, doing so in the memoir "Sur l'équation séculaire de la lune," where he shows that the apparent acceleration can be explained by the combined effects of the sun and the variations in the earth's orbital eccentricity. This removes the last of the then-recognized difficulties in matching the observed planetary and satellite motions with the expectations from Newtonian theory.

HEALTH SCIENCES (False Teeth)

Nicolas Dubois de Chémant (1753–1824, French) describes his patented porcelain teeth in *Dissertation sur les avantages des nouvelles dents, et retaliers artificiels, incorruptibles et sans odeur.* Dubois de Chémant manufactures the teeth through a process modified from one invented in 1776 by an apothecary named Duchâteau.

PHYSICS (Fluid Dynamics)
EARTH SCIENCES (Fluid Flow)

Joseph Lagrange (1736–1813, French) in his *Mécanique analytique* (1788—PHYSICS) develops the procedure of describing fluid flow by tracing the paths of particles over time (Lagrangian flow). This contrasts with the "Eulerian" method of describing fluid flow, developed earlier by Leonhard Euler (1761—PHYSICS).

PHYSICS (Mechanics)

Joseph Lagrange (1736–1813, French) mathematically develops the science of the mechanics of points and rigid bodies in his *Traité de mécanique analytique* [Analytical mechanics]. Emphasizing the use of pure analysis, he refers to physical processes only rarely, uses no geometry, and presents no diagrams. He assimilates many results from the past half century, including his own work in the calculus of variations (1760—MATHEMATICS), analytical results of Leonhard Euler, Jean le Rond d'Alembert, and others, and d'Alembert's Principle in mechanics (1743—PHYSICS).

1788–1804

BIOLOGY (Serpents; Fish)

Bernard Lacépède (1756–1825, French) completes the 44-volume *Histoire Naturelle* (1749–1788—BIOLOGY), begun by Georges Buffon, with eight volumes on serpents and fishes.

1789

ASTRONOMY (Saturn; Enceladus, Mimas)

William Herschel (1738–1822, German-English) discovers Enceladus and Mimas, two inner satellites of Saturn.

BIOLOGY (Descriptive)

Gilbert White (1720–1793, English) publishes *The Natural History and Antiquities of Selborne,* presenting a selection of his daily notes over many years on the natural phenomena and plant and animal life of his home town of Selborne, England. Particular attention is given to the habits and habitats of various bird and other animal species.

BIOLOGY (Plant Classification)

Antoine Laurent de Jussieu's (1748–1836, French) *Genera plantarum* presents a classification of plants based on groupings into natural families such as the grasses, lilies and palms. The classification, an outgrowth of unpublished work begun by his uncle Bernard de Jussieu (1699–1777, French) while arranging plants as director of the Royal Gardens at Trianon, Versailles, is an early attempt at classifying plants under a so-called "natural" classification system.

CHEMISTRY (Acids)

Claude Berthollet (1748–1822, French) establishes that oxygen is not a constituent of either hydrocyanic acid or hydrogen sulfide, thereby refuting the belief expressed by Antoine Lavoisier in his *Traité élémentaire de chimie* (1789—CHEMISTRY) that all acids contain oxygen. Since neither hydrocyanic acid nor hydrogen sulfide is a strong acid, these are not considered crucial countering examples. A more serious challenge to the Lavoisier theory is given later by Joseph Gay-Lussac and Louis Thenard (1809—CHEMISTRY).

CHEMISTRY (Atomic Theory)

Segments of the book *Comparative View of the Phlogistic and Anti-Phlogistic Theories* by William Higgins (1762/1763–1825, Irish) anticipate the atomic theory later developed by John Dalton (1803—CHEMISTRY); however, the book is not widely distributed, and Dalton probably never sees it. Higgins also anticipates later chemical symbolism developed by Jöns Jacob Berzelius when he uses first letter abbreviations for many elements. He furthermore writes simple formulae for chemical compounds: for instance, I-D represents water, a combination of Inflammable air (hydrogen) and Dephlogisticated air (oxygen).

CHEMISTRY (Conservation of Matter)

Antoine Lavoisier, in chapter 13 of *Traité élémentaire de chimie* (1789—CHEMISTRY), publishes the first widely read explicit statement of the Law of Conservation of Matter. He does so in a section on fermentation, stating that both in the laboratory and in nature the same quantity of matter exists after an operation as before. The principle had been implicitly assumed by Lavoisier in his previous researches and also by Joseph Black and Henry Cavendish in portions of their work. Here Lavoisier explicitly asserts its fundamental importance for chemical experiments. With it, the chemical balance becomes an important tool in chemical investigations; and indeed Lavoisier moves beyond his predecessors in consistent use of the balance.

CHEMISTRY (Quantification; Nomenclature; Elements)

Publication of Antoine Lavoisier's (1743–1794, French) *Traité élémentaire de chimie présenté dans un ordre nouveau et d'après les découvertes modernes* [Elementary treatise on chemistry presented in a new systematic order, containing all the modern discoveries], a work sometimes described as marking the start of chemistry as a science, and classed with Charles Darwin's *Origin of Species* in biology (1859—BIOLOGY) and Isaac Newton's *Principia mathematica* in physics (1687—ASTRONOMY). Lavoisier explains his reasons for rejecting the phlogiston concept, elaborates his oxygen theory of combustion (1778—CHEMISTRY), states and uses the Law of Conservation of Matter (1789—CHEMISTRY), encourages quantification, and details the revised system of chemical nomenclature first set out in 1787, with terms derived from Greek and Latin. Dephlogisticated air becomes "oxygen" (oxys = acid, gennao = I beget), the calx of a metal its "oxide," inflammable air "hydrogen" (hydro = water, gennao = I beget), and so on. The book is apparently the first to present a list of known elements, an element being defined by Lavoisier as a substance which has so far proven incapable of being decomposed into component parts. Among the substances included in his list of elements are light and heat (or caloric) as well as 23 substances which continue in the twentieth century to be regarded as elements. Lavoisier clearly opposes the ancient four-element (air, earth, fire, water) hypothesis.

CHEMISTRY (Uranium)

Martin Klaproth (1743–1817, German) discovers an unknown metal in pitchblende, which had formerly been thought to consist of zinc and iron. Although unable to isolate the new metal, Klaproth names it "uranium" after the recently discovered planet Uranus (1781—ASTRONOMY).

EARTH SCIENCES (Volcanoes)

Abraham Werner (1749–1817, German) conjectures (wrongly) that volcanoes are geologically recent phenomena arising from the combustion of underground coal and other inflammable substances.

1790

ASTRONOMY (Meteorites)

The Barbotan meteorite lands in France and is witnessed and recorded by city officials. The chemist Claude Berthollet (1748–1822, French) joins others in decrying the insertion of such "folk tales" in an official record.

CHEMISTRY (Nomenclature)

Vicente Telles (1764–1804, Portuguese) encourages acceptance in Portugal of the revised chemical nomenclature of Louis Guyton de Morveau, Antoine Lavoisier, Claude Berthollet, and Antoine Fourcroy (1787—CHEMISTRY) by incorporating it in his *Elementos de chimica.*

HEALTH SCIENCES (Gangrene)

Publication of Charles White's (1728–1813, English) "Observations on Gangrenes and Mortifications."

PHYSICS (Metric System)

Under instruction from the French National Assembly, a committee is created by the Paris Academy of Sciences to consider the problem of the lack of standardization in the many existent systems of weights and measures. Among the committee members are Pierre Simon de Laplace, Joseph Lagrange, Antoine Lavoisier, and Gaspard Monge. In 1791 the committee proposes a length standard of one meter as one ten millionth of a quadrant of the earth's circumference, and a mass standard of one gram as the mass of a cubic centimeter of water at 4°C. By the time of the disbanding of the committee in 1799, the metric system of weights and measures is officially accepted.

1791

BIOLOGY (Animal Electricity; Frogs)

Luigi Galvani (1737–1798, Italian), in "De viribus electricitatis in motu musculari commentarius," publishes descriptions and results of his experiments, begun in 1780, on animal electricity and the muscles and nerves of frogs. The work, originally published in journal form, is republished in book form, as *De viribus electricitatis,* in 1792.

BIOLOGY (Observational)

William Bartram (1739–1823, American) publishes *Travels Through North and South Carolina, Georgia, East and West Florida,* describing his four years of travel in the American south and especially his observations on the plant and animal life, the customs of the Indians, and a range of natural phenomena.

CHEMISTRY (Equivalent Proportions)

Jeremias Richter (1762–1807, German) proposes the Law of Equivalent Proportions, stating that if amount x of substance A combines chemically with amount y of substance B, and amount z of substance C also combines with amount y of B, then amount x of substance A will combine with amount z of substance *C*. Extensive research follows in the construction of tables of equivalent weights, presenting the relative weights with which various elements combine chemically.

EARTH SCIENCES (Mineral Veins)

Abraham Werner (1749–1817, German) publishes a *Neue Theorie von der Entstehung der Gänge* [New theory of the formation of veins], applying to mineral veins his general theory of the former existence of a universal ocean from which the primitive mountains and the earth's minerals and strata were formed by precipitation (1787—EARTH SCIENCES). Veins are said to have formed as fissures in the earth's crust filled from above by chemical precipitation from the universal ocean, with the middle part of the vein commonly formed later than the external part. The theory is accepted at the time by many supporters of the Neptunist school (1787—EARTH SCIENCES), but is rejected along with many other facets of Neptunist theory within the next half century.

PHYSICS (Electroscope)

Alessandro Volta (1745–1827, Italian) improves the condensing electroscope by adding two metal plates with a layer of insulating lacquer between them. He thereby increases the sensitivity by approximately two orders of magnitude. The new electroscope is a significant advance over the earlier use of pith balls and straws to detect electric charges.

PHYSICS (Meter)

In order to establish a metric length, astronomers Jean Delambre (1749–1822, French) and Pierre Méchain (1744–1804, French) measure precisely the meridian arc from Dunkirk to Barcelona.

1792

CHEMISTRY (General)

Publication of Antoine de Fourcroy's (1755–1809, French) *Philosophie chimique*.

CHEMISTRY (Quantification)

Jeremias Richter (1762–1807, German) attempts to quantify the forces of affinity in the first volume of his *Anfangsgrunde der Stochyometrie* [Outlines of stoichiometry, or the art of measuring chemical elements]. Richter determines the amounts of sulfuric, muriatic, and nitric acid required to neutralize specific amounts of the bases alumina, ammonia, baryta, lime, magnesia, potash, and soda. He implicitly uses the Law of Constant Composition (later stated formally by Joseph Proust [1797—CHEMISTRY]).

PHYSICS (Temperature)

Thomas Wedgwood (1771–1805, English) asserts that the same temperature is required for any body to become red hot.

1793

BIOLOGY (Pollination)

In *Secret of Nature Displayed,* Christian Sprengel (1750–1816, German) describes the process of pollination, particularly the accessory influence of wind and insects in cross-pollination, and presents observational evidence demonstrating that even with hermaphrodite flowers, having both stamens and pistils, the pollen is carried to another flower rather than self-fertilizing, self-fertilization being prevented in some such flowers by having the stamens and pistils mature at different times.

METEOROLOGY (General)

John Dalton (1766–1844, English) publishes *Meteorological Observations and Essays,* presenting tabulated data on atmospheric temperature, pressure, humidity, wind, precipitation, thunder, and the aurora borealis, plus essays on such topics as the trade winds, the aurora borealis, atmospheric water vapor, and evaporation. His description of water vapor as existing within the atmosphere in a state independent from the other gases anticipates his more general Law of Partial Pressures (1801—METEOROLOGY/CHEMISTRY).

SUPPLEMENTAL (French Academy of Sciences)

The French Academy of Sciences is abolished by the French Revolutionary Convention.

1794

ASTRONOMY (Meteorites)

Ernst Chladni (1756–1827, German) suggests that meteorites are not terrestrial but cosmic in origin.

BIOLOGY (Evolution)

Publication of Erasmus Darwin's (1731–1802, English) *Zoonomia,* in which he asserts that species evolve over time, through both the inheritance of acquired characteristics and the preferential survival of those competing species which are most suited to prevailing conditions. He supports his conclusion that all life forms descended from a single source with such evidence as the similarity in form of various organisms and the known modifications of animal populations through selective breeding. Jean Baptiste Lamarck's later theory (1809—BIOLOGY) is fuller, but omits the mechanism of competition. Both E. Darwin and Lamarck believe that each organism has an inner force leading it to a higher stage.

CHEMISTRY (Nomenclature)

Samuel Mitchill (1764–1831, American) encourages adoption of the new 1787 chemical nomenclature of Louis Guyton de Morveau, Antoine Lavoisier, Claude Berthollet, and Antoine Fourcroy (1787—CHEMISTRY), becoming the best known American chemist to do so thus far.

CHEMISTRY (Yttria)

Johan Gadolin (1760–1852, Finnish) analyzes a mineral deposit found by Carl Arrhenius (1757–1824, Swedish) in 1787 and discovers in it the new earth yttria. The original deposit had been named ytterite by Arrhenius but is later renamed gadolinite.

EARTH SCIENCES (Glacial History)

James Hutton (1726–1797, Scottish) visits the Jura mountains and decides that erratic boulders far downstream from current glaciers suggest the former existence of more extensive glaciation, adding to the evidence earlier presented by Bernard Kuhn (1787—EARTH SCIENCES).

HEALTH SCIENCES (Color Blindness)

John Dalton (1766–1844, English), himself suffering from red-green color blindness, publishes an early scientific description of the condition in the paper "Extraordinary Facts Relating to the Vision of Colours."

MATHEMATICS (Least Squares)

Carl Friedrich Gauss (1777–1855, German) invents the method of least squares, in which a magnitude x is estimated from a set of observations by minimizing the sum of the squares of the deviations of the measured values from x. Gauss fails to publish the method until 1809, by which time it has been invented and published independently by Adrien Legendre (1806—MATHEMATICS).

SUPPLEMENTAL (Guillotining of Lavoisier)

France's most famous chemist, Antoine Lavoisier (1743–1794, French), is guillotined during the French Revolution because of his part-time role as a tax

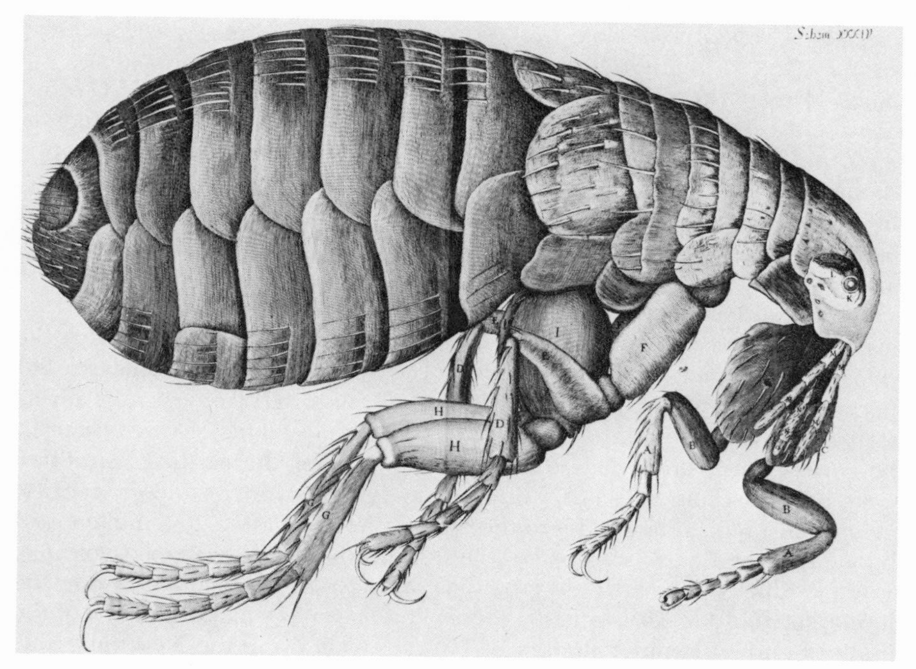

FIGURE 9. *A flea intricately depicted from microscopic observations by Robert Hooke. The original illustration, included as a foldout plate in Hooke's* Micrographia *(see 1665—*BIOLOGY *entry), is two feet long. (By permission of the Houghton Library, Harvard University.)*

collector for a period during the old regime. When his preeminent place in the scientific community is mentioned during his trial, the President of the Court is reputed to have replied: "The Republic has no need of men of science."

1795

BIOLOGY (Archetypes)
SUPPLEMENTAL (Scientific Methods: Intuition)

Johann Wolfgang von Goethe (1749–1832, German) asserts that there exist two archetype plans for the structures of all living bodies, one for the plant world and one for the animal world. Goethe advocates a holistic approach toward science, emphasizing intuition and a concern for the whole rather than a separation into parts.

CHEMISTRY (Titanium)

Martin Klaproth (1743–1817, German) discovers the element titanium but is unsuccessful in his attempts to isolate it.

EARTH SCIENCES (Geology; Uniformitarianism)

James Hutton (1726–1797, Scottish) expands his earlier memoir (1785—EARTH SCIENCES) into the two-volume *Theory of the Earth, With Proofs and Illustrations,* founding (some say) modern geology. Hutton counters the Catastrophe Theory that the earth's form derived largely from major catastrophic events with the Uniformitarian principle that the geological forces slowly changing the earth's crust are operating in the present in the same manner that they operated in the past. Based on this principle and extensive observations of the Scottish countryside, including the many angular unconformities where one series of strata rests on the upturned edges of another, he concludes that the widespread layered strata were deposited as sediments of an earlier sea but then were raised above the level of the sea and distorted from their relatively horizontal layerings into their present configurations by major convulsions of the earth's crust due largely to the internal heat of the earth. Unstratified rocks, including mineral veins, whinstone in dykes, porphyry, and granite, are seen as having been injected into the crust from below while in a molten condition. Thus both aqueous processes and subterranean (for instance, volcanic) processes due to the earth's internal heat are recognized as important in the development of the earth's crustal rocks. This becomes a major doctrine of the Plutonist and Vulcanist followers of Hutton, countering the Neptunist doctrine that the aqueous origin is overwhelming (1787—EARTH SCIENCES). Periods dominated by deposition alternate with violent upheavals, with a continual modification of the surface through the pounding of waves, river and stream erosion, chemical and mechanical disintegration, and, at the ocean bottom, through the consolidation of deposited sediments by subterranean heat. All combines to suggest an eternal geological cycle, with no evidence of a beginning or an end.

EARTH SCIENCES (Granite)

James Hutton in his *Theory of the Earth* (1795—EARTH SCIENCES) asserts that granite is of igneous origin and that although it generally underlies other rocks, it is often younger rather than older than the overlying strata, as it has been intruded upward from below. This sharply contrasts with Abraham Werner's conception of granite as the first chemical precipitate from the primeval universal ocean (1787—EARTH SCIENCES).

EARTH SCIENCES (Glacial History; Erratic Boulders)

James Hutton suggests in his *Theory of the Earth* (1795—EARTH SCIENCES) that the granite erratics in the Jura Mountains were transported to their present positions by glaciers.

EARTH SCIENCES (Valleys)

James Hutton in his *Theory of the Earth* (1795—EARTH SCIENCES) suggests that valleys are in large part created and contoured by the rivers and streams running through them.

EARTH SCIENCES (Volcanoes)

James Hutton in his *Theory of the Earth* (1795—EARTH SCIENCES) asserts that volcanoes are manifestations of the internal heat of the earth and are not, as believed by Abraham Werner, relatively recent phenomena due to the combustion of inflammable substances (1789—EARTH SCIENCES). Hutton contends instead that volcanoes arise where the internal molten materials break through to the earth's surface.

MATHEMATICS (Euclid's Parallel Postulate)

John Playfair (1748–1819, Scottish) reformulates the Parallel Postulate of Euclid, declaring that given a line l and a point P not on the line, then in the plane containing P and l there is exactly one line through P which does not intersect l. This formulation gains wide acceptance.

MATHEMATICS (Geometry)

Gaspard Monge (1746–1818, French) publishes *Feuilles d'analyse appliquée à la géometrie* on descriptive geometry following a delay of 32 years during which the French military forbade publication due to the presumed relevance of the work to fortifications. Monge uses two planar projections to represent solids, an important innovation for the development of architecture and mechanical drawing in the nineteenth century.

MATHEMATICS (Quadratic Reciprocity)

Carl Friedrich Gauss (1777–1855, German) independently discovers and first proves the law of quadratic reciprocity, a result suspected earlier by both Leonhard Euler and Adrien Legendre. Called by Gauss the *theorema aureum* or "gem of arithmetic," this law states that for primes p and q, if p and q are both of the form $4n + 3$, then either there exists an integer x such that p divides $x^2 - q$ or there exists an x such that q divides $x^2 - p$, but not both, whereas if either p or q is instead of the form $4n + 1$, then if there exists an x such that p divides $x^2 - q$ there also exists a y such that q divides $y^2 - p$. Gauss's development of the concept of congruence (1801—MATHEMATICS) later simplifies the statement of such laws.

PHYSICS (Metric System)

The metric system becomes the official system of weights and measures in France.

1796

ASTRONOMY (Black Holes)

In *Exposition du système du monde* (1796—ASTRONOMY), Pierre Simon de Laplace, like John Michell before him (1784—ASTRONOMY), hints at the possible existence of black holes, calculating that no light could escape from a body with the earth's density and a radius 250 times that of the sun.

ASTRONOMY (Evolution of the Solar System; Spiral Nebulae)

Pierre Simon de Laplace (1749–1827, French) presents, in *Exposition du système du monde,* an hypothesis regarding the origin and evolution of the solar system. Laplace postulates an initial rotating mass, which, upon cooling, condensed in size, with segments breaking off to form the planets. This "Nebular Hypothesis" serves to explain various observed aspects of planetary motions: the fact that the planets all revolve around the sun in the same direction and all the satellites revolve around the planets in the same direction; the fact that the planetary orbits and even the satellite orbits are all in nearly the same plane; and the fact that the sun, the planets, and the satellites with observed rotations all rotate about their axes in the same direction. Furthermore, he hypothesizes that the spiral nebulae are whirlpools of gas, cosmically close to the solar system, each likely eventually to form an individual new star or planet.

BIOLOGY (Evolution; Archetypes)

Geoffroy Saint-Hilaire (1772–1844, French) publishes his conviction that not all life forms were created at the beginning, many species instead developing from common earlier types. He does not believe, however, that species are still undergoing modification. Among his suggestions is that all animals derive from a single archetype plan and have the same basic organs, although specific organs might be exaggerated in some animals and reduced or eliminated in others.

CHEMISTRY (Diamonds)

Smithson Tennant (1761–1815, English) establishes that diamonds are composed exclusively of carbon.

EARTH SCIENCES (Folding)

In the final volume of his *Voyages dans les Alpes* (1779–1796—EARTH SCIENCES), Horace de Saussure (1740–1799, Swiss) concludes that the prominent anticlinal folds in the mountains bordering Lake Lucerne probably resulted from folding of strata which had originally been horizontal, or, with lesser probability, from an upward force originating from below. He also concludes that the limestones constituting these mountains were not formed by crystallization from the primeval ocean, as he had earlier believed, but by mechanical deposition.

HEALTH SCIENCES (Smallpox Vaccination)

Edward Jenner (1749–1823, English) develops a vaccination against the widely dreaded disease of smallpox and thereby lays the foundations of modern immunology. The vaccination consists of a cowpox virus, found effective by Jenner after experimenting and investigating the validity of widespread rumors that dairymaids who had had cowpox never contracted smallpox. The vaccination technique proves safer than the Turkish and Chinese inoculation techniques brought to England in 1717. Jenner successfully vaccinates an eight-year-old boy who, upon being inoculated with variolous matter from a pustule of a smallpox victim two months later, remains healthy.

MATHEMATICS (Geometry)

Analysis of the roots of the equation $(x^p - 1)/(x - 1) = 0$ leads Carl Friedrich Gauss (1777–1855, German) to discover a method of constructing, with straightedge and compass alone, the regular 17-sided polygon. This feat is considered by some as the first significant advance in Euclidean geometry since Euclid's original text, written c.300 B.C.

MATHEMATICS (Natural Numbers)

Carl Friedrich Gauss (1777–1855, German) determines that every natural number is the sum of three triangular numbers, a triangular number being a number in the sequence 0, 1, 3, 6, 10, 15. . . . More completely, the nth triangular number is the sum of the first $n - 1$ integers: $0 + 1 + 2 + 3 + \ldots + (n - 1)$.

MATHEMATICS (Negative Numbers; Complex Numbers)

The continuing reluctance of many to accept negative and complex numbers is reflected in William Frend's (1757–1841, English) *Principles of Algebra,* where he accepts as roots of equations only those that are positive and real, and where he criticizes others for not doing likewise.

c.1796

EARTH SCIENCES (Geological Column)

Abraham Werner (1749–1817, German) revises his subdivisions of the geological column (1787—EARTH SCIENCES) by adding a set of Transitional Rocks between the Primitive and Floetz Rocks and by further subdividing the major divisions.

1796–1814

MATHEMATICS (Gauss's Diary)

Carl Friedrich Gauss (1777–1855, German) keeps a mathematical diary recording very brief statements of 146 discoveries or results obtained by him during this time period, starting with his discovery of a method of constructing the regular 17-sided polygon (1796—MATHEMATICS). This extremely concise, 19-page diary is finally published posthumously in 1901.

1797

ASTRONOMY (Comets)

Heinrich Olbers (1758–1840, German) develops a successful and acclaimed method for calculating the orbit of a comet.

CHEMISTRY (Constant Composition)

Joseph Proust (1754–1826, French) proposes the Law of Constant Composition (or Definite Proportions); that is, that any particular chemical compound always has the same relative weights of its constituent elements. Proust proceeds to distinguish a chemical compound from a simple mixture of elements.

EARTH SCIENCES (Ocean Circulation)

Benjamin Thompson (Count Rumford, 1753–1814, American) describes experiments leading to his discovery of convection currents in the essay "On the propagation of heat in fluids." He proceeds to examine the possible role of convection in the world ocean, concluding from his own theoretical work and from the 1751 deep water temperature measurements of Henry Ellis (1751—EARTH SCIENCES) that the existence of cold water at depth in the tropics implies a meridional circulation transporting deep water from the polar regions toward the equator. Water cooled at the surface in high latitudes gets denser and hence descends. This cold, dense water spreads along the sea floor toward the equator, forcing a countercurrent from equator to pole at the surface. Thompson furthermore supplements this large-scale model of ocean circulation with an explanation of how in high latitudes a smaller scale vertical circulation can take place which results in significant heat transfer to the atmosphere: cold winds cool surface waters, which consequently descend, initiating a vertical circulation that brings to the surface warmer waters which in turn transfer heat to the atmosphere, cool, and descend, continuing the vertical circulation pattern.

MATHEMATICS (Calculus)

Joseph Lagrange (1736–1813, French) publishes *Théorie des fonctions analy-*

tiques, a work aimed at rigorizing calculus. Lagrange attempts to eliminate all infinitely small values, limits, and fluxions, and to place the subject on a fully algebraic base dealing exclusively with finite quantities, although allowing infinite series. The development is flawed in several aspects, including the assumption that all functions have power series expansions and the failure to examine sufficiently the issue of the convergence of the series used, but it remains a major advance toward the creation of a theory of functions of a real variable. Lagrange uses extensively Taylor series expansions (1715–1717—MATHEMATICS) and tries to establish that all functions $f(x)$ have such expansions and that these can be obtained purely algebraically.

MATHEMATICS (Calculus; Negative Numbers)

Publication of Lazare Nicolas Carnot's (1753–1823, French) *Reflections on the Metaphysics of Infinitesimal Calculus.* The book includes a criticism of the use of negative numbers for anything other than as an aid in calculations.

MATHEMATICS (Complex Numbers)

Caspar Wessel (1745–1818, Norwegian) presents the first instance of what becomes the standard geometric representation of complex numbers, employing the *x*-axis as the axis of reals and the *y*-axis as the axis of imaginaries (see 1685—MATHEMATICS for forerunner). Wessel fully explains the Argand diagram, later named for Jean Robert Argand (see 1806—MATHEMATICS).

PHYSICS (Fluid Dynamics; Venturi Principle)

Giovanni Venturi (1746–1822, Italian) discovers that when water passes through a constricted area its velocity increases and its pressure falls, both returning to normal after the water emerges from the constricted space.

1797–1812

CHEMISTRY (Combustion)

Experiments by Vasily Petrov (1761–1834, Russian) help confirm Antoine Lavoisier's theory of combustion based on oxygen (1778—CHEMISTRY) and help refute the phlogiston theory.

1798

ASTRONOMY (Star Catalog)

Caroline Herschel (1750–1848, English) publishes an *Index to Flamsteed's Observations of the Fixed Stars,* revising the star catalog of John Flamsteed (1725—ASTRONOMY) and indexing every star in the earlier work.

BIOLOGY (Fixity of Species)

Georges Cuvier (1769–1832, French) uses remains found during Napoleon's Egyptian campaign as evidence for the concept of the fixity of species. Cuvier believes species to be basically fixed from creation, although allowing some changes, particularly the sudden disappearance of individual species, to occur during cataclysmic events. Such an anti-evolutionist position is supported by the apparent identical structure of birds preserved for 3,000 years in Egyptian tombs and birds found currently living along the Nile. Cuvier publishes "Tableau élémentaire de l'histoire naturelle des animaux."

BIOLOGY (Population Pressure)

In *An Essay on the Principle of Population as it Affects the Future Improvements of Society,* Thomas Robert Malthus (1766–1834, English) asserts that since population increases more rapidly than food supply, it must eventually be limited by war, famine, disease, sexual abstinence, or other such means. In the meantime, during periods when population is not thus checked, wages will tend to sink to a subsistence level.

CHEMISTRY (Beryllium)

Nicolas Vauquelin (1763–1829, French) discovers beryllium, a new earth found in the mineral beryl (not isolated until 1828). Vauquelin originally names it "glucina," meaning "sweet," due to the taste of the salts formed from it.

CHEMISTRY (Chromium)

Nicolas Vauquelin (1763–1829, French) discovers and succeeds in isolating chromium, naming it for its many-colored compounds. Vauquelin obtains chromium trioxide by evaporating the filtrate from crocite (Siberian red lead) which had its lead removed by precipitation with hydrochloric acid. By means of intense heating followed by cooling, he then reduces the chromium trioxide to chromium.

CHEMISTRY (Liquid Ammonia)

Louis Guyton de Morveau (1737–1816, French) succeeds in liquefying ammonia by using a mixture of ice and calcium chloride to cool ammonia gas to $-44°C$.

EARTH SCIENCES (Mass of the Earth)

Henry Cavendish (1731–1810, English) uses a method suggested in mid century by John Michell (1724–1793, English) for determining the earth's mass.

A rod with two small lead balls on either side is suspended with fine wire. Two large lead spheres are then moved near the lead balls, and the gravitational attraction is calculated by the amount of turning of the rod. From this the earth's mass is calculated, which, when divided by the accepted value for the earth's volume, yields an average earth density of 5.5 grams cm^{-3}. This is a marked improvement over the 4.5 gm cm^{-3} value obtained by Nevil Maskelyne (1774—EARTH SCIENCES).

EARTH SCIENCES (Volcanism; Basalt)

James Hall (1761–1832, Scottish) demonstrates experimentally that lavas when cooled quickly can be fused into glass, which then, if remelted and cooled slowly, can transform back into a more stony substance. He thereby offers an explanation for the occurrence in some dykes of a vitreous outer portion in contact with the surrounding rock (the rock having been relatively cold as the original lava rose into it, hence causing rapid cooling) and a more crystalline structure in the central portions. His experiments further establish that the Scottish basalts are likely remains of ancient lava flows.

HEALTH SCIENCES (Smallpox)

Edward Jenner (1749–1823, English) describes his smallpox vaccination, including his 1796 success with an eight-year-old boy (1796—HEALTH SCIENCES), in *Inquiry into the Causes and Effects of Variolae Vaccinae.* Also included are descriptions of several case studies and much anecdotal material on cowpox and smallpox in human sufferers. The smallpox vaccine gains widespread use almost immediately.

MATHEMATICS (Algebra)

Joseph Lagrange (1736–1813, French) publishes, in *Traité de la résolution des équations numériques de tous degrés* [Treatise on the resolution of numerical equations of all degrees], a proof that every algebraic equation has a root.

MATHEMATICS (Number Theory)

Adrien Legendre (1752–1833, French) determines the number of ways an integer can be represented as a sum of two squares and also proves that all positive integers not of the form $4^h(8k + 7)$ can be represented as a sum of three squares. In the two-volume *Essai sur la théorie des nombres* (published 1797–1798), he simplifies and extends the number theory developments of Joseph Lagrange (1773—MATHEMATICS), presents the first systematic treatment of ternary quadratics, makes extensive use of the law of quadratic reciprocity (1795—MATHEMATICS), and conjectures that, as n approaches infinity, the number of primes less than n approaches $n/(ln\ n - 1.08366)$.

PHYSICS (Gravitational Constant)

Henry Cavendish (1731–1810, English) makes the first calculation of the gravitational constant *G* (see above experiment [1798—EARTH SCIENCES] for the mass of the earth).

PHYSICS (Heat)

In "An Inquiry Concerning the Source of the Heat Which Is Excited by Friction," Benjamin Thompson (Count Rumford, 1753–1814, American) provides observational evidence that heat is a form of energy rather than a substance. He had been led to the hypothesis that friction is an inexhaustible source of heat while considering the boring of cannon at Munich's military arsenal and had proceeded to experiment with brass guns at the arsenal. The experiments confirm the hypothesis, justifying his conclusion that heat is not a material substance as others had believed. He goes on to equate heat to motion.

1799

BIOLOGY (Comparative Human Anatomy)

Charles White (1728–1813, English) compares skull features and other characteristics among various human races in *An Account of the Regular Gradation in Man.*

CHEMISTRY (Diamond)

Diamond is converted to graphite, then carbonic acid by Louis Guyton de Morveau (1737–1816, French), who perceives the diamond as partially oxidized carbon.

CHEMISTRY (Law of Constant Composition)

Joseph Proust (1754–1826, French) shows that copper carbonate always has the same proportions of copper and carbon, thereby illustrating the Law of Constant Composition. Over the next nine years Proust further confirms this law through his analysis of numerous additional compounds, repeatedly refuting in scientific papers the opposing position, advocated particularly by his contemporary Claude Berthollet (1748–1822, French), that a compound's composition could vary considerably, depending, for instance, on the method of preparation. By 1808, due largely to the efforts of Proust, the Law of Constant Composition is well established.

EARTH SCIENCES (Stratigraphy)

William Smith (1769–1839, English) constructs a table indicating the ordered stratigraphy underlying English soil, from the Coal stratum to the Chalk stratum, plus the characteristic fossils to be found in the individual strata. The table appears in manuscript form, with the title "Tabular View of the Order of Strata in the Vicinity of Bath with Their Respective Organic Remains." Smith's identification of particular sets of fossils in particular strata allows the correspondence of strata from different regions even in locations where erosion, faults, and other adjustments have eliminated some of the stratigraphic layers.

MATHEMATICS (Fundamental Theorem of Algebra)

Carl Friedrich Gauss (1777–1855, German) presents in his doctoral dissertation *Demonstratio nova theorematis omnem functionem algebraicam rationalem integram unius variabilis in factores reales primi vel secundi gradus revolvi posse* [A new proof that every rational integral function of one variable can be resolved into real factors of the first or second degree] a new (many say the first), more rigorous proof of the Fundamental Theorem of Algebra that every polynomial can be factored into linear and quadratic real factors. Since an equivalent result was first conjectured by Albert Girard in 1629, many proofs have been attempted, though all previous ones are regarded as unsatisfactory (including some of Gauss's 1799 attempts).

MATHEMATICS (Higher-Degree Equations)

Paolo Ruffini (1765–1822, Italian) sets out to prove the impossibility of solving the general equation of degree equal to or greater than five using only the operations of addition, subtraction, multiplication, division, and extraction of *n*th roots. Equivalently, he seeks to prove the impossibility of a solution by radicals of such an equation. He presents some initial progress in this direction in his *Teoria generale dell equazioni* in 1799, prior to further advances (1813—MATHEMATICS).

MATHEMATICS (Projective Geometry)

Publication of Gaspard Monge's (1746–1818, French) *Traité de géométrie descriptive,* where, as in his 1795 work, he projects three-dimensional objects into two two-dimensional planes.

PHYSICS (Heat)

Reputedly, through experiments with pieces of ice rubbed together in a closed system at the freezing point of water, Humphry Davy (1778–1829, English) supports Benjamin Thompson's concept of heat as mechanical energy of particles (1798—PHYSICS). (Like the picture of Galileo dropping balls from the Tower of Pisa, this event may be apocryphal.)

1799–1806

MATHEMATICS (Calculus)

In a continuation of earlier work (1797—MATHEMATICS), Joseph Lagrange (1736–1813, French) publishes *Calcul des fonctions,* where he attempts to reconstruct calculus without using infinitesimals or limits, hoping thereby to place the subject on sounder footing. He relies heavily on power series.

1799–1825

ASTRONOMY (Celestial Mechanics)

Pierre Simon de Laplace (1749–1827, French) publishes his five-volume work *Mécanique céleste* [Celestial mechanics]. Included is a detailed mathematical development of the solar system's gravitational mechanics, with a derivation that the system is stable and that the supposed irregularities in the velocities of Saturn, Jupiter, and the moon are periodic and self-regulating. Among the specifics, Laplace presents his theory that the ring system of Saturn is composed of numerous narrow, although solid rings (1785—ASTRONOMY), plus discussions of the precession of the equinoxes, the flattening of the earth, and the level of the tides.

ASTRONOMY (Cosmology)

Pierre Simon de Laplace (1749–1827, French) includes in his *Mécanique céleste* (1799–1825—ASTRONOMY) a qualitative model of the evolution of the solar system. According to the model, the sun formed upon the cooling and contracting of an original gas mass, from which nebulous rings broke off and eventually condensed one by one to form the planets. The planets too had rings broken from them which in general condensed into satellites but in the case of Saturn remained as rings.

1800

BIOLOGY (Definition)

Karl Burdach (1776–1847, German) introduces the term "biology," using it in a restricted sense to denote the combined morphological, physiological, and psychological study of human beings. Two years later, a broader definition is given by Gottfried Treviranus and Jean Baptiste Lamarck (1802—BIOLOGY).

BIOLOGY (Tissues)

Marie-Françoise-Xavier Bichat (1771–1802, French) classifies tissues into 21 types and publishes *Traité des membranes* [Treatise on membranes] and *Recherches physiologiques sur la vie et la mort* [Physiological researches on life and death].

CHEMISTRY (Electrolysis; Water)

Within weeks of the announcement of the invention of the voltaic cell (1800—PHYSICS), William Nicholson (1753–1815, English) and Anthony Carlisle (1768–1840, English) succeed in generating hydrogen and oxygen gases by passing an electric current through water. They thus establish the usefulness of the converse of the voltaic cell: using electricity to produce chemical reactions. The results are published as an "Account of the New Electrical or Galvanic Apparatus of Sig. Alex. Volta, and Experiments performed with the Same."

CHEMISTRY (Nitrous Oxide)
HEALTH SCIENCES (Anesthesiology)

Humphry Davy (1778–1829, English) discovers nitrous oxide. While testing the gas, he finds it produces a giddy, intoxicated feeling if inhaled. Breathing it becomes popular, and it acquires the denotation "laughing gas." Davy suggests its use during surgery as an anesthetic, but the idea is not adopted until Horace Wells does so in 1844 (1844—HEALTH SCIENCES).

CHEMISTRY (Phlogiston)

Still unconvinced by Antoine Lavoisier's theory of combustion and the compound nature of water (1783—CHEMISTRY), Joseph Priestley (1733–1804, English) publishes the *Doctrine of Phlogiston Established and the Composition of Water Refuted.*

CHEMISTRY (Platinum)

William Wollaston (1766–1828, English) develops a revised process for making platinum malleable (1782—CHEMISTRY), a process which gains wide usage, particularly in the production of laboratory apparatus.

HEALTH SCIENCES (Water Purification)

Chlorine—then known as oxymuriatic acid (see 1811—CHEMISTRY)—is used medically by Louis Guyton de Morveau (1737–1816, French) and William Cruikshank (1745–1800, English) to purify water.

MATHEMATICS (Double Periodicity)

The scientific diary of Carl Friedrich Gauss (1777–1855, German) indicates that he discovered general functions with two distinct periods in the years 1797–1800. However, he never publishes the work, and a quarter century passes before the concept of double periodicity is rediscovered by Niels Abel (1825—MATHEMATICS).

METEOROLOGY (Wet and Dry Bulb Hygrometer)

John Leslie (1766–1832, Scottish) devises a wet and dry bulb hygrometer and explains the theory behind it. The instrument consists of a U-shaped tube partially filled with liquid and closed by bulbs at both ends, one of which is covered with wet muslin. Evaporative cooling causes the air to contract on the side of the tube with wet muslin, so that the level to which the liquid rises on that side can be used as a measure of the evaporation and hence of the humidity of the air.

PHYSICS (Infrared Radiation)

William Herschel (1738–1822, German-English) moves a thermometer along the color spectrum produced by a prism in order to investigate the heating properties of light of different colors. By moving the thermometer beyond the red end of the spectrum he discovers an unexpected continued heating effect. He consequently suggests that invisible energy is radiated in the region beyond red, thereby discovering infrared radiation. Herschel publishes the results in *An Investigation of the Powers of Prismatic Colours to Heat and Illuminate Objects.*

PHYSICS (Voltaic Pile)

Alessandro Volta (1745–1827, Italian) announces his recent invention of the first battery, the voltaic pile, consisting of alternating slices of copper, zinc, and blotting paper soaked in brine, and capable of producing a continuous flow of electricity. Although at times awkward and messy, the voltaic pile is a significant advance, providing the first method of artificially producing a reasonably steady electric current. As a result, the demand for frogs' legs for electrical experiments quickly declines, as researchers begin using the new voltaic pile (or voltaic cell) almost immediately.

1801

ASTRONOMY (Asteroids)

On January 1, Giuseppe Piazzi (1746–1826, Italian) discovers the asteroid Ceres. This is the first of thousands of asteroids to be discovered between the orbits of Mars and Jupiter and is felt by some to be the missing planet needed to fill the void in the sequence of Bode's Law (1772—ASTRONOMY). It has a diameter of roughly 770 km. Piazzi views Ceres for 41 days, then due to illness misses some observations and is unable to relocate the asteroid until December 31, when aided by the calculations of Carl Friedrich Gauss (1777–1855, German). Gauss determines the orbit from the 41 days of data.

ASTRONOMY (Star Catalog)

In *Uranographia,* Johann Bode (1747–1826, German) publishes a catalog of 17,240 stars and nebulae, along with numerous star maps.

ASTRONOMY (Star Catalog)

Joseph de Lalande (1732–1807, French) catalogs 47,000 stars in *Histoire céleste française.*

BIOLOGY (Evolution)

Jean Baptiste Lamarck (1744–1829, French), in *Systême des animaux sans vertèbres,* first publishes his views on evolution. Later enlarged upon (1809—BIOLOGY), these views include the doctrine that species—including man—are descended from other species, and that as animals adjust to their environment they can acquire new organs and characteristics which can be passed through inheritance to their offspring. Lamarck suggests that all change in the organic as well as the inorganic world results from natural law rather than from the intervention of God.

CHEMISTRY (Affinity)

Claude Berthollet (1748–1822, French) publishes his "Recherches sur les lois de l'affinité" [Researches into the laws of chemical affinity] on the nature of chemical affinity and the fact that many factors influence chemical reactions, including mass, solubility, volatility, elasticity, efflorescence, gravitation, and reciprocal affinity between substances. He compares chemical affinity to the force of gravity.

CHEMISTRY (Columbium)

Charles Hatchett (1765–1847, English) discovers the metal columbium (niobium) while examining minerals in the British Museum.

CHEMISTRY (Gas Laws)

Independently of Jacques Charles (1787—CHEMISTRY), John Dalton (1766–1844, English) empirically determines that as a gas is heated its volume increases proportionately with its temperature.

CHEMISTRY (Vanadium)

Andrés Manuel del Río (1764–1849, Mexican) discovers the metal vanadium, found in the brown lead mineral *Plomo pardo de Zimapan,* though later, in 1805, he accepts the misinterpretation of others that vanadium is not distinct from chromium. Vanadium is rediscovered in 1831.

EARTH SCIENCES (Crystallography)

René-Just Haüy (1743–1821, French) advances the study of mineralogy and helps found the science of crystallography with his four-volume *Traité de minéralogie,* in which he demonstrates the constancy and distinctiveness of aspects of the geometric form of individual crystals.

HEALTH SCIENCES (Astigmatism)

Thomas Young (1773–1829, English) explains astigmatism as a result of the irregular curvature of the eye's cornea.

HEALTH SCIENCES (Mental Illness)

Philippe Pinel (1745–1826, French) encourages humane treatment of the mentally insane and endeavors to place the study of mental illness on a more empirical basis in his *Traité médico-philosophique sur l'aliénation mentale ou la manie*.

MATHEMATICS (Calculus)

Joseph Lagrange (1736–1813, French) publishes *Leçons sur le calcul des fonctions* both as a commentary on and a supplement to his *Théorie des fonctions analytiques* (1797—MATHEMATICS).

MATHEMATICS (Number Theory; Congruence)

Carl Friedrich Gauss's (1777–1855, German) *Disquisitiones arithmeticae* [Arithmetical researches] introduces, among other things, the concept of congruence, whereby a set of elements is separated into disjoint classes by means of an equivalence relation. Gauss develops the congruence concept in the book's first sections and then uses it to unify and extend the work of his predecessors on arithmetical divisibility and to create a coherent theory of quadratic forms, including a proof of the law of quadratic reciprocity, called by Gauss the *theorema aureum* or "gem of arithmetic" (1795—MATHEMATICS). He also generalizes his earlier work on constructing the regular polygon of 17 sides (1796—MATHEMATICS), presenting a universal criterion for determining which regular n-sided polygons can be constructed with only straightedge and compass and which cannot. Finally, he includes a rigorous proof of the long-known Fundamental Theorem of Arithmetic that any positive integer can be uniquely (except for order) expressed as a product of primes.

METEOROLOGY/CHEMISTRY (Atmospheric Composition; Law of Partial Pressures)

John Dalton (1766–1844, English) tries to explain why the gases of the atmosphere remain mixed instead of segregating with the heaviest element at the bottom. He accepts Isaac Newton's conception of a gas as composed of mutually repulsive particles but modifies this to contend that in a mixture of two or more gases, such as the atmosphere, the particles of one gas are not mutually repulsive to those of another but instead behave independently. This leads to his Law of Partial Pressures, first formally enunciated in a paper in 1801, that the total pressure of the atmosphere (or more generally of any gaseous mixture) equals the sum of the pressures exerted by the individual gases, each of which exerts its pressure independently of the others. Further elaborations of his work in meteorology lead Dalton to an atomic theory in chemistry (CHEMISTRY entries for 1802 and 1803).

PHYSICS (Gravitation)

Well before Albert Einstein (1911—PHYSICS), whose work is from a very different perspective, Johann von Soldner (1776–1833, German) uses classical Newtonian gravitation theory to calculate the deflection expected for a ray of light passing by the sun, based on the assumption that light consists of Newtonian particles traveling at speed *c*.

PHYSICS (Ultraviolet Radiation)

Spurred by William Herschel's discovery of invisible radiation detectable beyond the red end of the spectrum (infrared radiation) (1800—PHYSICS), Johann Ritter (1776–1810, German) seeks evidence for invisible radiation beyond the violet end of the spectrum and succeeds in discovering ultraviolet radiation through the darkening of silver-chloride soaked paper.

1801–1804

PHYSICS (Wave Theory of Light; Principle of Interference)

Thomas Young (1773–1829, English) revives the wave theory of light propagation through his discovery of the principle of interference—stating that darkness is produced when two waves of light are exactly out of step and light amplification occurs when two waves are in step—and his analysis and quantitative studies of interference phenomena. Although he acknowledges Isaac Newton repeatedly, he is strongly attacked by some of his countrymen, particularly Henry Brougham, for opposing the corpuscular philosophy which is associated with Newton's name. Young rejects the particle theory of light in favor of his explanation of light as a wave in the universal ether, the latter explanation being felt to better account for the observed interference phenomena and for the fact that the speed of light appears to be independent of the temperature and other characteristics of the light source. Young publishes the following three papers stating his position and evidence for it: "On the Theory of Light and Colours" (1802), "An Account of Some Cases of the Production of Colours not hitherto described" (1802), and "Experiments and Calculations Relative to Physical Optics" (1804).

1802

ASTRONOMY (Asteroids)

Heinrich Olbers (1758–1840, German) discovers the asteroid Pallas, and Carl Friedrich Gauss (1777–1855, German) calculates its orbit.

ASTRONOMY (Navigation)

Nathaniel Bowditch (1773–1838, American) publishes the *New American Practical Navigator.*

BIOLOGY (Classification)

Jean Baptiste Lamarck (1744–1829, French) divides the invertebrates into ten classes and the vertebrates into four (fishes, reptiles, birds, mammals), placing primary emphasis on the nervous system. He then arranges these classes along a strictly linear scale.

BIOLOGY (Definition)

Gottfried Treviranus (1776–1837, German) and Jean Baptiste Lamarck (1744–1829, French), in separate works, broaden the definition of "biology" from Karl Burdach's usage (1800—BIOLOGY) to now have it signify the study of life in general. Both Treviranus and Lamarck show dismay with the heavy eighteenth-century emphasis on cataloging and classification, and both desire to initiate a new study (biology), not simply rename an existent one.

CHEMISTRY (Atomic Theory)

After puzzling over the failure of the various gases in the atmosphere to separate from one another (1801—METEOROLOGY), John Dalton (1766–1844, English) concludes that different elements are composed of essentially different atoms and that Isaac Newton's belief that atoms within a gas repel each other is valid only for atoms of the same element.

CHEMISTRY (Gas Laws: Charles' Law)

Joseph Gay-Lussac (1778–1850, French) obtains, independently of the work of Jacques Charles (1787—CHEMISTRY) and John Dalton (1801—CHEMISTRY), the relationship that a gas at constant pressure expands in volume proportionally to the temperature (known as Charles' Law or the Charles-Gay-Lussac Law).

CHEMISTRY (Photography)

Thomas Wedgwood (1771–1805, English) publishes *An Account of a Method of Copying Painting Upon Glass, and of Making Profiles by the Agency of Light Upon Nitrate of Silver,* following experiments done by himself and Humphry Davy (1778–1829, English) in which paintings on silver-nitrate-coated glass were used as negatives and an image was made on an aligned plate upon exposure to light. Since Wedgwood and Davy did not know how to "fix" the image, further exposure to light continued to darken and finally destroy the image plate. (The action of light on silver chloride had been known in the sixteenth century.)

CHEMISTRY (Spectroscopy)

In examining a candle flame through a prism, William Wollaston (1766–1828, English) notes the discontinuous nature of the spectrum produced. Having sharpened the prism's resolving power, Wollaston is able to distinguish dark absorption lines, thereby increasing the usefulness of spectroscopy as a scientific tool.

CHEMISTRY (Symbolism)

Thomas Thomson (1773–1852, Scottish) symbolizes individual minerals by the first letters of their names and combinations by a string of such letters written in order of percent composition.

CHEMISTRY (Tantalum)

The element tantalum is discovered by Anders Ekeberg (1767–1813, Swedish) while analyzing a sample of tantalite.

EARTH SCIENCES (Hydrogeology)

Jean Baptiste Lamarck (1744–1829, French) publishes *Hydrogéologie,* a small volume on the past and present effects of water in modifying the earth's surface.

EARTH SCIENCES (Ocean Waves)

Franz Gerstner (1756–1832) publishes a theory of surface waves in the open ocean.

EARTH SCIENCES (Uniformitarianism)

In *Illustrations of the Huttonian Theory of the Earth,* John Playfair (1748–1819, Scottish) explains the geological principles set forth by James Hutton in his *Theory of the Earth* (1795—EARTH SCIENCES), making them basically more comprehensible to most readers and including essays on different aspects of the system, some elaborated with Playfair's own observations and reflections.

EARTH SCIENCES (Valleys)

John Playfair (1748–1819, Scottish), reiterating elements of James Hutton's theory (1795—EARTH SCIENCES), refutes the explanation that valleys with central streams formed during cataclysmic upheavals and only later had streams or rivers flow through them, arguing instead that the valleys are carved by the streams, which thereby determine their shapes and distributions.

METEOROLOGY (Water Vapor)

John Dalton (1766–1844, English) suggests that water vapor is a gas, thereby refuting earlier conceptions of fire-filled bubbles of water (1666—METEOROLOGY). Dalton asserts that water vapor mixes nonchemically with the other atmospheric gases to form air.

1802–1803

EARTH SCIENCES (Volcanoes)

Reversing himself from a strict Wernerian viewpoint, Leopold von Buch (1774–1853, German) concludes from studies of the extinct volcanoes of the Auvergne district of central France that basalt can be volcanic in origin, that volcanic forces can break through masses of granite from below, that granite can undergo a series of changes to become lava, and that the domite hills without craters, such as the Puy de Dôme, were pushed upward after the granite was softened by heated vapors. He qualifies the non-Wernerian aspects of the conclusions with the caution that the volcanoes of Auvergne may be unique or nearly so.

1803

ASTRONOMY (Meteorites)

Jean Baptiste Biot (1774–1862, French), commissioned by the French Institut National to travel to the area of l'Aigle, France, and collect evidence to refute the recent rumors about stones falling from the sky, reports that he believes the many self-proclaimed witnesses.

CHEMISTRY/PHYSICS (Atomic Theory)

John Dalton (1766–1844, English) revives and sharpens the atomic theory put forth in the fourth century B.C. by the Greeks Leucippus and Democritus. The theory asserts that all material substances are made of atoms, which combine in varied proportions. A major contrast between the Greek and revised versions is that the Greeks had used shape to distinguish atoms, whereas Dalton uses weight. Convinced from his work in meteorology (1801—METEOROLOGY; 1802—CHEMISTRY) that all atoms cannot have the same weight, Dalton hypothesizes that the atoms of one element do have the same weight but that this weight differs from that of the atoms of any other element. Dalton attempts to determine relative weights, and does prepare the first table of atomic weights, but does so with erroneous assumptions regarding the numbers of atoms combining to form compounds. Specifically, if two constituents combine in only one way, it is assumed that the compound consists of equal numbers of atoms of each constituent. Thus water is interpreted as HO (rather than H_2O), from which oxygen is calculated to have an atomic weight eight times that of hydrogen (instead of 16 times) to correspond to the known weight ratio of the two elements in water. The table of atomic weights is appended to a paper "On the Absorption of Gases by Water" read to the Manchester Philosophical Society in 1803 and printed in 1805.

CHEMISTRY (Cerium)

Jöns Jacob Berzelius (1779–1848, Swedish) and Wilhelm Hisinger (1766–1852, Swedish) discover the element cerium, independently discovered by Martin Klaproth (1743–1817, German).

CHEMISTRY (Electrolysis)

While experimenting with the voltaic pile, Jöns Jacob Berzelius (1779–1848, Swedish) and Wilhelm Hisinger (1766–1852, Swedish) find that during electrolytic decomposition of salts, the bases accumulate at the negative pole and the acids at the positive pole, thus suggesting that acids and bases have opposite electrical charges.

CHEMISTRY (Gas Laws: Henry's Law)

William Henry (1774–1836, English) discovers that when a gas is absorbed in a liquid the weight of the gas dissolved is directly proportional to the pressure of the gas over the liquid. Formally stated in 1808, this is later termed Henry's Law. It contributes directly to the atomic theory of John Dalton (1803—CHEMISTRY/PHYSICS), who extends the law to mixtures of gases, in conjunction with his own Law of Partial Pressures (1801—METEOROLOGY).

CHEMISTRY (Palladium; Rhodium)

William Wollaston (1766–1828, English) discovers the elements palladium and rhodium, obtaining both from crude platinum ore. Palladium is named after the newly discovered asteroid Pallas (1802—ASTRONOMY).

CHEMISTRY (Platinum)

Robert Hare (1781–1858, American) volatilizes platinum.

CHEMISTRY (Reversibility)

Claude Berthollet (1748–1822, French) expands his *Recherches sur les lois de l'affinité* (1801—CHEMISTRY) into his *Essai de statique chimique* [Essay on chemical statics], the two volumes together helping to lay the foundations of an understanding of chemical reactions. A notable advance is his recognition and discussion of the reversibility of chemical reactions. Berthollet's recognition that the mass of the reacting compounds influences the reaction is a step toward the Law of Mass Action later formulated by Cato Guldberg and Peter Waage (1867—CHEMISTRY).

MATHEMATICS (Projective Geometry)

Projective geometry experiences a revival with the publication of Lazare Nicolas Carnot's (1753–1823, French) *Géométrie de position.* Carnot encourages development of "pure" or synthetic geometry and attempts to demonstrate it to be as powerful as the analytic geometry of René Descartes (1637—MATHEMATICS).

METEOROLOGY (Cloud Classification)

Luke Howard (1772–1864, English) classifies clouds into: (1) wispy, high "cirrus" clouds, (2) lumpy, low-level "cumulus" clouds, and (3) horizontal sheets of "stratus" clouds. He also uses "nimbus" to refer to a cloud in which water vapor is condensing to rain, hail, or snow.

PHYSICS (Mechanics)

Publication of Lazare Nicolas Carnot's (1753–1823, French) *Principes fondamentaux de l'équilibre et du mouvement.*

1804

ASTRONOMY (Asteroids)

Discovery of the asteroid Juno.

BIOLOGY (Photosynthesis)

Nicholas de Saussure (1767–1845, Swiss) publishes *Recherches chimiques sur la végétation* [Chemical researches on vegetation], asserts the importance of carbon dioxide and soil nitrogen to green plants, and demonstrates that plants absorb water. He shows that plants receive their carbon from atmospheric carbon dioxide, not from the soil as earlier theorists had supposed.

CHEMISTRY (Atomic Theory; Multiple Proportions)

John Dalton (1766–1844, English) formulates his Law of Multiple Proportions, stating that two elements which combine to form more than one compound do so in a manner such that the weights of one element combining with unit weight of the other element are in simple whole number ratios.

CHEMISTRY (Osmium; Iridium)

Smithson Tennant (1761–1815, English) separates two new metals—osmium and iridium—from a black powder remaining after dissolving crude platinum in dilute aqua regia. He names osmium for its odor and iridium for the varied colors of its salts.

METEOROLOGY (Atmospheric Composition)

Joseph Gay-Lussac (1778–1850, French), accompanied by physicist Jean Baptiste Biot (1774–1862, French), makes two balloon ascents to determine the composition of the atmosphere at various heights and the variation in the earth's magnetic field. He reaches an altitude of just over 7,000 meters.

PHYSICS (Radiation)

John Leslie (1766–1832, Scottish) publishes *An Experimental Inquiry into the Nature and Propagation of Heat,* in which he establishes that the absorptivity of a surface equals its emissivity and that the reflectivity of a surface decreases as its emissivity increases. He also shows that as heat is radiated from a point on a surface, the intensity of the radiation from that point in any unobstructed direction varies proportionally to the sine of the angle to the surface.

PHYSICS (Torque)

Louis Poinsot (1777–1859, French) introduces the concept of "torque" in his *Éléments de statique* [Elements of statics].

1804–1811

EARTH SCIENCES (Fossils)

James Parkinson (1755–1824, English) publishes the three-volume *Organic Remains of a Former World,* providing an illustrated introduction to the scientific description of fossils. Volume 1 (1804) covers the plant kingdom, volume 2 (1808) fossil zoophytes, and volume 3 (1811) fossil amphibia, echini, insects, mammals, shells, and starfish. To reconcile the long geological time scale indicated by the fossil record with the theological notion of a seven-day Creation, Parkinson accepts, along with others, the notion that each of the seven Creation "days" corresponds to an extended period of time.

1805

CHEMISTRY (Morphine)

Friedrich Sertürner (1783–1841, German) isolates crystalline morphine from opium, originally naming the new substance morphium. The name is soon changed to morphine.

EARTH SCIENCES (Volcanism)

James Hall (1761–1832, Scottish) experimentally confirms the assumption of James Hutton that high pressures within the earth's crust markedly affect the impact of volcanic heat on rocks subjected to it. Hall shows that even if exposed to extremely high temperatures carbonate of lime can be fused without loss of carbonic acid if the incident pressure is sufficiently great.

METEOROLOGY (Winds; Beaufort Scale)

Francis Beaufort (1774–1857, Irish) devises the Beaufort scale of wind velocities, with magnitudes ranging from 0 for calm conditions to 12 for winds of hurricane intensity.

MATHEMATICS (Computers)

In a technological innovation later of importance to the development of computers, Joseph Jacquard (1752–1834, French) revolutionizes the textile industry by the Jacquard attachment to the loom. The attachment automates the weaving of fabrics through use of a series of cards with punched holes, forerunners of the punched cards later used for input to early computers.

1805–1834

METEOROLOGY/EARTH SCIENCES (Descriptive)

Alexander von Humboldt (1769–1859, German) advances the sciences of both geography and meteorology with the publication of 23 volumes describing his 1799–1804 expedition with A. J. A. Bonpland to South and Central America: *Voyage de Humboldt et Bonpland.* Humboldt examines the relation between air temperature and altitude in the Andes, measures the variation with latitude in magnetic intensity, and studies tropical storms, volcanic activity, plant distributions, meteor showers, and the use of guano for fertilizer.

1806

BIOLOGY (Intercellular Space)

Giovan Amici (1786–1868, Italian) studies the spaces between plant cells and determines their importance for conduction of gases within the plant.

CHEMISTRY (Catalysts)

Charles Désormes (1777–1862, French) and Nicholas Clément (1779–1841, French) provide an explanation of the ability of saltpeter to quicken the production of sulfuric acid, a property earlier recognized and used empirically by Joshua Ward (1740s—CHEMISTRY). Désormes and Clément's explanation: the saltpeter reacts with burning sulfur to release the gas nitric oxide. The nitric oxide molecules gain oxygen atoms from the air, creating nitrogen dioxide, which then loses oxygen to the sulfur dioxide, forming sulfur trioxide, which reacts with water to form sulfuric acid. In this way the nitric oxide transfers oxygen from the air to the sulfur dioxide, without itself becoming significantly depleted.

CHEMISTRY (Organic; Classification)

Antoine de Fourcroy (1755–1809, French) progresses toward a more-complete classification of organic substances by subdividing into about 20 categories each those substances with a vegetable origin and those with an animal origin.

MATHEMATICS (Argand Diagram)

Independently of the earlier work of Caspar Wessel (1797—MATHEMATICS), Jean Robert Argand (1768–1822, French) develops the Argand diagram for representing complex numbers and presents the method in "Essai sur une manière de représenter les quantités imaginaires dans les constructions géométriques" [Essay on a manner of representing imaginary quantities in geometric constructions].

MATHEMATICS (Calculus; Continuity)

André-Marie Ampère (1775–1836, French) "proves" that at any point where a function is continuous it also has a derivative. This erroneous result (erroneous in the context of later, more formal definitions of continuity and derivative), along with supposed proofs, is stated in texts throughout the first half of the nineteenth century, reflecting the unsatisfactory state of rigor and particularly the imprecise definitions of function and functional concepts.

MATHEMATICS (Statistics, Least Squares)

In the supplement "On the method of least squares" to his volume *New Methods for Determination of a Comet's Orbit,* Adrien Legendre (1752–1833, French) presents the method of least squares for obtaining the "best possible" value of a measured magnitude from a set of observations which differ due to random measurement errors. The desired value x is that value which minimizes the sum of the squares of the deviations of the observed values from x. This work is done independently of Carl Friedrich Gauss, who discovered the method earlier (1794—MATHEMATICS) but failed to publish until 1809.

1807

ASTRONOMY (Asteroids)

Heinrich Olbers (1758–1840, German) discovers the asteroid Vesta and conjectures from its orbit and the orbits of other known asteroids that the asteroids are the remaining fragments of a former planet.

CHEMISTRY (Atomic Theory)

Thomas Thomson (1773–1852, Scottish) helps popularize John Dalton's atomic theory (1803—CHEMISTRY) with his five-volume *System of Chemistry.*

CHEMISTRY (Electrochemistry; Potassium, Sodium)

Humphry Davy (1778–1829, English) isolates potassium and sodium, doing so through electrolysis with the most powerful electric battery of the time (a voltaic pile, 1800—PHYSICS). The two metals are isolated at the negative electrode. Davy thereby establishes the ability of electrochemistry to isolate highly active elements which had proved unresponsive to traditional chemical techniques.

The high reactivity of the two elements later proves useful in isolating other elements, such as silicon and zirconium (1824—CHEMISTRY) and titanium (1825—CHEMISTRY).

MATHEMATICS (Differential Geometry)

Gaspard Monge (1746–1818, French) publishes a collection of his papers on the calculus of surfaces and curves in *Application de l'analyse à la géométrie*, reputedly the first book devoted to differential geometry.

MATHEMATICS (Fourier Analysis)

Jean Joseph Fourier (1768–1830, French) first announces his theorem that a wide class of periodic phenomena can be fully described by sums of sine and cosine functions. Joseph Lagrange and others criticize the work for lack of rigor, but Fourier pursues the development and later publishes an expanded theory (1822—MATHEMATICS).

PHYSICS (Heat)

Thomas Young (1773–1829, English) concludes from work on radiation from hot bodies and on infrared heating that heat, like light, is more likely a wave vibration than a material substance.

PHYSICS (Optics; Color Vision)

Thomas Young (1773–1829, English) asserts the Young-Helmholtz theory of color vision.

1808

ASTRONOMY (Satellites of Jupiter)

Jean Delambre (1749–1822, French) publishes "Nouvelles tables écliptiques des satellites de Jupiter, d'après la théorie de M. Laplace, et la totalité des observations faites depuis 1662, jusqu'à l'an 1802" in *Tables astronomiques publiées par le Bureau des Longitudes de France*.

BIOLOGY (Bones)

Antoine de Fourcroy (1755–1809, French) and Nicolas Vauquelin (1763–1829, French) analyze ox bones and find an unexpected manganese content.

CHEMISTRY (Atomic Theory)

John Dalton (1766–1844, English) publishes *A New System of Chemical Philosophy*, in which he elaborates his atomic theory first developed five years earlier (1803—CHEMISTRY). Two cornerstones are the Law of Constant Composition

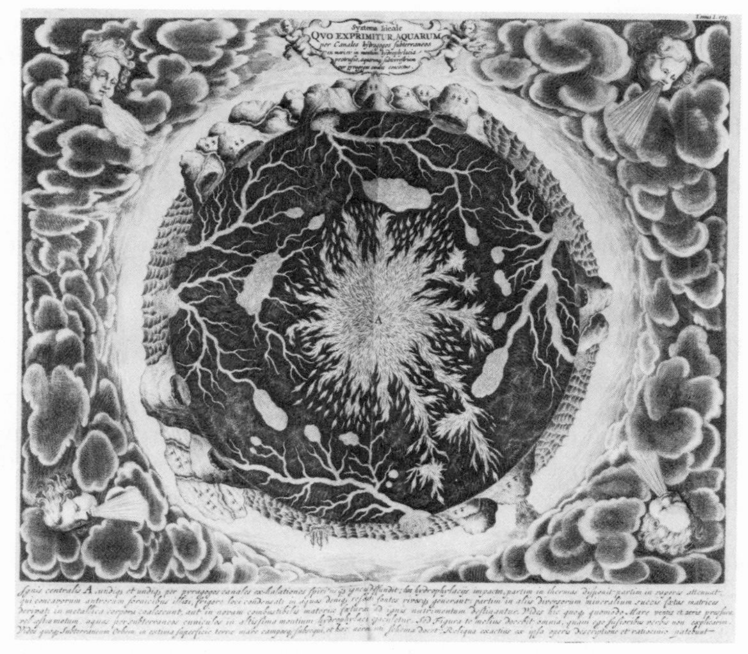

FIGURE 10. *A cross-sectional view of the earth's interior structure as envisioned by Athanasius Kircher and presented in his* Mundus subterraneus *[The subterranean world] (see 1665—*EARTH SCIENCES *entry). A great central fire reaches outward through fissures and other pathways, resulting in volcanic eruptions at spots on the earth's surface and elsewhere in the heating of underground water and consequent formation of hot springs. (By permission of the Houghton Library, Harvard University.)*

(1799—CHEMISTRY) and the Law of Multiple Proportions (1804—CHEMISTRY). Proceeding by rules of "maximum simplicity," he assumes that two elements which combine to form compounds do so such that if only one compound is formed that compound has equal numbers of atoms of each element and if more than one compound is formed then one compound will have equal numbers of each element and a second compound will have exactly twice as many atoms of one element as of the other. Dalton symbolizes elements by circles variously filled in—for example, oxygen is symbolized by an empty circle, hydrogen by a circle with an enclosed dot, gold by a circle with an enclosed G—and compounds by juxtaposing in symmetrical fashion the symbols for the constituent elements. The symbols help provide a visualization of the atomic concepts.

CHEMISTRY (Atomic Theory)

Thomas Thomson (1773–1852, Scottish) and William Wollaston (1766–1828, English) present experimental confirmation of the atomic theory's Law of Multiple Proportions when they, independently, prepare pairs of salts formed from the same elements, one salt of the pair having twice as much base as the other. The work intrigues Jöns Jacob Berzelius (1779–1848, Swedish), who proceeds to add extensive further confirmation over the next several years.

CHEMISTRY (Barium)

Humphry Davy (1778–1829, English) isolates barium, strontium, calcium, magnesium, and boron. Efforts to isolate barium in particular had been made in the 1700s but without success; Davy's achievement is through use of the voltaic pile, invented in 1800.

CHEMISTRY (Boric Acid)

Joseph Gay-Lussac (1778–1850, French) and Louis Thenard (1777–1857, French) treat boric acid with potassium and conclude from the results that boric acid is a combination of oxygen and an unknown combustible substance. Later they recompose the boric acid from the two constituents.

CHEMISTRY (Formulae)

Thomas Thomson (1773–1852, Scottish), in a paper on oxalic acid, introduces quantified chemical symbolism for compounds, a compound with, for instance, two parts oxygen (w) and one part carbon (c) being denoted by $2w + c$.

CHEMISTRY (Gas Laws)

William Henry (1774–1836, English), in *Epitome of Chemistry,* states that when a gas is absorbed in a liquid the weight of the gas dissolved and the pressure of the gas over the liquid are directly proportional (Henry's Law, discovered in 1803).

CHEMISTRY (Sulfur)

After careful experimentation, Humphry Davy (1778–1829, English) erroneously concludes that sulfur is not an element itself but a compound of other elements, including oxygen and hydrogen (see 1809—CHEMISTRY).

EARTH SCIENCES (Geognosy)

Robert Jameson (1774–1854, Scottish) helps spread the geological teachings of Abraham Werner (see, for instance, 1787—EARTH SCIENCES) with his *Treatise on Geognosy.* The term "geognosy," popularized by Werner, refers to the science of the earth and especially the distribution and nature of the mineral and rock layers composing it.

EARTH SCIENCES (Paleontology; Stratigraphy)

Georges Cuvier (1769–1832, French) and Alexandre Brongniart (1770–1847, French) describe the formations of the Seine basin and their numerous strata, helping establish thereby the basic principles of paleontological stratigraphy, including the use of fossils for identifying individual strata. They discuss the Chalk formation at the bottom, with its weak stratification and distinctive fossils, the Plastic Clay formation with its abrupt transition from the Chalk and its lack of fossils, and the Sand and Calcaire Grossier formation with many bands of limestone and marl.

PHYSICS (Polarization of Light)

Étienne Malus (1775–1812, French) discovers that light can be polarized by reflection as well as by passage through a crystal.

1809

ASTRONOMY (Planetary Orbits)

Carl Friedrich Gauss (1777–1855, German), in *Theoria motus corporum celestium* [Theory of the motions of heavenly bodies], summarizes his calculation of the orbits of Ceres (1801—ASTRONOMY) and Pallas (1802—ASTRONOMY), and presents a general procedure for calculating the orbits of asteroids, planets, satellites, and comets from limited observational data.

BIOLOGY (Evolution)

Jean Baptiste Lamarck (1744–1829, French) publishes *Philosophie zoologique* [Zoological philosophy], in which he develops a clear statement of organic evolution, including examples and theory. His earlier linear scale of beings (1802—BIOLOGY) he now views as an evolutionary scale reflecting the major steps through which the simplest organisms have evolved into the human species. Two natural laws are postulated to govern evolution: (1) an organ is strengthened by frequent use, weakened—sometimes to the point of disappearing—by infrequent use; (2) changes in individuals which are common to both sexes are passed on to future generations. Snakes are believed to have lost their presumed original four legs through disuse, and giraffes to have gained extraordinarily long necks through habitual upward stretchings toward the branches and leaves of trees; hence, there is the fundamental presumption of the inheritance of acquired characteristics. Changes in the species result from changes in the environment; the varied paths of evolution have been irregular due to the varied environments encountered. The work, though influential, is widely criticized. Opponents attempt to demonstrate the noninheritability of acquired characteristics by observing the offspring of rats whose tails have been cut off prior to copulation; Lamarck replies that the absence of the tail is not transmitted to the offspring because it occurs through accident rather than infrequent use. Lamarck's animal classification is based primarily on the nervous system and secondarily on the respiratory and circulatory systems.

CHEMISTRY (Acids)

Joseph Gay-Lussac (1778–1850, French) and Louis Thenard (1777–1857,

French) study muriatic acid but reject their own evidence that it contains no oxygen (see CHEMISTRY entries for 1789 and 1810).

CHEMISTRY (Columbium; Tantalum)

William Wollaston (1766–1828, English) erroneously concludes that columbium and tantalum are identical.

CHEMISTRY (Gay-Lussac Law)

Joseph Gay-Lussac (1778–1850, French) asserts that when two gases combine chemically they do so such that the volumes involved are in the ratio of small whole numbers. John Dalton, who emphasizes weights, not volumes, rejects the Gay-Lussac suggestion, later widely accepted and termed the Gay-Lussac Law.

CHEMISTRY (Sulfur)

Joseph Gay-Lussac (1778–1850, French) and Louis Thenard (1777–1857, French) establish that sulfur is an element, refuting earlier work of such chemists as Humphry Davy, who had concluded that sulfur is a compound (1808—CHEMISTRY).

EARTH SCIENCES (Volcanoes)

After extensive observations in Italy, the Canary Islands, and the volcanic regions of central France, Leopold von Buch (1774–1853, German) puts forward his theory of "Craters of Elevation" to account for the form of volcanic mountains. He suggests that the Puy de Dôme and other smooth, dome-shaped mountains in central France, as well as the more obviously volcanic mountains with distinctive craters, were all formed by molten material from the earth's interior being thrust upward. In some cases the molten material solidified retaining its form; in other cases it burst through the peak, creating lava streams and a normal crater.

MATHEMATICS (Gaussian Distribution; Least Squares)

Carl Friedrich Gauss (1777–1855, German) inserts an explanation of his method of least squares (1794—MATHEMATICS) at the end of his *Theoria motus* (1809—ASTRONOMY). Further, in the book's third section he includes the equation of the bell-shaped Gaussian distribution curve.

PHYSICS (Light)

Thomas Young (1773–1829, English) applies the wave theory of light to the phenomena of refraction and dispersion.

1809–1854

PHYSICS (Mechanics of Flight)

A series of experiments and publications on the mechanics of flight establish George Cayley (1773–1857, English) as the founder of modern aerodynamics. He calculates power-to-weight ratios required at various speeds, distinguishes lift and drag, experiments with wing design, steering rudders, and air screws.

1810

BIOLOGY (Australian Flora)

Robert Brown (1773–1858, Scottish) describes the flora of Australia in *Prodromus florae novae Hollandiae et insulae van Diemen* after analyzing the specimens he collected during an 1801–1805 expedition.

CHEMISTRY (Acids)

Humphry Davy (1778–1829, English) shows that oxygen is not a constituent of muriatic (or hydrochloric) acid, thereby destroying the earlier belief, held by Antoine Lavoisier and others, that all acids contain oxygen (1789—CHEMISTRY).

CHEMISTRY (Chlorine)

Humphry Davy (1778–1829, English) establishes—to the satisfaction of many but not all—that chlorine (then known as oxymuriatic acid; see 1811—CHEMISTRY) is an element.

CHEMISTRY (Organic)

Joseph Gay-Lussac (1778–1850, French) and Louis Thenard (1777–1857, French) analyze sugar, starch, gum, and selected acids and resins, determining their percent compositions by mixing with potassium chlorate, inserting in a heated tube, and analyzing the resulting gases.

CHEMISTRY (Preservation)

Nicholas Appert (1750–1841, French) publishes *L'art de conserver pendant plusieurs années toutes les substances animals et végétales* [The art of conserving over several years all animal and vegetable substances], a book describing a method developed by Appert over the previous two decades for preserving food by placing it in champagne bottles, corking and sealing, and then submerging it in boiling water. The key elements are the application of heat, to destroy the ferments causing spoilage, and the exclusion of air.

EARTH SCIENCES (Uplift)

Leopold von Buch (1774–1853, German) announces that a large part of Sweden is slowly rising, evidence coming from the widespread recognition by inhabitants along the east coast that local bays and gulfs are withdrawing over time. Von Buch concludes that the land must be rising, as the probability of sea level lowering is far less.

HEALTH SCIENCES (Homeopathy)

Samuel Hahnemann (1755–1843, German) founds homeopathy with the book *Organon of the Rational Art of Healing.*

MATHEMATICS (Journals)

Joseph Gergonne (1771–1859, French) founds the influential, though short-lived *Annales de mathématiques pures et appliquées,* allegedly the first privately funded mathematical periodical and one of the first periodicals specifically devoted to mathematics. It ceases publication in 1832.

PHYSICS (Nature of Light)

In the two-volume *Zur Farbenlehre,* Johann Wolfgang von Goethe (1749–1832, German) attacks Isaac Newton's theory of light (1704—PHYSICS) and presents a more psychologically oriented examination of color.

1810–1819

BIOLOGY (Brain Anatomy)

Franz Gall (1758–1828, Austrian) and Johann Spurzheim (1776–1832, German), in the four-volume work *Anatomie et physiologie du systeme nerveux* [Anatomy and physiology of the nervous system], show that nerve fibers compose the brain's white matter and introduce the concept that the various mental processes are localized in different regions of the brain. Control of speech, for instance, is determined to be centered in the frontal lobes. Gall also founds phrenology, claiming that an individual's character and mental abilities are reflected in and can be determined from the precise shape of the skull, this shape revealing that of the underlying cerebrum, the sizes of whose individual regions (relative to the "normal" for that region) supposedly correlate strongly with the powers of the corresponding mental processes.

1810–1820

CHEMISTRY (Atomic Theory)

Jöns Jacob Berzelius (1779–1848, Swedish) analyzes over 2,000 inorganic compounds to determine the weight ratios of the various constituent elements and in so doing adds experimental evidence to the atomic theory. His analyses show that John Dalton's Law of Multiple Proportions (1804—CHEMISTRY) is soundly based experimentally; and he emphasizes that only the atomic hypothesis provides a reasonable theoretical basis for the law.

1811

BIOLOGY (Brain; Nervous System)

Charles Bell (1774–1842, Scottish) asserts in his *Idea of a New Anatomy of the Brain* that different parts of the brain undertake different functions and that the specific functions of each of the various divisions of the peripheral nerves derive from the part of the brain connected to that division. He describes an experiment in which he cut the spinal nerve and found the back muscles to undergo convulsions when the anterior roots were touched but not when the posterior roots were touched. This contributes later to the development of the Bell-Magendie Law (1822—BIOLOGY), but is used here to support Bell's contention that the cerebellum is the locus of involuntary nervous functions whereas the cerebrum is the locus of voluntary nervous functions. (Bell believed the posterior root filaments to be connected to the cerebellum and the anterior root filaments to be connected to the cerebrum.)

CHEMISTRY (Avogadro's Hypothesis)

In "Essai d'une manière de déterminer les masses relatives des molécules élémentaires des corps, et les proportions selon lesquelles elles entrent dans ces combinaisons" [Essay on a manner of determining the relative masses of the elementary molecules of bodies, and the proportions in which they enter into these compounds], Amedeo Avogadro (1776–1856, Italian) postulates that, for a given temperature, pressure and volume, all gases have the same number of "molecules." This requires the innovative—and initially controversial—concept that atoms of the same element can become bound together, a molecule of nitrogen gas, for instance, being composed of pairs of nitrogen atoms. Later termed Avogadro's Hypothesis, Avogadro arrives at it after contemplating the Gay-Lussac Law (1809—CHEMISTRY) and deciding that such an hypothesis is the simplest explanation of the law. Among the consequences of accepting the Avogadro Hypothesis is the determination that water is composed of twice as many hydrogen molecules as oxygen molecules since, upon being separated into the two gases, the volume of the resulting hydrogen is twice that of the resulting oxygen. The masses of hydrogen and oxygen in water being in the ratio 1:8, the conclusion is therefore that the molecular weight of oxygen is 16 times that of hydrogen. John Dalton and other early supporters of the atomic theory had conceived of water as HO rather than the H_2O now advanced by Avogadro and were reluctant to accept the possibility of single-element, multi-atom molecules because of the belief that like atoms repel each other (1802—CHEMISTRY). The Avogadro Hypothesis receives little acknowledgment for half a century but becomes widely accepted afterwards (see CHEMISTRY entries for 1860 and 1865).

CHEMISTRY (Chlorine)

Humphry Davy (1778–1829, English) summarizes results from his experiments on oxymuriatic acid in the paper "On Some of the Combinations of Oxymuriatic Gas and Oxygen, and on the Chemical Relations of these Principles to Inflammable Bodies" and there suggests the name "chlorine" for the misnamed "oxymuriatic acid."

CHEMISTRY (Iodine)

Iodine is discovered by Bernard Courtois (1777–1838, French) while burning algae in order to obtain sodium and potassium compounds. Preliminary examination of the substance leads Courtois to suspect iodine to be a new element, after which he turns it over to Charles Désormes (1777–1862, French) and Nicholas Clément (1779–1841, French) for a more thorough examination.

EARTH SCIENCES (Peruvian Coastal Current)

Alexander von Humboldt (1769–1859, German) speculates that the coldness of the northward flowing current along the coast of Peru, later termed the Humboldt current, derives from advection of Antarctic waters. (This explanation loses much of its appeal when it is observed later in the century that the temperature in the current sometimes decreases downstream for considerable distances. Urbain de Tessan suggests that the cold waters result from upwelling rather than surface advection [1844—EARTH SCIENCES].)

EARTH SCIENCES (Stratigraphy)

Georges Cuvier (1769–1832, French) and Alexandre Brongniart (1770–1847, French) publish their *Essai sur la géographie minéralogique des environs de Paris, avec une carte géognostique et des coupes de terrain* on the formations in the Paris region, including those of the Chalk, Plastic Clay, Coarse Limestone, Silicious Limestone, Gypsum, Marine Marls, Sandstone Without Shells, Marine Sandstone, Burrstones Without Shells, Marls and Burrstones With Shells, and Alluvial Clay. This work expands upon an earlier work (1808—EARTH SCIENCES) by incorporating observations from the three intervening years.

PHYSICS (Optics; Brewster's Law)

David Brewster (1781–1868, Scottish) formulates Brewster's Law that the maximum polarization of the reflected component of a beam of light occurs when the angle of incidence is the inverse tangent of the refractive index n of the surface from which the light is reflected; that is, when $\tan i = n$.

1812

CHEMISTRY (General)

Publication of Humphry Davy's (1778–1829, English) *Elements of Chemical Philosophy.*

CHEMISTRY (Glucose)

Konstantin Kirchhoff (1764–1833, Russian) produces glucose by boiling starch in water with a small amount of sulfuric acid, the acid serving as a catalyst.

EARTH SCIENCES (Catastrophism)

Georges Cuvier (1769–1832, French) attaches to his *Recherches sur les ossemens fossiles* (1812—EARTH SCIENCES) a *Discours préliminaire* [Preliminary discourse on the theory of the earth], later published separately, in 1826, as *Discours sur les révolutions de la surface du globe* [Discourse on the revolutions of the surface of the globe]. In this work Cuvier asserts from his studies of the stratigraphy of the Paris basin (EARTH SCIENCES entries for 1808 and 1811) that the earth has experienced in the past many sudden and violent catastrophes. During these there have been mass extinctions of species, with new species arising in subsequent periods. Cuvier rejects the speculation that species have evolved from one to another, finding no evidence for such an evolution in the fossil record.

EARTH SCIENCES (Fossils)

Publication of Georges Cuvier's (1769–1832, French) *Recherches sur les ossemens fossiles* [Researches on fossil bones], helping to establish Cuvier, through his systematic analyses of these and other fossil remnants, as a founder of comparative vertebrate paleontology. Cuvier extrapolates from the fossil remains to speculate on the actual appearance of the former plant or animal, and establishes that many of the remains belong to species which are now extinct.

HEALTH SCIENCES (Appendicitis)

James Parkinson (1755–1824, English) describes appendicitis and attributes the cause of death in appendicitis cases to perforation. Although the condition had been described earlier (1554—HEALTH SCIENCES), this is reputedly the first description in English and the first identification of perforation as the cause of death.

HEALTH SCIENCES (Mental Illness)

Publication of Benjamin Rush's (1746–1813, American) *Medical Inquiries and*

Observations Upon the Diseases of the Mind, one of the first modern works on mental disorders.

MATHEMATICS (Differential Notation)

George Peacock (1791–1858, English), John Herschel (1792–1871, English), and Charles Babbage (1792–1871, English) found the Analytical Society at Trinity College, Cambridge, to reform the notation and teaching of calculus, particularly encouraging replacement of the fluxions of Isaac Newton (MATHEMATICS entries for 1664–1666 and 1669) with the differential notation of Gottfried Wilhelm Leibniz (1673–1676—MATHEMATICS), the latter having greatly assisted Continental mathematicians over the preceding century.

MATHEMATICS (Infinite Series)

Carl Friedrich Gauss (1777–1855, German) presents a rigorous investigation of the convergence of the hypergeometric series, determining which restrictions on the parameters of the series would insure that the series converges.

MATHEMATICS (Probability Theory)

Pierre Simon de Laplace (1749–1827, French) becomes the first to make extensive use of analysis in the study of probabilities, doing so in his work *Théorie analytique des probabilités.* By applying infinitesimal analysis to what had been discrete mathematics, Laplace greatly increases the power of probability theory. In addition to analyzing games of chance, geometrical probabilities, Bernoulli's theorem (1713—MATHEMATICS), and Adrien Legendre's theory of least squares (1806—MATHEMATICS), he here introduces the Laplace transform.

PHYSICS (Forces)

In *Ansicht der Chemischen Naturgesetze durch die neueren Entdeckungen gewonnen,* Hans Oersted (1777–1851, Danish) attempts to establish the identity of all forces of matter, suggesting that one basic force gives rise to such varied effects as light, heat, electricity, and magnetism. An expanded, French version of the work is published in 1813 as *Recherches sur l'identité des forces chimiques et électriques* [Researches on the identity of chemical and electrical forces].

PHYSICS (Mechanical Universe; Mathematical Determinism)

Pierre Simon de Laplace (1749–1827, French) suggests that complete knowledge of the masses, positions, and velocities of all particles at any instant would enable precise calculation of all past and future events.

1812–1816

MATHEMATICS (Logic)

Publication of Georg Hegel's (1770–1831, German) *Science of Logic.*

1813

BIOLOGY (Evolution)

William Wells (1757–1817, American-English) presents to the Royal Society a paper entitled "An Account of a White Female, Part of Whose Skin Resembled that of a Negro," containing a clear statement of the principle of natural selection with respect to humankind. Wells states that domesticated animals can be improved by human-directed selective breeding and that similarly the varieties of humans have developed in nature by a natural selection process suiting the individual species to the countries they inhabit.

CHEMISTRY (Nitrogen Trichloride)

Pierre Dulong (1785–1838, French) discovers the explosive nitrogen trichloride, although not without sacrifice, as he loses an eye and two fingers during the investigation.

CHEMISTRY (Symbolism)

Jöns Jacob Berzelius (1779–1848, Swedish) publishes the first paper incorporating his emerging chemical symbolism, later to become generally accepted in the scientific literature. The paper, entitled "Experiments on the Nature of Azote, of Hydrogen, and of Ammonia and Upon the Degrees of Oxidation of Which Azote is Susceptible," incorporates a symbolism whereby an element is generally represented by the first letter of its Latin name, or, in the event of elements with the same first letter, by the first two letters. The use of letters, earlier employed by Thomas Thomson (CHEMISTRY entries for 1802 and 1808), has a clear typographical advantage over the symbols of Jean Hassenfratz and Pierre Adet (1787—CHEMISTRY) and the circles of John Dalton (1808—CHEMISTRY).

EARTH SCIENCES (Stratigraphy)

Jean Julien d'Omalius d'Halloy (1783–1875, Belgian) extends the work of Georges Cuvier and Alexandre Brongniart on the Tertiary formations of the Seine basin (1808—EARTH SCIENCES) and describes and maps the major subdivisions of the Cretaceous and Jurassic series.

MATHEMATICS (Higher-Order Equations)

Paolo Ruffini (1765–1822, Italian) proves, although without complete rigor, the impossibility of solving by radicals the general equation of degree equal to or greater than 5.

PHYSICS (Potential)

Siméon Poisson (1781–1840, French) derives the Poisson equation for the potential of a point within a continuous distribution of matter, $\nabla^2 v = (\partial^2 v/\partial x^2) +$

$(\partial^2 v/\partial y^2) + (\partial^2 v/\partial z^2) = -4\pi\rho$ (ρ = density), although does not provide a rigorous proof. The result generalizes Laplace's equation, $\nabla^2 v = 0$, which is valid for the special case of zero density.

1814

ASTRONOMY (Spectroscopy)

After constructing an exceptional prism, Joseph Fraunhofer (1787–1826, German) observes and notes dark absorption lines in the solar spectrum. He identifies eight prominent lines, later studying them in detail and mapping 576 lines of varying distinction. He labels the major lines with letters of the alphabet, and finds that the (reflected) light from the moon, Venus, and Mars contains the same absorption lines as those in the solar spectrum, while light from other stars contains lines differing from those in the solar spectrum.

CHEMISTRY (Atomic Weights)

Jöns Jacob Berzelius (1779–1848, Swedish), having undertaken an extensive program to determine the relative atomic weights of the known elements, publishes initial values for some of those elements, preparatory to his first atomic weight table, published in 1818.

CHEMISTRY (Avogadro's Hypothesis)

André Ampère (1775–1836, French) independently arrives at Avogadro's Hypothesis (1811—CHEMISTRY).

CHEMISTRY (Iodine)

Joseph Gay-Lussac (1778–1850, French) publishes his classic research on iodine.

CHEMISTRY (Isomers)

Joseph Gay-Lussac (1778–1850, French) suggests that the arrangement of elements within a compound might be important, so that two compounds with the same complement of elements, in the same proportions, might have very different properties. This prediction is confirmed in 1823 with analyses of silver cyanate and silver fulminate. The term "isomer" is introduced later (1830—CHEMISTRY).

CHEMISTRY (Symbolism)

Publication of Jöns Jacob Berzelius's (1779–1848, Swedish) *Theory of Chemical Proportions and the Chemical Action of Electricity,* in which he complements his discussion of the theory by an expanded version of his emerging chemical symbolism (1813—CHEMISTRY), continuing to use the first one or two letters of the Latin name to symbolize an element, and symbolizing compounds quantitatively, with a plus sign between constituent parts. Thus water is written $2H + O$.

EARTH SCIENCES (Mineral Classification)

Jöns Jacob Berzelius (1779–1848, Swedish) provides a classification of minerals based on chemical composition in his *New System of Mineralogy*.

MATHEMATICS (Operator)

The notion of a mathematical "operator" is developed by François Servois (1767–1847, French).

MATHEMATICS (Planimeter)

J. H. Hermann creates the first planimeter, a mechanical instrument to measure the area of a planar figure.

1814–1815

HEALTH SCIENCES (Toxicology)

Mathieu Orfila (1787–1853, Spanish) helps found toxicology, and especially the study of its medical-legal implications, with his *Traité de toxicologie générale* [Treatise on general toxicology; 1814] and his two-volume *Traité des poisons* [Treatise on poisons; 1814–1815].

1815

ASTRONOMY (Comets)

Heinrich Olbers (1758–1840, German) discovers Olbers's comet, later found to have a period of 73 years.

CHEMISTRY (Cyanogen; Radicals)

Joseph Gay-Lussac (1778–1850, French) discovers cyanogen and describes its properties, preparation, and various compounds in the paper "Recherche sur l'acide prussique." He finds (and recognizes its importance) that when the gas combines with hydrogen it forms a compound which acts in certain reactions as though it were an individual element. His identification of the cyano radical is the first recognized identification of an organic radical.

EARTH SCIENCES (Erratic Boulders; Glacial History)

Jean Pierre Perraudin (1767–1858, Swiss), a mountaineer and guide in the Swiss Alps, speculates from his observations of the movement of large boulders by glaciers in the Swiss Alps, that the erratic boulders found scattered in the area could have arrived at their present locations through glacier transport. Hence he conjectures that local glaciers formerly extended to greater distances.

EARTH SCIENCES (Geological Mapping; Stratigraphy)

William Smith (1769–1839, English), among the founders of the study of stratigraphy, publishes "A Geological Map of England and Wales, with Part of Scotland; exhibiting the Collieries, Mines, and Canals, the Marshes and Fen Lands originally overflowed by the Sea; and the Varieties of Soil, according to the Variations of the Substrata; illustrated by the most descriptive Names of Places, and of Local Districts; showing also the Rivers, Sites of Parks, and Principal Seats of the Nobility and Gentry; and the opposite coast of France." The complete map measures 8′9″ by 6′2″, is hand painted with 20 colors, uses a scale of 5 miles to the inch, and is accompanied by a 50-page explanatory memoir. It is the first attempt to map in such detail the stratigraphy of an entire country, and the memoir indicates the many practical uses that such a map can have, for mining, agriculture, road building, finding underground water, and other purposes. Smith's work, following years of direct observation, indicates a widespread regularity in the strata underlying English soil. He includes in the accompanying memoir his earlier table of English strata and their characteristic fossils (1799—EARTH SCIENCES).

EARTH SCIENCES (Volcanism)

Nicholas Desmarest (1725–1815, French) leaves at his death an unpublished, elaborately detailed topographic map of the volcanic geology of Auvergne, begun in 1764. (A preliminary, reduced version was published in Desmarest's 1774 memoir on the volcanic origin of basalt.)

MATHEMATICS (Permutation Groups)

Augustin Louis Cauchy (1789–1857, French) investigates what are now called permutation groups (see 1770—MATHEMATICS), discovering several basic theorems.

1816

CHEMISTRY (Atomic Theory; Prout's Hypothesis)

William Prout (1785–1850, English), extrapolating from the evidence that most elements seem to have atomic weights that are whole number multiples of the atomic weight of hydrogen, hypothesizes that each element is in some manner a combination of hydrogen atoms.

CHEMISTRY (Catalysts)

Humphry Davy (1778–1829, English) finds that certain organic vapors can combine with oxygen more readily when in the presence of platinum or other metals.

EARTH SCIENCES (Stratigraphy; Fossils)

William Smith (1769–1839, English) publishes a thin volume entitled *Strata Identified by Organized Fossils Containing Prints on Coloured Paper of the Most Characteristic Specimens in each Stratum,* describing and illustrating the characteristic fossils found in the various strata (or formations) depicted in his "Geological Map of England" (1815—EARTH SCIENCES). He explains that his early observational conclusion that identical fossil species are found in a given stratum even at locations widely separated in space (1799—EARTH SCIENCES) helped him frequently in subsequent investigations to distinguish one stratum from another.

HEALTH SCIENCES (Stethoscope)

René Laënnec (1781–1826, French) invents the stethoscope, using first a cylindrical roll of paper, then a wood cylinder to detect chest noises and help diagnose disease.

MATHEMATICS (Fundamental Theorem of Algebra)

In his third of four major proofs of the Fundamental Theorem of Algebra (1799—MATHEMATICS), Carl Friedrich Gauss (1777–1855, German) employs complex integrals and details of complex number theory.

MATHEMATICS (Non-Euclidean Geometry)

Carl Friedrich Gauss (1777–1855, German) vaguely suggests, in a book review, the possibility of a consistent geometry not satisfying the Euclidean parallel postulate. Although he never publishes further on the subject, he apparently develops many consequences of denying the parallel postulate and over the years becomes increasingly convinced that that postulate is unprovable from the other, less-questioned Euclidean postulates. Contemporaneously, Nikolai Lobachevsky and János Bolyai carry out independent developments of non-Euclidean geometry and proceed to publish them (MATHEMATICS entries for 1829 and 1832).

PHYSICS (Elastic Surfaces)
MATHEMATICS (Applied)

Sophie Germain (1776–1831, French) is awarded the *prix extraordinaire* of the French Academy of Sciences for a memoir she entered in a contest to formulate a mathematical theory of elastic surfaces. Germain's memoir treats vibrations of plane and curved elastic surfaces. She later publishes an enlarged version (1821—PHYSICS).

PHYSICS (Optics)

David Brewster (1781–1868, Scottish) invents the kaleidoscope.

1816–1822

BIOLOGY (Classification; Evolution)

In his seven-volume *Histoire naturelle des animaux sans vertèbres* [Natural history of animals without vertebrae] on invertebrate zoology, Jean Baptiste Lamarck (1744–1829, French) strongly modifies the scale of being he had formulated earlier (1802—BIOLOGY). Having failed to discover an evolutionary link between the invertebrate and vertebrate classes, he replaces the strictly linear scale of 1802 with a branched arrangement.

1817

BIOLOGY (Chlorophyll)

Pierre Pelletier (1788–1842, French) and Joseph Caventou (1795–1877, French) name and isolate chlorophyll.

BIOLOGY (General)

Publication of Georges Cuvier's (1769–1832, French) *Le Règne animal distribué d'après son organisation.*

CHEMISTRY (Cadmium)

Cadmium is discovered by Friedrich Strohmeyer (1776–1835, German).

CHEMISTRY (Lithium)

Lithium is discovered by Johan Arfwedson (1792–1841, Swedish), who tries unsuccessfully to isolate it.

CHEMISTRY (Meteorite Composition)

In examining the Pallas meteorite from Siberia, André Laugier (1770–1832, French) detects chromium and sulfur in addition to the previously reported iron and nickel.

CHEMISTRY (Morphine)

Friedrich Sertürner (1783–1841, German) describes a series of salts formed from morphine, a substance first isolated by him in impure form 12 years earlier (1805—CHEMISTRY). His determination now of the alkaline nature of morphine marks an important step in the beginnings of alkaloid chemistry. Its isolation is the first isolation of an alkali with a vegetable origin.

EARTH SCIENCES (Fossils)

William Smith (1769–1839, English) describes, in *Stratigraphical System of Or-*

ganised Fossils, with reference to the Specimens of the Original Collection in the British Museum Explaining Their State of Preservation and Their Use in Identifying the British Strata, the state and stratigraphical origin of approximately 700 fossil specimens that he had collected from England and Wales and deposited in the British Museum. Emphasizing the organization of fossils according to strata and the usefulness of this organization for identifying strata, he includes a table of the fossils, the strata in which they are found, and the continuity of these strata.

EARTH SCIENCES (Geological Mapping)

William Maclure (1763–1840, American) completes a geological map and description of the eastern United States (west to about 94°W), presenting the broad distribution of the major geologic formations at a scale of 120 miles to an inch. In spite of several errors in rock classifications, the map is notable for its attempt to present the geology of so vast a region.

EARTH SCIENCES (Mineral Classification)

Abraham Werner (1749–1817, German) leaves at his death a manuscript presenting his final system of mineral classification, updated from its initial presentation four decades earlier (1774—EARTH SCIENCES). The four main classes into which he places and subdivides 317 "minerals" are the Earthy Minerals, Saline Minerals, Combustible Minerals, and Metallic Minerals.

HEALTH SCIENCES (Parkinson's Disease; Shaking Palsy)

In an *Essay on the Shaking Palsy,* James Parkinson (1755–1824, English) describes a degenerative brain disorder characterized by involuntary trembling motions and lessened muscular power. Based on his observation of a variety of palsied conditions, he shows that symptoms previously assumed characteristic of distinct diseases reflect instead a single disease, referred to as "shaking palsy" by him, though later termed Parkinson's Disease. Commonly it begins with trembling in mid-life, then leads to increasing physical, though not mental, degeneration.

MATHEMATICS (Calculus)

Bernard Bolzano (1781–1848, Czechoslovakian) publishes *Rein analytischer Beweis,* centered on a proof that if f and g are continuous functions and $f(a) < g(a)$ while $f(b) > g(b)$, then there exists an x between a and b such that $f(x) = g(x)$. The book is particularly noted for Bolzano's attempt to develop his results without the use of infinitesimals. His definition of continuous function, not involving infinitesimals, later becomes standard (although reworded). Approximately, Bolzano defines $f(x)$ as continuous at x if for arbitrarily large N there is an $\varepsilon > 0$ such that $|f(x + \Delta x) - f(x)| < (1/N)$ whenever $\Delta x < \varepsilon$. The function is then a "continuous function" if it is continuous at each value x in its domain. In proving the book's central theorem Bolzano also develops results regarding conditions of convergence and existence of greatest lower bounds.

MATHEMATICS (Geometry; Arithmetic)

In a letter to Heinrich Olbers, Carl Friedrich Gauss (1777–1855, German) asserts that the principles of arithmetic are a priori but those of geometry are not. (Most earlier commentators had believed both to be a priori; by the end of the century, many believe neither to be a priori.)

1817–1859

EARTH SCIENCES (Geography)

Carl Ritter (1779–1859, German), in his 19-volume *Die Erdkunde, im Verhältniss zur Natur und zur Geschichte des Menschen, oder allgemeine vergleichende Geographie als sichere Grundlage des Studiums und Unterrichts in physikalischen und historischen Wissenschaften* [The science of the earth in relation to nature and the history of mankind; or, general comparative geography as the solid foundation of the study of, and instruction in, the physical and historical sciences], attempts to place the study of geography on a scientific foundation, emphasizing not the traditional, sometimes massive encyclopedic collections of individual facts but rather the understanding of the interconnections among the seemingly diverse, empirically obtained geographical data.

1818

ASTRONOMY (Encke's Comet)

Johann Encke (1791–1865, German) determines the orbit of Encke's comet.

ASTRONOMY (Star Catalog)

Friedrich Bessel (1784–1846, German) catalogs 3,222 stars in *Fundamenta Astronomiae*.

CHEMISTRY (Atomic Weights; Symbolism)

Jöns Jacob Berzelius (1779–1848, Swedish) publishes a table of relative atomic weights, based on a standard 100 for oxygen. His is the first table to present the approximate relative weights accepted in the twentieth century. In that respect it is a major advance over the table of atomic weights constructed by John Dalton (1803—CHEMISTRY), who had based his determinations on the assumption that compounds have the same number of atoms of each constituent. Berzelius also publishes molecular weights of 2,000 compounds and further develops his symbolism for writing chemical elements and chemical formulae (see also CHEMISTRY entries for 1813 and 1814). Berzelius symbolizes the elements by the first one or two letters of the Latin name and symbolizes compounds by juxtaposing the element symbols, superscripted with the number of atoms involved if greater than one; for example, carbon dioxide is symbolized as CO^2. His is basically the symbolism retained in the twentieth century, although the superscripts of Berzelius have been replaced by subscripts (1834—CHEMISTRY).

CHEMISTRY (Lithium)

Humphry Davy (1778–1829, English) isolates a small amount of the newly discovered metal lithium (1817—CHEMISTRY).

CHEMISTRY (Selenium)

Jöns Jacob Berzelius (1779–1848, Swedish) discovers selenium, though at first mistaking it for tellurium. The discovery is made while manufacturing sulfuric acid by burning distilled sulfur. Berzelius also discovers the elements silicon (1823—CHEMISTRY), thorium (1829—CHEMISTRY), cerium (1803—CHEMISTRY), and zirconium (1824–CHEMISTRY).

CHEMISTRY (Strychnine)

Pierre Pelletier (1788–1842, French) and Joseph Caventou (1795–1877, French) discover strychnine, the first alkali of vegetable origin to be discovered after morphine (see CHEMISTRY entries for 1805 and 1817).

EARTH SCIENCES (Glacial Striations; Glacial History)

Mountaineer Jean Pierre Perraudin (1767–1858, Swiss) notes the similarity in the striations being made by glaciers in currently glaciated regions to the numerous systematic linear scars on hard rocks in currently unglaciated regions and suggests from this and his earlier speculations of former glacial transport of erratic boulders (1815—EARTH SCIENCES) that glaciers once covered what are currently unglaciated locations, some of the glaciers extending at least 5 km further than at present and filling the valley of the Val de Bagnes. Perraudin obtains a convert in Ignatz Venetz (1829—EARTH SCIENCES).

EARTH SCIENCES (Reefs)

Adelbert von Chamisso (1781–1838, German) completes a four-year study of reef structures.

EARTH SCIENCES (Sea Floor)

John Ross (1777–1856, English) leads an oceanographic expedition to Baffin Bay, where he collects samples of the deep sea floor. His reports of finding worms in the mud from the ocean floor are basically ignored as impossible.

PHYSICS (Wave Theory of Light)

Augustin Fresnel (1788–1827, French) publishes in his *Mémoire sur la diffraction de la lumière* [Memoir on the diffraction of light] the first of a series of calculations demonstrating the ability of a transverse wave theory of light to account for the details of observed reflection, refraction, interference, polarization, and diffraction patterns, such as the appearance of diffraction fringes as light spreads around objects. These calculations help convert many scientists to a wave theory of light and make a strong case that the waves are transverse. (Thomas Young, who had earlier revived a wave theory [1801–1804—PHYSICS], had considered the undulations to be longitudinal, although had recently, in 1817, suggested the possibility of a small transverse component. By that time, however, Fresnel had begun to consider the undulation itself to be transverse.)

1819

ASTRONOMY (Meteorite Composition)

M. Stromeyer detects cobalt in several meteorite samples.

CHEMISTRY (Molar Heat Capacities; Specific Heats)

Pierre Dulong (1785–1838, French) and Alexis Petit (1791–1820, French) find that many substances have molar heat capacities of approximately 6 cal/mole °C, implying, through the constancy of the molar heat capacities, that the specific heats tend to be inversely proportional to the atomic weights. The result is announced in the paper "Recherches sur quelques points importants de la théorie de la chaleur" [Researches on several important points of the theory of heat].

EARTH SCIENCES (Geological Mapping)

George Greenough (1778–1855, English) publishes a geological map of England and Wales at a scale of 11 miles to an inch, providing more detail than William Smith had (1815—EARTH SCIENCES) for the formations older than the coal stratum.

HEALTH SCIENCES (Disinfection)

Chlorine begins to be used for purposes of disinfection.

HEALTH SCIENCES (Stethoscope)

René Laënnec (1781–1826, French) describes his stethoscope (1816—HEALTH SCIENCES) and its uses and potentials for disease diagnosis in his two-volume *Traité de l'auscultation médiate* [Treatise on mediate auscultation].

MATHEMATICS (Number Theory)

By showing that the composite number 341 divides $2^{341} - 2$, Pierre Sarrus (1798–1861, French) proves that Fermat's Little Theorem (1640—MATHEMAT-

ICS) cannot be strengthened to: for whole numbers p and $n > 1$, p divides $n^p - n$ if and only if p is prime.

PHYSICS (Wave Theory of Light)

Dominique-François Arago (1786–1853, French) and Augustin Fresnel (1788–1827, French) describe experiments demonstrating that light vibrates transversely to its direction of forward movement, doing so in "Mémoire sur l'action que les rayons de lumière polarisée exercent les uns sur les autres" [Memoir on the action that the rays of polarized light exert on one another].

1820

CHEMISTRY (Law of Isomorphism)

Following work done in 1819, Eilhard Mitscherlich (1794–1863, German) publishes the principle of the Law of Isomorphism, stating that similar crystalline forms reflect analogous chemical formulae. The principle aids Jöns Jacob Berzelius (1779–1848, Swedish) in determining atomic weights of various elements.

CHEMISTRY (Substitution Reactions)

In what is recognized as the first planned substitution reaction, Michael Faraday (1791–1867, English) succeeds in replacing the hydrogen atoms of ethylene (then known as "olefiant gas") by chlorine, creating C_2Cl_6 and C_2Cl_4, the first two carbon/chlorine compounds.

EARTH SCIENCES (Geology and Religion)

William Buckland (1784–1856, English), in his inaugural lecture at Oxford, entitled "The Connexion of Geology with Religion Explained," states the conviction that the study of geology should be directed toward confirming religious beliefs, particularly regarding the creation of the world and the occurrence of Noah's Flood.

EARTH SCIENCES (Sea Water)

Alexander Marcet (1770–1822, Swiss-English) induces from observation that all sea water contains the same ingredients in roughly the same proportions.

MATHEMATICS (Probability Theory; Determinism)

Publication of the third edition of Pierre Simon de Laplace's (1749–1827, French) *Theory of Probability*. This edition becomes one of the classics on the statistical view of understanding nature. Although philosophically Laplace himself maintains a deterministic concept of nature, within a century the statistical view leads others to reject the deterministic philosophy and seek instead for a "most probable" behavior.

PHYSICS (Electromagnetism)

Hans Oersted (1777–1851, Danish) discovers the magnetic action of electricity as he examines the deflection of a compass needle by an electric current moving along a nearby wire. Oersted concludes that there exists a magnetic field surrounding the current and publishes a description of his experiments and results in the paper "Experimenta circa efficaciam conflictus electrici in acum magneticam" [Experiments on the effect of a current of electricity on the magnetic needle].

PHYSICS (Electromagnetism)

André Ampère (1775–1836, French) reflects on the magnetic action of electricity as recently demonstrated by Hans Oersted (1820—PHYSICS) and concludes that a magnet consists of closed electrical currents. He experiments with two current-carrying wires and demonstrates their mutual interaction, then coils the wires into helices and notes the magnetlike mutual attraction and repulsion of the two helices, confirming his emerging theory that in fact magnetism is electricity in motion.

PHYSICS (Galvanometer)

Johann Schweigger (1779–1857, German) invents the galvanometer, making use of the magnetic effect produced by an electric current, discovered earlier in the year by Hans Oersted (1820—PHYSICS), in order to construct an instrument to measure electric current. Schweigger's galvanometer consists of a magnetic needle with a coil of wire wound around it. As current flows through the wire, the magnetic needle is deflected, and the strength of the current becomes obtainable through measuring the angle of deflection.

PHYSICS (Law of Biot and Savart)

Following Hans Oersted's discovery of the magnetic action of electricity (1820—PHYSICS), Jean Baptiste Biot (1774–1862, French) and Félix Savart (1791–1841, French) determine that the intensity of the magnetic field is inversely proportional to the distance to the conductor. They obtain this result, later termed the Law of Biot and Savart, by measuring the rate of oscillation of a magnetic dipole placed successively at various distances from a current-carrying wire.

1820–1830

MATHEMATICS (Differential Equations; Existence Theorems)

Augustin Louis Cauchy (1789–1857, French) recognizes the need for determining whether differential equations of various types have solutions, irrespective of whether the solutions have been found. He develops a set of existence theorems, later modified by Rudolph Lipschitz (1876—MATHEMATICS).

1821

ASTRONOMY (Uranus)

Alexis Bouvard (1767–1843, French) publishes observations on the orbit of the planet Uranus, including distinct irregularities compared to the expected ellipse. (The irregularities are later explained through the discovery of Neptune [1846—ASTRONOMY].)

MATHEMATICS (Analysis)

With the publication of *Cours d'analyse algébrique* [Course on algebraic analysis], Augustin Louis Cauchy (1789–1857, French) lays the foundation of the classical theory of functions of a real variable. He adds rigor to the calculus based on a new, sounder theory of limits than had existed previously, being careful to define precisely such terms as "function," "derivative," "limit," and "continuity," and to distinguish divergent from convergent series. His definitions of "limit" and "continuity" adjust the previous meanings somewhat, increasing their utility. He proceeds to define the derivative $f'(x)$ of a function $f(x)$ as the limit as i approaches 0 of $[f(x + i) - f(x)]/i$. Cauchy also discovers and proves both the root test and the ratio test for convergence of a series.

PHYSICS (Conductivity)

Humphry Davy (1778–1829, English) determines that the conductivity of metals varies inversely with the temperature. This variation receives a practical application half a century later when William Siemens uses it to invent the platinum-resistance thermometer (see 1871—PHYSICS).

PHYSICS (Elastic Surfaces)
MATHEMATICS (Applied)

Sophie Germain (1776–1831, French) publishes an enlarged version of her prize-winning essay on elastic surfaces (1816—PHYSICS), under the title *Remarques sur la nature, les bornes et l'étendue de la question des surfaces élastiques et équation générale de ces surfaces.* She describes a general vibrating elastic surface with a fourth-order partial differential equation.

PHYSICS (Electromagnetism; Lines of Force)

Michael Faraday (1791–1867, English) discovers electromagnetic rotation and reports the discovery in the paper "On Some New Electro-Magnetical Motions, and on the Theory of Magnetism." Employing a magnet and a wire with a flowing current, he causes each separately to rotate round the other and furthermore concludes that a current-carrying wire is surrounded by a circular "line" of magnetic force. This is his first published mention of the "line of force" concept.

1822

BIOLOGY (Bell-Magendie Law)

Extending the work of Charles Bell on the functional specificity of the nerve roots (1811—BIOLOGY), François Magendie (1783–1855, French) formulates and demonstrates the Bell-Magendie Law that the anterior roots of the spinal cord control movement while the posterior roots control sensation.

EARTH SCIENCES (Crystallography)

Posthumous publication of René-Just Haüy's (1743–1821, French) two-volume *Traité de Cristallographie suivi d'une application de cette science à la détermination des espèces minérales et d'une nouvelle méthode pour mettre les formes crystallines en projection.*

MATHEMATICS (Computers)

Charles Babbage (1792–1871, English) produces a primitive Difference Engine capable of calculating simple functional values to six decimal places, a central purpose being the construction of tables of logarithmic and trigonometric functions.

MATHEMATICS (Foundations)

In *Versuch eines vollkommen consequenten Systems der Mathematik* [Study of a complete consistent system of mathematics], Martin Ohm (1792–1872, German) attempts to place fractions and negative numbers on a solid logical basis, assuming the natural numbers a priori, and to reduce analysis to arithmetic.

MATHEMATICS (Fourier Analysis)

The sixth section of Jean Joseph Fourier's *La Théorie analytique de la chaleur* (1822—PHYSICS) includes the basic elements of Fourier analysis, elaborated from his initial statement of the theory (1807—MATHEMATICS). Fourier establishes that any continuous function can be represented as a sum of sine and cosine curves.

MATHEMATICS (Poncelet-Steiner Theorem)

Jean Victor Poncelet in *Traité des propriétés projectives des figures* (1822—MATHEMATICS), suggests that, with the exception of the construction of circular arcs, any construction which can be made with a straightedge and compass can be made with only a straightedge if one fixed circle and its center are provided.

This is later proved by Jakob Steiner (1833—MATHEMATICS) and called the Poncelet-Steiner Theorem.

MATHEMATICS (Projective Geometry)

Jean Victor Poncelet (1788–1867, French) publishes *Traité des propriétés projectives des figures,* a classic synthetic treatment of geometry. Poncelet is the first major mathematician to encourage development of projective geometry as a separate branch of mathematics, and he includes in his *Traité* such concepts as those of cross ratio, involution, perspectivity, and projectivity, plus develops a theory of polygons which are circumscribed to one conic while being inscribed in another.

PHYSICS (Critical Temperatures)

Charles Cagniard de la Tour (1777–1859, French) introduces the concept of "critical" temperature, above which value a liquid will vaporize irrespective of the pressure, and presents experimentally determined critical-state temperatures and pressures for various liquids in the paper "Exposé de quelques résultats obtenus par l'action combinée de la chaleur et de la compression sur certains liquides, tel que l'eau, l'alcool, l'éther sulfurique et l'essence de pétrole" [Report of some results obtained by the combined action of heat and compression on certain liquids, such as water, alcohol, sulfuric ether, and gasoline].

PHYSICS (Heat Conduction)
MATHEMATICS (Theory of Functions; Calculus)

Jean Joseph Fourier (1768–1830, French) develops his theory of heat conduction in *La Théorie analytique de la chaleur* [The analytic theory of heat], making extensive application of mathematics to physical problems. The work becomes a key source in later research on the theory of functions and on the foundations of calculus.

PHYSICS (Mechanics; Dimensions)

Jean Joseph Fourier (1768–1830, French) suggests using mass, time, and length as fundamental dimensions and representing velocity, acceleration, and other variables in terms of them.

PHYSICS (Thermoelectricity)

Thomas Seebeck (1770–1831, Estonian) discovers thermoelectricity, showing that a temperature differential between the junctions of two different metals in a closed circuit can create a continuously flowing electric current. He publishes an account of the discovery in the paper "Magnetische Polarisation der Metalle und Erze durch Temperatur-Differenz."

1823

BIOLOGY (Stomach Acid)

The gastric juices of the stomach are found by William Prout (1785–1850, English) to contain free hydrochloric (or muriatic) acid, a fact he demonstrates in 1824 before the Royal Society through separate experiments on three of four equal parts of the digested contents of a stomach.

CHEMISTRY (Catalysis)

Johann Döbereiner (1780–1849, German) spurs the study of catalysis by his invention of an automatic lighter, in which a strip of platinum foil serves as an instigator, causing a jet of hydrogen to react with oxygen and catch fire.

CHEMISTRY (Fats; Purity)

Michel Chevreul (1786–1889, French) explains the composition of fats in his *Recherches chimiques sur les corps gras d'origine animale.* He shows that a fat is a compound of glycerol with an organic acid and that the creation of soaps from saponification of fats results from the replacement of glycerol by an inorganic base. This is one of the first works addressing the issue of the fundamental structure of a large class of compounds. Chevreul also introduces the constancy of the melting point as a criterion for the purity of a substance, doing so for the crystalline substances obtained by treating soaps with acid.

CHEMISTRY (Isomers)

Justus von Liebig (1803–1873, German) prepares silver fulminate, a compound with the same chemical composition but very different properties from silver cyanate, a separate compound analyzed by Friedrich Wöhler (1800–1882, German). Joseph Gay-Lussac (1778–1850, French) notes the identical composition, and refers back to his earlier predictions of the existence of such pairs of compounds (1814—CHEMISTRY). He thus identifies the concept of isomerism, although not the term.

CHEMISTRY (Liquid Chlorine)

Michael Faraday (1791–1867, English) liquefies chlorine, requiring a temperature of 239 K, and describes his procedure and some of the properties of liquid chlorine in the paper "On Fluid Chlorine."

CHEMISTRY (Liquid Hydrogen Bromide)

Hydrogen bromide is liquefied, requiring a temperature of 206 K.

CHEMISTRY (Silicon)

Jöns Jacob Berzelius (1779–1848, Swedish) discovers silicon.

CHEMISTRY (Waterproofing)

Charles Macintosh (1766–1842, Scottish) discovers waterproofing by spreading dissolved rubber between two sheets of cotton.

EARTH SCIENCES (Noah's Flood)

William Buckland (1784–1856, English) interprets fossil evidence from over 20 European caves plus large boulders and irregular deposits of gravel, sand, and clay in Great Britain as supporting the biblical concept of an ancient universal flood, presenting his analysis in *Reliquiae diluvianae; Observations on the Organic Remains Contained in Caves, Fissures, and Diluvial Gravel, and on Other Geological Phenomena, Attesting the Action of a Universal Deluge*. He estimates the timing of the flood at 5,000–6,000 years before the present.

MATHEMATICS (Non-Euclidean Geometry)

János Bolyai (1802–1860, Hungarian) develops a consistent non-Euclidean geometry by denying Euclid's fifth postulate and replacing it by the assumption that there exists more than one line parallel to a given line through a given point (hyperbolic geometry). Bolyai writes a description of his geometry as an appendix to a book by his father. Unfortunately, publication of the book is delayed almost a decade (1832—MATHEMATICS), by which time non-Euclidean geometry is independently discovered and published by Nikolai Lobachevsky (MATHEMATICS entries for 1826 and 1829).

MATHEMATICS (Observational Errors)

In *Theoria combinationis observationum erroribus minimus obnoxiae,* Carl Friedrich Gauss (1777–1855, German) summarizes and generalizes his earlier work on the theory of observational errors. Included is a more mathematically rigorous development of the method of least squares than appeared in his initial published version (1809—MATHEMATICS), plus some of his subsequent work in mathematical statistics.

PHYSICS (Fluid Flow; Navier-Stokes Equations)

Claude Navier (1785–1836, French) derives the Navier-Stokes equations for fluid flow, basing his derivation on the existence of a force acting between two fluid particles and dependent on the distance separating the particles and the relative velocities. George Stokes independently derives the equations two decades later (1845—PHYSICS).

1823–1830

BIOLOGY (Fertilization)

In a series of studies on plant fertilization, Giovan Amici (1786–1868, Italian) and Adolphe Brongniart (1801–1876, French) determine the crucial role of the stigma and pollen tube.

1824

ASTRONOMY (Earth-Sun Distance)

Johann Encke (1791–1865, German) uses a parallactic method to compute the distance from the earth to the sun, obtaining a value of 95.5 million miles.

BIOLOGY (Fertilization)

Jean Louis Prévost (1790–1850, Swiss) and Jean Baptiste Dumas (1800–1884, French) argue that the sperm, contrary to widespread opinion, is not superfluous but rather is a vital element in the process of fertilization.

CHEMISTRY (Photography; Daguerreotypes)

Louis Daguerre (1789–1851, French) produces photographic plates of the type later termed "daguerreotypes." He proceeds to develop the process over the next 15 years, being joined by Joseph Niepce (1765–1833, French) for the period 1829–1833.

CHEMISTRY (Silicon; Zirconium)

Jöns Jacob Berzelius (1779–1848, Swedish) isolates silicon and zirconium, the latter following many unsuccessful attempts by others. Berzelius succeeds in the zirconium isolation by heating a mixture of potassium and potassium zirconium fluoride in an iron tube protected by a platinum crucible.

HEALTH SCIENCES (Anesthesiology)

Henry Hickman (1800–1830, English) administers carbon dioxide to an animal prior to operating on it, in order to induce temporary sleep.

MATHEMATICS (Algebra)

Niels Abel (1802–1829, Norwegian), independently of Paolo Ruffini (1813—MATHEMATICS), proves the impossibility of solving by radicals the general equation of degree equal to or greater than five. This work inspires Evariste Galois's further work on solvability (1830–1832—MATHEMATICS).

MATHEMATICS (Bessel Coefficients)

Friedrich Bessel (1784–1846, German) systematizes the theory of a class of coefficients—later known as Bessel coefficients—encountered frequently in physical investigations over the preceding decades.

MATHEMATICS (Inversion)

Jakob Steiner (1796–1863, Swiss-German) introduces the inversive transformation whereby point P inside a circle of radius r and center C is transformed to a point P' outside the circle such that P' lies on the line through C and P and the product of the distances CP and CP' is r^2. More formally, point (x,y) in a plane with coordinate system centered at C is transformed to point $(x',y') = (r^2x/(x^2 + y^2), r^2y/(x^2 + y^2))$. Hence points on the circle are transformed to themselves and the transformation creates a one-to-one correspondence between the points outside the circle and the non-C points inside the circle. Steiner does not publish his results but others independently create the inversive transformation and it becomes one of the first nonlinear transformations to be studied in depth.

PHYSICS (Thermodynamics)

The first substantial theory of heat engines is presented by Sadi Carnot (1796–1832, French) in his *Réflexions sur la puissance motrice du feu* [Reflections on the motive power of fire]. The key element in the generation of power is the temperature difference between two bodies and the consequent passage of heat (caloric) from the body with the higher temperature to the body with the lower temperature. Carnot does not believe that heat is converted into work but rather that work is done as the heat passes from high to low temperature. Gases are advantageous in developing power from heat because they can be heated, cooled and compressed easily. Carnot develops the concepts of a reversible engine and of a reversible (or Carnot) cycle, and he states the result that for two given temperatures, all reversible engines operating between bodies at these two temperatures have the same efficiency and that this efficiency exceeds, or at worst equals, that of any irreversible engine operating between the two bodies.

1825

ASTRONOMY (Celestial Mechanics)

The final volume of the five-volume *Mécanique céleste* of Pierre Simon de Laplace (1749–1827, French) is published (see 1799–1825—ASTRONOMY).

CHEMISTRY (Aluminum)

Hans Oersted (1777–1851, Danish) isolates aluminum, doing so by heating alumina and carbon, bringing them in contact with chlorine to produce anhy-

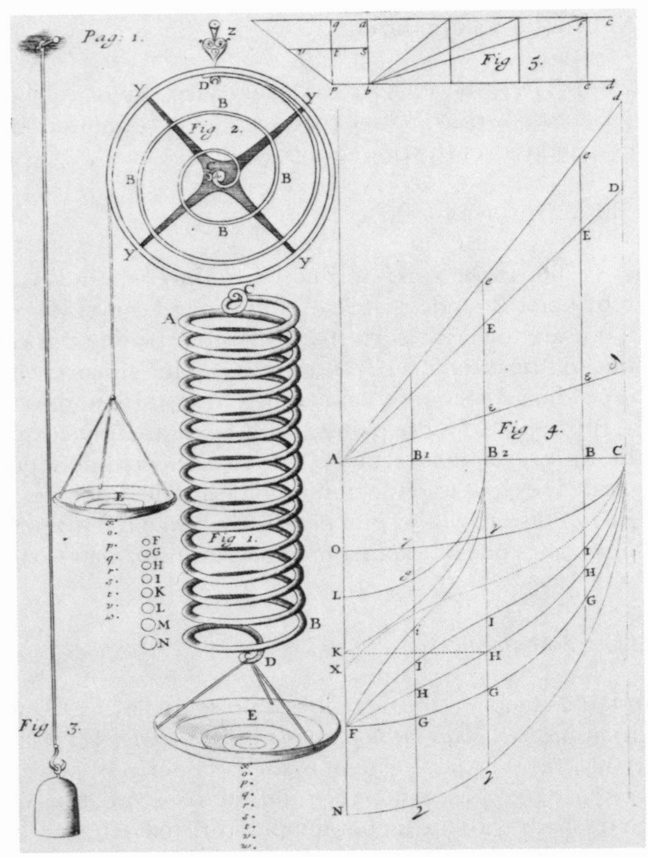

FIGURE 11. *Apparatus used by Robert Hooke to determine the relationship between stress and strain in springs, along with curves drawn demonstrating that relationship and particularly the linear relation later referred to as Hooke's Law (see 1678—PHYSICS entry). The illustrations come from Hooke's* De potentia restitutiva; or, Of Spring, Explaining the Power of Springing Bodies *(see 1678—PHYSICS entry). (By permission of the Houghton Library, Harvard University.)*

drous aluminum chloride, then warming the latter with potassium amalgam to produce aluminum amalgam, and finally distilling the aluminum amalgam in a vacuum.

CHEMISTRY (Benzene)

Michael Faraday (1791–1867, English) discovers and isolates benzene (C_6H_6).

CHEMISTRY (Bromine)

Carl Löwig (1803–1890, German) isolates bromine, but does not publish mention of the isolation until later, subsequent to Antoine Balard's publication on the properties of bromine (1826—CHEMISTRY).

CHEMISTRY (Isobutylene; Isomers)

Michael Faraday (1791–1867, English) discovers isobutylene, a hydrocarbon having the same chemical composition as olefiant gas but different chemical properties. This provides further support for the existence of isomerism, first discovered through analyses of silver cyanate and silver fulminate (1823—CHEMISTRY).

CHEMISTRY (Limelight)

Thomas Drummond (1797–1840, English) and Goldsworthy Gurney (1798–1875, English) produce limelight, claimed by Drummond to be 83 times brighter than any previous artificial light. The light is produced by heating lime in a flame from alcohol burned in oxygen-rich air. Drummond reflects the glow from the lime by a parabolic mirror, resulting in an intense beam of light. Tried briefly in street lighting, limelight is later used mainly in lighthouses and theater spotlighting.

CHEMISTRY (Titanium)

Jöns Jacob Berzelius (1779–1848, Swedish) succeeds in isolating titanium through reactions between titanium oxide and potassium.

EARTH SCIENCES (Volcanoes; Islands)

Leopold von Buch (1774–1853, German), in *Physikalische Beschreibung der canarischen Inseln,* describes his *Ehrebungs* or Upheaval hypothesis on the creation of volcanoes, and classifies all volcanoes into either central or line volcanoes, depending on whether the volcanic force broke out in one spot or at several points along a fissure or other linear (or nonlinear) path. Extrapolating from conclusions drawn during studies of the Canary Islands, he suggests that most and perhaps all ocean islands are volcanic islands thrust upward from the ocean floor by suboceanic fires.

HEALTH SCIENCES (Astigmatism)

George Airy (1801–1892, English) constructs cylindrical-spherical eyeglasses to correct for astigmatism.

MATHEMATICS (Biquadratic Reciprocity)

Carl Friedrich Gauss (1777–1855, German) complements his law of quadratic reciprocity (1795—MATHEMATICS) with a law of biquadratic reciprocity, using complex number theory and his newly introduced Gaussian complex integers $a + bi$ with a, b rational.

MATHEMATICS (Complex Variables)

Augustin Louis Cauchy (1789–1857, French) initiates the theory of functions of a complex variable, introducing the Cauchy integral theorem with residues in his *Mémoire sur les intégrales définies, prises entre des limites imaginaires.*

MATHEMATICS (Elliptic Functions)

Niels Abel (1802–1829, Norwegian) inverts the elliptic integral, thereby creating elliptic functions, and discovers the double periodicity of these functions (published 1827).

METEOROLOGY (Gulf Stream Effect)

Edward Sabine (1788–1883, English) suggests that the mild winter of 1821–1822 in Europe resulted from an anomalously strong Gulf Stream in that year, extending to the European coast instead of dissipating well westward in the North Atlantic.

PHYSICS (Electricity)

Georg Ohm (1789–1854, German) publishes results of a sequence of experiments on the magnitude of current flow through a variety of test wires connected to a thermopile. When presented more mathematically by Ohm two years later, the results include a statement of Ohm's Law (see 1827—PHYSICS).

PHYSICS (Electromagnet)

William Sturgeon (1783–1850, English) receives an award for important improvements in the electromagnet, increasing the magnetic effect by placing an iron bar with an insulating shellac coating in a solenoid. The magnetic force is further increased later when Joseph Henry (1797–1878, American) insulates the wires as well.

1825–1826

MATHEMATICS (Projective Geometry; Duality)

The principle of duality in geometry is given its first clear expression in the work of Joseph Gergonne (1771–1859, French), who presents the dual of Desargues' Theorem (1648—MATHEMATICS), interchanging the terms "point" and "line." Gergonne's work precipitates an outpouring of publications listing pairs of such dual theorems, in which the terms "point" and "line" are interchanged.

1825–1832

MATHEMATICS (Elliptic Integrals)

Adrien Legendre (1752–1833, French) publishes, in three volumes, a systematic account of his theory of elliptic integrals: *Traité des fonctions elliptiques et des intégrales eulériennes.*

1826

ASTRONOMY (Olbers's Paradox)

Heinrich Olbers (1758–1840, German) asks why, if the number of stars is infinite and the stars are distributed roughly uniformly in space, the night sky remains basically dark. Although asked before by Jean-Philippe Loÿs de Cheseaux (1744—ASTRONOMY), the question becomes known as Olbers's Paradox.

BIOLOGY (Evolution)

In a paper on Spongilla, Robert Grant (1793–1874, Scottish) briefly indicates his belief that some species are descended from other species and that during the descent improvements can occur.

CHEMISTRY (Aniline)

Otto Unverdorben (1806–1873, German) becomes the first to prepare aniline.

CHEMISTRY (Atomic Weights)

Jöns Jacob Berzelius (1779–1848, Swedish) produces a new, revised version of his table of atomic weights (see 1818—CHEMISTRY). In this version, most elements are presented with atomic weights very close to those accepted in the twentieth century, although silver, sodium, and potassium are each listed with values roughly twice those accepted now.

CHEMISTRY (Bromine)

Antoine Balard (1802–1876, French) announces his discovery of bromine and describes its properties, compounds, and major natural sources. The discovery had occurred unexpectedly in 1824 while Balard studied the flora of a salt marsh. Balard's prediction that the atomic weight of bromine will fall approximately halfway between the weights of chlorine and iodine is verified by the later determination of atomic weights by Jöns Jacob Berzelius. (See also 1825—CHEMISTRY.)

CHEMISTRY (Nobili's Rings)

Leopoldo Nobili (1784–1835, Italian) describes the colorful rings formed electrolytically with lead and salt solutions, later referred to as Nobili's rings.

CHEMISTRY (Osmosis)

Henri Dutrochet (1776–1847, French) quantitatively examines osmosis, determining that the pressures involved during the diffusion of solutions are proportional to the solution concentrations.

EARTH SCIENCES (Geological Mapping)

Leopold von Buch (1774–1853, German) maps the geology of Germany in a 24-sheet map.

MATHEMATICS (Calculus)

Augustin Louis Cauchy (1789–1857, French) expands his *Cours d'analyse* (1821—MATHEMATICS) in *Applications du calcul infinitésimal à la géométrie.*

MATHEMATICS (Convergence)

Niels Abel (1802–1829, Norwegian) investigates the general binomial series, and constructs a logarithmic test for the convergence of a series.

MATHEMATICS (Journals)

August Crelle (1780–1855, Germany) founds the influential and continuing *Journal für die reine und angewandte Mathematik* [Journal for pure and applied mathematics], devoted primarily to pure mathematics and commonly referred to as Crelle's *Journal.*

MATHEMATICS (Non-Euclidean Geometry)

Nikolai Lobachevsky (1792–1856, Russian) first announces his hyperbolic geometry, replacing Euclid's parallel postulate by a postulate assuming more than one parallel line to a given line through a given external point.

PHYSICS (Galvanometer)

Leopoldo Nobili (1784–1835, Italian) invents a high-sensitivity galvanometer.

1827

ASTRONOMY (Binary Stars)

Félix Savary (1797–1841, French) shows that binary star pairs obey the motion patterns expected from Isaac Newton's law of gravitation (1687—ASTRONOMY).

BIOLOGY (Mammalian Ovum)

Karl von Baer (1792–1876, Russian), who discovered the mammalian ovum, publishes *Epistola de ova mammalium et hominis generis.*

BIOLOGY (Microscopes)

The first achromatic microscopes appear, eliminating the color distortion plaguing earlier microscopes. Among the inventors is Giovan Amici (1786–1868, Italian).

CHEMISTRY (Aluminum)

Friedrich Wöhler (1800–1882, German) isolates aluminum, although only in very small quantities. The procedure uses potassium to decompose anhydrous aluminum chloride.

CHEMISTRY/HEALTH SCIENCES (Food Classification)

William Prout (1785–1850, English) classifies foods into carbohydrates, fats, proteins, and water (using the terms saccharinous, oleaginous, and albuminous for the first three classifications respectively).

CHEMISTRY (Photography; Camera)

The first camera image is produced on a metal plate by Joseph Niepce (1765–1833, French).

HEALTH SCIENCES (Epilepsy)

Louis Bravais (French) describes hemiplegic epilepsy in his "Recherches sur les symptômes et le traitement de l'épilepsie hémiplégique." Later described in detail by John Hughlings Jackson (1863—HEALTH SCIENCES), this particular type of epilepsy is sometimes called "Bravais-Jacksonian epilepsy" and sometimes simply "Jacksonian epilepsy."

MATHEMATICS (Barycentric Calculus)

In the work *Der barycentrische calcul,* August Möbius (1790–1868, German) introduces homogeneous coordinates and creates barycentric calculus. Möbius identifies each point of a plane by three coordinates representing the respective amounts of mass to be placed at the three vertices of a given triangle such that the point identified is located at the center of gravity. (The coordinates are not unique, but their ratios are.) As a result, the equations of curves and surfaces become homogeneous. Möbius identifies the following transformations: congruence, similarity, collineation, and affine transformations.

MATHEMATICS (Elliptic Functions)

Niels Abel (1802–1829, Norwegian) publishes his invention of elliptic functions and discovery of their double periodicity (see 1825—MATHEMATICS).

MATHEMATICS (Quadratic Differential Forms; Curvature)

The first systematic study of quadratic differential forms in two variables is presented by Carl Friedrich Gauss (1777–1855, German) in *Disquisitiones generales circa superficies curvas* [General investigations of curved surfaces], where he develops the "intrinsic geometry" of a surface, using curvilinear coordinates. A central theme of the work is the curvature of surfaces, and Gauss introduces the concept of the radius of curvature at an individual point, this being defined for planar figures as the radius of the circle tangent to the curve at the point. For surfaces in three dimensions, the Gaussian curvature is the inverse of the product of the minimum and maximum radii of curvature at the given point for the curves intersecting the surface on planes through the normal line at the point. The theory culminates in Gauss's *theorema egregium* stating that the total curvature of a surface with squared linear element $(ds)^2 = E(du)^2 + F(du)(dv) + G(dv)^2$ depends only on the coefficients E, F, G, and their derivatives.

PHYSICS (Brownian Motion)

Robert Brown (1773–1858, Scottish) observes a microscopic random motion associated with pollen in water and soon discovers that the same random motion occurs with suspended inorganic particles as well. Termed "Brownian motion," this is finally explained decades later with the kinetic theory. Brown in the meantime publishes (in 1828) a pamphlet entitled and providing "A Brief Account of Microscopical Observations made in the Months of June, July, and August, 1827, on the Particles Contained in the Pollen of Plants; and on the General Existence of Active Molecules in Organic and Inorganic Bodies."

PHYSICS (Electromagnetism; Ampere's Law)

André Ampère (1775–1836, French) publishes his *Mémoire sur la théorie mathématique des phénomènes électrodynamiques uniquement déduite de l'expérience* [Memoir on the mathematical theory of electrodynamic phenomena deduced solely from experiment] in which he investigates mathematically the relations between electric currents, presents laws of electrical action, along with experimental verification, and elaborates his theory of magnets as assemblages of electric currents (1820—PHYSICS). Among the laws stated is an inverse square law for the mutual action of two magnetic elements, analogous to the inverse square law for gravitation devised by Isaac Newton (1687—ASTRONOMY/PHYSICS). The new law, later termed Ampère's Law, states that the attractive and repulsive forces between two magnetic elements are directed along the straight line joining the elements and vary in intensity in inverse proportion to the square of the distance between the elements.

PHYSICS (Ohm's Law)

In *Die galvanische Kette, Mathematisch bearbeitet* [The galvanic circuit investigated mathematically], Georg Ohm (1789–1854, German) presents a statement and theoretical development of Ohm's Law for the conduction of an electric current: if a voltage v is applied through a resistance r, then the resultant current i satisfies $i = v/r$. The law was determined from experiments sending a current through a variety of test wires (see 1825—PHYSICS) and from application of Jean Joseph Fourier's ideas on heat conduction (1822—PHYSICS) to electricity.

1827–1839

BIOLOGY (Birds)

John James Audubon (1785–1851, American) publishes a multivolume collection of over one thousand paintings of American bird life, *The Birds of America* (1827–1838), plus a five-volume accompanying text, *Ornithological Biography* (1831–1839). For the text, Audubon obtains the scientific assistance of William MacGillivray (1796–1852, Scottish).

1828

ASTRONOMY (Nebulae)

Caroline Herschel (1750–1848, English), much of whose career as an astronomer was spent assisting her brother William Herschel, publishes *Reduction and Arrangement in the Form of a Catalogue in Zones of all the Star Clusters and Nebulae Observed by Sir William Herschel.*

BIOLOGY (Fish)

Karl von Martius (1794–1868, German) and Louis Agassiz (1807–1873, Swiss-American) publish an analysis and classification of Amazon fish.

BIOLOGY (Fossil Vegetables; Classification)

Adolphe Brongniart (1801–1876, French) identifies four successive major periods of vegetation, from the Carboniferous to the Tertiary, in his *Prodrome d'une histoire des végétaux fossiles* and classifies the vegetable kingdom into the following six classes: Agamae, cellular cryptogams, vascular cryptogams, gymnosperms, monocotyledonous angiosperms, dicotyledonous angiosperms.

CHEMISTRY (Alcohol; Radicals)

Jean Baptiste Dumas (1800–1884, French) and Polydore Boullay (1806–1835, French) examine various reactions involving alcohol and explain them on the basis of alcohol's being a hydrate of ethylene (then regarded as C_4H_4, with alcohol as $C_4H_4 \cdot H_2O$, carbon being believed to have an atomic weight of 6 instead of 12). This work adds to earlier work, for example that of Joseph Gay-Lussac with cyanogen (1815—CHEMISTRY), in suggesting the importance of radicals in chemical reactions.

CHEMISTRY (Beryllium)

Beryllium is isolated independently by Friedrich Wöhler (1800–1882, German) and Antoine Bussy (1794–1882, French).

CHEMISTRY (Organic Chemistry)

Friedrich Wöhler (1800–1882, German) prepares the organic compound urea ($CO[NH_2]_2$) by evaporating the inorganic compound ammonium cyanate (NH_4OCN). This is the first artificial creation of a product which is created in nature within a living being. It thus weakens the foundations of the theory of vitalism, which states that organic compounds require a "vital force," although it by no means destroys such a theory, since the ammonium cyanate itself derived from an organic substance. Indeed, the concepts of vitalism remain popular among individual groups for years. Wöhler's work also results in the broadening of organic chemistry to include the study of all carbon compounds, whether created naturally or artificially, and it provides chemists with another example of isomerism (CHEMISTRY entries for 1814 and 1823), urea and ammonium cyanate having identical chemical compositions but different chemical properties.

PHYSICS (Electricity; Magnetism)
MATHEMATICS (Calculus; Green's Theorem)

George Green (1793–1841, English) publishes his "Essay on the Application of Mathematical Analysis to the Theories of Electricity and Magnetism," one of the first mathematical analyses of electromagnetic concepts. The paper includes the theorem later termed Green's Theorem, which reduces selected volume integrals to surface integrals. It remains essentially unknown until 1846, when William Thomson has it reprinted in August Crelle's *Journal für Mathematik* (1826—MATHEMATICS). The theorem is discovered and announced independently, also in 1828, by Michel Ostrogradski (1801–1861, Russian), and it is known as Ostrogradski's Theorem in the Soviet Union.

PHYSICS (Optics)

In *Theory of Systems of Rays,* William Rowan Hamilton (1805–1865, Irish) unifies the study of optics through the principle of "varying action." He also correctly predicts the later experimental discovery of conical refraction.

1828–1831

MATHEMATICS (Homogeneous Coordinates)

In *Analytische-geometrische Entwickelungen,* Julius Plücker (1801–1868, German) develops in depth the usage of homogeneous coordinates (see 1827—MATHEMATICS).

1828–1837

BIOLOGY (Embryology)

In one of the founding works of modern embryology, the two-volume *History of the Development of Animals,* Karl von Baer (1792–1876, Russian) presents his theory of how the embryo, including organs and tissues, develops out of the single fertilized egg cell. He shows the early development of the embryo to be similar for wide classes of animals.

1829

CHEMISTRY (Forerunner to the Periodic Table)

Johann Döbereiner (1780–1849, German) reports the phenomenon of element triads, in which an element with atomic weight midway between two other elements has properties similar to or intermediate to the other two elements. In particular, bromine is roughly midway between chlorine and iodine, strontium between calcium and barium, and selenium between sulfur and tellurium. (Later, when atomic numbers are understood, these nine elements are found to have atomic numbers 35, 17, 53, 38, 20, 56, 34, 16, and 52 respectively.) Considered little more than amusing at the time, the element triads are understood more fully with the development of the periodic table 40 years later (1869—CHEMISTRY).

CHEMISTRY (Thorium)

Jöns Jacob Berzelius (1779–1848, Swedish) discovers thorium, naming it after the Scandinavian god Thor. He succeeds in isolating the element through generating a reaction between thorium oxide and potassium.

EARTH SCIENCES (Climatology; Glaciology)

Following several-years' examination of the glaciers of the Swiss Alps in light of Jean Perraudin's ideas on possible earlier greater glacial extent (EARTH SCIENCES entries for 1815 and 1818), Ignatz Venetz (1788–1859, Swiss) announces to the Swiss Society of Natural Sciences his thesis that the glaciers of the Alps once covered the Jura Mountains and extended northward beyond the mountainous areas into the plains. He uses the current distribution of erratic boulders and moraines to support his theory, in particular suggesting that several ridges of debris three miles downstream from the current Flesch glacier are moraines deposited in the past by that glacier.

EARTH SCIENCES (Niagara Falls)

Robert Bakewell (1768–1843, English) notes that Niagara Falls is formed from a drift deposit and calculates the approximate time of the original deposition as 10,000 years before present, basing his calculations on the length of the gorge and the estimated 3-feet/year upstream recession of the waterfall in response to the slow erosion of the rock layer by the flowing water. The drift is considered to have been deposited during the biblical Noah's Flood.

EARTH SCIENCES (Petrology: Nicol Prism)
PHYSICS (Optics: Nicol Prism)

William Nicol (c.1768–1851, Scottish) describes the construction and use of the Nicol Prism, a device for creating a beam of plane polarized light, subsequently widely used by petrographers in microscopic examinations of minerals and rocks. The prism consists of two pieces of calcareous spar (or Iceland spar) cemented together with Canada balsam.

HEALTH SCIENCES (Hydropathy)

Vincenz Priessnitz (1801–1851, German) develops hydropathy, using water to treat physical and mental ailments.

MATHEMATICS (Elliptic Functions)

Carl Jacobi's (1804–1851, German) *Fundamenta nova theoriae functionum ellipticarum* details his work on elliptic functions and further develops and simplifies Niels Abel's concepts (MATHEMATICS entries for 1825 and 1827) of doubly periodic functions and inversion of elliptic integrals. Jacobi here emphasizes the double periodicity of the elliptic functions, only later, in 1835–1836, founding the theory of elliptic functions on the theta function (quasi doubly periodic functions defined by infinite series) instead.

MATHEMATICS (Non-Euclidean Geometry)

Nikolai Lobachevsky (1792–1856, Russian) publishes a description of his non-Euclidean geometry (see 1826—MATHEMATICS). His, like the contemporary but unpublished work of Carl Friedrich Gauss (1816—MATHEMATICS), is a hyperbolic geometry, with more than one line parallel to a given line through a given external point.

MATHEMATICS (Statistics; Census Data)

Adolphe Quetelet (1796–1874, Belgian) constructs the first statistical breakdown of a national census, examining for the Belgian census the correlation of death with age, sex, season, occupation, and economic status.

PHYSICS (Least Constraint)

Carl Friedrich Gauss (1777–1855, German) presents, in the paper "On a New General Principle of Mechanics," his principle of least constraint: the motion of a system of points which are influenced both by each other and by outside conditions is such as to maximize the agreement with free motion, given the existent constraints.

1830

BIOLOGY (Blood)

Ferdinand Wurzer finds manganese to be a normal constituent of human blood.

BIOLOGY (Fossil Zoology)
EARTH SCIENCES (Fossils)

Leopold von Buch (1774–1853, German) includes in "Über die Ammoniten in den älteren Gibirgs-Schichten" a plea for increased zoological study of fossils, as a full understanding of organic beings will require a study of all forms which ever lived, not just those now living. He cites particularly the pioneering work done by Georges Cuvier c.1802 relating the now-extinct ammonites to other animal forms.

CHEMISTRY (Isomers)

Jöns Jacob Berzelius (1779–1848, Swedish) introduces the term "homosynthetic" or "isomeric" for sets of compounds having the same chemical constitution but different chemical properties, a phenomenon first discovered in 1823, with additional examples having been identified since that time (CHEMISTRY entries for 1825 and 1828).

EARTH SCIENCES (Geological Epochs)

Charles Lyell (1797–1875, English) introduces a three-part division of geologic time into the Eocene, Miocene, and Pliocene epochs.

EARTH SCIENCES (Mountain Formation)

Élie de Beaumont (1798–1874, French) publishes *Recherches sur quelques-unes des révolutions de la surface du globe* [Researches on some of the revolutions which have taken place on the surface of the globe], in which he upholds the theory that each mountain range is catastrophic in origin and that the origin of mountains is due basically to the cooling of the earth and the resultant radial contraction of its crust.

MATHEMATICS (Line Coordinates)

Julius Plücker (1801–1868, German) creates the concept of line coordinates and with it proves the principle of duality. Plücker considers the collection of all lines tangent to a curve, pointing out that this collection determines the shape of the curve as much as the collection of points does. He refers to the family of tangents as a "line curve."

MATHEMATICS (Number Theory)

Adrien Legendre (1752–1833, French) expands his *Essai sur la théorie des nombres* (1798—MATHEMATICS) into a two-volume third edition entitled *Théorie des nombres* and including recent advances, particularly by Carl Friedrich Gauss.

MATHEMATICS (Projective Geometry)

Julius Plücker (1801–1868, German) presents a separate set of homogeneous coordinates from those of August Möbius (1827—MATHEMATICS) by taking a fixed triangle and setting the coordinates of any point *P* in the plane equal to the perpendicular distances to the three sides of the triangle.

SUPPLEMENTAL (Science Fraud)

Charles Babbage (1792–1871, English), later noted especially for his pioneering work on computers (MATHEMATICS entries for 1822 and 1833), writes *Reflections on the Decline of Science in England,* in which he includes an identification and naming of different types and levels of fraud. Among these are "trimming," where a too-high and a too-low observation are both adjusted to obtain the same average but appear closer to the desired mean; "cooking," where there is selective reporting of the favorable observations; and, most seriously, "forging," where observations are unjustifiably fabricated.

1830–1832

MATHEMATICS (Galois Theory)

Building on and moving beyond earlier results of Niels Abel (1824—MATHEMATICS), Evariste Galois (1811–1832, French) resolves the general problem of the solvability by radicals of algebraic equations. He associates with each equation a group, later termed the Galois group of the equation, and proves that the equation is solvable if and only if the Galois group has a series of maximal normal subgroups with composition factors which are prime numbers. For each degree greater than four, there exist equations with Galois groups which do not satisfy this property, and hence the general equation of degree five or greater is not solvable by radicals. Galois's investigations into the theory of algebraic equations form the basis of what later becomes Galois theory. He coins the term "group," invents Galois imaginaries, generalizes Fermat's "Little Theorem" (1640—MATHEMATICS), and initiates the study of finite fields, although his career is cut short at age 20 when he is killed in a duel.

1830–1833

EARTH SCIENCES (Uniformitarianism)

Charles Lyell's (1797–1875, English) three-volume *The Principles of Geology, being an Attempt to Explain the Former Changes of the Earth's Surface by Reference to Causes now in Operation* expands upon the uniformitarian theories introduced by James Hutton (EARTH SCIENCES entries for 1785 and 1795) and lays the basis of later geology. Presenting and analyzing the available evidence of geologic changes, Lyell opposes the Catastrophe Theory that the earth's present form is largely the result of infrequent violent events and instead offers the Principle of Uniformity, maintaining that forces operating on the earth's surface in the past are essentially the same as those operating now, and that, given enough time, these forces can indeed account for the observable state of the earth.

1830–1842

SUPPLEMENTAL (Positivism)

Auguste Comte (1798–1857, French) founds Positivism, a philosophy stressing that the seeker after knowledge should aim to describe observable phenomena rather than to delve into metaphysical explanations of causes. Comte details his ideas in his *Cours de philosophie positive* [The course of positive philosophy], where he includes a concept of the evolution of human society governed by natural laws.

1831

BIOLOGY (Cell Nucleus)

Robert Brown (1773–1858, Scottish) discovers the cell nucleus.

BIOLOGY (Evolution)

In *Naval Timber and Arboriculture,* Patrick Matthew publishes views on the origin of species very similar in basic outline to those later developed at length by Charles Darwin (1859—BIOLOGY). However, Matthew's work is little known and has little influence on the subsequent development of the theory of evolution.

CHEMISTRY (Chloroform)

Samuel Guthrie (1782–1848, American), Justus von Liebig (1803–1873, German), and Eugène Soubeiran (1793–1858, French) independently discover chloroform, and Soubeiran publishes a description in "Recherches sur quelques combinaisons du chlore." Guthrie and Liebig both publish articles the following year on the discovery, Guthrie's entitled "New Mode of Preparing a Spiritous Solution of Chloric Ether" and Liebig's entitled "Ueber die Verbindungen, welche durche die Einwirkung des Chlors auf Alkohol, Aether, ölbildenes Gas und Essiggeist entstehen."

CHEMISTRY (Magnesium)

In the paper "Sur le radical métallique de la magnésie," Antoine Bussy (1794–1882, French) describes a new method of isolating magnesium, by heating magnesium chloride and potassium, which has allowed him to isolate the first substantial amount of the substance. (Humphry Davy had prepared a minute quantity in 1808.)

CHEMISTRY (Vanadium)

Nils Sefstrom (1787–1845, Swedish), while examining a piece of iron because of its unusual softness, rediscovers vanadium, a metal first discovered in 1801 but then misinterpreted as chromium.

EARTH SCIENCES (Gulf Stream)

Recognition is given to a theory worked out earlier by James Rennell (1742–1830, English) hypothesizing that the Gulf Stream is a downhill flow of water, caused by the trade winds through the mechanism of producing a pile of water in the Gulf of Mexico and along the North American coast.

EARTH SCIENCES (Magnetic North Pole)

James Ross (1800–1862, English) locates the north magnetic pole in Boothia Peninsula during a four-year Arctic expedition.

EARTH SCIENCES (Petrology: Thin Slices)

William Nicol (c.1768–1851, Scottish) describes his newly developed method of preparing thin slices of petrified wood in the article "Process of Preparing Fossil Plants for the Microscope." The method, later used extensively for minerals and rocks, begins with cutting a thin slice of the material, grinding it, polishing it, and cementing it to a piece of glass. The exposed surface is then further ground until acquiring the required amount of transparency.

MATHEMATICS (Complex Numbers)

Carl Friedrich Gauss (1777–1855, German) presents a unified theory of complex numbers, constructed largely as a result of his investigation into a particular problem in diophantine analysis, that being to determine the conditions that primes p, q must satisfy in order that $x^4 = qy + p$ or $z^4 = pw + q$ be solvable in integers. Gauss provides the complex numbers with a formal mathematical development, deducing their properties from the postulates of arithmetic after defining $a + bi$ as the number couple (a,b), and then appropriately defining equality $[(a,b) = (c,d)$ if $a = c$ and $b = d]$, addition $[(a,b) + (c,d) =$

$(a+c,b+d)$], and multiplication [$(a,b) \times (c,d) = (ac-bd,ad+bc)$]. He also represents the numbers geometrically, identifying each complex number with a point on the plane. However, he fails to publish the number-couple work, which is independently developed and published by William Rowan Hamilton (1837—MATHEMATICS).

MATHEMATICS (Duality)

In *Analytische-geometrische Entwickelungen,* Julius Plücker (1801–1868, German) develops and generalizes the concept of duality, having noted that interchanging the parameters and variables in the equation of a line results in the equation of a point.

MATHEMATICS (Negative Numbers; Complex Numbers)

Augustus De Morgan (1806–1871, English), in *On the Study and Difficulties of Mathematics,* objects to using either negative numbers or complex numbers.

MATHEMATICS (Prime Numbers)

Carl Friedrich Gauss's work in complex number theory (1831—MATHEMATICS) provides a new concept of "prime" number, whereby some prime numbers in real integer arithmetic, such as 3, remain prime but others become composite by virtue of being factorable into two complex numbers with integer coefficients, such as $5 = (1 + 2i)(1 - 2i)$.

MATHEMATICS (Ternary Quadratics)

Ludwig Seeber (1793–1855, German) solves the problem of reduction of forms for ternary quadratics, as Joseph Lagrange had done for binary quadratics (1773—MATHEMATICS).

METEOROLOGY (Storms)

William Redfield (1789–1857, American) publishes "Remarks on the Prevailing Storms of the Atlantic Coast of the North American States," in which he presents a theory of the rotary motion of storms, asserting that the winds in storms are directed counterclockwise around a center which is moving in the direction of the prevailing winds. (Later it is realized that storms in the Southern Hemisphere revolve clockwise, due to the Coriolis effect [1835—METEOROLOGY].)

PHYSICS (Electromagnetic Induction)

Michael Faraday (1791–1867, English) discovers electromagnetic induction. Using an iron ring with two coils of copper wire wrapped around opposite sides, Faraday finds that when a current is sent through one coil, a magnetic needle under the second coil is deflected whenever the strength of the current in the first coil is varied. Faraday explains that the current magnetizes the iron ring, which then induces the current in the second coil. The induced current is in the same direction as the original current when that current is increasing, and in the opposite direction when the original current is decreasing. His description is unmathematical and stresses the physical existence of lines of magnetic force and similar lines of electric force. He rejects the notion of action at a distance. When Faraday successfully generates electricity from magnetism—complementing the 1820 results of Hans Oersted, which revealed the magnetic effects of electricity (1820—PHYSICS)—it follows ten years of attempts and he is careful to show that the electricity produced indeed has the characteristics of electricity produced by other means. Faraday develops an electric generator far superior to any previous ones, creating a continuous flow of electricity from heat and mechanical energy.

PHYSICS (Magic Lyre)

Charles Wheatstone (1802–1875, English) invents the "Magic Lyre," which reproduces sounds through boards connected by a rod.

1831–1836

EARTH SCIENCES (Natural History)
BIOLOGY (Observational)

Charles Darwin (1809–1882, English) is the naturalist aboard H.M.S. *Beagle* for a five-year voyage, during which time he studies the geology and biology of the areas visited, including the coasts of South America and the islands of the Pacific. In addition to examining living plants and animals, he examines fossils of many species now extinct.

1832

CHEMISTRY (Organic Compounds; Benzoyl Radical)

A means of classifying organic compounds is provided in Justus von Liebig (1803–1873, German) and Friedrich Wöhler's (1800–1882, German) paper "Researches on the Radical of Benzoic Acid," where the importance of radicals to organic chemistry is illustrated by compounds containing the benzoyl radical C_7H_5O (written by Liebig and Wöhler as $C_{14}H_{10}O_2$). The work illustrates the applicability to organic chemistry of various methods developed originally in studies in inorganic chemistry. Shortly thereafter many chemists begin seeking common radicals in other organic compounds.

EARTH SCIENCES (Atlantic Currents)

The posthumous *An Investigation of the Currents of the Atlantic Ocean and of Those Which Prevail Between the Indian Ocean and the Atlantic Ocean* by James Rennell (1742–1830, English) presents the most comprehensive overview yet published of the water circulation in the Atlantic Ocean.

EARTH SCIENCES (North American Fossils)

Timothy Conrad (1803–1877, American) publishes *Fossil Shells of the Tertiary Formations of North America.*

EARTH SCIENCES (Paleoclimatology; Ice Ages)

Reinhard Bernhardi (German) publishes a little-read paper suggesting the former existence of a polar ice cap with ice extending to southern Germany and depositing the erratic boulders and mounds of rock fragments now scattered over the landscape. He further indicates that Europe has experienced both noticeably warmer and noticeably colder periods than the present.

HEALTH SCIENCES (Hodgkin's Disease)

Thomas Hodgkin (1798–1866, English) describes what is later called "Hodgkin's disease," a cancer of the lymph nodes which often spreads to the spleen and liver.

MATHEMATICS (Biquadratic Reciprocity)

Carl Friedrich Gauss (1777–1855, German) proves the law of biquadratic reciprocity.

MATHEMATICS (Hyperelliptic Functions; Abelian Functions)

Carl Jacobi (1804–1851, German), in reviewing Adrien Legendre's third supplement to his *Traité des fonctions elliptiques et des intégrales eulériennes,* suggests renaming Legendre's "hyperelliptical transcendental functions" as "Abelian transcendental functions" after Niels Abel. Jacobi, whose work on hyperelliptic integrals helps lead to the extensive nineteenth-century development of the theory of abelian functions of *n* variables, suggests, in partial analogy to doubly periodic functions, that hyperelliptic integrals can be inverted to hyperelliptic functions through a generalization of elliptic theta functions.

MATHEMATICS (Non-Euclidean Geometry)

János Bolyai's (1802–1860, Hungarian) description of his non-Euclidean, hyperbolic geometry (see 1823—MATHEMATICS), is finally published as a 26-page appendix, "Appendix scientiam spatii absolute veram exhibens" [The science of absolute space], to the first volume of his father Farkas Bolyai's (1775–1856, Hungarian) *Tentamen* (see 1832–1833—MATHEMATICS).

MATHEMATICS (Projective Geometry)

In *Systematische Entwickelung der Abhängigkeit geometrischer Gestalten von einander,*

Jakob Steiner (1796–1863, Swiss-German) unifies the methods of projective geometry and applies them to a wide variety of problems.

METEOROLOGY (Clouds)

Luke Howard (1772–1864, English) advances beyond simple cloud classification (1803—METEOROLOGY) with his "Essay on the Modifications of Clouds."

PHYSICS (Electromagnetic Self-Induction)

Joseph Henry (1797–1878, American) discovers the principle of electromagnetic self-inductance.

PHYSICS (CGS System of Measurement)
EARTH SCIENCES (Geomagnetism)

In *Intensitas vis magneticae terrestris ad mensuram absolutam revocata* [Determination of the strength of terrestrial magnetism in absolute units], Carl Friedrich Gauss (1777–1855, German) presents new, more precise methods of determining the earth's magnetic field and introduces an expanded system of units, using Coulomb's Law to extend to magnetism and electrostatics the standardized use of length, mass, and time as basic units in mechanics. This leads, with later extensions by Wilhelm Weber (1804–1891, German), to the widespread acceptance of the CGS (centimeter/gram/second) system of measurements in the 1880s.

c.1832

EARTH SCIENCES (Stratigraphy)

Adam Sedgwick (1785–1873, English) discovers volcanic rocks intercalated among the Cambrian and Lower Silurian marine sediments of North Wales.

1832–1833

MATHEMATICS (Foundations)

In *Tentamen juventutem studiosam in elementa matheseos* [Essay on the elements of mathematics for studious youths], Farkas Bolyai (1775–1856, Hungarian) attempts to place geometry (volume 1), analysis, algebra, and arithmetic (volume 2) on a rigorous foundation.

1833

ASTRONOMY (Binary Stars)

John Herschel (1792–1871, English) publishes a star catalog that includes a list of binary stars extended from an earlier list of his father William Herschel and which additionally includes nebulae.

BIOLOGY (Digestion)

In *Experiments and Observations on the Gastric Juice, and the Physiology of Digestion,* William Beaumont (1785–1853, American) describes a series of observations and experiments on digestion carried out over a nine-year period with a soldier wounded in the abdomen in 1822 and living thereafter with a direct passage outward from the stomach allowing food and gastric juice to be siphoned from the stomach or directly inserted into it. Beaumont constructs tables, for instance, on the times needed to digest various foods, and also addresses the broader issue of whether the gastric juice is strictly a chemical solvent or whether some "vital force" is needed for the digestive process. Following experiments in which the gastric juice accomplishes digestion both inside and outside the stomach, Beaumont concludes that the process is purely chemical.

CHEMISTRY (Creosote)

Karl Reichenbach (1788–1869, German) discovers creosote.

CHEMISTRY (Electrolysis)

Michael Faraday (1791–1867, English) analyzes the process of electrolysis, that is, the process of producing chemical changes with an electric current, and concludes that the extent of electrolytic decomposition of a substance is proportional to the strength and time of application of the current.

CHEMISTRY (Symbolism)

Edward Turner (1796–1837, English) accelerates the acceptance of the chemical symbolism of Jöns Jacob Berzelius (CHEMISTRY entries for 1813 and 1814) by including it in the fourth edition of his *Elements of Chemistry*. Turner had incorporated practically no symbolism in the 1831 third edition, but finds it increasingly difficult to explain some of the modern results without symbols.

EARTH SCIENCES (Erratic Boulders)

Charles Lyell (1797–1875, English), in the 1833 volume of his *Principles of Geology* (1830–1833—EARTH SCIENCES), argues that the erratic boulders found in various locations throughout Europe were deposited by boulder-laden icebergs during the Great Flood (the biblical Noah's Flood). This extends the widely accepted explanation that the erratics were deposited during the Flood by providing a mechanism by which the boulders could be transported.

MATHEMATICS (Algebra)

In "Report on the Recent Progress and Present State of Certain Branches of Analysis," George Peacock (1791–1858, English) distinguishes arithmetical algebra, where the symbols represent only the positive integers, and the more general symbolic algebra. He sets forth the principle that results established as valid for all arithmetical algebra will be valid also for symbolic algebra. Later he is able to derive the principle from other axioms.

MATHEMATICS (Computers)

Charles Babbage (1792–1871, English) conceives a design for an "Analytical Engine" which he then works on for almost four decades. A major forerunner to twentieth-century computers, the Analytical Engine uses two sets of punched cards, one for operations and one for variables. The idea was derived partly from the punched cards of the Jacquard attachment to the loom (1805—MATHEMATICS). Augusta Ada Byron (Countess of Lovelace, 1815–1852, English) writes a set of instructions for the machine to compute Bernoulli numbers. This is considered the first computer program, although development of the Analytical Engine never reaches the stage of allowing the program to be run.

MATHEMATICS (Geometry)

Ludwig Magnus (1790–1861, German) combines two quadratic transformations to obtain a quartic transformation, with straight lines corresponding to quartic curves.

MATHEMATICS (Poncelet-Steiner Theorem)

Jakob Steiner (1796–1863, Swiss-German) proves Jean Poncelet's conjecture (1822—MATHEMATICS) that, with the exception of circular arcs, all constructions possible with straightedge and compass are also possible with straightedge alone given one fixed circle and its center, doing so in his *Die geometrischen Konstruktionen ausgeführt mittelst der geraden Linie und eines festen Kreises, als Lehrgegenstand auf höheren Unterrichtsanstalten und zur praktischen Benützung*.

MATHEMATICS (Series; Conditional Convergence)

Augustin Louis Cauchy (1789–1857, French), warning of the dangers of casually rearranging terms in a series, provides an example of a conditionally convergent series with different sums depending on the order of the terms.

PHYSICS (Electromagnetism)
MATHEMATICS (Applied Topology)

In one of the earliest scientific applications of topology, Carl Friedrich Gauss (1777–1855, German) develops the concepts of linked and nonlinked circuits for studies in electromagnetism.

PHYSICS (Least Action)

William Rowan Hamilton (1805–1865, Irish) warns that, however useful and important Pierre de Maupertuis's Principle of Least Action (1744—PHYSICS) has become in physics, it should not be regarded as necessarily valid, as economy is indeed not basic for all natural happenings.

PHYSICS (Telegraphy)

Making practical use of Hans Oersted's and Michael Faraday's recent discoveries in electromagnetism (see PHYSICS entries for 1820 and 1831), Carl Friedrich Gauss (1777–1855, German) and Wilhelm Weber (1804–1891, German) construct an electric telegraph for use over the 1.25 mile distance between Gauss's house and an observatory.

1833–1835

MATHEMATICS (Irrational Numbers)

William Rowan Hamilton (1805–1865, Irish) presents, in two papers to the Royal Irish Academy (published later as "Algebra as the Science of Pure Time"), one of the first attempts at analyzing the basis of the irrational numbers. He presents a theory of both the rationals and irrationals based on algebraic number couples and the notion of time. Irrationals are defined as partitions of rationals, in the manner of Dedekind cuts (1872—MATHEMATICS).

1833–1840

BIOLOGY (Comparative Anatomy)

Richard Owen (1804–1892, English) publishes the five-volume *Descriptive and Illustrated Catalogue of the Physiological Series of Comparative Anatomy.*

1833–1844

EARTH SCIENCES (Paleontology; Fossil Fishes)

Publication of Louis Agassiz's (1807–1873, Swiss-American) *Recherches sur les poissons fossiles,* a five-volume study of over a thousand fossil fishes.

1834

CHEMISTRY (Carbolic Acid; Phenol)

Friedlieb Runge (1794–1867, German) discovers carbolic acid, or phenol, and prepares it by distilling coal.

CHEMISTRY (Gas Laws; Equation of State)

Émile Clapeyron (1799–1864, French) combines the gas laws of Robert Boyle (1662—CHEMISTRY) and of Jacques Charles (1787—CHEMISTRY) to arrive at the equation of state for an ideal gas.

CHEMISTRY (Gas Laws; Diffusion)

Thomas Graham (1805–1869, Scottish) formulates Graham's Law, asserting that the ratio of the speeds at which two different gases diffuse is inverse to the ratio of the square roots of the gas densities. Thus oxygen, being sixteen times as dense as hydrogen, diffuses only one fourth as rapidly.

CHEMISTRY (Law of Substitution)

Jean Baptiste Dumas (1800–1884, French) formulates the Law of Substitution after demonstrating that hydrogen can be replaced in an organic compound by halogens. The concept is then further developed by Auguste Laurent (1807–1853, French) as he creates various reactions whereby hydrogen is replaced by chlorine, bromine, and nitric acid.

CHEMISTRY (Lithium; Strontium; Spectroscopy)

William Talbot (1800–1877, English) asserts that dark lines observed by David Brewster (1781–1868, Scottish) in the spectrum of light shining through nitrous acid vapors are caused by the absorption of light by those vapors. Talbot also distinguishes lithium from strontium through spectroscopy.

CHEMISTRY (Radicals)

Jean Baptiste Dumas (1800–1884, French) notes instances in which the chemical properties of an organic radical are not significantly altered even after an electropositive element is replaced by an electronegative one. This suggests that in organic chemistry, the chemical properties might not be strongly dependent on electrical properties (see 1840—CHEMISTRY).

CHEMISTRY (Radicals)

Justus von Liebig (1803–1873, German) examines alcohol and ether, regarding alcohol as the hydrate of the ethyl radical and ether as the oxide of the ethyl radical.

CHEMISTRY (Symbolism)

Justus von Liebig (1803–1873, German) revises the symbolism for chemical compounds by replacing Jöns Jacob Berzelius's superscripts (1818—CHEMISTRY) by subscripts. Thus CO^2 becomes CO_2. He also rejects the earlier use of a bar to denote doubling, so that C^2H^4 with a bar through it becomes C_4H_8.

EARTH SCIENCES (Glacial History)

In a talk before the Swiss Society of Natural Sciences, Johann de Charpentier (1786–1855, German) outlines evidence supporting the claims of Ignatz Venetz (1829—EARTH SCIENCES) that Alpine glaciers have extended to lower elevations in the past.

EARTH SCIENCES (Ocean Currents)

William Redfield (1789–1857, American) rejects the generally accepted viewpoint that ocean currents are caused exclusively by winds, suggesting that, analogously to atmospheric winds, ocean currents may be driven at least in part by density differences within the water.

MATHEMATICS (Analysis)

Bernard Bolzano (1781–1848, Czechoslovakian) devises a function which is continuous throughout an interval but has no derivative at any point on that interval. As this work is overlooked for decades, the credit for first devising such a function is often given to Karl Weierstrass (1861—MATHEMATICS).

MATHEMATICS (Computers)

Inspired by an article on the computing Difference Engine of Charles Babbage (1822—MATHEMATICS), Pehr Scheutz (1785–1873, Swedish) constructs an improved and operable version, although is unsuccessful in securing significant support or recognition until the 1850s (1853—MATHEMATICS).

MATHEMATICS (Periodicity)

Carl Jacobi (1804–1851, German) proves that there are no triply periodic single-valued functions of one variable and that the ratio of the periods of any doubly periodic single-valued function of one variable cannot be real.

MATHEMATICS (Projective Geometry)

Julius Plücker (1801–1868, German), in *System der analytischen Geometrie*, creates canonical forms for the general curve of order *n* and completely classifies curves of the third order.

PHYSICS (Electric Discharge Speed)

Charles Wheatstone (1802–1875, English) measures a conductor's electric discharge speed, using quarter-mile lengths of wire, a revolving mirror, and three spark gaps to determine the lengths of intervals between sparks. The result is an estimated velocity of electricity of 250,000 miles/sec, 30% greater than the velocity of light. The high result is later attributed at least in part to the looped rather than straight wire.

PHYSICS (Lenz's Law)

Heinrich Lenz (1804–1865, Russian) formulates Lenz's Law that the direction of a current produced by electrodynamic induction is always such as to oppose by its electromagnetic action the flux change which gave rise to it. Lenz describes the law, along with experimental confirmation, in the paper "Ueber die Bestimmung der Richtung der durch elektrodynamische Vertheilung erregten galvanischen Ströme."

PHYSICS (Peltier Effect)

Jean Peltier (1785–1845, French) discovers experimentally that a junction between two dissimilar metals tends to absorb heat when an electric current is passed across it in one direction but to lose heat when the current is passed in the opposite direction. The thermoelectric cooling and heating of the junctions is later termed the Peltier effect.

1834–1837

ASTRONOMY (Southern Skies)

John Herschel (1792–1871, English) observes the Southern Hemisphere stars from the Cape of Good Hope, recording 2,000 previously unrecorded binary star systems and over 1,700 previously unrecorded nebulae. He also observes a nova in 1837 which increases in brightness until 1843, at which point its brightness is comparable to that of Sirius.

1835

EARTH SCIENCES (Delta Advance)

Charles Beke (1800–1874, English) reveals geological evidence for the advance of the Tigris-Euphrates delta.

EARTH SCIENCES (Glaciology)

In *Sur la cause probable du transport des blocs erratiques de la Suisse,* Johann de Charpentier (1786–1855, German), like Ignatz Venetz before him (1829—EARTH SCIENCES), advocates the position that erratic granite boulders in Switzerland were transported to their present positions by glaciers, which must therefore have extended further at some point in the past than at present.

EARTH SCIENCES (Stratigraphy: Silurian and Cambrian Systems)

Adam Sedgwick (1785–1873, English) and Roderick Murchison (1792–1871, Scottish) publish a joint paper "On the Silurian and Cambrian Systems" from Murchison's ongoing research and definitions of the Silurian System in South Wales (see also 1838—EARTH SCIENCES) and Sedgwick's ongoing research and definitions of the Cambrian System in North Wales. Serious disputes later arise between Sedgwick and Murchison resulting from further studies that indicate an overlap between the two systems as originally defined.

METEOROLOGY (Winds; Coriolis Effect)

Gaspard de Coriolis (1792–1843, French) theorizes in "Mémoire sur les équations du mouvement relatif des systèmes de corps" [Memoir on the equations of relative movement of systems of bodies] on the motion of objects relative to a separate moving surface. Later, particularly with the work of William Ferrel (1855—METEOROLOGY), the analysis is used to explain, in conjunction with the earth's rotation, the general patterns of trade winds and the rotations of hurricanes and tornadoes. Such researches later lead to the naming of the effect of the earth's rotation on motions of terrestrial bodies the Coriolis effect. This effect results in a rightward deflection in the Northern Hemisphere and a leftward deflection in the Southern Hemisphere, so that, for example, as the winds move toward the low pressure at the center of a hurricane, they are deflected to the right in the Northern Hemisphere, leading to counterclockwise rotation of Northern Hemisphere hurricanes, and to the left in the Southern Hemisphere, leading to clockwise rotation of Southern Hemisphere hurricanes.

PHYSICS (Projectile Motion)

Heinrich Magnus (1802–1870, German) discovers the adjustment in a projectile's path due to its rotation, later termed the "Magnus effect."

SUPPLEMENTAL (Religion and Science: Copernican Theory)

The Index of forbidden books published every several years by the Catholic Church fails to include the books teaching the Copernican theory of the earth's revolution around the sun (1543—ASTRONOMY), books initially placed on the list in 1616. However, there is no pardon of Galileo Galilei, who had been condemned for his support of the theory (1633—SUPPLEMENTAL).

1836

BIOLOGY (Enzymes: Pepsin)

Theodor Schwann (1810–1882, German) discovers and isolates pepsin, a digestive enzyme in the human stomach, and reports his results in the paper "Ueber das Wessen des Verdauungsprocesses." Though the ability of hydrochloric acid to break up or digest food had been known previously, Schwann shows that the existence of pepsin greatly increases the power of the stomach juices to dissolve meat.

BIOLOGY (Yeast)

Yeast is found by Charles Cagniard de la Tour (1777–1859, French) and Theodor Schwann (1810–1882, German) to be a living organism.

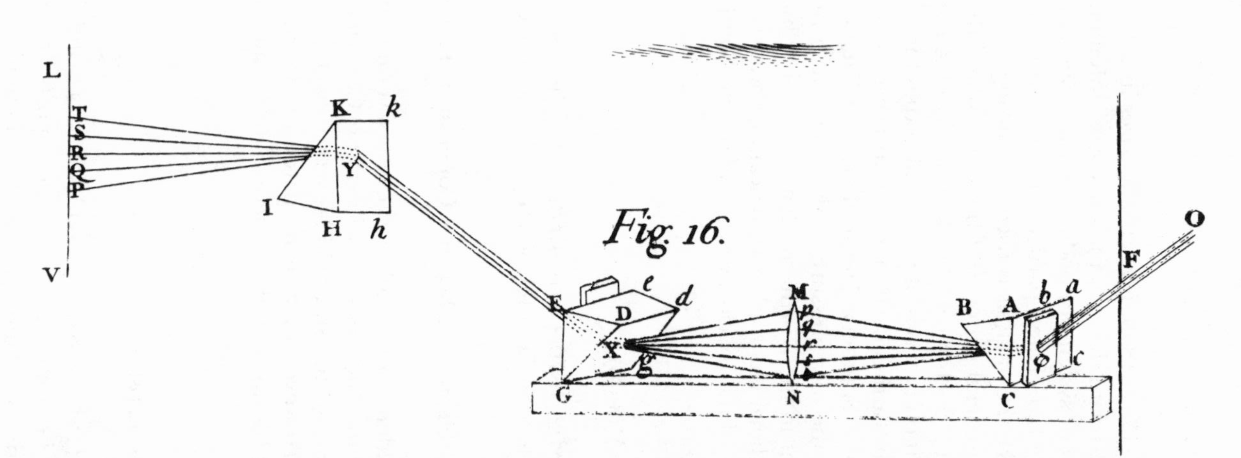

FIGURE 12. *A schematic representation by Isaac Newton illustrating the nature of white light in his* Opticks *(see 1704—*PHYSICS *entry). White light from 0 incident on a prism ABC is dispersed to the various colors of the rainbow, from which a properly placed lens MN is able to merge it again to reform white light at X. A second prism DEG refracts the white light to Y, where a third prism again separates it to the colors of the rainbow, confirming the similarity of the white light at Y to the initial white light at 0. (By permission of the Houghton Library, Harvard University.)*

CHEMISTRY (Acetylene)

Edmund Davy (1785–1851, English) discovers and prepares acetylene.

CHEMISTRY (Catalysis)

Jöns Jacob Berzelius (1779–1848, Swedish) examines various examples in the literature of catalytic-type reactions, concluding that a common force exists and introducing the terms "catalysis" and "catalytic force."

EARTH SCIENCES (Glaciology)

Johann de Charpentier (1786–1855, German) and Ignatz Venetz (1788–1859, Swiss) convince Louis Agassiz (1807–1873, Swiss-American) that many features in the currently unglaciated landscape were formed in the past by glaciers. Once convinced, Agassiz assimilates the evidence, develops a broader-scale theory, and moves toward publication and vigorous advocation of the theory (see EARTH SCIENCES entries for 1837 and 1840).

EARTH SCIENCES (Observational)

Charles Darwin (1809–1882, English) publishes *A Naturalist's Voyage Around the World*, a first account of his observations from the 1831–1836 voyage of the *Beagle*.

HEALTH SCIENCES (Speech)
BIOLOGY (Brain)

Marc Dax (d.1837, French) concludes from his work with brain damaged patients, none of whom had speech loss unless the brain's left hemisphere was damaged, that the brain is asymmetric with respect to its control of various functions and that it is the left hemisphere which controls speech. The work is not widely read, and Paul Broca rediscovers the concepts, probably independently, later in the century (1861—BIOLOGY).

MATHEMATICS (Existence Proofs)

Augustin Louis Cauchy (1789–1857, French) expands upon his existence proofs for solutions of differential equations (1820–1830—MATHEMATICS) by providing limits within which the solutions are certain to converge.

MATHEMATICS (Journals)

Joseph Liouville (1809–1882, French) founds and edits the *Journal de mathématiques pures and appliquées*. Still active, this journal continues to be referred to as Liouville's *Journal*.

PHYSICS (Battery)

By lessening polarization, John Daniell (1790–1845, English) improves the voltaic electric cell (1800—PHYSICS) and replaces it with a "Daniell cell" which more reliably produces a steady current.

1837

BIOLOGY (Blood; Respiration)

Heinrich Magnus (1802–1870, German) analyzes gases in the blood and shows that a higher concentration of oxygen exists in the blood flowing in arteries than in that flowing in veins, thereby suggesting that respiration takes place in the tissues.

BIOLOGY (Microbes)

Theodor Schwann (1810–1882, German) publishes his discovery that meat goes bad only due to subvisible animals (microbes).

BIOLOGY (Photosynthesis)

Henri Dutrochet (1776–1847, French) studies the absorption by plants of atmospheric carbon dioxide and concludes that carbon dioxide is absorbed only by those plant cells that contain green pigment and only in the presence of light.

CHEMISTRY (Organic)

In the paper "Note on the present state of organic chemistry," Jean Baptiste Dumas (1800–1884, French) and Justus von Liebig (1803–1873, German) claim to have found the key to systematically organizing organic chemistry, basing research not on the elements, as in inorganic chemistry, but on the radicals. The radicals in organic chemistry act analogously to the elements in mineral chemistry, with the same general principles of combination and reaction.

EARTH SCIENCES (Climatology; Ice Ages)

Louis Agassiz (1807–1873, Swiss-American) presents his famous "Discourse at Neuchâtel," a talk to the Swiss Society of Natural Sciences during which he first outlines his thesis of an earlier period of widespread ice-age conditions extending from the North Pole to the Mediterranean and Caspian Seas. Although suggestions of greater glacial extent in the past had been made earlier by others (see EARTH SCIENCES entries for 1787, 1795, 1815, 1818, 1829, 1832, and 1835), these suggestions failed to gain major recognition until the advocacy of Agassiz, and Agassiz's theory is much grander, hypothesizing a full-fledged ice cap, with an immense polar ice sheet covering Europe south to the Mediterranean. Agassiz supports his contention with observations from the Jura Mountains in Switzerland, explaining by means of the presumed ice motions the erratic granite boulders found scattered over the limestone, the polished rock surfaces, and the systematic striations on the exposed bedrock. Agassiz adopts the term "ice age" coined in 1836 by Karl Schimper (as *Eiszeit*) and hypothesizes widespread annihilation of animal life during this period.

EARTH SCIENCES (Devonian Rocks)

William Lonsdale (1794–1871, English) concludes from studies of fossil remains that the greywacke and limestone of South Devonshire fall between the Silurian and Carboniferous formations. Adam Sedgwick (1785–1873, English) and Roderick Murchison (1792–1871, Scottish) instead initially conclude on the basis of lithology that these rocks belong within the Cambrian System, although they become converted to Lonsdale's position after examining the fossil evidence (see also 1839—EARTH SCIENCES).

MATHEMATICS (Calculus of Variations)

Carl Jacobi (1804–1851, German) extends the work of Adrien Legendre (1786—MATHEMATICS) on determination of criteria for distinguishing maximal from minimal solutions in the calculus of variations. In so doing he offers a geometrical interpretation and introduces the conjugate point.

MATHEMATICS (Complex Numbers)

William Rowan Hamilton (1805–1865, Irish) presents a rigorous treatment of complex numbers as real number pairs as he develops a formal algebra of such pairs, including the multiplicative rule that $(a,b)(c,d) = (ac-bd, ad+bc)$. The development, combining and extending earlier presentations (1833–1835—MATHEMATICS), appears in the paper "Algebraic Couples with a Preliminary Essay on Time."

MATHEMATICS (Convergent Series)

Gustav Peter Dirichlet (1805–1859, German) proves that the sum of an absolutely convergent series is not altered by a reordering of the terms in the series.

MATHEMATICS (Functions)

Gustav Peter Dirichlet (1805–1859, German) defines a function of a real variable as a table or correspondence between two sets of numbers.

MATHEMATICS (Geometrical Constructions)

Pierre Wantzel (1814–1848, French) derives necessary and sufficient conditions on the coefficients of an algebraic equation to ensure that the solution is constructible with straightedge and compass alone. In the process he proves both that the cube cannot be doubled and that certain angles cannot be trisected exclusively by means of straightedge and compass.

MATHEMATICS (Geometry)

Julius Plücker (1801–1868, German) examines the general problem of the intersections of curves of degrees *m* and *n*.

MATHEMATICS (Poisson Distribution)

Siméon Poisson (1781–1840, French) includes in his *Recherches sur la probabilité des jugements* a description of the limiting case of the binomial distribution later termed the Poisson distribution. With the binomial distribution describing the distribution of outcomes of n trials where each trial has a probability p of success and $q = 1 - p$ of failure, the Poisson distribution is the limiting case as n increases indefinitely but p decreases proportionately so that np remains constant.

MATHEMATICS (Convergence)

Augustin Louis Cauchy (1789–1857, French) constructs the Cauchy integral test for the convergence of a series.

PHYSICS (Heat)

Carl Friedrich Mohr (1806–1879, German) rejects the view that heat is a material substance, insisting instead that heat is an oscillatory motion of the particles within a substance, and that it, like electricity, light, magnetism, and cohesion, is simply one form in which force can appear.

PHYSICS (Telegraphy)

A patent is granted for an electric telegraph invented by Charles Wheatstone (1802–1875, English) and William Cooke (1806–1879, English). The Daniell cell (1836—PHYSICS) is used as a source of electricity and the Sturgeon electromagnet (1825—PHYSICS) as a key element in the recording device.

1837–1848

EARTH SCIENCES (Thematic Geographical Atlases)

Heinrich Berghaus (1797–1884, German) prepares the *Berghaus Physikalischer Atlas,* a set of 93 maps intended as a supplement to Alexander von Humboldt's *Kosmos* (1845–1862—SUPPLEMENTAL). The maps present up to date information on anthropogeography, biogeography, climatology, geology, geomagnetism, hydrography, and zoogeography.

1838

ASTRONOMY (Stellar Parallax)

Friedrich Bessel (1784–1846, German) measures the heliocentric parallax of 61 Cygni to be 0.31″. This is the first authenticated measurement of a star's heliocentric parallax and hence of its distance from the solar system. Using the 186 million mile diameter of the earth's orbit as a base line and the 0.31″ parallax, Bessel calculates a distance of 11 light years.

BIOLOGY (Cell Theory)

Theodor Schwann (1810–1882, German) publishes his *Microscopical Researches into the Accordance in the Structure and Growth of Animals and Plants,* where he elaborates the cell theory, asserting that all plant and animal life consists of cells, that each cell contains a nucleus and a surrounding membrane, and that cells grow and divide. Acknowledging the work of Jacob Schleiden with plant cells (1838—BIOLOGY), Schwann shows that there exists a cell structure in animals which is similar in many respects to the cell structure in plants. He emphasizes the importance of the cell nucleus, coins the term "metabolism," and discovers the cells forming the nerve sheaths, later termed "Schwann cells." He compares the formation of cells—an organic process—to crystallization—an inorganic process.

BIOLOGY (Ferns)

Publication of William Hooker's (1785–1865, English) *Genera filicum* on ferns.

BIOLOGY (Plant Cells; Cell Nucleus)

Jacob Schleiden (1804–1881, German) describes his researches on the development of plant cells, arguing the existence of cell nucleii or cytoblasts, upon which new cells develop and grow.

EARTH SCIENCES (Stratigraphy: Silurian System)

Roderick Murchison (1792–1871, Scottish) extends stratigraphical studies further backward in time as he describes the stratigraphy and associated fossils underlying the Old Red Sandstone of South Wales in *The Silurian System.* He defines the Silurian formations, separating them to the Upper Silurian, with the Ludlow rocks and Wenlock limestone and shale, and the Lower Silurian, with the Caradoc sandstones and Llandeilo flags. Within a few years, the Silurian system is identified and examined in France, Turkey, Scandinavia, and the United States.

HEALTH SCIENCES (Teeth)

Horace Wells (1815–1848, American) publishes *An Essay on Teeth; Comprising a Brief Description of Their Formation, Diseases and Proper Treatment.*

1839

ASTRONOMY (Stellar Parallax; Stellar Distances)

Thomas Henderson (1798–1844, Scottish) announces that his 1832 observations from the Cape of Good Hope of the star Alpha Centauri reveal a stellar parallax of 1.16″. From this he determines the distance to Alpha Centauri to be four light years. (Though not known at the time, Alpha Centauri is the star closest to the solar system. Its parallax is considerably larger than that of 61 Cygni announced by Friedrich Bessel in late 1838, but by failing to announce his interpretation of his 1832 measurements until 1839 Henderson loses priority in the search for a stellar parallax.)

CHEMISTRY (Electroplating)

Patents for electroplating are taken out by Carl Jacobi (1804–1851, German) and Werner Siemens (1816–1892, German).

CHEMISTRY (Fuel Cell)

William Grove (1811–1896, English) constructs a fuel cell, placing test tubes of hydrogen and oxygen gas over two platinum strips in a glass vessel containing dilute sulfuric acid, with a connecting wire. The bottom half of each platinum strip is in contact with the acid, the top half with the hydrogen or oxygen gas.

CHEMISTRY (Lanthanum)

Carl Mosander (1797–1858, Swedish) discovers lanthanum.

CHEMISTRY (Photographic Negatives)

William Talbot (1800–1877, English) produces the first photographic negative.

CHEMISTRY (Photography; Daguerreotypes)

Louis Daguerre (1789–1851, French) announces the invention of the daguerreotype, having worked on improving the images since his initial discovery of the photographic process (1824—CHEMISTRY).

CHEMISTRY (Vulcanization)

Charles Goodyear (1800–1860, American) develops a process for vulcanization of rubber.

EARTH SCIENCES (Devonian System)

Adam Sedgwick (1785–1873, English) and Roderick Murchison (1792–1871, Scottish), expanding upon the earlier work of William Lonsdale in South Devonshire (1837—EARTH SCIENCES), term as "Devonian" the South Devonshire rocks characterized by fossils of marine fauna and determined to fall between the Silurian and Carboniferous formations. They then travel to Germany, seeking and finding rocks of the newly defined Devonian System.

EARTH SCIENCES (Geology; Volcanic History)

Charles Maclaren (1782–1866, Scottish) publishes a *Sketch of the Geology of Fife and the Lothians,* including a detailed examination of the structure of the Arthur's Seat and Pentlands hills and the volcanic history revealed within them.

EARTH SCIENCES (Ice Ages)

Timothy Conrad (1803–1877, American) supports Louis Agassiz's contention that ice formerly extended well beyond its present coverage (1837—EARTH SCIENCES) and extends the original contention by describing erratic boulders, striations, and polished rocks in western New York State, suggesting that these indicate former more extensive glaciation in North America just as Agassiz's evidence from the Alps indicates former more extensive glaciation in Europe.

HEALTH SCIENCES (Dentistry)

Chapin Harris (1806–1860, American) publishes the popular dental textbook *The Dental Art, a Practical Treatise on Dental Surgery.*

MATHEMATICS (Geometry; Plücker's Equations)

Julius Plücker (1801–1868, German) presents in *Theorie der algebraischen Curven* Plücker's equations, relating the order and class of a curve to the numbers of double points, double tangents and points of inflection. These equations contribute directly to further advances by Plücker, Arthur Cayley, and others in geometry and the theory of algebraic functions.

MATHEMATICS (Number Theory)

Publication of Gustav Peter Dirichlet's (1805–1859, German) *Recherches sur diverses applications de l'analyse infinitésimale à la théorie des nombres,* noted in particular for its central use of the limiting process.

PHYSICS (Photovoltaic Effect)

Alexandre Becquerel (1820–1891, French) discovers the photovoltaic effect, finding a 0.1 volt difference of potential produced upon shining a light on one of two electrodes of the same material immersed in an electrolyte.

PHYSICS (Potential Theory)

Carl Friedrich Gauss (1777–1855, German) publishes *Allgemeine Lehrsätze in*

Beziehung auf die imp verkehrten Verhältnisse des Quadrats der Entfernung wirkenden Anziehungs-und Abstossungs-kräfte [General theorems on attractive and repulsive forces which act according to the inverse square of the distance], helping inaugurate rigorous development of potential theory. Gauss includes a rigorous proof of Poisson's equation (1813—PHYSICS).

1839–1840

ASTRONOMY (Moon)

The first telescopic photographs of the moon are taken by John Draper (1811–1882, American).

1839–1843

EARTH SCIENCES (Observational Oceanography)

James Ross (1800–1862, English) leads a ship-based expedition to determine the earth's magnetic properties in the south polar seas, and to accumulate oceanographic data from dredging, sounding, and temperature measurements. In the process, Ross discovers the Ross Sea and Victoria Land, Antarctica.

1840

BIOLOGY (Blood)
PHYSICS (Conservation of Energy)

Robert Mayer (1814–1878, German), working as a ship's doctor near Java, finds that the blood in the crew's veins is brighter red than venous blood in colder climates. He hypothesizes that more oxygen is in the blood in tropical climates because less is needed for food combustion to maintain body heat. Body heat is presumed to derive from the food's chemical energy, chemical energy ("force") being converted thereby into heat energy ("force"). Such ideas lead Mayer to the law of conservation of energy (1842—PHYSICS).

BIOLOGY (Physiology)

Johannes Peter Müller (1801–1858, German) completes his *Handbuch der Physiologie des Menschen* [Handbook of human physiology], publishing the second of two volumes, the first having been published in 1834.

CHEMISTRY (Hess's Law; Chemical Reactions)

Germain Henri Hess (1802–1850, Russian) formulates Hess's Law that the amount of heat generated or absorbed in a sequence of chemical reactions is determined by the starting and ending conditions of the sequence and is not dependent upon the number of intermediate steps. This is later recognized as a special case of the law of conservation of energy.

CHEMISTRY (Ozone)

Christian Schönbein (1799–1868, Swiss) discovers ozone.

CHEMISTRY (Structural Types; Nomenclature)

Jean Baptiste Dumas (1800–1884, French) advances a theory that the chemical properties of an organic compound are determined by its structure and not, as seems to be the case with inorganic compounds, by its electrical properties. He illustrates the concept with the example that the basic qualitative properties of acetic acid are retained even after replacing three fourths of the hydrogen by chlorine. As a consequence, he calls for a revision of chemical nomenclature for organic compounds, basing terms on common features, not on elemental composition.

EARTH SCIENCES (Geological Mapping)

A geological map of France, begun in 1825 and constructed under the supervision of Élie de Beaumont (1798–1874, French) and Pierre Armand Dufrénoy (1792–1857, French), is completed at a scale of 1:500,000. The map is based on the principles of George Greenough's map of England and Wales (1819—EARTH SCIENCES), which had led to the initiation of the French project in 1822 with de Beaumont and Dufrénoy spending time in England between 1822 and 1824 studying the English map and its construction. This was followed by nine years of field work in France, 1825–1834, collecting the necessary data for the construction of the French map. De Beaumont and Dufrénoy supplement the map with a text on the history, composition, and physical description of French geological features.

EARTH SCIENCES (Glaciology; Paleoclimatology)

Publication of Louis Agassiz's (1807–1873, Swiss-American) *Études sur les glaciers* [Studies on glaciers], an early description of glacial motion and deposits, presenting the results of Agassiz's many glacial expeditions and elaborating his theory, first presented in 1837, of a past ice age (1837—EARTH SCIENCES). Later in the year Agassiz announces in an address to the British Association for the Advancement of Science that he believes that the ice sheets not only covered northern Europe but northern North America and Asia as well. He also obtains two important converts by convincing William Buckland, who then convinces Charles Lyell, that various deposits in Britain are of glacial origin. All three make presentations supporting the glacial theory at a meeting of the London Geological Society.

HEALTH SCIENCES (Basedow's Disease)

Karl von Basedow (1799–1854, German) describes exophthalmic goiter, later termed Basedow's disease.

HEALTH SCIENCES (Germ Theory)

Jacob Henle (1809–1885, German) presents an early version of the germ theory of disease in his work *Pathologische Untersuchungen* [Pathological investigations], in which he argues that contagion is due to living matter which can reproduce, can be transmitted by contact or through the atmosphere, and can live as a parasite on the diseased body. He classifies diseases as contagious, miasmatic, and miasmatic-contagious, malaria, for instance, being miasmatic, syphilis and rabies being contagious, and smallpox and puerperal fever being miasmatic-contagious. To establish the specific cause for an individual disease he indicates the necessity to (1) show that the disease only occurs in conjunction with the suspected cause, (2) isolate the suspected cause, and (3) test the isolated suspected cause to see that it produces the disease.

c.1840

MATHEMATICS (Infinite Classes)

Bernard Bolzano (1781–1848, Czechoslovakian) distinguishes between denumerable and nondenumerable infinite classes, the first being capable of being placed in one-to-one correspondence with the integers, and indicates that the set of real numbers is indeed nondenumerable.

1840–1842

SUPPLEMENTAL (Knowledge)

In *The Positive Philosophy,* Auguste Comte (1798–1857, French) develops his thesis that each branch of knowledge passes through three stages: the theological, where supernatural beings are thought to produce all phenomena; the metaphysical, where natural phenomena are believed to result from fundamental energies; and the positive or scientific, where the search is not for absolutes but for a study, based on reason, observation, and experimentation, of the laws relating phenomena.

1841

BIOLOGY (Algae)

Publication of William Henry Harvey's (1811–1866, Irish) *A Manual of the British Algae.*

CHEMISTRY (Didymium)

Carl Mosander (1797–1858, Swedish) treats lanthanum with dilute nitric acid and extracts what he believes to be a new element, named didymium. Didymium is considered a pure earth until in 1885 it is decomposed by Auer von Welsbach.

CHEMISTRY (Uranium)

Uranium is isolated by Eugène Peligot (1811–1890, French).

EARTH SCIENCES (Earth's Shape)

Friedrich Bessel (1784–1846, German) calculates the ellipticity of the earth, obtaining a value of 1/299.

EARTH SCIENCES (Niagara Falls)

Charles Lyell (1797–1875, English) suggests that the drift forming Niagara Falls is a glacial deposit. After revising Robert Bakewell's estimate (1829—EARTH SCIENCES) of the recession rate of the Falls from 3 to 1 feet/year, he estimates that the ice left the deposit and began to retreat approximately 30,000 years ago.

EARTH SCIENCES (Paleoclimatology; Sea Level)

Charles Maclaren (1782–1866, Scottish), assuming Louis Agassiz's theory of a former ice age (1837—EARTH SCIENCES), with an ice sheet extending southward to 35°N and averaging one mile in thickness, approximates that sea level was 800 feet lower during the ice age than it is now.

HEALTH SCIENCES (Hypnotism)

James Braid (1795–1860, Scottish) revives mesmerism as a therapeutic technique and coins the term "hypnotism."

MATHEMATICS (Determinants; Jacobians)

Carl Jacobi (1804–1851, German) develops the theory of determinants, including mention of the "functional determinant" or "Jacobian" in his paper "Deformatione et proprietatibus determinantium." Later in the year he publishes a further development specifically of the theory of Jacobians in the memoir "De determinantibus functionalibus." He shows, for instance, that the Jacobian of a set of *n* functions in *n* variables is identically zero if and only if the functions are mutually dependent. Although functional determinants had been used earlier, they are later named after Jacobi due to his extensive development of their properties.

MATHEMATICS (Isoperimetry)

Jakob Steiner (1796–1863, Swiss-German) completes the two-part paper "Ueber Maximum und Minimum bei den Figuren in der Ebene, auf der Kugelfläche und im Raume überhaupt," in which he states and provides five separate geometric methods of proving the isoperimetric theorem that the circle is the planar figure of maximum area for a given perimeter, a theorem for which he had provided several proofs in the 1830s. The major gap in Steiner's proofs is that he assumes a priori that an area-maximizing curve exists. This gap is filled in the 1870s by Karl Weierstrass as he proves the existence of the maximizing curve. (In Steiner's fifth proof he rephrases the theorem to state that for a given area, the planar figure with the shortest perimeter is the circle).

METEOROLOGY (Storms; Adiabatic Lapse Rate)

James Espy (1785–1860, American) publishes a monograph on *The Philosophy of Storms.* Although the two concepts are not inherently contradictory, heated controversy occurs between the exponents of Espy's emphasis on vertical convection and condensation as the primary drivers of a storm and exponents of William Redfield's theory of the rotary motion of storms (1831—METEOROLOGY). Espy determines the important role that the release of latent heat (during condensation in clouds) has in providing energy for initiating and prolonging storms. Espy also, c.1841, invents and uses an instrument, the "nepheloscope," to simulate cloud behavior and to measure adiabatic lapse rates, under both wet (with condensation) and dry (without condensation) conditions, an adiabatic lapse rate being the rate of temperature decrease with altitude of an air parcel rising without gain or loss of heat.

PHYSICS (Joule's Law)

James Joule (1818–1889, English) asserts that the rate at which heat is evolved in a current propagated along a metallic conductor is proportional to the product of the resistance and the square of the intensity of the current. Later termed Joule's Law, it appears along with experimental evidence in the paper "On the Heat Evolved by Metallic Conductors of Electricity and in the Cells of a Battery During Electrolysis."

1842

CHEMISTRY (Agricultural)

John Lawes (1814–1900, English) develops superphosphate, the first commercial artificial fertilizer. Superphosphate is made by combining sulfuric acid with various phosphates to increase their solubility. Lawes uses animal bones as the phosphate source until 1847, then uses mineral phosphates.

EARTH SCIENCES (Coral Reefs)

Charles Darwin (1809–1882, English) publishes *The Structure and Distribution of Coral Reefs* (volume 1 of the *Geology of the Voyage of the Beagle*), in which he presents extensive evidence leading to the major conclusion that the three main types of coral reefs—fringing reefs, barrier reefs, and atolls—are related as successive stages of a single process based on the slow subsidence of portions of the earth's crust.

EARTH SCIENCES (Ice Ages)

Joseph Adhémar (1797–1862, French) postulates that the earth's ice ages have occurred regularly in response to the 22,000-year cycle of the precession of the equinoxes (1754—ASTRONOMY), providing thereby supposedly the first attempt at an astronomical explanation of the ice ages. Adhémar suggests that the key variable is the length of the winter and therefore that the Northern and Southern Hemisphere ice ages have occurred alternately each 11,000 years. During the warm periods he sees the possibility of catastrophic floods, and his book is entitled *Les révolutions de la mer, déluges périodiques* [Revolutions of the sea, periodic deluges].

EARTH SCIENCES (Mountains)

William Rogers (1805–1881, American) and Henry Rogers (1809–1866, American) describe their extensive investigations of the geology, structure, and origin of the Appalachian Mountains in *On the Physical Structure of the Appalachian Chain, as exemplifying the Laws which have regulated the Elevation of Mountain Chains Generally.* They attribute features of the wavelike structure of the mountains to transformations and movements within the presumed molten rock beneath the earth's crust.

HEALTH SCIENCES (Anesthesiology)

Crawford Long (1815–1878, American) uses sulfuric ether as an anesthetic while operating on a neck tumor. (However, this remains generally unknown until he publishes an account in 1849, three years after a public demonstration of the use of ether anesthesia by William Thomas Morton.)

MATHEMATICS (Isoperimetry)

Jakob Steiner (1796–1863, Swiss-German) proves that for any given perimeter, the triangle with the greatest area is the equilateral triangle.

PHYSICS (Conservation of Energy)

In "Remarks on the Forces of Inorganic Nature," Robert Mayer (1814–1878, German) publishes his ideas regarding the conservation of energy, initiated largely from his earlier speculations on the conversion of chemical to heat energy within the human body (1840—BIOLOGY). Mayer develops conversions among potential energy (termed "falling force"), motion and heat, and he calculates that raising a weight 365 feet increases its potential energy by an amount equivalent to raising the temperature of the same weight of water from 0° to 1°C. The paper does not receive much attention, but within the next few years other researchers, notably James Joule, Hermann von Helmholtz, and Ludwig Colding, independently postulate the principle of the conservation of energy and it then rapidly gains adherents.

PHYSICS (Doppler Effect)

Johann Christian Doppler (1803–1853, Austrian) determines that the frequency of a received sound wave depends upon the movement of the source, with recession of the source from the receiver decreasing the frequency and with approach toward the receiver increasing it. He thereby discovers and details the basic principles of the Doppler effect (the drop in pitch heard as a moving sound source passes a listener).

1843

ASTRONOMY (Earth's Orbital Variations)

After ten years of calculations based on Isaac Newton's law of gravitation (1687—ASTRONOMY) and the orbits of the seven known planets, Urbain Le-Verrier (1811–1877, French) develops formulas for calculating the changes in the earth's orbit and shows that the orbital eccentricity has varied between about 0 and 6% over the past 100,000 years.

ASTRONOMY (Sunspots)

Samuel Schwabe (1789–1875, German) discovers an 11-year periodicity in sunspots—the sunspot cycle. Though the data base at this point is limited to a 17-year tally of sunspot counts, the hypothesized 11-year periodicity is substantiated as the data record is extended.

BIOLOGY (Chorda Tympani)

In "On the Chorda Tympani," Claude Bernard (1813–1878, French) describes this small segment of the facial nerve and demonstrates its importance for taste and the secretion of saliva.

CHEMISTRY (Erbium)

Carl Mosander (1797–1858, Swedish) discovers the metal erbium.

CHEMISTRY (Formulae)

Charles Gerhardt (1816–1856, French) adjusts the writing of formulae for select chemical compounds so that, for instance, water becomes H_2O instead of the H_4O_2 used by Justus von Liebig (1803–1873, German) and others.

CHEMISTRY (Radicals)

Robert Bunsen (1811–1899, German) reports success in isolating a radical, specifically the free cacodyl radical, obtained as a product of a reaction between cacodyl chloride and zinc .

EARTH SCIENCES (Oceanic Abyssal Theory)

Edward Forbes (1815–1854, English) suggests that the cold and darkness in the deep ocean might prohibit the existence of marine life below a depth of about 300 fathoms. Although evidence to the contrary has already been reported (for example, 1818—EARTH SCIENCES), the "Abyssal" or "azoic" theory becomes widely accepted over the next decades.

EARTH SCIENCES (Paleozoic System)

Adam Sedgwick (1785–1873, English) proposes classifying all Paleozoic rocks into one system, with three major subdivisions, the first including the Cambrian and Silurian series, the second including the Devonian series, and the third including the Carboniferous and Permian series.

HEALTH SCIENCES (Clinical Medicine)

Thomas Watson (1792–1882, English) publishes a two-volume collection of lectures on clinical medicine, *Lectures on the Principles and Practice of Physic*. His suggestion in volume 2 that operating physicians wear rubber gloves is reputedly the first such suggestion.

HEALTH SCIENCES (Digestion)

Publication of Claude Bernard's (1813–1878, French) *On the Gastric Juice and its Function in Digestion,* a work previewing Bernard's later extensive comparative investigations into the processes involved in the digestion of fats versus carbohydrates versus proteins.

HEALTH SCIENCES (Puerperal Fever)

In a paper entitled "The Contagiousness of Puerperal Fever," Oliver Wendell Holmes (1809–1894, American) suggests that puerperal fever may be transmitted from patient to patient by attending physicians and encourages basic preventive measures against such transmission: after conducting postmortem sections or examining cases of puerperal fever, physicians should change their clothes and wash their hands in calcium chloride before visiting women in childbed. The paper excites strong opposition from Philadelphia physicians.

MATHEMATICS (*N*-Dimensional Geometry)

Arthur Cayley (1821–1895, English) invents *n*-dimensional geometry by applying the language of geometry to systems of equations in any number of variables.

MATHEMATICS (Quaternions)

Following years of attempts to develop a three-dimensional analogue to the complex numbers, William Rowan Hamilton (1805–1865, Irish) discovers that the commutative law is not required for a self-consistent algebra and creates a consistent noncommutative algebra, the algebra of quaternions. A quaternion can be viewed as a number of the form $a + bi + cj + dk$, with a, b, c, d reals, $i^2 = j^2 = k^2 = -1$, $ij = -ji = k$, $jk = -kj = i$, and $ki = -ik = j$. Hamilton divides the quaternions into a real or scalar part, a, and a complex or vector part, $bi + cj + dk$. He determines some of the chief properties of vectors while developing the theory of quaternions, but vector analysis itself is developed in more generality by Hermann Grassmann (1844—MATHEMATICS).

MATHEMATICS (Series)

Augustin Louis Cauchy (1789–1857, French) creates a theory of divergent series.

PHYSICS (Mechanical Equivalent of Heat)

James Joule (1818–1889, English) publishes his first determination of the mechanical equivalent of heat, obtaining it by comparing the heat generated by a magneto-electric machine under various conditions with the mechanical force required for turning the apparatus.

PHYSICS (Optics; Refraction)

In *Dioptrische Untersuchungen* [Investigations on the refraction of light], Carl Friedrich Gauss (1777–1855, German) extends the study of refraction to lenses which are no longer restricted to having negligible thickness. He shows geometrically that the same formulae used for single lenses with negligible thickness can be used for determining the refraction near the principal axis of any lens or system of lenses.

1843–1844

BIOLOGY (Evolution)

Samuel Haldeman (1812–1880, American) publishes a review of the current beliefs regarding whether or not species have developed and modified over time.

1844

ASTRONOMY (Binary Stars)

Friedrich Bessel (1784–1846, German) announces the postulated existence of unseen companion stars to both Sirius and Procyon based on periodic perturbations in their motions. In the case of Sirius, he calculates a 50-year period in the perturbations and an amplitude of 11 arc-seconds. (Actual telescopic sighting of these companion stars comes in 1862 for Sirius [1862—ASTRONOMY] and in 1896 for Procyon.)

ASTRONOMY (Solar Spectrum)

John Draper (1811–1882, American) makes the first successful daguerreotype picture of the sun's diffraction spectrum.

BIOLOGY (Chemical Decomposition of Plants)

Jean Baptiste Boussingault (1802–1887, French) and Jean Baptiste Dumas (1800–1884, French) show that plants decompose various compounds, reducing carbonic acid to carbon, water to hydrogen, ammonium hydroxide to ammonium, and nitric acid to nitrogen.

BIOLOGY (Evolution)

Robert Chambers (1802–1871, Scottish) advances a theory of biological evolution in his widely read but anonymous *Vestiges of the Natural History of Creation*. He expresses an Actualistic point of view, agreeing with the Uniformitarians that history should, as much as possible, be described by processes still operating, but disagreeing with their assumption that the earth exists in a near steady state. The book helps focus attention on basic issues regarding the evolution hypothesis.

BIOLOGY (Evolution)

Charles Darwin (1809–1882, English) writes a 231-page statement expressing his theory of natural selection as a mechanism for evolution. Over the next decade and a half he records numerous favorable and unfavorable pieces of evidence, before publishing his more complete statement in *The Origin of Species* (1859—BIOLOGY).

CHEMISTRY (Atomic Theory)

Michael Faraday (1791–1867, English) expresses his preference for the atomic concepts of Rudjer Bošković (1758—CHEMISTRY) over those of John Dalton (1803—CHEMISTRY), claiming to show in the paper "Speculation Touching Electrical Conduction and the Nature of Matter" that our observational knowledge of electrical conduction is compatible only with Boscovichean atoms.

CHEMISTRY (Ruthenium)

Ruthenium is discovered by Karl Klaus (1796–1864, Estonian).

EARTH SCIENCES (Upwelling)

Urbain de Tessan (1804–1879, French), in his report of the voyage of the *Vénus, Voyage autour du monde sur la frégate "La Vénus," pendant les années 1836–1839* [Voyage around the world on the frigate *La Vénus* during the years 1836–1839], mentions upwelling of deep water along the Peruvian coast and the resulting low temperatures of the coastal waters off South America.

EARTH SCIENCES (Volcanic Islands)

Charles Darwin (1809–1882, English) publishes an account of the volcanic islands studied along the route of the 1831–1836 voyage of the *Beagle.* Entitled *Geological Observations on the Volcanic Islands visited during the voyage of HMS Beagle, together with some brief notices on the geology of Australia and the Cape of Good Hope,* this is volume 2 of the *Geology of the Voyage of the Beagle,* volume 1 having been published in 1842 on coral reefs.

HEALTH SCIENCES (Anesthesiology)

Dentist Horace Wells (1815–1848, American) demonstrates the use of nitrous oxide as an anesthetic and begins using it in his dental practice, after having had it administered to himself successfully by colleague John Riggs while having a wisdom tooth removed.

MATHEMATICS (Algebra)

The theory of binary cubics is initiated with the work of Ferdinand Eisenstein (1823–1852, German), which includes additionally the first development of an algebraic covariant.

MATHEMATICS (Geometry of Numbers)

Ferdinand Eisenstein (1823–1852, German) helps initiate the theory of the geometry of numbers with his formula for the number of lattice points on and within a circle.

MATHEMATICS (Logical Operators)

George Boole (1815–1864, English) develops the calculus of operators.

MATHEMATICS (*N*-Dimensional Space; Vectors)

Hermann Grassmann (1809–1877, German), independently of Arthur Cayley (1843—MATHEMATICS), arrives at the notion of *n*-dimensional space. The Grassmann theory of "extended magnitude" (published as *Die lineale Ausdehnungslehre, ein neuer Zweig der Mathematik* [The theory of linear extension, a new branch of mathematics]) can be seen as a substantial generalization of quaternions, creating an algebra of vectors. Grassmann develops not only the concept of noncommutative multiplication, as had William Rowan Hamilton (1843—MATHEMATICS), but also the possibility of nonassociativity.

MATHEMATICS (Transcendental Numbers)

Joseph Liouville (1809–1882, French) invents a method for constructing a set of transcendental numbers, all of the form $(a_1/10) + (a_2/10^{2!}) + (a_3/10^{3!}) + \ldots$, where the a_i are single digit integers. These are the first numbers proved to be transcendental. Liouville also makes some limited progress toward proving e and e^2 to be transcendental by showing that neither of them is the root of any quadratic equation with integral coefficients, but the full proof that e cannot be the root of any polynomial equation with integral coefficients is not obtained until the work of Charles Hermite (1873—MATHEMATICS).

1844–1847

BIOLOGY (Antarctic Flora)

Joseph Hooker (1817–1911, English) publishes *Antarctic Flora.*

1845

ASTRONOMY (Asteroids)

The discoveries of Astrea and Hébé by Karl Hencke (1793–1866, German) mark the first discoveries of asteroids since the discovery of Vesta (1807—ASTRONOMY), the fourth to be discovered (the first three being Ceres, Pallas, and Juno, in 1801, 1802, and 1804 respectively). Within the next half century, however, over 300 asteroid discoveries are made.

ASTRONOMY (Neptune)

Careful calculations of the differences between the expected and observed motions of Uranus lead John Couch Adams (1819–1892, English) and Urbain LeVerrier (1811–1877, French) independently to postulate the existence and calculate the mass and orbit of an unknown planet perturbing the Uranus orbit. LeVerrier encourages the astronomer Johann Galle to search for such a planet, a search soon successfully completed with the discovery of Neptune (1846—ASTRONOMY).

ASTRONOMY (Spiral Nebulae)

William Parsons (Lord Rosse, 1800–1867, English-Irish) discovers the spiral structure of individual nebulae, using a 72-inch reflector telescope that he had constructed over the previous 18 years.

ASTRONOMY (Sunspots)

Jean Bernard Léon Foucault (1819–1868, French) and Armand Fizeau (1819–1896, French) photograph sunspots.

CHEMISTRY (Low-Temperature Chemistry)

Ethylene is liquefied, requiring a temperature of 169 K.

EARTH SCIENCES (Continental Drift)

George Windsor Earl reads a paper to the Royal Geographical Society presenting his thesis that Sumatra, Java, and Borneo were formerly separated from the Asiatic continent by only a very shallow sea, resulting in the wide similarity among their flora and fauna. Similarly, Earl asserts that the marsupials on New Guinea and Australia suggest that these two islands were also separated by a much shallower sea in the past than in the present.

EARTH SCIENCES (Ocean Tides)

George Airy (1801–1892, English) charts high water locations in the North Sea at various times of day.

MATHEMATICS (Bertrand's Postulate)

Joseph Bertrand (1822–1900, French) conjectures that for any integer $n > 3$, there exists a prime between n and $2n - 2$, inclusive. Known as Bertrand's postulate, this is later proved by Pafnuty Chebyshev (1850—MATHEMATICS).

MATHEMATICS (Invariance)

Arthur Cayley (1821–1895, English) generalizes aspects of the theory of determinants into a theory of algebraic invariants and completes the *Theory of Linear Transformations*.

METEOROLOGY (Isotherms; Continentality)

Alexander von Humboldt (1769–1859, German) constructs a world map of average annual temperatures, accumulating data from various stations around the world and using it to plot lines of equal temperature (isotherms). Humboldt notes the effect of large land masses and develops the concept of "continentality," whereby, due to the relative thermal inertia of the oceans, isotherms on seasonal maps depart markedly from latitude circles, with large continental areas being colder in winter and warmer in summer than ocean areas at the same latitude.

PHYSICS (Diamagnetism; Paramagnetism)

Michael Faraday (1791–1867, English) classifies materials as "paramagnetic" or "diamagnetic" depending on whether they align themselves in a magnetic field along or across the lines of magnetic force. The first, paramagnetic materials such as iron, are drawn into the more intense region of the field while the second, diamagnetic materials such as bismuth, move away from the intense magnetism.

PHYSICS (Fluid Flow; Navier-Stokes Equations)

George Stokes (1819–1903, Irish) independently derives the Navier-Stokes equations first derived by Claude Navier (1823—PHYSICS). Stokes accomplishes the derivation using a continuous fluid model rather than a molecular one.

PHYSICS (Light)

Michael Faraday (1791–1867, English) discovers that polarized light will be rotated by a longitudinal magnetic field.

c.1845

EARTH SCIENCES (Ice Ages)

Joshua Trimmer (1795–1857, English) finds two distinct layers of glacial till on an East Anglican cliff and concludes that, at least in Great Britain, there were at least two separate ice age glaciations.

1845–1850

MATHEMATICS (Actuarial Statistics)

Carl Friedrich Gauss (1777–1855, German) applies probability theory to the problem of determining pension plans for widows, deriving key methods for later actuarial practices.

1845–1862

SUPPLEMENTAL (Cosmos)

Alexander von Humboldt (1769–1859, German) publishes the popular, wide-ranging *Kosmos, Entwurf einer physischen Weltbeschreibung* [Cosmos, a sketch of a physical description of the universe], a five-volume work covering and attempting to unify numerous studies, including the astronomical laws of celestial space, geophysics, man's place in the universe, and the historical and current status of man's efforts to portray nature pictorially and to describe it scientifically.

1846

ASTRONOMY (Neptune)

Johann Galle (1812–1910, German), at the Berlin Observatory, locates the planet (Neptune) predicted independently by John Couch Adams and Urbain LeVerrier (1845—ASTRONOMY). The planet is only 52 seconds of arc from LeVerrier's predicted position for the date of the discovery (September 23, 1846).

BIOLOGY (Evolution)

Publication of a short paper by Jean Julien D'Omalius d'Halloy (1783–1875, Belgian) presenting the opinion that many species have probably developed through descent from others rather than through a separate creation.

BIOLOGY (Zoophytes; Corals)
EARTH SCIENCES (Subsidence)

James Dana (1813–1895, American) classifies and describes the physiology and ecology of 261 actinoid zoophytes and 483 coral zoophytes, many previously unknown, in *Zoophytes,* one of many formal expedition reports from a 1838–1842 around-the-world United States exploring expedition under the command of Charles Wilkes. While en route in 1839 Dana heard of Charles Darwin's theory of the interrelatedness of the three major coral structures and the importance of subsidence to their development (1842—EARTH SCIENCES) and proceeded, with his continuing observations, especially in the Fijis, to provide evidence for and refinement of the theory. He concludes from evidence of subsidence on 200 Pacific islands, that the subsidence basically increased from south to north or northeast as far as the Hawaiian Islands.

CHEMISTRY (Nitroglycerine)

Ascanio Sobrero (1812–1888, Italian) prepares the colorless, explosive liquid nitroglycerine.

EARTH SCIENCES (Denudation)

Andrew Ramsay (1814–1891, Scottish) provides stratigraphic evidence in the memoir "On the Denudation of South Wales and the Adjacent Counties of England" that these regions have experienced large-scale denudation. He also suggests that water is able to produce large-scale marine planation on landmasses which are stationary or subsiding, thereby complementing earlier suggestions of extensive dissective marine action on rising landmasses.

EARTH SCIENCES (Earthquakes)

Robert Mallet (1810–1881, Irish) publishes a paper "On the Dynamics of Earthquakes" in which he attempts to reduce the enormous amount of scattered earthquake-related observations to a theory of earthquake motion based on the laws of mechanics. He defines earthquake motion as "the transit of a wave of elastic compression . . . through the surface and crust of the earth. . . ."

EARTH SCIENCES (Paleoclimatology; Pleistocene)

Edward Forbes (1815–1854, English) suggests the term "Pleistocene"—earlier used for the Newer Pliocene—for the post-Pliocene period, and speculates that this is the period of deposition of the deposits recently identified as glacial drift, following the ice age theory of Louis Agassiz and others (1837—EARTH SCIENCES). (Charles Lyell in 1839 had proposed the term "Pleistocene" as a synonym for the Newer Pliocene, but after Forbes it gains general usage with his suggested post-Pliocene meaning.)

EARTH SCIENCES (Volcanism)

James Dana (1813–1895, American) uses the low fusibility of feldspar and the general vertical circulatory motion of a boiling fluid, with the fluid rising in the center and descending on the sides, to explain the tendency in many volcanic regions to have solid feldspathic rocks in the center surrounded by basaltic lavas.

HEALTH SCIENCES (Anesthesiology)

Dentist William Thomas Morton (1819–1868, American) demonstrates the use of ether as an anesthetic, first while removing an ulcerated tooth and later while assisting John Warren (1778–1856, American) at an operation to remove a neck tumor.

MATHEMATICS (Galois Theory)

Joseph Liouville (1809–1882, French) accomplishes the difficult task of penetrating and editing some of the manuscripts left by Evariste Galois (1830–1832—MATHEMATICS) and publishes them in Liouville's *Journal* (1836—MATHEMATICS), thereby bringing them to the attention of the mathematical community.

MATHEMATICS (Ideal Numbers; Prime Factorization)

Ernst Kummer (1810–1893, German) extends the notion of "prime" numbers in the integers to "ideal" numbers in general algebraic domains of rationality, preserving the existence of a unique factorization into prime factors, the prime factors in the more general domains being the ideal numbers.

METEOROLOGY (Gulf Stream Effect)
EARTH SCIENCES (Gulf Stream)

Edward Sabine (1788–1883, English) reiterates his earlier suggestion on the impact of the Gulf Stream on European weather (1825—METEOROLOGY) in the

The like Eclipse having not for many Ages been seen in the Southern Parts of Great Britain, I thought it not improper to give the Publick an Account thereof, that the suddain darkness, wherein the Starrs will be visible about the Sun, may give no Surprize to the People, who would, if unadvertized, be apt to look upon it as Ominous, and to Interpret it as portending evill to our Sovereign Lord King George and his Government, which God preserve. Hereby they will see that there is nothing in it more than Natural, and nomore than the necessary result of the Motions of the Sun and Moon; And how well these are understood will appear by this Eclipse.

According to what has been formerly Observed, compared wth our best Tables, we conclude ye Center of ye Moon's shade will be very near ye Lizard point, when it is about 5 min: past Nine at London; and that from thence in Eleven minutes of time, it will traverse ye whole Kingdom, passing by Plymouth, Bristol, Glocester, Daventry, Peterborough & Boston, near wch it will leave ye Island: On each side of ye Tract for about 75 Miles, the Sun will be Totally darkned, but for less & less Time, as you are nearer those limits, wch are represented in ye Scheme, passing on ye one side near Chester, Leeds, and York; and on ye other by Chichester, Gravesend, and Harwich.

At London we Compute the Middle to fall at 13 min: past 9 in ye Morning, when tis dubious whether it will be a Total Eclips or no, London being so near ye Southern limit. The first beginning will be there at 7 min: past Eight, and ye end at 24 min: past Ten. The Ovall figure shews ye space ye Shadow will take up at ye time of the Middle at London; And its Center will pass on to ye Eastwards with a Velocity of nearly 30 Geographical Miles in a min: of Time.

NB. The Curious are desired to Observe it, and especially the duration of Total Darkness, with all the care they can; for therby the Situation and dimensions of the Shadow will be nicely determin'd; and by means therof we may be enabled to Predict the like Appearances for ye future, to a greater degree of certainty than can be pretended to at present, for want of such Observations.

By their humble Servant Edmund Halley.

Sold by I. Senex, at the Globe in Salisbury Court, near Fleetstreet, who makes, and sells ye Newest and Correctest Maps, and Globes of 3, 9, 12, and 16 Inches Diameter, at moderate Prices

Sold also by William Taylor at the Ship in Paternoster Row — Entered in the Hall Book paper "On the Cause of Remarkably Mild Winters Which Occasionally Occur in England," pointing out particularly the mild 1845–1846 winter. Sabine suggests that the reason the Gulf Stream (1770—EARTH SCIENCES) extends farther in some years than in others is its variable initial velocity at the Florida Straits, with higher velocities, and a consequent more extended Gulf Stream, produced by more intense trade winds driving more water into the Gulf of Mexico, this water causing an unusually large difference in sea level between the Gulf and the Atlantic and thus leading to a stronger outflow through the Florida Straits. He thus indicates a correlation between the Atlantic trade winds in summer and European weather the following winter, and relates oceanic anomalies in one area (the Gulf of Mexico) to subsequent atmospheric anomalies elsewhere (Europe). Several decades later, the search for similar relationships intensifies.

PHYSICS (Electrodynamometer)

Wilhelm Weber (1804–1891, German) introduces the electrodynamometer for measuring currents, particularly alternating currents. The instrument contains a fixed coil suspended in the field of a movable coil with the strength of the forces between the currents in the two coils reflecting the current amount.

PHYSICS (Light; Ether)

Michael Faraday (1791–1867, English) in "Thoughts on Ray Vibrations" suggests that no ether or other vibrating medium is necessary for the transmission of light, because instead a line of particles could transmit energy by vibrating transversely to the direction of a strain placed on that line.

1846–1848

CHEMISTRY(Molecular Structure)

While studying the crystallization of tartrates, Louis Pasteur (1822–1895, French) discovers molecular asymmetry and demonstrates the existence of optical isomers (1823—CHEMISTRY). This work is among the earliest dealing with the three-dimensional structure of molecules and is often felt to mark the origin of stereochemistry. Pasteur presents pairs of mirror-imaged crystals, these deriving from optically isomeric salts.

FIGURE 13. *The path of the moon's shadow across England as predicted by Edmond Halley for the total solar eclipse of April 22, 1715 (see 1715—*ASTRONOMY *entry). The path as observed on the day of the eclipse followed the predicted path almost precisely. (By permission of the Houghton Library, Harvard University.)*

1846–1864

BIOLOGY (Ferns)

Publication of William Hooker's (1785–1865, English) five-volume *Species filicum* on ferns, extending his earlier *Genera filicum* (1838—BIOLOGY).

1847

BIOLOGY (Gorilla)

Thomas Savage (1804–1880, American) and Jeffries Wyman (1814–1874, American) announce, in the article "Notice of the External Characters and Habits of *Troglodytes gorilla,* a New Species of Orang from the Gaboon River," a second species of African orang, naming it the "gorilla." Distinguishing the gorilla from the other African orang, the *Troglodytes niger,* they describe the animal's physical appearance, motions, habits, dwellings, and bone structure.

EARTH SCIENCES (Ice Ages)

Edouard Collomb (1801–1875, French) reports the existence, in the French Vosges mountains, of two layers of glacial till separated by stream deposits. This suggests some glacier retreat, although conceivably minor, between two periods of greater glacial extent.

EARTH SCIENCES (Mountain Ranges)

James Dana (1813–1895, American) suggests that many mountain ranges, as well as continents, were formed not by a force from below causing an irruption of igneous matter, but by contraction resulting from the cooling of the earth. He specifies the Appalachian and Rocky Mountain ranges as two arising from effects of contraction.

EARTH SCIENCES (Wind and Current Charts)

Matthew Maury (1806–1873, American) publishes the first of his *Wind and Current Charts* for the world oceans, this initial set dealing with the North Atlantic.

HEALTH SCIENCES (Anesthesiology: Chloroform)

James Simpson (1811–1870, Scottish) introduces chloroform as an anesthetic and administers it by dropping the chloroform onto a towel or handkerchief. John Snow (1813–1858, English) quickly adopts usage of chloroform, although rejects Simpson's method of administering it, instead inventing apparatus for administering exact percentages of chloroform in air. The two methods remain in use for the remainder of the nineteenth century, with advocates on each side denouncing the alternate method.

HEALTH SCIENCES (Anesthesiology: Ether)

John Snow (1813–1858, English), premier English anesthetist, publishes *On Ether* describing the properties of the drug, the apparatus invented by Snow for administering it, and various practical considerations regarding its safe administration. He divides anesthesia into five stages.

HEALTH SCIENCES (Puerperal Fever)

Ignaz Semmelweis (1818–1865, Hungarian) determines that puerperal fever (childbed fever) is brought to the woman in labor by the hands and instruments of examining physicians and can be eliminated through a thorough cleansing, in a solution of water with chloride of lime, of the hands, instruments, and other items brought in contact with the patient.

MATHEMATICS (Convergence)

George Stokes (1819–1903, English) creates the concept of uniform convergence of a series, a concept independently recreated in 1848 by Philipp von Seidel (1821–1896, German).

MATHEMATICS (Complex Numbers)

Augustin Louis Cauchy (1789–1857, French) objects to the use of complex or imaginary numbers and finds a method of eliminating i (the square root of negative 1) by constructing residues to the modulus $x^2 + 1$. These residues have the formal properties of the complex number system, with x replacing i.

MATHEMATICS (Logic)

Publication of Augustus De Morgan's (1806–1871, English) *Formal Logic.* De Morgan introduces the concept that all types of relations must be dealt with in the study of logic, not only the "to be" relations of Aristotelian logic.

MATHEMATICS (Logic; Boolean Algebra)

Publication of George Boole's (1815–1864, English) *The Mathematical Analysis of Logic,* where, for the first time, algebraic operations are systematically applied to logic. This initiates the study of Boolean algebra.

MATHEMATICS (Number Theory)

Ferdinand Eisenstein (1823–1852, German) determines arithmetically the number of ways an integer can be represented as a sum of six or eight squares.

MATHEMATICS (Projective Geometry)

Karl Christian von Staudt (1798–1867, German) attempts to eliminate length and congruence, magnitude and number from projective geometry in his *Geometrie der Lage* [Geometry of position].

MATHEMATICS (Topology)

Although some research in topology can be traced back to René Descartes (1639—MATHEMATICS), the systematic study of topology as a branch of geometry is said to begin with the publication of Johann Listing's (1806–1882, German) *Studies in Topology.*

PHYSICS (Conservation of Energy)

Hermann von Helmholtz (1821–1894, German) publishes the paper "Ueber die Erhaltung der Kraft" [On the conservation of energy], in which he applies the principle of the conservation of energy to various physical problems and mathematically discusses the equivalence of that principle to the assumption that the natural forces between points act along the lines connecting the points and have intensities dependent solely on the distances between the points.

PHYSICS (Conservation of Energy)

James Joule (1818–1889, English) lectures on the conservation of energy and asserts the essential equivalence and interconvertibility of heat, mechanical, electrical, and chemical energy. His ideas derived initially from studies of the heat produced by an electric current, but expanded to include additional forms of energy. Joule asserts that the "living force" (kinetic energy) of a volume can be converted from or to heat but cannot be destroyed. A given quantity of heat or of mechanical energy always produces the same amount of "living force." Joule presents the numerical conversion factors for mechanical to kinetic to heat energy, obtained from his experiments with an electromagnetic engine. Arranging a set of falling weights to produce a stirring of paddles in a container of water, Joule determines the amount of mechanical energy required to produce a given amount of heat energy, as calculated from the temperature increase of the water. He thereby determines the mechanical equivalent of heat: a 772-lb weight falling 1 foot raises the temperature of 1 lb of water by 1°F. Further experimentation leads him to conclude that a given expenditure of work of any kind can always be arranged to produce a set quantity of heat energy. In this way he establishes that heat is a form of energy.

1847–1894

EARTH SCIENCES (Paleontology)

James Hall (1811–1898, American) publishes the 13-volume *Paleontology of New York* on the Silurian and Devonian periods of the Paleozoic era in the state of New York.

1848

ASTRONOMY (Doppler-Fizeau Effect)

Armand Fizeau (1819–1896, French) adapts the Doppler Effect, originally discovered for sound waves (1842—PHYSICS), to light, finding that light waves exhibit a similar effect, with the light from a receding source containing a red shift in its spectrum. Termed the Doppler-Fizeau Effect, this becomes in the 1920s the basis for the theory of an expanding universe (ASTRONOMY entries for 1928 and 1929).

ASTRONOMY (Saturn; Hyperion)

George Bond (1825–1865, American) and William Lassell (1799–1880, English) discover Hyperion, the eighth satellite of Saturn to be discovered. Hyperion follows an eccentric orbit between the satellites Titan and Iapetus.

ASTRONOMY (Saturn's Rings; Roche Limit)

Édouard Roche (1820–1883, French) suggests that Saturn's rings are composed of millions of pieces created by tidal disruption of a liquid moon, based on calculations of how close a satellite could be to a planet the size of Saturn without being tidally disrupted. The distance he calculates (roughly 1.5 Saturn radii) for the outer limit of the region where the tidal forces would exceed the internal gravitation of a liquid moon approximates the distance from Saturn of the outer edge of the main Saturn rings. This limit is later termed the Roche limit.

ASTRONOMY (Solar Energy)

Robert Mayer (1814–1878, German) hypothesizes that meteorites and asteroids falling into the sun provide the mechanism for maintaining the sun's heat.

BIOLOGY (Archetype Vertebrate)

Richard Owen (1804–1892, English), in *On the Archetype and Homologies of the Vertebrate Skeleton*, describes the supposed archetype vertebrate, consisting of a connected series of similar vertebra which in different animals are modified to form skulls, tails, and other body parts.

CHEMISTRY (Crystals)
MATHEMATICS (Group Theory)

Auguste Bravais (1811–1863, French) begins examination of the rotations and translations of crystals into themselves, in the process advancing the studies both of crystalline structure and of group theory.

HEALTH SCIENCES (Peritonitis; Appendix)

Henry Hancock (1809–1880, English) reports a successful operation for peritonitis in "Disease of the Appendix Caeci Cured by Operation."

1849

CHEMISTRY (Amines)

Charles Wurtz (1817–1884, French) discovers methyl amine and ethyl amine, two primary amines formed from ammonia, by replacing hydrogen by methyl or ethyl respectively. August von Hofmann (1818–1892, German) then prepares many additional amines and examines more fully their relationship to ammonia.

HEALTH SCIENCES (Cholera)

John Snow (1813–1858, English) publishes the 31-page pamphlet *On the Mode of Communication of Cholera,* in which he rejects earlier miasma theories of cholera transmission and asserts instead that cholera is transmitted by a parasitic microorganism in polluted drinking water. A much enlarged (162 pages) second edition published in 1855 includes discussions of 1832, 1849, and 1854 London epidemics, with tabulated data on cholera deaths by district and water supply, and street maps showing the locations of crucial water pumps (see also 1854—HEALTH SCIENCES).

MATHEMATICS (Geometry; Solomon's Seal)

Arthur Cayley (1821–1895, English) and George Salmon (1819–1904, Irish) show that every three-dimensional cubic surface contains exactly 27 straight lines, real or imaginary (referred to as "Solomon's seal").

MATHEMATICS (Number Theory)

While trying to prove Fermat's Last Theorem (1637—MATHEMATICS), Ernst Kummer (1810–1893, German) factors $x^p + y^p$ and extends the theory of Gaussian complex numbers (1831—MATHEMATICS).

MATHEMATICS (Number Theory)

Charles Hermite (1822–1901, French) initiates the arithmetical theory of bilinear forms.

PHYSICS (Heat)

James Thomson (1822–1892, English) uses Sadi Carnot's theory of heat (1824—PHYSICS) to theoretically predict how much the melting point of ice should be lowered by pressure. William Thomson (1824–1907, English) then verifies the predicted values experimentally. This partial verification of the Carnot theory delays temporarily the shift toward the theory of James Joule (1847—PHYSICS), involving the conservation of energy, in which heat is converted to other energy forms.

PHYSICS (Light)

George Stokes's (1819–1903, English) *Dynamical Theory of Diffraction* adds further support to the wave theory of light.

PHYSICS (Speed of Light)

Armand Fizeau (1819–1896, French) measures the speed of light in air, later complementing this by the measurement of the somewhat slower speed in water.

1850

ASTRONOMY (Saturn's Rings)

William Bond (1789–1859, American) and George Bond (1825–1865, American) discover a ring of Saturn inward of the B ring. The near transparency of the new ring lessens support for the theory that the rings are each continuous solid bodies.

CHEMISTRY (Hydrolysis)

Ludwig Wilhelmy (1812–1864, German) presents one of the earliest mathematical equations describing a chemical process, doing so for the hydrolysis of cane sugar and obtaining $dz/dt = -kz$ for the change in sugar concentration z with respect to time t, k being a constant.

MATHEMATICS (Bertrand's Postulate)

Pafnuty Chebyshev (1821–1894, Russian) proves Joseph Bertrand's postulate (1845—MATHEMATICS) that for any integer $n > 3$ there exists a prime between n and $2n - 2$.

MATHEMATICS (Continued Fractions)

Carl Jacobi (1804–1851, German) generalizes continued fractions, using simultaneous difference equations.

MATHEMATICS (Infinite Classes)

In the posthumously published *Paradoxien des Unendlichen* [Paradoxes of the infinite], where he discusses properties of infinite sets and distinguishes between actual and potential infinites, Bernard Bolzano (1781–1848, Czechoslovakian) defines a class to be “infinite” if it is equivalent to a subset of itself and to be “finite” otherwise.

MATHEMATICS (Number Theory)

Ferdinand Eisenstein (1823–1852, German) presents criteria for the solvability of binomial congruences $x^n \equiv r \bmod m$ for n prime.

MATHEMATICS (Number Theory)

Pafnuty Chebyshev (1821–1894, Russian) proves that the number of primes less than or equal to n, for any integer $n > 1$, is at least $7n/(8 \log n)$ and no greater than $9n/(8 \log n)$. (See 1896—MATHEMATICS, prime number theorem.)

METEOROLOGY (Global Surface Winds)

Matthew Maury (1806–1873, American) draws a generalized plot of surface winds over the globe, showing a region of calm at the equator, later known as the doldrums, northeast trades north of the doldrums to about 30°N and southeast trades south of the doldrums to about 30°S, a belt of mid-latitude calms in each hemisphere, later known as Horse Latitudes, prevailing westerlies in the high mid-latitudes poleward of the mid-latitude calms, and polar calms at the highest latitudes. The mid-latitude westerlies are depicted as southwesterly in the Northern Hemisphere and northwesterly in the Southern Hemisphere.

PHYSICS (Heat; Second Law of Thermodynamics)

Rudolf Clausius (1822–1888, German) partially reconciles the research on the theory of heat done by Sadi Carnot (1824—PHYSICS) with that done by James Joule (1847—PHYSICS), showing that Carnot’s basic experimental results are not inconsistent with Joule’s interpretation that heat is destroyed and converted into other energy forms. In the paper “Ueber die bewegende Kraft der Wärme,” he follows Joule in rejecting the premise, basic to the earlier caloric theory, that heat is conserved, and he begins to develop the subject of thermodynamics, basing it on two central laws: the law of conservation of energy (the first law of thermodynamics), and the law that, in the absence of external constraints, the net flow of heat between two bodies will be from the warmer to the cooler one (the second law of thermodynamics). The latter, formulated by Clausius at this time, is later restated to assert that the entropy of a system never decreases.

PHYSICS (Stokes’ Law)

George Stokes (1819–1903, English) publishes results of investigations of the motion of pendula through viscous media and based on that work formulates Stokes’ Law on the terminal velocity of objects falling through fluids.

1851

ASTRONOMY (Solar Corona)

August Busch (1804–1855, German) observes a total eclipse of the sun and photographs the sun's corona.

BIOLOGY (Evolution)

Within a paper on the pathology of inflammation, a paper little read outside the medical field, Henry Freke postulates that all organic beings have descended from the same primordial form.

CHEMISTRY (Crystallography)

Publication of Auguste Bravais's (1811–1863, French) *Études cristallographiques,* where he applies group theory to the study of the structure of crystals.

CHEMISTRY/PHYSICS (Kinetic Theory of Gases)

James Joule (1818–1889, English) contributes to the emerging kinetic theory of gas pressure with his calculation of the velocity of gaseous hydrogen molecules in the paper "Some Remarks on Heat and the Constitution of Elastic Fluids." He calculates a molecular velocity of 6,225 feet per second for hydrogen gas at a temperature of 60°F and a barometric pressure of 30 inches.

EARTH SCIENCES (Earth's Rotation; Foucault Pendulum)

Jean Bernard Léon Foucault (1819–1868, French) constructs the Foucault pendulum and with it demonstrates experimentally that the earth rotates. The pendulum consists of a 62-lb iron ball attached at the end of a 200-ft wire hung from the ceiling of the Paris Pantheon and allowed to swing freely. Relative to the earth, the pendulum's plane of oscillation is clockwise, implying—from Newton's result that a freely swinging pendulum has a fixed plane of oscillation—that the floor of the room is turning counterclockwise, in response to the rotation of the earth.

HEALTH SCIENCES (Ophthalmoscope)

Hermann von Helmholtz (1821–1894, German) invents the ophthalmoscope.

MATHEMATICS (Cauchy-Riemann Equations; Riemann Surface)

Georg Riemann (1826–1866, German) creates a theory of functions of a complex variable $u + iv = f(x + iy)$, with x and y reals, in his doctoral dissertation *Grundlagen für eine allegemeine Theorie der Functionen einer veränderlichen complexen Grosse* [Foundations for a general theory of functions of a complex variable]. In clarifying his usage of the term "complex function," he introduces what are to become known as the Cauchy-Riemann equations: $\partial u/\partial x = \partial v/\partial y$ and $\partial u/\partial y = -\partial v/\partial x$. He establishes the existence of a function with which any simply connected region in the real plane can be transformed into any such region in the complex plane and vice versa, and he introduces the concept of the n-sheeted Riemann surface, in which n planes are connected so as to create a single-valued aspect to a complex function which had been n-valued in the normal Gaussian plane. He thereby inserts topological considerations into the study of complex functions.

PHYSICS (Ether)

Armand Fizeau (1819–1896, French) carries out an experiment to show the ether-drag effect, concluding from the results that if the ether exists, then its motion is not affected by the moving objects within it.

PHYSICS (Temperature Scales)

In "On the Dynamical Theory of Heat" (1851—PHYSICS), William Thomson proposes the Kelvin or Absolute temperature scale, with degree intervals equivalent to those on the Centigrade scale but with the zero point at −273.7°C [later −273.15°C], the temperature at which, theoretically, from the classical point of view, gases will have contracted to zero volume and all motion will have ceased. Thomson is led to the concept of an absolute temperature scale by consideration of Sadi Carnot's reversible heat engine (1824—PHYSICS), which, when operated between two isotherms and two adiabats, provides the basis for the definition of an absolute scale.

PHYSICS (Thermodynamics)

In a presentation entitled "On the Dynamical Theory of Heat," William Thomson (1824–1907, English) further develops the emerging study of thermodynamics. Thomson emphasizes that although energy is conserved (the first law of thermodynamics), there also tends to be a dissipation of the amount of energy actually available to do organized work (second law of thermodynamics).

1852

BIOLOGY (Evolution)

In the lengthy essay "A Theory of Population, deduced from the general law of animal fertility," Herbert Spencer (1820–1903, English) contrasts the Biblical creation theory with the theory that organic beings have developed through the ages, arguing that species have altered over the centuries because of changing environmental circumstances. Anticipating aspects of the evolutionary theory later developed by Charles Darwin (1859—BIOLOGY), Spencer extends Thomas Malthus's concept of population pressure (1798—BIOLOGY) to animal societies in general, concludes that species of necessity engage in a struggle for survival, and coins the phrase "survival of the fittest."

CHEMISTRY (Joule-Thomson Effect)

James Joule (1818–1889, English) and William Thomson (1824–1907, English) show that a gas expanding into a vacuum without the addition of external work indeed does undergo a change in temperature, in spite of theoretical speculations that it would not do so. The temperature change occurs due to the internal work required to overcome the attractive forces between molecules. For all gases except hydrogen the temperature decreases during the expansion. This effect is subsequently termed the Joule-Thomson effect.

CHEMISTRY (Valency)

The concept of valency emerges with the work of Edward Frankland (1825–1899, English) on metallo-organic compounds. In the paper "On a New Series of Organic Bodies containing Metals," Frankland points out that atoms have limited capacities for forming compounds and specifies, for instance, that in combining with organic radicals, one atom of nitrogen, phosphorus, arsenic, or antimony will combine with three or five radicals, whereas one atom of zinc, mercury, or oxygen will combine with two radicals. Frankland refers to the "combining power" of the elements rather than the "valency," a term which is coined later.

EARTH SCIENCES (Ice Ages)

Alexander von Humboldt (1769–1859, German) criticizes Joseph Adhémar's theory that the earth's ice ages have occurred in response to the precession of the equinoxes (1842—EARTH SCIENCES). Humboldt claims that Adhémar erred in basing his theory on the number of hours of daylight at a given location, when, Humboldt believes, the crucial variable is instead the annual amount of received solar radiation, a variable which is not altered by the precession cycle according to d'Alembert's precessional calculations (1754—ASTRONOMY).

EARTH SCIENCES (Lake Bonneville)

Howard Stansbury (American) infers from water marks in the flat lands around the Great Salt Lake in Utah that at some time in the past there was a very much larger lake (or sea) in this region, extending hundreds of miles.

EARTH SCIENCES (Mountain Formation)

Élie de Beaumont (1798–1874, French) further elaborates, in *Notice sur les systèmes de montagnes,* his theory of mountain formation based on the upheaval of mountains due to the diminishing volume of the earth (see also 1830—EARTH SCIENCES), the latter being perceived as arising from the slow dissipation of the heat contained in the earth when its crust was in a state of fusion. Mountain chains arise in the regions of the crust undergoing transverse compression. Believing that all mountain ranges of the same age belong to a single system and generally run parallel to each other, he identifies 21 such systems and a pentagonal network of their intersections.

MATHEMATICS (Four-Color Problem)

Francis Guthrie (d.1899) conjectures that four colors will suffice to color any planar map in a manner such that no two countries with a common boundary of positive length are colored identically. (This conjecture attracts numerous unsuccessful attempts at proofs over the succeeding century.)

MATHEMATICS (Topology)

L. Schlafli (Swiss) generalizes Leonhard Euler's formula relating the number of vertices, faces, and edges in two-dimensional polyhedra, $N_0 - N_1 + N_2 = 2$ (1752—MATHEMATICS), to n-dimensional space.

PHYSICS (Gyroscope)

Jean Bernard Léon Foucault (1819–1868, French) invents the gyroscope.

1852–1855

EARTH SCIENCES (Ice Ages)

Robert Chambers (1802–1871, Scottish) supports Louis Agassiz's conception of a widespread former glaciation (EARTH SCIENCES entries for 1837 and 1840) with several papers presenting evidence from superficial deposits and striated rocks of the former glaciation of Scotland.

1853

ASTRONOMY (Solar Energy; Solar Age)

Hermann von Helmholtz (1821–1894, German) postulates that the sun's energy is generated from a gravitational collapse of its mass, and calculates that this requires, for the known solar energy output, shrinkage of only a few hundred feet a year. He estimates the age of the sun at 20–30 million years and the expected remaining life span at 10 million years.

CHEMISTRY (Spectroscopy)

Anders Angström (1814–1874, Swedish) shows that the rays emitted by an incandescent gas have the same refrangibility as the rays absorbed by the same gas.

EARTH SCIENCES (Observational Oceanography)

Largely on the initiative of Matthew Maury (1806–1873, American), an international conference is convened at Brussels, where it is decided that every maritime vessel should collect data on winds, currents, temperatures, and other atmospheric and oceanographic parameters at regular intervals. The data from the resulting continuing international cooperation forms the basis for much of the increased oceanographic knowledge obtained over the succeeding century.

EARTH SCIENCES (Slate Cleavage)

Henry Sorby (1826–1908, English) seeks to determine the origin of slate cleavage by microscopically examining slate rocks and other rocks of similar composition but not possessing cleavage. He concludes that the particles composing slate rocks have been rearranged, while the rocks have undergone significant changes in their dimensions, with the major cleavage lines parallel to the direction of greatest elongation and perpendicular to the direction of greatest compression.

HEALTH SCIENCES (Anesthesiology)

The use of anesthesia gains considerable respectability when John Snow (1813–1858, English) administers chloroform to Queen Victoria during the birth of Prince Leopold, although in some communities substantial religious and medical prejudice against anesthesiology remains for decades.

MATHEMATICS (Algebra)

Leopold Kronecker (1823–1891, German) constructs transcendental solutions to the general cubic and quartic equations.

MATHEMATICS (Computers)

Pehr Scheutz (1785–1873, Swedish), aided by Charles Babbage (1792–1871, English) and funds from the Swedish Academy, completes an improved version of his Difference Engine. In contrast to the earlier, smaller version (1834—MATHEMATICS), this version leads to honors, medals, and monetary awards for Scheutz.

MATHEMATICS (π)

William Shanks (1812–1882, English) calculates π to 707 decimal places.

MATHEMATICS (Quaternions)

In *Lectures on Quaternions,* William Rowan Hamilton (1805–1865, Irish) extends his use of number pairs for the complex numbers (1837—MATHEMATICS) to four dimensions and develops his theory of quaternions (1843—MATHEMATICS), including extensive applications to geometry and spherical trigonometry.

1853–1861

ASTRONOMY (Sun's Rotation; Sunspots)

Richard Carrington (1826–1875, English) examines the sun's rotation through a series of sunspot observations. He notices the faster movement of sunspots in the lower solar latitudes and concludes that the photosphere undergoes a systematic drift.

1854

ASTRONOMY (Cosmology: Heat Death)

Hermann von Helmholtz (1821–1894, German) uses the second law of thermodynamics (1850—PHYSICS) to postulate the eventual "heat death" of the universe, after the continual passage of heat from warm to cool bodies has finally resulted in a universally uniform temperature.

BIOLOGY (Plant Constitution)

Alexander Braun (1805–1877, German) discovers that zinc occurs naturally in plants.

CHEMISTRY (Second Law of Thermodynamics)

Rudolf Clausius (1822–1888, German) restates the second law of thermodynamics (see 1850—PHYSICS), asserting that in any irreversible change the entropy of the system will increase.

CHEMISTRY (Silicon; Aluminum)

Henri Deville (1818–1881, French) creates the first crystalline silicon and perfects an electrolytic process for obtaining metallic aluminum from sodium aluminum chloride.

CHEMISTRY (Spectroscopy)

David Alter (1807–1881, American) concludes from experiment that each element has a characteristic spectrum.

EARTH SCIENCES (Ocean Bathymetry)

Matthew Maury (1806–1873, American) creates apparently the first bathymetric contour map of a large oceanic region, plotting the bottom topography of the North Atlantic at 1,000 fathom intervals on his "Bathymetrical Map of the North Atlantic Basin".

HEALTH SCIENCES (Cholera)

Through quantitative investigation of the current London cholera epidemic, John Snow (1813–1858, English) verifies his earlier thesis (1849—HEALTH SCIENCES) that the disease is spread through the water supply rather than the air as commonly believed. He finds a fatality rate several times higher in regions supplied by water from the fecal-contaminated portion of the Thames River than in regions supplied by water from upstream. The particularly polluted water from the Broad Street pump resulted in 500 deaths within a few hundred yards of this pump over a 10-day period. Snow argues that cholera is spread by a living, water-borne cell or germ and that fatalities could be reduced by boiling water and decontaminating soiled linen. He publishes his analysis in the 1854 paper "On the Communication of Cholera by Impure Thames Water" and in the 1855 enlarged edition of his earlier pamphlet *On the Mode of Communication of Cholera* (1849—HEALTH SCIENCES).

MATHEMATICS (Analysis)

Georg Riemann (1826–1866, German) creates a function of one variable which is continuous at all irrational values of that variable but discontinuous at all rational values.

MATHEMATICS (Calculus)

George Stokes (1819–1903, English) reduces selected surface integrals to line integrals.

MATHEMATICS (Calculus; Riemann Integral)

Georg Riemann (1826–1866, German) in his *Habilitationsschrift* (or probationary essay) for becoming a Privatdozent, examines necessary and sufficient conditions for the expansion of a function in a trigonometric or Fourier series, and in doing so redefines the definite integral so that it no longer requires continuity as had an earlier definition by Augustin Louis Cauchy (1821—MATHEMATICS). If lower and upper limits (as the length of the largest x-axis division approaches 0) are calculated for any bounded real function $f(x)$ on the interval $[x_0, x_n]$ and if these two limits are identical, then that value is the Riemann integral. Soon researchers present examples of bounded functions having Riemann integrals but not being differentiable. (The *Habilitationsschrift* is not actually published until 1867, when it appears posthumously.)

MATHEMATICS (Group Theory; Cayley's Theorem)

Arthur Cayley (1821–1895, English) presents the first technical definition of a group (defined above under 1770—MATHEMATICS), listing postulates and emphasizing that a group is determined by the rules governing its elements. He represents a group by its multiplication table and states Cayley's Theorem, that every group is isomorphic to a subgroup of the permutation group A(S) for some S.

MATHEMATICS (Logic)

In *An Investigation of the Laws of Thought, on Which Are Founded the Mathematical Theories of Logic and Probabilities,* George Boole (1815–1864, English) places logic on a mathematical basis, including an axiomatic structure, and thereby contributes greatly to the founding of symbolic logic. Boole aims at expressing the laws of human reasoning in symbolic form and at developing a calculus appropriate to the symbolic language.

MATHEMATICS (Riemannian Spaces; Non-Euclidean Geometry)

Georg Riemann (1826–1866, German) completes a classic work on the foundations of geometry (not actually published until 1867): *Über die Hypothesen welche der Geometrie zu Grunde liegen* [On the hypotheses which lie at the foundations of geometry], in which he argues that the postulates of Euclidean geometry are not self-evident truths but empirically determined axioms which can be replaced by alternative axioms. Urging a greatly broadened concept of geometry as the study of n-dimensional manifolds, or sets of ordered n-tuples $(x_1, x_2, \ldots, x_n)$, he generalizes Carl Friedrich Gauss's quadratic differential forms in two variables (1827—MATHEMATICS) to quadratic differential forms in n variables, creating a distance measure generalized from the Pythagorean Theorem. The distance between points $(x_1, \ldots, x_n)$ and $(x_1+x_1', \ldots, x_n+x_n')$ is the square root of the sum of all $g_{ij}x_i'x_j'$ where $1 \leq i \leq j \leq n$ and the g_{ij} are functions of $x_1, \ldots, x_n$. Each specific set of g_{ij}'s defines a separate "space." Later terminology labels any space with a system of measurement defined by such a formula as a Riemannian space. Riemann also generalizes Gauss's definition of the curvature of a surface to the curvature of an n-dimensional space, with the curvature defined fully as a function of the g_{ij}. Riemann's theory encompasses as specific subcases Euclidean geometry (in which $n = 2$ or 3, and $g_{ij} = 1$ when $i = j$ but $g_{ij} = 0$ when $i \neq j$) plus the non-Euclidean hyperbolic geometries of Gauss, Nikolai Lobachevsky, and János Bolyai (MATHEMATICS entries for 1816, 1823, 1826, 1829, 1832). It also founds elliptic (or spherical) geometry, where no two lines are parallel and the sum of the angles of a triangle exceeds 180°. For an example of elliptic geometry Riemann defines lines to be great circles on a sphere. His work greatly accelerates the acceptance of non-Euclidean geometries, and he goes so far as to entertain the possibility that the new geometries could be as relevant to the physical world as the standard geometry of Euclid.

1854–1857

MATHEMATICS (Hermitian Forms)

Charles Hermite (1822–1901, French) introduces Hermitian forms, later to prove of importance for quantum theory.

1855

CHEMISTRY (Fertilizers)

Justus von Liebig (1803–1873, German) undertakes a detailed study of chemical fertilizers, thereby becoming, by some accounts, the founder of agricultural chemistry *(Die Grundsatze der Agrikulturchemie).*

CHEMISTRY (Lithium)

Although lithium had been discovered in 1817 and isolated in 1818, in 1855 Robert Bunsen (1811–1899, German) and Augustus Matthiessen (1831–1870, English) prepare the first quantity of sufficient size to allow a thorough investigation of its properties. The preparation is done by heating lithium chloride and passing an electric current through it.

CHEMISTRY (Wurtz Reaction)

Charles Wurtz (1817–1884, French) develops the Wurtz Reaction, whereby hydrocarbons are synthesized through reactions between alkyl halides and sodium.

EARTH SCIENCES (Laurentian and Huronian Systems)

William Logan (1798–1875, Canadian) and T. Sterry Hunt (1826–1892, American) describe the geography and geology of Canada, in particular the rocks of the Laurentian formation centered on the Laurentian Mountains, and the Huronian System found under the Silurian rocks in the vicinity of Lakes Huron and Superior. The studies of Logan and Hunt on the "Laurentian" and "Huronian" systems (terms coined by Logan) are among the first detailed studies of pre-Cambrian rocks.

EARTH SCIENCES (Oceanography)
METEOROLOGY (General)

Publication of Matthew Maury's (1806–1873, American) *Physical Geography of the Sea,* a compilation of Maury's oceanographic and meteorological writings that becomes popular in Europe and America as an oceanography textbook. It includes his bathymetrical chart of the North Atlantic (1854—EARTH SCIENCES) and is one of the first works to deal with the ocean and atmosphere as a united system.

HEALTH SCIENCES (Addison's Disease)

Thomas Addison (1793–1860, English) recognizes the disease of the adrenal glands later known as Addison's disease.

METEOROLOGY (Atmospheric Circulation)

William Ferrel (1817–1891, American) examines pressure data compiled by the U.S. Navy's chief hydrographer Matthew Maury (1806–1873, American) along with known wind patterns to construct a schematic picture of the general circulation of the atmosphere. Ferrel suggests three major vertical circulation cells in each of the Northern and Southern Hemispheres: (1) in the Hadley Cell air rises near the equator and descends near 30° latitude, (2) in the mid-latitude (Ferrel) cell air rises near 60° latitude and descends near 30° latitude, (3) in the high latitude cell air rises near 60° and descends at the pole. Ferrel applies the studies of Gaspard de Coriolis (1835—METEOROLOGY) to the atmosphere, suggesting that moving air is deflected to the right in the Northern Hemisphere and to the left in the Southern Hemisphere (later termed Ferrel's Law), thereby explaining the easterly trade winds in low latitudes, as the near-surface air of the Hadley Cell is deflected from its 30°-to-equator path, and the westerly winds in the mid-latitudes as the surface air of the Ferrel Cell is deflected from its low-to-high latitude path. Ferrel is similarly able to explain the counterclockwise rotation of cyclones in the Northern Hemisphere and the clockwise rotation of cyclones in the Southern Hemisphere.

METEOROLOGY (Weather Maps)

On February 16, M. LeVerrier (French) suggests to the French emperor that a network of weather stations is desirable to provide forewarning of impending disasters, and three days later presents a weather map of France constructed from observations at 10 stations. The idea of constructing such maps spreads rapidly.

PHYSICS (Electromagnetism)

In the paper "Faraday's Lines of Force," James Clerk Maxwell (1831–1879, Scottish) represents Faraday's conceptions of electricity and magnetism mathematically and interconnects the treatment of electromagnetism through lines of force with the treatment through action at a distance. Maxwell also conceives a mechanical model of the lines of force concept, with beads of magnetic vortices alternating with beads of electric vortices.

1856

BIOLOGY (Fermentation; Pasteurization)

Louis Pasteur (1822–1895, French) describes his recent researches on the problem of excessive fermentation in the article "Recherches sur la Putrefaction." In an effort to assist the French wine industry, plagued by a widespread souring of fermenting wine, Pasteur had determined that the excess fermentation could be eliminated either by boiling the liquid to destroy active yeast or by filtering out microorganisms. He consequently introduces the process of "pasteurization" whereby a liquid is heated to 55°C, destroying undesired yeast without effecting a major chemical change in the wine. Pasteur formulates a germ theory of fermentation, which he expands over the next two decades by examining vinegars and beers as well as wines. The work contributes to his rejection of the theory of spontaneous generation.

CHEMISTRY (Dyes)

The synthetic dye industry emerges from research in organic chemistry when William Perkin (1838–1907, English) creates the artificial aniline dye, mauve, by synthesis from impure aniline, a component of coal tar, during an attempt to synthesize the drug quinine. Perkin proceeds to produce mauve commercially from benzene, also a component of coal tar, after resolving major production problems. The method consists of nitration, followed by reduction in acid using either iron or tin. Production of aniline red and aniline blue quickly follow. Although mauve is many times mentioned as the first synthetic dye, picric acid had been prepared in 1771 and aurin (or rosolic acid) was discovered in 1834 and observed to produce red colors.

CHEMISTRY (Steel)

Henry Bessemer (1813–1898, English) produces steel through the Bessemer process, a process also being developed independently by William Kelly (1811–1888, American).

METEOROLOGY (Dynamic)
EARTH SCIENCES (Oceanography)

In "An Essay on the Winds and Currents of the Ocean," William Ferrel (1817–1891, American) presents a general scheme for the large-scale circulation of the atmosphere and suggests that ocean currents are approximately geostrophic, that is, resulting from the pressure gradient and Coriolis forces and therefore flowing parallel to the isobars.

1856–1857

CHEMISTRY (Photography)

The art of photography becomes open to amateurs through the development of the dry collodion process.

1856–1860

MATHEMATICS (Geometry)

In *Beiträge zur Geometrie der Lage,* a revised edition of his earlier *Geometrie der Lage* (1847—MATHEMATICS), Karl Christian von Staudt (1798–1867, German) attempts to show that analytic methods are superfluous for geometry. However, the complications introduced have the result of convincing many of the practical if not absolute need for analytic methods. Von Staudt tries to geometrize real and complex numbers, introducing complex projective spaces of one, two, and three dimensions. Imaginary elements are presented as double elements of elliptic involutions.

1857

ASTRONOMY (Asteroid Gaps)

Daniel Kirkwood (1814–1895, American) points out the existence of "gaps" in the distribution of the mean distances from the sun of the known asteroids, a discovery which he formally publishes and elaborates nine years later (1866—ASTRONOMY).

ASTRONOMY (Saturn's Rings)

Following a sequence of computations on the nature of Saturn's rings, James Clerk Maxwell (1831–1879, Scottish) revises the ringlet hypothesis (1799–1825—ASTRONOMY) of Pierre Simon de Laplace by suggesting that numerous small individual bodies with independent orbits compose the ringlets. Maxwell's calculations show that gravitational forces would tear apart even a very thin ring if it were a continuous solid body. (Observational evidence supporting Maxwell appears late in the century when James Keeler and William Campbell show that the inner rings rotate faster than the outer ones, as gravitational theory would demand for individual satellites [1895—ASTRONOMY].)

CHEMISTRY/PHYSICS (Kinetic Theory of Gases; Evaporation)

In the paper "Ueber die Art der Bewegung, welche wir Wärme nennen," Rudolf Clausius (1822–1888, German) establishes mathematically that the heat in a gas cannot be accounted for exclusively by translational motion of the molecules and asserts that molecules have rotational and vibrational motion as well as translational motion. He consequently rejects the contentions that the translational kinetic energy is conserved during molecular collisions and that all molecules have equal, constant velocities. His allowance for differing molecular velocities enables him to offer a new explanation of evaporation, asserting that the molecules able to overcome the attractive forces of the liquid and "escape" to the gaseous state are those with high velocities (and hence high kinetic energies). Hence evaporation produces a loss of energy in the liquid and a decrease in temperature.

EARTH SCIENCES (Continental Drift)

Contained within the broadscale, somewhat mystical view of nature expressed in his *Key to the Geology of the Globe: An Essay*, Richard Owen (1810–1897, American) includes the speculation that the continents have drifted apart from each other. Although discussed at the time, the essay is largely forgotten in the following decades.

MATHEMATICS (Topology)

Georg Riemann (1826–1866, German) publishes his theory of Abelian functions, makes increased use of topological methods in the theory of functions of a complex variable, and introduces the use of cross cuts to define the *n*-fold connectivity of a surface. A cross cut is a cut that begins and ends at the edge of a surface.

METEOROLOGY (Winds)

In his "Note sur le rapport de l'intensité et de la direction du vent avec les écarts simultanés du baromètre," Christoph Buys Ballot (1817–1890, Dutch) asserts that actual winds are approximately geostrophic, that is, such as to balance the pressure gradient and Coriolis forces.

1857–1858

EARTH SCIENCES (Observational Oceanography)

Several oceanographic discoveries are made as the crew members of the British ship *Cyclops* perform soundings in the North Atlantic along the planned telegraphic cable route between Europe and North America. Among the discoveries are the "Telegraph Plateau" on the mid-Atlantic Ridge and the widespread distribution of the Globigerina ooze sediment along the mid-Atlantic Ridge.

MATHEMATICS (Matrices)

Arthur Cayley (1821–1895, English) invents and develops the algebra of matrices, deriving the notion from determinants and from simultaneous linear equations.

1858

BIOLOGY (Evolution)

Alfred Wallace (1823–1913, English) sends to Charles Darwin an 11-page paper ("On the Tendency of Varieties to Depart Indefinitely from the Original Type") presenting ideas on the evolution of species that are close to those developed independently, and in great detail, by Darwin over the previous two decades. This leads to the joint presentation to the London Linnean Society in July of Wallace's paper and Darwin's "On the Tendency of Species to Form Varieties and on the Perpetuation of Varieties and Species by Natural Means of Selection." More importantly, it spurs Darwin to complete his *Origin of Species* for publication in 1859. Wallace's ideas are founded largely on studies in Brazil and the East Indies.

CHEMISTRY/PHYSICS (Kinetic Theory of Gases; Mean Free Path Length)

Rudolf Clausius (1822–1888, German) introduces and develops the concept of the mean free path length of a molecule in the paper "Ueber die mittlehre Länge der Wege." The work is undertaken partly in response to Buys Ballot's argument that gaseous diffusion would have to proceed almost instantaneously if molecules really moved at the velocities indicated by the emerging but controversial kinetic theory of gases. Clausius responds that this is prevented by the tremendous number of collisions amongst the molecules. He determines an equation for the mean free path length L of a molecule based on the average distance λ between molecules and the radius ρ of the repulsive sphere about each molecule: $L = \rho\lambda^3/(4\pi\rho^3/3)$, the denominator being the volume of a molecule's collision sphere.

CHEMISTRY (Organic; Carbon)

In the paper "The Constitution and Metamorphoses of Chemical Compounds and the Chemical Nature of Carbon," August Kekule (1829–1896, German) establishes two major facts of organic chemistry: carbon has four "affinity units" (that is, its valence is 4), and carbon atoms can chemically combine with one another. He further shows that certain compounds can be easily transformed one to another because the structure of the core carbon atoms remains the same. He not only offers an explanation of isomerism but predicts the numbers of isomers of specific kinds.

CHEMISTRY (Symbolic Formulae)

Archibald Couper (1831–1892, Scottish) publishes symbolic formulae for chemical compounds, illustrating the linking of constituent atoms by dotted lines connecting the respective atomic symbols. Although the dotted lines are later replaced by solid lines, the general scheme gains gradual acceptance.

EARTH SCIENCES (Petrology: Rock Origins)

Henry Sorby (1826–1908, English) shows, in the memoir "On the Microscopical Structure of Crystals, Indicating the Origin of Minerals and Rocks," that microscopic study of rocks and minerals can reveal important details of their mode of formation. Through analysis of the size and content of their cavities, he surmises whether individual crystals were deposited from solution in water or from igneous fusion, and approximates the relative rates of formation and the relative temperatures and pressures during formation.

HEALTH SCIENCES (Anesthesiology)

John Snow (1813–1858, English) publishes *On Chloroform and Other Anaesthetics*, describing various anesthetics and the apparatuses and procedures for administering them.

FIGURE 14. *An illustration from a catalog of Johann Scheuchzer's fossil collection (see 1716—*EARTH SCIENCES *entry), depicting Scheuchzer and a student in the then popular pastime of fossil collection. Scheuchzer argued strongly for the organic origin of fossils (see 1709—*EARTH SCIENCES *entry) in the ongoing debate concerning whether fossils are the remains of once living beings. (By permission of the Houghton Library, Harvard University.)*

HEALTH SCIENCES (Cellular Pathology)

Publication of Rudolph Virchow's (1821–1902, German) *Cellularpathologie* [Cellular pathology], helping to establish him as a founder of the subject. Virchow extends cell theory (1838—BIOLOGY) to diseased areas of the body and analyzes disease as warfare on the cell level.

MATHEMATICS (Group Theory)

Thomas Kirkman (1806–1895, English) makes one of the earliest attempts to determine all permutation groups of given orders.

MATHEMATICS (Quintic Equations)

Charles Hermite (1822–1901, French) constructs transcendental functions appropriate for solving the general quintic equation.

MATHEMATICS (Topology; Möbius Band)

August Möbius (1790–1868, German) discovers the Möbius band, a surface with only one side, which, when cut along the middle, remains as one piece.

Johann Listing (1806–1882, German) also, and independently, discovers one-sided surfaces.

PHYSICS (Hydrodynamics)

Hermann von Helmholtz (1821–1894, German) publishes a mathematical theory of vortex motion, constituting the first detailed analysis of fluid motion not constrained to being irrotational.

SUPPLEMENTAL (Evolution)

Herbert Spencer (1820–1903, English) writes that evolution can be applied to every science, including astronomy, biology, and human history.

1859

ASTRONOMY (Saturn's Rings)

Publication of James Clerk Maxwell's (1831–1879, Scottish) paper "On the Stability of the Motion of Saturn's Rings."

ASTRONOMY (Solar Composition)

Gustav Kirchhoff (1824–1887, German) reasons that the dark Fraunhofer lines found in the solar spectrum (1814—ASTRONOMY) are produced because gases in the outer solar atmosphere absorb radiation at the wavelengths of those lines. Upon comparing the solar spectrum with laboratory spectra of individual elements, Kirchhoff and Robert Bunsen (1811–1899, German) establish that the specific absorption lines in the solar spectrum suggest the existence of hydrogen, nickel, iron, sodium, calcium, and magnesium in the sun. They thereby present the first evidence regarding the chemical composition of a star. Over the next three years, Kirchhoff further determines, again spectroscopically, that the sun contains barium, chromium, copper, and zinc as well.

ASTRONOMY (Solar Flares)

The first record of a solar flare is made by Richard Carrington (1826–1875, English), who observes a stream of light jump from the sun's surface and then relax about five minutes later. Carrington hypothesizes the cause to be the impact of a large meteor.

ASTRONOMY/CHEMISTRY (Spectroscopy)

Gustav Kirchhoff (1824–1887, German) asserts conditions under which dark line, bright line, or continuous spectra can be expected: a hot solid, liquid or gas under heavy pressure exhibits a continuous spectrum unless separated from the viewer by a cool gas under low pressure, in which case a dark-line (or absorption) spectrum appears; a hot gas under low pressure exhibits a bright-line (or emission) spectrum.

ASTRONOMY (Vulcan)

Peculiarities in the orbit of Mercury lead Urbain LeVerrier (1811–1877, French) to postulate the existence of a planet Vulcan between Mercury and the sun. (No such planet is discovered; Albert Einstein explains the peculiarities with the theory of relativity [1916—PHYSICS / ASTRONOMY].)

BIOLOGY (Evolution)

Thomas Huxley (1825–1895, English) presents a lecture, entitled "Persistent Types of Animal Life," in which he argues that at any time the existent species of plants and animals are the result of a gradual modification of previous species.

BIOLOGY (Evolution)

Charles Darwin (1809–1882, English) publishes his major work presenting his theory of the evolution of species: *On the Origin of Species by Means of Natural Selection.* In a sustained central argument, Darwin presents a wealth of data to support his thesis that species evolve over time and that the evolution is guided by a "natural selection" process similar to the artificial selection common in the breeding of animals and plants. He had begun his first notebook on the topic in 1837, after becoming convinced of the reality of evolution through reflection on the fossils he had collected and the species he had observed, particularly in South America and the Galapagos Islands, during the voyage of the *Beagle* (1831–1836—EARTH SCIENCES). He hypothesized the "natural selection" mechanism a year later after reading Thomas Malthus's *Essay on Population* (1798—BIOLOGY) and contemplating the struggle for existence within and between species necessitated by the tendency for geometrical population increases within species. He then developed (and modified) the theory over the next 20 years. Darwin emphasizes that variations occur in nature, and that population increases, if left alone, will always eventually lead to competition within and between species. As a result, the profitable variations tend to be preserved and inherited, while the unprofitable ones tend to be destroyed. This natural selection leads to the extinction of some and further evolution of other species. Among the types of evidence detailed by Darwin are instances of natural mimicry, the similarity of embryos of different species, the geographical distributions of various animals, and fossil evidence, younger strata containing fossils of more advanced life forms than older strata. One of Darwin's central postulates is that, given enough time, large effects can result from small, gradual changes. Hence he inserts into biology a principle similar to the uniformitarianism inserted into geology by James Hutton (EARTH SCIENCES entries for 1785 and 1795) and Charles Lyell (1830–1833—EARTH SCIENCES). Although the work is particularly noted for the mechanism of natural selection and the evidence amassed in support of the concept of evolution, Darwin includes also the possibility that among the variations on which natural selection acts are some which are characteristics acquired through use or disuse and subsequently inherited. This idea of the inheritance of acquired characteristics had been a central element in the earlier evolutionary theory of Jean Baptiste Lamarck (1809—BIOLOGY).

BIOLOGY (Evolution)

In the first part of his *Introduction to the Australian Flora*, Joseph Hooker (1817–1911, English) presents the doctrine that plant species have descended and changed over time. The many observations included lend support to some of the tenets of Charles Darwin's more substantial work (1859—BIOLOGY).

CHEMISTRY/PHYSICS (Kinetic Theory of Gases)

His interest having been aroused by his work on the rings of Saturn (1859—ASTRONOMY), James Clerk Maxwell (1831–1879, Scottish) carries out the first extensive mathematical development of the kinetic theory of gases. He derives, theoretically, a law for the statistical distribution of molecular speeds in a gas and relates the results to the second law of thermodynamics (1850—PHYSICS): although it is not impossible that heat could pass from a low temperature gas to a high temperature gas, it is statistically highly improbable.

CHEMISTRY (Spectroscopy)

Gustav Kirchhoff (1824–1887, German) and Robert Bunsen (1811–1899, German) invent the Kirchhoff-Bunsen spectroscope, whereby the bright lines in the spectra of elements can be accurately mapped, and determine that, under sufficiently high pressure, all incandescent solids and liquids will produce continuous spectra. Kirchhoff hypothesizes that under normal conditions each chemical element produces a characteristic spectrum, that this spectrum can be obtained even from a very small quantity of the element, and that it is not affected by other elements which might be present. These hypotheses allow the discovery of new elements through spectroscopic analysis.

EARTH SCIENCES (Thin Sections)

Henry Sorby (1826–1908, English) advocates to the Geological Society of France increased use of thin sections in the study of rocks and minerals (see EARTH SCIENCES entries for 1831, 1853, 1858), as such sections reveal the structure and composition of the rocks, and throw light on their mode of origin and changes over time. Sorby indicates, for instance, that thin sections have revealed innumerable cavities in granitic rocks, with these cavities enclosing water and saline solutions.

MATHEMATICS (Geometry)

Arthur Cayley (1821–1895, English) uses cross ratios and an "absolute" conic to create a projective equivalent to metric distances, thereby reducing metric geometry to a subcase of projective geometry.

MATHEMATICS (Riemann's Hypothesis; Zeta Function)

Georg Riemann (1826–1866, German) publishes a memoir "Ueber die Anzahl der Primzahlen unter einer gegebenen Grösse" [On the number of primes under a given magnitude] applying complex number theory to the problem of devising a formula for the number of primes less than any given number. In the process he investigates the infinite series called the "zeta function": $1 + (1/2^s) + (1/3^s) + (1/4^s) + \ldots$, where s is a complex number $a + bi$. He states without proof a formula for the number of roots of the zeta function between 0 and T (a formula finally proved in 1905), and he presents the hypothesis that if the zeta function equals 0 for an $s = a + bi$ with $0 < a < 1$, then $a = 1/2$. This hypothesis, later termed Riemann's Hypothesis, becomes a point of departure for much research in the theory of prime numbers over the next hundred years.

PHYSICS (Radiation)

Gustav Kirchhoff (1824–1887, German) announces that, for any emitting body, the coefficient of emission and the coefficient of absorption are in a ratio that is a function only of wavelength and temperature.

PHYSICS (Thermodynamics)

William Rankin (1820–1872, English) publishes the popular *Manual of the Steam Engine and Other Prime Movers* to educate engineers regarding thermodynamic principles.

1860

ASTRONOMY (Solar Prominences)

Warren de la Rue (1815–1889, English) takes wet-plate photographs of the moon-blocked sun during a total solar eclipse, and from them discovers solar prominences.

BIOLOGY (Evolution)

Louis Agassiz (1807–1873, Swiss-American), along with other noted scientists, criticizes Charles Darwin's *Origin of Species* (1859—BIOLOGY), claiming that species have not evolved but that each has been separately and divinely created.

CHEMISTRY (Acetylene)

Marcellin Berthelot (1827–1907, French) rediscovers and names acetylene (1836—CHEMISTRY). His work over the next several years includes the determination of many of the properties of acetylene plus the use of acetylene as a starting point in synthesizing styrene, naphthalene, acenaphthene, ethylene, methane, and benzene.

CHEMISTRY (Avogadro's Hypothesis)

At the international Karlsruhe conference (1860—CHEMISTRY), Stanislao Cannizzaro (1826–1910, Italian) revives the work of Amedeo Avogadro, particularly Avogadro's Hypothesis and the important distinction between atoms and molecules (1811—CHEMISTRY). Cannizzaro employs Avogadro's Hypothesis in the straightforward determination of molecular weights of gaseous compounds by comparing the weight of a volume of the gas to that of an equal volume of hydrogen. From molecular weights he proceeds to atomic weights. The work leads quickly to wide acceptance of the formerly ignored Avogadro Hypothesis. Cannizzaro also establishes the usefulness of atomic weights in determining the formulae of organic compounds.

CHEMISTRY (Cesium)

Robert Bunsen (1811–1899, German) and Gustav Kirchhoff (1824–1887, German) demonstrate the value of their spectroscope (1859—CHEMISTRY) by heating an ore, finding unknown spectral lines (in the blue region), and thereby discovering the metal cesium.

CHEMISTRY (Karlsruhe Congress)

An international conference is held at Karlsruhe to unravel the current difficulties in chemical nomenclature and symbolism and atomic weight and valency calculations. The 140 chemists attending from throughout Europe aim at standardization and better definition of important chemical concepts, a major item being the definitions of "atom" and "molecule," words often used synonymously in the previous half century.

CHEMISTRY/BIOLOGY (Vitalism)

Marcellin Berthelot (1827–1907, French), in the work *Chimie organique fondée sur la synthèse*, seriously undermines the remaining support for the theory of vitalism when he provides multiple examples of artificial synthesis of organic compounds from carbon, nitrogen, hydrogen, and oxygen.

MATHEMATICS (Logic)

In "Syllabus of a Proposed System of Logic," the fourth of five papers on syllogisms (1846–1860), Augustus De Morgan (1806–1871, English) initiates the study of the logic of relations.

PHYSICS (Black Body)

Gustav Kirchhoff (1824–1887, German) introduces the concepts of black body and emissivity. A black body is a surface that absorbs all radiation incident upon it. The emissivity of a body is the ratio of the intensity of radiation emitted by the body to the intensity of radiation which would be emitted by a black body at the same temperature.

1860s

CHEMISTRY (Kinetic Theory of Gases)

James Clerk Maxwell (1831–1879, Scottish) develops the kinetic theory of gases, including a rigorous statistical analysis assuming a gas is a collection of rapidly moving, perfectly elastic molecules. Although developed for nonexistent "ideal" gases, with molecules of zero size and no intermolecular attraction, the theory has application to a wide range of real gases and accounts for the known pressure/temperature/volume relationships. Ludwig Boltzmann (1844–1906, Austrian) independently develops a similar theory.

MATHEMATICS (Algebra)

Benjamin Peirce (1809–1880, American) develops the theory of linear associative algebras (published in 1881).

MATHEMATICS (Rationals; Irrationals)

In a series of lectures, Karl Weierstrass (1815–1897, German) develops the properties of the rational numbers from those of the naturals through use of number couples. Accepting the integers and their properties as given, he asserts that no additional axioms are needed to develop the full real number system. Both the positive rationals and the negative integers are defined as couples of natural numbers and the negative rationals as couples of negative integers. Irrational numbers become aggregates of rational numbers, or, approximately, ordered infinite sequences of rational numbers.

1861

BIOLOGY (Evolution)

Henry Freke, in his book, *On the Origin of Species by Means of Organic Affinity* further elaborates his ideas of evolution, introduced a decade earlier in a medical journal (1851—BIOLOGY). He emphasizes in the preface the importance of the fact that he and Charles Darwin (1859—BIOLOGY) have arrived at the same basic conclusion—that all organic species have descended from one primordial germ—but have done so in different ways, Darwin emphasizing analogy and Freke emphasizing induction. Freke assures his readers that nothing in his work denies the biblical record.

BIOLOGY (Evolution; Archaeopteryx)

The fossil *Archaeopteryx* is discovered in Bavaria and proclaimed the "missing link" in the evolutionary sequence from reptile to bird. The fossil remains indicate teeth, wings with claws, and a long tail with a vertebrated structure.

BIOLOGY (Speech Center)

Paul Broca (1824–1880, French) localizes the brain's speech center in right-handed individuals as in the left portion of the frontal cortex, and hypothesizes that it will be found to be in the brain's right hemisphere for left-handed individuals. His conclusions are based in large part on examination of the brain of a recently deceased 51-year-old male who had lost the ability to speak (except for the one syllable "tan") at age 30. Broca's description of the case and conclusion that the speech center probably resides in the third left frontal convolution are reported in the paper "Remarques sur le siège de la faculté du langage articulé, suivie d'une observation d'aphémie (perte de la parole)." Although Marc Dax had earlier also concluded an asymmetry of the human brain with respect to speech (1836—HEALTH SCIENCES) and although the specifics of Broca's conclusions are later disputed, the basic concepts gain widespread attention due to Broca's research.

CHEMISTRY (Organic Chemistry)

August Kekule (1829–1896, German) introduces the modern definition of organic chemistry as the study of carbon compounds, whether produced artificially or naturally.

CHEMISTRY (Rubidium)

Robert Bunsen (1811–1899, German) and Gustav Kirchhoff (1824–1887, German) announce the discovery of the metal rubidium, found in lepidolite through use of the spectroscope (1859—CHEMISTRY) and identification of a previously unknown red line in the spectrum.

CHEMISTRY (Symbolism; Benzene)

Joseph Loschmidt (1821–1895, Austrian) devises symbolic formulae for several hundred compounds, that for benzene comprising one large circle (carbon) with six symmetrically attached small circles (hydrogen). Announcement of Loschmidt's symbolic structure for benzene precedes by several years the more famous similar structure announced by August Kekule (1865—CHEMISTRY) but gains little recognition.

CHEMISTRY (Thallium)

William Crookes (1832–1919, English) discovers thallium while spectroscopically examining the residues of a sulfuric acid plant, hoping to find tellurium. No tellurium lines are found, but the previously unknown green line leads to the naming of the new element (thallium = green branch).

HEALTH SCIENCES (Puerperal Fever)

Ignaz Semmelweis (1818–1865, Hungarian) publishes *Die Aetiologie, der Begriff und die Prophylaxis des Kindbettfiebers* [The etiology, concept, and prophylaxis of puerperal fever], the first half of which contains the extensive statistical data he had compiled over the previous 15 years on the occurrence and mortality rates of childbed fever in various hospitals; plus the manner in which he analyzed the data and formulated his conclusion that childbed fever is communicated to a woman in labor through decomposed matter on the unwashed or improperly washed hands or instruments of examining physicians or assistants. The second half of the book reviews his correspondence on the subject and the published literature supporting and opposing his explanations. The strength of his belief that thousands continue to die because his practices are not being followed contributes to the vehemence of his attacks in the second half of the book on the opponents of his doctrine.

MATHEMATICS (Analysis)

By presenting the equation of a curve which is continuous but has no tangent at any point, Karl Weierstrass (1815–1897, German) establishes that a continuous function is not necessarily differentiable. He presents the example in classroom lectures in 1861, then more formally in a published paper in 1872. Bernard Bolzano's earlier similar work (1834—MATHEMATICS) had gone essentially unnoticed.

PHYSICS (Heat Conduction)
MATHEMATICS (Differential Forms; Riemann-Christoffel Tensor)

Georg Riemann (1826–1866, German) submits the essay *Über eine Frage der Wärmeleitung* [On a question in the conduction of heat] to the French Academy. In it he develops the theory of quadratic differential forms, later basic to the study of relativity, and applies the theory to the conduction of heat in a solid. While working on the problem of heat conduction, he invents the Riemann-Christoffel tensor, later used in the development of tensor calculus.

1862

ASTRONOMY (Sirius B)

Alvan Clark (1804–1887, American) and his son Alvan Graham Clark (1832–1897, American) discover Sirius B, the companion star to Sirius, while testing a lens for a new refractor telescope. The star is located in the position predicted by Friedrich Bessel (1844—ASTRONOMY).

ASTRONOMY (Solar Atmosphere)

Anders Angström (1814–1874, Swedish) discovers, through spectroscopy, the presence of hydrogen in the sun's atmosphere.

ASTRONOMY (Star Catalogs)

Friedrich Argelander (1799–1875, German) completes his tabulation of the positions and brightnesses of over 324,000 Northern Hemisphere stars, listing stars up to the ninth magnitude. The results are published in *Bonner Durchmusterung.*

BIOLOGY (Fertilization)

Charles Darwin (1809–1882, English), in *Fertilisation of Orchids,* details the process of cross fertilization, the importance of insects in transferring pollen, and the methods evolved in the orchid to attract the essential insects.

BIOLOGY (Starch; Carbon Assimilation)

Julius von Sachs (1832–1897, German) establishes that the starch found in green plant cells is produced by photosynthesis from the carbon dioxide absorbed from the atmosphere.

CHEMISTRY (Isopropyl)

Charles Friedel (1832–1899, French) prepares isopropyl, the first secondary alcohol.

CHEMISTRY (Periodic Table Forerunner)

Noting the periodic recurrence of various properties of the known atomic elements, geologist Alexandre-Émile Béguyer de Chancourtois (1820–1886, French) arranges the elements on a cylinder, in order of atomic weight, so that, for example, lithium, sodium and potassium are on one perpendicular while oxygen, sulfur, selenium, and tellurium are on another.

CHEMISTRY (Thallium)

Claude Lamy (1820–1878, French) isolates thallium.

EARTH SCIENCES (Experimental Geology: Silicious Rocks)

Gabriel Daubrée (1814–1896, French) performs controlled laboratory experiments on the formation and destruction of silicates and other naturally occurring materials. He shows, for instance, the strong effects of superheated water, even in small quantities, causing silicates to crystallize at temperatures well below their normal points of fusion. He regards this influence of water as probably crucial in the natural crystallization of silicious rocks.

EARTH SCIENCES (Geology)

James Dana (1813–1895, American) publishes his *Manual of Geology.*

EARTH SCIENCES (Valleys; Lakes)

Andrew Ramsay (1814–1891, Scottish) publishes the papers "The Excavation of the Valleys of the Alps" and "On the Glacial Origin of Certain Lakes in Switzerland, the Black Forest, . . . and Elsewhere." Ramsay suggests that rivers initially created many of the Alpine valleys, and that glaciers created many large basins. Many of the Alpine valleys initially created by rivers were further modified not only by rivers but also by glaciers, which tended to widen and deepen the existent valleys.

HEALTH SCIENCES (Germ Theory)
BIOLOGY (Spontaneous Generation)

Louis Pasteur (1822–1895, French) amasses experimental evidence against the theory of the spontaneous generation of organisms and develops the theory that infection is spread by germs, extending his germ theory of fermentation (1856—BIOLOGY) to a germ theory of disease. Proceeding over the next several decades on the basis of this germ theory, he searches for and develops effective inoculations against several specific diseases, including chicken cholera (1880—BIOLOGY), anthrax (1882—HEALTH SCIENCES), and rabies (1885—HEALTH SCIENCES).

MATHEMATICS (Tensor Calculus)

In an amplified version of his earlier theory of extension (1844—MATHEMATICS), Hermann Grassmann (1809–1877, German) develops the algebra of tensor calculus, including quaternions, determinants, and matrices as special cases. The work is little appreciated until it is applied to general relativity (1916—PHYSICS).

PHYSICS (Acoustics)

Hermann von Helmholtz (1821–1894, German) details his researches in acoustics in the book *Tonempfindungen* [Sensations of tone].

SUPPLEMENTAL (Knowledge; Evolution)

In *First Principles,* the first volume of a projected ten-volume work, Herbert Spencer (1820–1903, English) distinguishes the unknowable from its knowable manifestations, or phenomena. He asserts that the manifestations evolve over time and that the various evolutions—for example, those in astronomy, geology, biology, psychology—are all really one Evolution proceeding everywhere in all facets of the universe.

1862–1869

BIOLOGY (Conchology)

Gwyn Jeffreys (1809–1885, Welsh) publishes the five-volume *British Conchology, or an Account of the Mollusca Which Now Inhabit the British Isles and the Surrounding Seas.*

1863

BIOLOGY (Brain)

Ivan Sechenov (1829–1905, Russian) describes the physiological bases of various psychic processes in his "Refleksy golovnogo mozga" [Reflexes of the brain].

CHEMISTRY (Indium)

While spectroscopically analyzing zinc ore after roasting and decomposing, Ferdinand Reich (1799–1882, German) and Hieronymus Richter (1824–1898, German) discover indium, naming it for a prominent line in the indigo portion of the spectrum.

EARTH SCIENCES (Ice Ages)

Archibald Geikie (1835–1924, Scottish) compiles accumulating evidence of the glacial origin of certain surficial deposits in Scotland, and presents the finding of plant remains between layers of glacial till as evidence that separate ice ages existed with substantial intervening periods of warm climate.

EARTH SCIENCES (Uniformitarianism; Prehistoric Man)

Publication of Charles Lyell's (1797–1875, English) *The Geological Evidence of the Antiquity of Man,* where he summarizes the known archaeological findings on prehistoric man, viewed in the context of uniformitarian geology.

HEALTH SCIENCES (Epilepsy)

John Hughlings Jackson (1835–1911, English) describes hemiplegic or focal epilepsy in the paper "Unilateral Epileptiform Seizures, Attended by Temporary Defect of Sight." Although the condition had earlier been noted by Louis Bravais (1827—HEALTH SCIENCES), Jackson's detailed description leads to the designation "Jacksonian epilepsy," or, less frequently, "Bravais-Jacksonian epilepsy."

MATHEMATICS (Cremona Transformations)

Luigi Cremona (1830–1903, Italian) generalizes the inversive transformation (1824—MATHEMATICS) to the general transformation sending point (x,y) in a plane to point $(R_1(x,y),R_2(x,y))$ where R_1 and R_2 are any rational algebraic functions. This birational transformation of the plane is later further generalized by Cremona to a birational transformation of three-dimensional space.

MATHEMATICS (Hypercomplex Number Systems)

Karl Weierstrass (1815–1897, German) proves that the only hypercomplex number systems with real coordinates and commutative multiplication such that $ab = 0$ implies $a = 0$ or $b = 0$ are (1) the algebra with one fundamental unit whose square is itself; and (2) the ordinary complex numbers (two fundamental units). (This work is not published until 1884.)

MATHEMATICS (Topology)

In *Theorie der elementaren Verwardschaft* [Theory of elementary relationships], August Möbius (1790–1868, German) examines the relationships between figures with one-to-one pointwise mappings such that neighboring points map to neighboring points. He also introduces the triangulation of polyhedra.

METEOROLOGY (Air Pollution Control)

The Alkali Act passes in Great Britain, forbidding industrial insertion of hydrogen chloride into the atmosphere and hence encouraging use and development of pollution-control methods.

METEOROLOGY (Greenhouse Effect)

John Tyndall (1820–1893, Irish) in an article "On Radiation Through the Earth's Atmosphere" describes the "greenhouse" effect whereby water vapor and other gases in the atmosphere absorb much of the thermal radiation emitted from the Earth but transmit most of the solar radiation entering the earth-atmosphere system from above.

METEOROLOGY (Weather Mapping)

Publication of Francis Galton's (1822–1911, English) "Meteorographica or Methods of Mapping the Weather."

1864

ASTRONOMY (Nebulae)

Publication of John Herschel's (1792–1871, English) *A General Catalogue of Nebulae,* later revised by Johann Dreyer to the widely used *New General Catalogue* or NGC (1888—ASTRONOMY).

CHEMISTRY (Butyl)

Aleksandr Butlerov (1828–1886, Russian) prepares tertiary butyl, the first tertiary alcohol.

CHEMISTRY (Law of Octaves)

John Newlands (1837–1898, English) orders the chemical elements according to atomic weight and notes that with only a few transpositions from this ordering there arises a rough periodicity in properties, with a period of 7. Hence, for instance, the eighth and fifteenth elements resemble the first. Newlands calls this the Law of Octaves and divides the elements into families according to the cyclical similarities. J. Lothar Meyer (1830–1895, German) independently engages in similar work, publishing *Die modernen Theorien der Chemie und ihre Bedeutung für die chemische statik,* in which he also supports the ideas of Stanislao Cannizzaro (1860—CHEMISTRY) and particularly Avogadro's Hypothesis (1811—CHEMISTRY).

CHEMISTRY (Symbolism)

A. Crum Brown (1838–1922, Scottish) devises structural formulae for chemical compounds, convenient particularly in distinguishing isomers.

EARTH SCIENCES (Ice Ages)

James Croll (1821–1890, Scottish) extends Joseph Adhémar's earlier hypothesis that the earth's ice ages are caused by changes in the earth's orbit (1842—EARTH SCIENCES), adding the 100,000-year cycle in the eccentricity of the earth's orbit to the precession of the equinoxes stressed by Adhémar. Using calculations from Urbain LeVerrier (1843—ASTRONOMY), Croll plots orbital changes over the past three million years, finds cyclical changes with long intervals of high eccentricity and long intervals of low eccentricity, and concludes that ice ages occur during periods of high eccentricity, alternating from the Northern to the Southern Hemisphere in response to the 22,000 year precession cycle. Wrongly believing the crucial factor to be minimum winter solar radiation, Croll postulates that when the eccentricity is high, the hemisphere whose winter occurs at the time of the earth's farthest distance from the sun will experience an ice age. The theory becomes widely discussed, with many researchers over the next three decades seeking field evidence in one direction or the other.

HEALTH SCIENCES (Neurology; Ophthalmoscope)

John Hughlings Jackson (1835–1911, English) initiates use of the ophthalmoscope for studying diseases of the nervous system.

1865

BIOLOGY (Genetics)

Gregor Mendel (1822–1884, Austrian) founds the science of genetics with the publication of the paper "Versuch uber Pflanzenhybriden" [Experiments in plant hybridization], presenting his laws of heredity supported by systematic tabulation of his extensive experimentation with the hybridization of pea plants, done from 1856 to 1863. Mendel concludes from his controlled cross-breeding of peas with contrasting heights, colors, and other observables, that certain characteristics are inherited in an undiluted fashion through indivisible units (genes). Dominant characteristics are transmitted unchanged to offspring; recessive characteristics become latent, to reappear in a 1:3 ratio in the second generation. For example, in crossing 6- to 7-foot pea plants with 0.75- to 1.5-foot plants, in the first generation of offspring all are tall while in the second generation 277 are dwarf and 787 tall, for a 1:2.84 ratio, thus showing tallness to be a dominant and dwarfness a recessive characteristic. Mendel further determines that characteristics controlling different elements of the plant—such as height and color—are transmitted independently of each other. (The paper, published in the *Proceedings of the Natural History Society of Brunn* and essentially unread for three decades, enters the mainstream of biology after being rediscovered in 1900 [1900—BIOLOGY].)

BIOLOGY (Silkworm Disease)

Louis Pasteur (1822–1895, French) begins examining the silkworm diseases devastating the French silk industry. He concludes by 1869 that the diseases are spread by microorganisms, isolates those organisms for two of the diseases, and shows how disease-free eggs and moths can be identified and used for breeding.

CHEMISTRY (Avogadro's Number)

Joseph Loschmidt (1821–1895, Austrian), calculating from the newly developed kinetic theory of gases (1860s—CHEMISTRY), obtains a value for the number of molecules in 22.4 liters of a gas, approximately 6×10^{23}. This is the first calculated estimate of Avogadro's number, resulting from the recent acceptance (1860—CHEMISTRY) of Avogadro's Hypothesis (1811—CHEMISTRY).

CHEMISTRY (Benzene Ring)

In the paper "Studies on Aromatic Compounds," August Kekule (1829–1896, German) announces his theory of the molecular structure of benzene, containing six carbon and six hydrogen atoms. The structure is based on the hexagonal "benzene ring," in which the six carbon atoms are linked alternately by single and double bonds. Kekule further proposes that all aromatic compounds have this ring of six carbon atoms as a common nucleus.

CHEMISTRY (Celluloid)

Alexander Parkes (1813–1890, English) creates celluloid, the first plastic.

EARTH SCIENCES (Paleoclimatology)

Charles Lyell (1797–1875, English) classifies the earth's history into the Precambrian, Paleozoic, Mesozoic, and Cenozoic Eras, and each era into periods, for instance, the current era, the Cenozoic, into Eocene, Miocene, Pliocene, post-Pliocene, and Recent. Lyell had been considering some such classification since 1830.

EARTH SCIENCES (Post Glacial Uplift)

Using evidence of the changing Scottish shorelines, Thomas Jamieson (1829–1913, Scottish) theorizes that the weight of former Northern Hemisphere ice sheets depressed the land downward during glaciation and resulted in a post glacial uplift after the removal of the ice. This provides an explanation for the puzzling existence of marine fossils above sea level.

EARTH SCIENCES (Sea Water)

Johan Forchhammer (1794–1865, Danish) succeeds, after extended efforts, in detecting boric acid in sea water.

HEALTH SCIENCES (Antiseptic Surgery)

The practice of antiseptic surgery is introduced by Joseph Lister (1827–1912, English) while attending a compound fracture of a leg. Lister uses carbolic acid as the antiseptic, sometimes diluted with water or olive oil. He also introduces the use of absorbable ligatures.

HEALTH SCIENCES (Experimental Medicine)

Claude Bernard (1813–1878, French) encourages the conversion of medicine to a more experimental science, with observation followed by hypothesis and experimentation, in his *Introduction à l'étude de la médecine expérimentale* [Introduction to the study of experimental medicine].

PHYSICS (Electromagnetic Theory of Light)

James Clerk Maxwell (1831–1879, Scottish) publishes his most influential paper, "On a Dynamical Theory of the Electromagnetic Field," read to the Royal

Society the previous December. The paper opens with a comment that the current mathematical theories dealing with electricity and magnetism have assumed the existence of action at a distance and have ignored the surrounding medium. Maxwell offers instead a field theory, assuming space to be filled with an ether that allows propagation of light. He derives equations that concisely express the major laws of electricity and magnetism, predicts electromagnetic waves of wavelength other than that of visible light, shows these waves to have a velocity equal to that of light, and shows light to be a form of electromagnetic wave. He provides a method for predicting the speed of light given the electric and magnetic properties of the medium through which it travels. He thus connects mathematically the disciplines of optics, electricity, and magnetism.

c.1865

MATHEMATICS (Weierstrass-Bolzano Theorem)

Karl Weierstrass (1815–1897, German) proves that if S is a bounded infinite set of points, then there exists a point P such that every neighborhood of P contains points of S. This result has since been termed the Weierstrass-Bolzano Theorem, in recognition of the dependence of Weierstrass's proof on methods earlier established by Bernard Bolzano.

1866

ASTRONOMY (Asteroid Gaps; Cassini Division)

Daniel Kirkwood (1814–1895, American) publishes an elaboration of his discovery of gaps in the distribution of the mean distances from the sun of the known asteroids (1857—ASTRONOMY), pointing out significant gaps at distances corresponding to periods of revolution 1/3, 2/5, and 2/7 ths the period of Jupiter. The lack of asteroids in such resonances with Jupiter contrasts with the existence of asteroids in resonance with Mars. Kirkwood also mentions that the prominent gap (the Cassini division) in the rings of Saturn between rings A and B occurs at the distance from Saturn corresponding to a period 1/3 that of the satellite Enceladus, suggesting a resonance gap in the rings similar to the resonance gaps in the asteroids.

ASTRONOMY (Nova)

William Huggins (1824–1910, English) undertakes the first spectroscopic examination of a nova, concluding, in part, that the nova is enveloped by hydrogen gas.

CHEMISTRY (Dynamite)

Alfred Nobel (1833–1896, Swedish) invents dynamite, using nitroglycerine and kieselguhr. Nobel had manufactured an explosive nitroglycerine-gunpowder mixture since 1863, but after several deaths in uncontrolled explosions, he experimented to develop a safer product, succeeding in the development of dynamite.

EARTH SCIENCES (Earth's Core)

Gabriel Daubrée (1814–1896, French) hypothesizes an iron core at the earth's center.

MATHEMATICS (Quaternions)

Posthumous publication of William Rowan Hamilton's (1805–1865, Irish) *Elements of Quaternions,* presenting hundreds of applications of quaternions to geometry, mechanics, and physics.

1867

BIOLOGY (Evolution)

Fleeming Jenkin (1833–1885, Welsh) refutes Charles Darwin's theory of evolution (1859—BIOLOGY) on the basis of the currently accepted mechanisms of heredity, whereby a child blends the characteristics of the two parents. Hence if a favorable chance variation occurs this will not lead to a new species but instead will quickly be muted, as the descendants in the first subsequent generation will inherit only 50% of the variation, those in the second generation will inherit only 25%, and so on.

CHEMISTRY (Law of Mass Action)

Cato Guldberg (1836–1902, Norwegian) and Peter Waage (1833–1900, Norwegian) establish the Law of Mass Action, asserting that, for a given temperature, a chemical reaction proceeds at a speed directly proportional to the concentration of the reacting substances. They present the theory in the pamphlet "Etudes sur les affinités chimiques" [Studies on chemical affinities].

EARTH SCIENCES (Paleoclimatology)

James Croll (1821–1890, Scottish) presents an astronomically based theory of the ice ages in the paper "On the Excentricity of the Earth's Orbit, and Its Physical Relations to the Glacial Epoch," extending his earlier work (1864—EARTH SCIENCES). He estimates that the last Glacial Epoch ended 80,000 years ago and suggests two positive feedbacks prolonging ice ages once they are initiated: (1) The high ice albedo results in more solar radiation being reflected from the earth's surface, thereby further increasing the cooling; (2) a Northern Hemisphere ice age would shift the warm equatorial currents southward sufficiently to reach the coast of South America southward of its easternmost point, thereby causing them to deflect leftward into the Southern Hemisphere instead of rightward into the Northern Hemisphere, hence resulting in cooler Northern Hemisphere waters.

HEALTH SCIENCES (Antiseptics)

Joseph Lister (1827–1912, English) publishes his first accounts of the importance of using antiseptics while treating wounds ("On a New Method of Treating Compound Fracture, Abscess etc., with Observations on the Conditions of Suppuration"; "On the Antiseptic Principle of the Practice of Surgery"). He is convinced through Louis Pasteur's germ theory (1862—HEALTH SCIENCES) that, contrary to common belief, suppuration is not a natural aid to the healing process, but is caused by germs from the hands and instruments of the physicians and from the air. To limit suppuration, Lister suggests applying a dressing of carbolic or phenic acid to the wound, cleaning the physician's hands, and sterilizing the instruments. Through a presentation of several case studies, Lister illustrates the variations called for in the treatment of specific wounds.

MATHEMATICS (Riemann Integral; Riemannian Spaces)

Posthumous publication of Georg Riemann's (1826–1866, German) two classic papers of 1854—his *Habilitationsschrift* on Fourier series, introducing the Riemann integral (1854—MATHEMATICS), and his *Über die Hypothesen welche der Geometrie zu Grunde liegen* on the foundations of geometry, including the introduction of Riemannian spaces (1854—MATHEMATICS).

METEOROLOGY (Aurora Borealis)

Anders Angström (1814–1874, Swedish) carries out the first investigation of the spectrum of the aurora borealis.

1868

ASTRONOMY (Sirius)

Observing that the spectral lines in the light spectrum from Sirius are shifted slightly toward the longer wavelengths (a red shift), William Huggins (1824–1910, English) applies the Doppler-Fizeau Principle (1848—ASTRONOMY) and calculates that Sirius is receding from the solar system at about 29 miles per second.

ASTRONOMY (Solar Composition)
CHEMISTRY (Helium)

Pierre Janssen (1824–1907, French) journeys to India for observation of a total solar eclipse and determination of the sun's chemical composition through spectroscopic examination of the chromosphere. He observes a previously unseen yellow spectral line near the D-line of sodium, also observed by J. Norman Lockyer (1836–1920, English), who interprets it as representing a new element and names it "helium" for Helios, the Greek god of the sun.

ASTRONOMY (Solar Prominences)

J. Norman Lockyer (1836–1920, English) and Pierre Janssen (1824–1907, French) independently devise a method of taking measurements of solar prominences in daylight with a spectroscope, no longer requiring an eclipse as in the initial discovery of prominences (1860—ASTRONOMY).

ASTRONOMY (Solar Spectrum)

Anders Angström (1814–1874, Swedish) lists 1,000 absorption lines in the solar spectrum.

BIOLOGY (Domestication; Evolution)

Publication of Charles Darwin's (1809–1882, English) *The Variation of Plants and Animals Under Domestication.* Darwin discusses the causes of variation and presents his Pangenesis hypothesis whereby the mechanism of heredity is believed to be by means of development in the child of particles derived from each cell of the parents. Among the many observations described, Darwin includes examples of apparent transmission of acquired characteristics, thereby lending some support to earlier evolutionary concepts of Jean Baptiste Lamarck (1809—BIOLOGY).

CHEMISTRY (Perfumes)

William Perkin (1838–1907, English) prepares coumarin from coal-tar derivatives. This is the first preparation of a natural perfume.

EARTH SCIENCES (Deep Sea Life; *Lightning* Expedition)

Wyville Thomson (1830–1882, Irish) and William Carpenter (1813–1885, Irish), skeptical of earlier suggestions that no life exists in the oceans beneath 300 fathoms (1843—EARTH SCIENCES), organize a deep-water dredging expedition in the waters north of Ireland and Scotland. Their skepticism is vindicated as dredging to 530 fathoms from H.M.S. *Lightning* at 55°36′N, 7°20′W yields crustaceans, echinoderms, molluscs, rhizopods, and sponges, as well as shells of globigerina, earlier known to be found on the ocean floor. In addition, they note unexpected temperature contrasts in deep waters, finding marked horizontal differences within relatively small regions. This apparent refutation of the widely held belief in a constant deep water temperature of 4°C leads Carpenter to organize a new voyage on H.M.S. *Porcupine* the following year (1869—EARTH SCIENCES).

EARTH SCIENCES (Sea Level; Paleoclimatology)

Charles Whittlesey (1808–1866, American) calculates that sea level was at least 350 feet lower at the timing of the maximum Northern Hemisphere glaciation than it is today, doing so by calculating an estimated volume of glacial ice from (1) available mappings of the horizontal extent of glaciation and (2) an average ice sheet thickness of one mile, from thickness estimates based on the altitude of glacial markings along mountain ranges which peaked above the top of the main glacial cover.

HEALTH SCIENCES/BIOLOGY (Brain)

To explain why patients damaged in one hemisphere of the brain are sometimes no longer able to speak, John Hughlings Jackson (1835–1911, English) speculates that the two hemispheres are not duplicates of each other but that instead one tends to dominate. For most people, the dominating hemisphere is on the left.

MATHEMATICS (Binary Forms)

Paul Gordan (1837–1912, German) establishes that every binary form has associated with it a finite complete system of invariants and covariants. His proof shows not just the existence of such associated systems, but constructive methods for computing them.

MATHEMATICS (Mathieu Functions)

Émile Mathieu's (1835–1890, French) research on the vibrations of an elliptic membrane lead to his introduction of elliptical cylindrical coordinates and a set of functions appropriate to them later termed "Mathieu functions."

MATHEMATICS (Non-Euclidean Geometry)

Eugenio Beltrami (1835–1899, Italian) demonstrates that the geodesics on a surface of constant negative curvature satisfy the postulates of plane hyperbolic geometry, the geometry formulated by Nikolai Lobachevsky (1829—MATHEMATICS) and János Bolyai (1832—MATHEMATICS), and that those on a surface of constant positive curvature satisfy the postulates of double elliptic geometry, formulated by Georg Riemann (1854—MATHEMATICS).

1868–1869

MATHEMATICS (Line Geometry)

In his *Neue Geometrie des Raumes* [New geometry of lines], Julius Plücker (1801–1868, German) elaborates the geometry of Cartesian three-dimensional space using straight lines, rather than points, as the basic elements. He also suggests other possibilities for basic elements, including planes, circles, and spheres.

1868–1878

CHEMISTRY (General)

Charles Wurtz (1817–1884, French) publishes the three-volume *Dictionnaire de chimie pure et appliquée* [*Dictionary of pure and applied chemistry*].

1869

BIOLOGY (Genetics)

Francis Galton (1822–1911, English) applies Charles Darwin's theory of evolution (1859—BIOLOGY) to man's mental inheritance. In *Hereditary Genius* he buttresses his contentions on the inherited nature of individual talent with much evidence and statistical analysis.

BIOLOGY (Geographical Zoology)

After analyzing 125,000 animal specimens collected from eight years of field work, Alfred Wallace (1823–1913, English) publishes *The Malay Archipelago, the Land of the Orangutan and the Bird of Paradise,* in which he develops his thesis that the animal life on the Malay Archipelago is divided into a western half similar to the animal life of India and an eastern half similar to that of Australia.

BIOLOGY (Growth Requirements)

Zinc is found by Jules Raulin (1836–1896, French) to be an essential element in the growth of Aspergillus. Soon it is found to be essential to numerous other plants and animals as well.

BIOLOGY (Soul)

In the paper "Has the Frog a Soul?" Thomas Huxley (1825–1895, English) describes a series of experiments on a frog the front part of whose brain has been eliminated. The frog still responds to certain stimuli, and Huxley asks whether the frog is an automaton and whether it has a soul. He suggests that examination of the latter question should shed light on the connections between the human body and soul.

CHEMISTRY (Atomic Volume Curve)

J. Lothar Meyer (1830–1895, German) plots the atomic volumes of the chemical elements against their atomic weights and connects the points to obtain a curve with pronounced peaks and valleys. Electronegative elements fall along ascending slopes of the curve while electropositive elements fall along descending slopes. The periodicity in atomic volume and electrochemical behavior revealed by this curve is matched by periodicities in other properties, such as fusibility, malleability, and volatility, when these or corresponding variables are plotted against atomic weight.

CHEMISTRY (Celluloid)

John Hyatt (1837–1920, American), with the assistance of his brother Isaiah, produces celluloid by dissolving pyroxyline and camphor in alcohol and heating the mixture under pressure.

CHEMISTRY (Color Photography)

Louis Ducos du Haroun (1837–1920, French) succeeds, using a trichrome process, in producing the first color photographs.

CHEMISTRY (Critical Temperature)

Following researches on the liquefaction of gases, Thomas Andrews (1847–1907, English) clarifies and develops the concept of critical temperatures (1822—PHYSICS), above which a gas cannot be liquefied, in the paper "On the Continuity of the Gaseous and Liquid States of Matter." He shows through his work with carbonic acid that above the critical temperature increased pressures can lead, without any discontinuous change, to a form of matter somewhere between the liquid and gaseous states.

CHEMISTRY (Periodic Table)

Dmitry Mendeleev (1834–1907, Russian) constructs the periodic table of the elements. Arranging the known elements basically in order of atomic weight and leaving occasional blank spaces, Mendeleev shows that elements with similar properties can be made to fall into columns. He predicts the discovery of elements to fill the gaps and proceeds to describe in detail the chemical properties of three of those elements. These three, eka-aluminum or gallium (CHEMISTRY entries for 1874 and 1875), eka-boron or scandium (1879—CHEMISTRY), and eka-silicon or germanium (1886—CHEMISTRY), are discovered, as predicted, within seventeen years and establish Mendeleev's fame. (Like John Newlands [1864—CHEMISTRY], Mendeleev finds it advantageous to reorder a few elements from a strict listing according to atomic weight. An explanation for why these changes yield stronger periodicities awaits the work of Henry Moseley on atomic numbers [1913—PHYSICS].)

CHEMISTRY (Periodic Table)

J. Lothar Meyer (1830–1895, German), independently of Dmitry Mendeleev (1869—CHEMISTRY), constructs a periodic table of elements, although he does not proceed to Mendeleev's predictions.

CHEMISTRY (Vanadium)

Vanadium is isolated by Henry Roscoe (1833–1915, English).

EARTH SCIENCES (Deep Sea Life; *Porcupine* Expedition)

Researchers on H.M.S. *Porcupine,* particularly Gwyn Jeffries (1809–1885, Welsh), Wyville Thomson (1830–1882, Irish), and William Carpenter (1813–1885, Irish), further confirm the two major findings of the *Lightning* expedition of the previous year (1868—EARTH SCIENCES): life exists in the deep oceans, and so do strong horizontal temperature variations. Dredging directed by Thomson at 47°38′N, 12°08′W yields marine animals at 2,435 fathoms, far surpassing the depths obtained during the *Lightning* expedition. Temperature measurements directed by Carpenter not only confirm the existence of horizontal contrasts found in the *Lightning* data but also indicate that these contrasts remain fairly constant with depth through long distances.

MATHEMATICS (Correlation Coefficient)

Francis Galton (1822–1911, English) introduces the correlation coefficient in his work *Hereditary Genius* (1869—BIOLOGY).

MATHEMATICS (Tensor Analysis)

Elwin Christoffel (1829–1900, German) determines necessary and sufficient conditions for transforming one quadratic form into another, thereby making significant progress toward the later theory of tensor calculus.

METEOROLOGY (Pressure; Winds)

Alexander Buchan (1829–1907, Scottish) maps monthly worldwide distributions of pressure and winds and publishes the paper "Mean Pressure and Prevailing Winds of the Globe." Isobars of pressure are plotted.

1870

ASTRONOMY (Sun)

Charles Young (1834–1908, American) verifies predictions made in 1861 by Gustav Kirchhoff that the outer layers of the sun's atmosphere produce a bright line spectrum, with the bright lines occurring at the locations of the dark lines in the normal solar spectrum. Young observes the spectral reversal during a solar eclipse. Kirchhoff's prediction derived from his interpretation (1859—ASTRONOMY) that the dark Fraunhofer lines (1814—ASTRONOMY) in the solar spectrum are produced because gases in the outer solar atmosphere absorb radiation at the specified wavelengths. These gases emit at the same wavelengths and hence produce the companion bright line spectrum.

BIOLOGY (Evolution)

Alfred Wallace (1823–1913, English) publishes *Contributions to the Theory of Natural Selection.*

BIOLOGY (Evolution)

Edward Cope (1840–1897, American) provides additional evidence for the theory of evolution in his *Systematic Arrangement of the Extinct Batrachia, Reptilia and Aves of North America.*

EARTH SCIENCES (Fjords)

Oscar Peschel (1826–1875, German) publishes a systematic study of several individual landform types, including fjords, islands, lakes, mountains, and valleys, in *Neue Probleme der vergleichenden Erdkunde als versuch einer Morphologie der Erdoberfläche.* In particular, he hypothesizes that fjords indenting the coastlines in high mid-latitudes were formed in the past by glacial activity.

EARTH SCIENCES (Loess)

Ferdinand von Richthofen (1833–1905, German) concludes that European, North American, and South American deposits of yellowish silt (loess) were deposited by wind during the last ice age, having rejected the thesis of a water deposition because of the homogeneity of the soil (lacking the stratification expected with water deposits), the high altitudes of some of the loess fields, and the findings of terrestrial fossils imbedded in the loess. Von Richthofen instead envisions that outwash streams from the edge of the ice sheet deposited silt which was then blown away by wind to cover nearby grasslands, an explanation which gains wide acceptance.

EARTH SCIENCES (Ocean Sediment Mapping)

Louis de Pourtalès (1823–1880, Swiss-American) maps the distribution of marine sediments off the east coast of the United States.

MATHEMATICS (Analysis)

Confidence in analytic methods lessens when Karl Weierstrass (1815–1897, German) proves invalid the earlier assumption that a particular equation of Pierre Simon de Laplace has a mathematical solution in the domain of continuous functions which would minimize a particular integral. The assumption of a solution, for instance by Georg Riemann in 1851, was based on the physical intuition that such a minimum should exist.

MATHEMATICS (Binary Forms)

Paul Gordan (1837–1912, German), in an extension of his work of 1868, establishes that not only does any binary form have associated with it a finite system of invariants and covariants, but any finite system of binary forms also has associated with itself such a system of invariants and covariants.

MATHEMATICS (Cayley's Theorem)

Camille Jordan (1838–1921, French) publishes his *Traité des substitutions et des équations algébriques* [Treatise on substitutions and algebraic equations], in which he proves Cayley's Theorem that every group of *n* elements is isomorphic to a subgroup of the permutation group on *n* symbols. Jordan refers to the work of Auguste Bravais (1848—CHEMISTRY) on the structure of crystals as evidence for the value of the theory of finite groups.

MATHEMATICS (Definition)

In the introduction to *Linear Associative Algebra* (1870—MATHEMATICS), Benjamin Peirce (1809–1880, American) defines mathematics as "the science which draws necessary conclusions."

MATHEMATICS (Function)

Hermann Hankel (1839–1873, German) presents his *Universitätsprogramme* in a paper subtitled "A Contribution to the Determination of the Concept of Function in General." He rejects earlier definitions of function as being either too general or not general enough and seeks a new definition which will be as broad as possible while still remaining conceptually useful. The result is a definition of function that requires the function and its differential quotients to be determined and finite for all except a finite number of values of the argument in any finite interval. Hankel also classifies linear functions as continuous or linear discontinuous functions and further classifies the latter as pointwise discontinuous or totally linearly discontinuous.

MATHEMATICS (Geometry; Consistency)

Felix Klein (1849–1925, German) establishes that if Euclidean geometry is consistent then non-Euclidean geometry is consistent as well.

MATHEMATICS (Hypernumbers)

In a short memoir on generalized hypernumbers, Benjamin Peirce (1809–1880, American) introduces the notions of character, direct units, and skew units.

MATHEMATICS (Linear Associative Algebras)

In the work *Linear Associative Algebra,* Benjamin Peirce (1809–1880, American) attempts to determine all linear associative algebras and to do so with a finite number of fundamental units. The book also presents a systematic study of hypercomplex numbers.

FIGURE 15. *An illustration from Jean Antoine Nollet's* Essai sur l'électricité des corps *[Essay on the electricity of bodies] (see 1746—*PHYSICS *entry) typifying the popular demonstrations of electrical phenomena in the mid-eighteenth century. A youth suspended on a board attached to the ceiling by silk cords is being electrified by induction, as a result of which his hand will attract the pieces of paper lying beneath it and the woman holding her finger to the youth's nose will draw a spark. (By permission of the Houghton Library, Harvard University.)*

MATHEMATICS (Linear Associative Algebras)

In an appendix to his father's *Linear Associative Algebra* (1870—MATHEMATICS), Charles Sanders Peirce (1839–1914, American) proves that the only linear associative algebras with real number coordinates and with $ab = 0$ implying $a = 0$ or $b = 0$ are the real numbers, the complex numbers, and the quaternions with real coefficients.

MATHEMATICS (Logic)

Charles Sanders Peirce (1839–1914, American) presents a notation, along the lines of Boolean algebra, for the logic of relations.

PHYSICS (Matter)

William Clifford (1845–1879, English) writes a paper "On the Space-Theory of Matter" and postulates that matter is simply "a manifestation of curvature in a space-time manifold."

1870–1875

CHEMISTRY (Periodic Classification)

Lewis Gibbes (1810–1894, American) constructs a "Synoptical Table of the Chemical Elements," which arranges the chemical elements into families according to valences. The work is not published until 1886, by which time the periodic tables of Dmitry Mendeleev (1869—CHEMISTRY) and J. Lothar Meyer (1869—CHEMISTRY) are well established. Like Mendeleev, Gibbes leaves blank spaces in his arrangement and predicts that elements will be discovered to fill them.

1870–1879

CHEMISTRY (Free Energy)

J. Willard Gibbs (1839–1903, American) applies the laws of thermodynamics to chemical reactions, developing the concept of "free energy," a quantity whose value cannot increase during a chemical reaction.

EARTH SCIENCES (Paleoclimatology; Lake Bonneville)

Grove Gilbert (1843–1918, American) examines fossils and superposed terraces in Utah and determines that the Great Salt Lake once extended over 20,000 square miles, the extended lake being termed Lake Bonneville. This confirms the earlier work of Howard Stansbury (1852—EARTH SCIENCES).

1871

BIOLOGY (Heart)

Henry Bowditch (1840–1911, American) reports that the contractions of the cardiac muscle increase in a steplike fashion in response to repeated uniform stimuli.

BIOLOGY (Human Evolution)

Charles Darwin (1809–1882, English) explicitly includes man as an evolved animal in his *The Descent of Man.*

BIOLOGY (Microorganisms)

Louis Pasteur (1822–1895, French) attributes the spoilage of beer to microorganisms.

BIOLOGY (Pterodactyl)

The first discovery of a pterodactyl in America is made by Othniel Marsh (1831–1899, American). The pterodactyl is an extinct flying reptile, with a long tail, teeth, and a membranous wing spreading from one finger to the side of the body.

EARTH SCIENCES (Corals)

Louis de Pourtalès (1823–1880, Swiss-American) publishes observationally based work on the corals of the deep sea.

MATHEMATICS (Non-Euclidean Geometry)

Felix Klein (1849–1925, German) uses Arthur Cayley's distance measure (1859—MATHEMATICS) to unify Euclidean and non-Euclidean geometries, showing that the various geometries can be represented in a manner so that the difference is represented fully in the distance function. Klein introduces the adjectives "parabolic," "elliptic," and "hyperbolic" for the respective geometries of Euclid, of Georg Riemann (1854—MATHEMATICS), and of Nikolai Lobachevsky (MATHEMATICS entries for 1826 and 1829), Carl Friedrich Gauss (1816—MATHEMATICS), and János Bolyai (MATHEMATICS entries for 1823 and 1832).

MATHEMATICS (Topology; Betti Numbers)

Enrico Betti (1823–1892, Italian) generalizes the connectivity numbers of Georg Riemann (1857—MATHEMATICS) to *n*-dimensional figures, creating what are now known as Betti numbers. The Betti number of a surface is the largest number of cross cuts which can be made without dividing the surface into more than one piece.

METEOROLOGY (Hurricanes)

Publication of Ludwig Colding's (1815–1888, Danish) *Les cyclones tropicaux* [Tropical cyclones].

PHYSICS (Electromagnetic Induction)

Hermann von Helmholtz (1821–1894, German) determines the propagation velocity of electromagnetic induction.

PHYSICS (Thermometry)

William Siemens (1823–1883, German) invents the platinum resistance thermometer, making use of the temperature-dependency of material conductivity (see the work of Humphry Davy under 1821—PHYSICS) to measure temperature.

1872

ASTRONOMY (Stellar Spectra)

Henry Draper (1837–1882, American) makes the first successful picture of the diffraction spectrum of a star other than the sun.

BIOLOGY (Emotions)

Charles Darwin (1809–1882, English) develops ideas of psychological evolution in *The Expression of the Emotions in Man and Animals.* His first extensive notes on this topic were made in 1839 as he recorded the development of the emotions exhibited by his first child. Later his observations expanded to emotions expressed by other humans and by domesticated animals.

EARTH SCIENCES (Subterranean Water)

Ludwig Colding (1815–1888, Danish) publishes the work *Des mouvement de l'eau souterraine* [On the movement of underground water].

HEALTH SCIENCES (Experimental Psychology)

Wilhelm Wundt's (1832–1920, German) *The Principles of Physiological Psychology* introduces the modern study of experimental psychology.

MATHEMATICS (Dedekind Cuts; Reals)

Publication of Richard Dedekind's (1831–1916, German) *Stetigkeit und irrationale Zahlen* [Continuity and irrational numbers], in which he presents a theory of the real number system based on Dedekind cuts, where, for any real number, the rational number line is divided into two classes, the class of rationals less than and the class of rationals greater than the given number. Any real number, rational or irrational, is thereby defined in terms of rationals. Dedekind proceeds to show how the cuts (A_1,A_2), (B_1,B_2) can be manipulated by such operations as addition and multiplication and establishes such properties as commutativity and associativity. In this work he also defines "infinite set," essentially stating that a set is infinite if it can be placed in one-to-one correspondence with a proper subset of itself.

MATHEMATICS (Functional Representations; Reals; Cauchy Sequences; Cantor Series)

Georg Cantor (1845–1918, German) publishes a proof that if a function can be represented by a trigonometric series then that representation is unique. In the course of the derivation, he introduces the concept of "derived point sets of the first species" and rigorously defines real numbers in terms of "fundamental" sequences of rationals $\{a_n\}$. Briefly, the first derived set of a point set P is the set $P^{(1)}$ of its limit points, the second derived set is the set $P^{(2)}$ of the limit points of $P^{(1)}$, . . . , and a set P is of the nth kind if $P^{(n)}$ has a finite number of points while $P^{(m)}$ has an infinite number of points for any $m < n$. The derived sets of the first species comprise all point sets of the rth kind, for finite r. The fundamental sequences, later termed Cauchy sequences, are sequences for which there exists an integer N with the property that for any integers $n > N$ and $m > 0$, the absolute value of $(a_{n+m} - a_n)$ is less than any preassigned $\varepsilon > 0$. Cantor assigns two Cauchy sequences $\{a_n\}$ and $\{b_n\}$ the same limit if $|a_n - b_n| < \varepsilon$ for sufficiently large n. A real number becomes an equivalence class of all Cauchy sequences with the same limit. In the same treatise Cantor shows that any positive real number can be represented by a series of the type $c_1 + (c_2/2!) + \ldots + (c_n/n!) + \ldots$, with $0 \leqslant c_n \leqslant n-1$, such a series later being termed a Cantor series.

MATHEMATICS (Geometry)

Felix Klein (1849–1925, German) characterizes various geometries according to the properties of figures remaining invariant under specific transformations. He announces his *Erlanger Programm* to study transformation-group equivalence for the purpose of unifying the study of the diverse geometries.

MATHEMATICS (Group Theory; Sylow's Theorem)

Ludvig Sylow (1832–1918, Norwegian) discovers the fundamental theorem of permutation groups that if p is a prime number and α is an integer such that p^α divides the order of the group but $p^{\alpha+1}$ does not divide the order of the group, then the group has a subgroup of order p^α.

MATHEMATICS (Heine-Borel Theorem)

While proving that any real function continuous on a closed bounded interval is uniformly continuous, Heinrich Heine (1821–1881, German) states and uses the following theorem: If S is a countably infinite set of subintervals of a closed interval $[a,b]$ and every point x in $[a,b]$ is an interior point (or, in the case of a and b, an endpoint) of at least one member of S, then there exists a finite subset S_1 of S such that every point in $[a,b]$ is an interior point of at least one member of S_1. The latter theorem is later named the Heine-Borel Theorem, following Emil Borel's recognition of its importance and statement and use of it as a separate theorem in 1895.

PHYSICS (Mach's Principle)

Ernst Mach (1838–1916, Czechoslovakian) refutes Newtonian concepts of absolute space and the inherent inertia of material bodies, arguing instead that all motion is relative to matter and that a totally isolated body would have no inertia. Albert Einstein in 1918, after the introduction of his theories of relativity (PHYSICS entries for 1905 and 1916) coins the term "Mach's Principle" for the concept that the inertia of a body and in fact the local structure of space and time are determined by the rest of the mass and energy in the universe.

1872–1876

EARTH SCIENCES (Observational Oceanography)

Researchers on board H.M.S. *Challenger* collect 1,441 water samples, 13,000 plant and animal specimens, and hundreds of sea floor deposits as the vessel voyages around the world to explore the physical and biological characteristics of the major ocean basins. Over the subsequent two decades, 50 volumes are published on the results of the voyage.

1873

ASTRONOMY (Moon)

Édouard Roche (1820–1883, French) hypothesizes that the moon formed by condensing from a cloud surrounding the earth.

BIOLOGY (Evolution)

Walter Bagehot's (1826–1877, English) *Physics and Politics* applies the theory of evolution by natural selection to the evolution of human customs and institutions.

CHEMISTRY (Critical Temperature)

Johannes van der Waals (1837–1923, Dutch) provides a molecular explanation of the known phenomenon that there exists a critical temperature for a gas above which it can exist only as a gas and below which it can be condensed to a two-phase system of gas and liquid (1822—PHYSICS; 1869—CHEMISTRY).

CHEMISTRY (Ideal Gas Law; Van der Waals Equation)

Johannes van der Waals (1837–1923, Dutch) generalizes the ideal gas law ($Pv = RT$, with p, v, T the pressure, volume per mole, and temperature of the gas, and R a universal gas constant) to adjust for the differences between real gases and ideal gases. The law, later termed van der Waals equation, contains constants a,b which depend on the gas under consideration: $(P + a/v^2)(v - b) = RT$. The constants a,b account for the finite rather than zero size of the gas molecules and for the fact that molecules experience weak forces of attraction (van der Waals forces).

CHEMISTRY (Thermodynamics)

J. Willard Gibbs (1839–1903, American) devises a method of representing thermodynamic properties geometrically and presents his techniques in the paper "Graphical Methods in the Thermodynamics of Fluids."

EARTH SCIENCES (Ice Ages; Interglacials)

Amos Worthen (1813–1888, American) adds to the accumulating evidence that there existed warm, interglacial periods between ice ages by reporting the existence of a layer of humus-rich soil between glacial till layers in Illinois. Such a soil is believed to form only with a warm climate supporting abundant plant life.

HEALTH SCIENCES (Leprosy)

Gerhard Hansen (1841–1912, Norwegian) discovers rod-shaped bodies, *Mycobacterium leprae,* in biopsy specimens from leprosy patients. He believes that these microorganisms, later termed Hansen's bacillus, are the causative agent of leprosy and proceeds over the next decade to experiment extensively with them. This is reputedly the first instance of an investigator's suggesting that microorganisms cause a specific chronic disease.

MATHEMATICS (Convergence)

Paul DuBois-Reymond (1831–1889, German) attempts, with partial success, to create a general theory of tests for convergence of series.

MATHEMATICS (Curl)

In his *Treatise on Electricity and Magnetism* (1873—PHYSICS), James Clerk Maxwell introduces the concept of the curl of a vector.

MATHEMATICS (*e*)

Charles Hermite (1822–1901, French) proves that the base of the natural logarithms, $e = 2.718 \ldots$, is transcendental. This is the first familiar number shown to be transcendental (see 1844—MATHEMATICS).

MATHEMATICS (Fourier Series)

Paul DuBois-Reymond (1831–1889, German) presents an example of a continuous function that cannot be represented by a Fourier series (MATHEMATICS entries for 1807 and 1822).

MATHEMATICS (Projective Geometry)

Felix Klein (1849–1925, German) eliminates Euclid's parallel postulate from projective geometry, as the quality of being parallel is not invariant under projections.

MATHEMATICS (Spectral Analysis)

Cleveland Abbe (1838–1916, American) presents the basic principles for use of optical techniques in two-dimensional spectral analysis.

PHYSICS (Electromagnetism)

James Clerk Maxwell (1831–1879, Scottish) expands his famed earlier paper (1865—PHYSICS) into the book *Treatise on Electricity and Magnetism,* where he fully develops the ideas of Michael Faraday (1831—PHYSICS) and his own extensions into a mathematical theory of the electromagnetic field.

PHYSICS (Photoconductivity)

Willoughby Smith discovers photoconductivity in selenium, finding that the metal's conductivity increases when exposed to light.

1873–1876

MATHEMATICS (Biquaternions)

William Clifford (1845–1879, English) generalizes the quaternions of William Rowan Hamilton (1843—MATHEMATICS) into biquaternions $q + aQ$, where q and Q are quaternions and $a^2 = 1$. Biquaternions developed from Clifford's interest in the geometry of motion and can be used for the study of motion in both Euclidean and non-Euclidean spaces.

METEOROLOGY (Radiometer)

William Crookes (1832–1919, English) invents the radiometer, an instrument for measuring radiation intensity.

1874

ASTRONOMY (Sunspots)

Walter Maunder (1851–1928, English) initiates the sunspot record of the Greenwich Observatory, taking daily photographs and tabulating various sunspot features, including the positions, number, area, and motions of sunspots.

CHEMISTRY (Gallium)

Paul Lecoq de Boisbaudran (1838–1912, French) discovers gallium, using the spark spectrum to do so. He names the element gallium in honor of France and then proceeds to isolate it and examine its properties.

CHEMISTRY (Stereochemistry; Carbon)

Jacobus Van't Hoff (1852–1911, Dutch) and Joseph LeBel (1847–1930, French) independently develop the theory of stereochemistry through their examinations of the asymmetric, three-dimensional molecular structure of carbon, both being stimulated in their studies by the earlier work of Louis Pasteur (1846–1848—CHEMISTRY).

EARTH SCIENCES (Paleoclimatology)

James Geikie's (1839–1915, Scottish) text *The Great Ice Age* provides an extensive treatment of the ice age issue. He argues that the findings of multiple till layers, often separated by peat, convincingly suggest multiple glaciations rather than a single major glacial advance, and uses this as evidence supporting the theory of James Croll of an astronomically driven ice age cycle (1864—EARTH SCIENCES).

HEALTH SCIENCES (Osteopathy)

Andrew Still (1828–1917, American) founds osteopathy. Deciding that all diseases originate from dislocations of the vertebrae, Still advocates treatment through massage and physical manipulation rather than drugs.

MATHEMATICS (Infinities; Algebraic Numbers; Reals; Set Theory)

Georg Cantor (1845–1918, German) distinguishes the sizes of infinite sets, defining two such sets to have the same "power" if and only if their elements can be put into one-to-one correspondence. (This definition is implicit in Cantor's 1874 paper, explicit later [1879—MATHEMATICS].) He then proceeds to prove that the set of all algebraic numbers has the same power as the set of natural numbers, that is, that it is "denumerable" or "countable," but the set of real numbers is not denumerable. He thereby further shows that transcendental numbers not only exist but are nondenumerably infinite. Cantor's paper presenting this work, "Über eine Eigenschaft des Inbegriffes aller reellen algebraischen Zahlen" [On a property of the collection of all real algebraic numbers], is considered the first formal publication on set theory.

MATHEMATICS (Logic)

Posthumous publication of Augustus De Morgan's (1806–1871, English) treatise *Formal Logic: Or, the Calculus of Inference, Necessary and Probable.*

PHYSICS (Electron)

Johnstone Stoney (1826–1911, Irish-English) deduces from (1) the electromagnetic nature of light and (2) the property of atoms of emitting characteristic spectra, that a type of electrical vibrator must exist within the atom to generate the light. He coins the term "electron" for this vibrator and asserts that the magnitude of its charge is the same as that on a hydrogen atom during electrolysis.

1875

BIOLOGY (Evolution)

In a letter to Francis Galton, Charles Darwin (1809–1882, English) indicates his growing conviction that evolution cannot be explained by natural selection alone but is adjusted by the inheritance of acquired characteristics.

BIOLOGY (Evolution)

Melchior Neumayr (1845–1890, German) supports Charles Darwin's theory of species evolution (1859—BIOLOGY) with evidence from a study of invertebrate fossils.

BIOLOGY (Fertilization)

With the aid of a microscope, Oscar Hertwig (1849–1922, German) becomes the first to observe the spermatozoon, or male germ cell, unite with the egg, or female germ cell. The observation is for the male and female germ cells of the sea urchin and is described by Hertwig in the article "Beiträge zur Kenntnis der Bildung, Befruchtung und Theilung des thierischen Eies," published in 1876.

BIOLOGY (Horticulture)

Luther Burbank (1849–1926, American) establishes a plant nursery in California, where he cultivates new varieties of grasses, grains, berries, fruits, and vegetables.

CHEMISTRY (Gallium)

Dmitry Mendeleev (1834–1907, Russian) establishes that gallium, discovered in 1874, is eka-aluminum, the missing element between aluminum and indium in Mendeleev's periodic table of the elements (1869—CHEMISTRY). The properties of gallium correspond well with those predicted, having an atomic weight of 69.9 and specific gravity of 5.94, compared to the predicted values of 68 and 5.9, and, as predicted, having the oxide formula Ga_2O_3, forming salts of the type GaX_2, slowly dissolving in acids and alkalies, and being unaffected by air.

EARTH SCIENCES (Bottom-Water Formation)

Joseph Prestwich (1812–1896, English) examines ocean temperature data recorded over the past 100 years and determines that at no location in the North Pacific does there exist significant sinking to depth (or bottom-water formation).

EARTH SCIENCES (Canyons)

In *Explorations of the Colorado River of the West,* John Wesley Powell (1834–1902, American) suggests that the Uinta canyons were created by rivers cutting through gradually rising rock.

EARTH SCIENCES (Graded Rivers)

Grove Gilbert (1843–1918, American) publishes a *Report on the Geology of the Henry Mountains,* a classic landform study. Gilbert is credited with developing the concept of graded rivers, in which an equilibrium has been reached between the slope of the land, the volume and velocity of the water, and the amount of detritus carried.

EARTH SCIENCES (Paleoclimatology; Ice Ages)

In *Climate and Time in Their Geological Relations: A Theory of Secular Changes of the Earth's Climate,* James Croll (1821–1890, Scottish) presents an astronomically based theory of the ice ages, expanded from his earlier works (EARTH SCIENCES entries for 1864 and 1867). He now adds the tilt of the earth's axis to his earlier consideration of the eccentricity of the orbit and the precession of the equinoxes. He feels that minimum tilt would be most conducive to ice ages, as it would result in the smallest amount of heat reaching the polar regions. As before, he believes the Northern and Southern Hemispheres undergo major glaciations alternately rather than simultaneously. The theory is received enthusiastically, though loses support in the 1880s and early 1890s, when geological evidence suggests that the Northern Hemisphere ice age lasted until much more recently than the peak calculated by Croll at 80,000 years before the present.

METEOROLOGY (Buys Ballot's Law)

Christoph Buys Ballot (1817–1890, Dutch) states that Northern Hemisphere winds blow in such a direction that high pressure will be on the right, low pressure on the left when a person stands with his back to the wind. The failure of the winds to blow directly from high to low pressure follows basically from the Coriolis effect, a concept developed by Gaspard de Coriolis (1835—METEOROLOGY) and applied to meteorology by William Ferrel (1855—METEOROLOGY).

METEOROLOGY (Supersaturation)

Paul Coulier (French) examines the phenomenon of supersaturation, determining through experiments with fog in an enclosed system that condensation does not always occur when cooling temperatures reach the saturation point. As the amount of fog produced in the enclosed system decreases with repeated heatings and coolings of the same mass of air, Coulier concludes that invisible dust particles initially serve as vital nucleii around which the water vapor condenses, and that after the dust settles to the ground, supersaturation can occur.

1875–1897

BIOLOGY (Indian Flora)

Joseph Hooker (1817–1911, English) publishes the seven-volume *Flora of British India.*

1876

ASTRONOMY (Solar Spectrum)

John Draper (1811–1882, American) takes the first photograph of the solar spectrum.

BIOLOGY (Criminal Psychology)

Cesare Lombroso (1836–1909, Italian) founds the science of criminal psychology through the publication of a pamphlet on the origin of criminal tendencies. Centering attention on the criminal rather than the crime, Lombroso develops the concept of the born criminal and encourages humane sentencing and treatment.

BIOLOGY (Fertilization)

In *The Effect of Cross- and Self-Fertilisation in the Vegetable Kingdom,* Charles Darwin (1809–1882, English) shows experimentally that cross fertilization increases the vigor and size of the offspring. He also describes the various means by which pollen is transported from one plant to another.

BIOLOGY (Biogeography; Evolution)

In the two-volume *Geographical Distribution of Animals,* Alfred Wallace (1823–1913, English) systematizes the science of biogeography and presents further evidence for the theory of evolution. He lists species of Irish elks, cave lions, rhinoceroses, elephants, hippopotamuses, and saber-toothed cats as extinct fauna formerly living in Europe, and species of large felines, horses, tapirs, llamas, mastodons, and ground sloths as extinct fauna from North America. He attempts a weather-based explanation of the apparent past massive extinction of large animals, claiming that these occurred during major ice ages, as a result of the cold, floods, and changes in sea level.

EARTH SCIENCES (Ice Ages)

J. J. Murphy in the paper "Glacial Climate and Polar Ice-Cap" rejects James Croll's earlier hypothesis that the combination of a long cold winter and short hot summer is conducive to glaciation, suggesting instead that glaciations are more likely under conditions of a long cool summer and a short mild winter, the cool summer preventing full melt of the snow cover accumulating during the mild winter.

HEALTH SCIENCES (Anthrax)

Robert Koch (1843–1910, German) demonstrates that bacilli are the cause of anthrax, determines that these organisms can be grown outside the animal body, and prepares pure cultures of the anthrax bacillus.

HEALTH SCIENCES/BIOLOGY (Brain)

John Hughlings Jackson (1835–1911, English) concludes from studies with a brain-damaged patient that perception is controlled from the rear of the right hemisphere of the brain.

MATHEMATICS (Differential Equations; Cauchy-Lipschitz Theorems)

Rudolph Lipschitz (1832–1903, German) further develops the 1820–1830 existence theorems of Augustin Louis Cauchy (1820–1830—MATHEMATICS), obtaining the Cauchy-Lipschitz Theorems.

PHYSICS (Applied Physics; Telephone)

Piecing together results obtained by himself and others, Alexander Graham Bell (1847–1922, American) applies the electromagnetic theory to long distance transmission of sound, inventing the telephone. The basic apparatus consists of a wire with each end wrapped around an iron bar and with a metal foil membrane placed close to each bar. Sound entering one membrane causes the membrane to vibrate, which alters the magnetic field around the first iron bar. This sends a current along the wire to the second iron bar and sets up a second fluctuating magnetic field. The second membrane then vibrates in the manner of the first and the sound is reproduced.

PHYSICS (Cathode Rays)

Eugen Goldstein (1850–1930, German) coins the term "cathode rays" for the invisible rays apparently emitted from the cathode of a tube filled with rarefied gas. The existence of the rays had been known at least since 1869.

PHYSICS (Philosophical; Pragmatism)

Charles Sanders Peirce (1839–1914, American) formulates the pragmatist principle that our conception of an object is contained entirely within our conception of the practical effects that the object might have.

PHYSICS (Research Laboratories)

Thomas Edison (1847–1931, American) sets up his "invention factory" at Menlo Park, New Jersey. This becomes the prototype of later industrial research centers. Among the 1876 inventions at the laboratory is a mimeograph machine.

1876–1878

CHEMISTRY (Chemical Thermodynamics; Phase Rule)

In the two-part memoir "On the Equilibrium of Heterogeneous Substances," J. Willard Gibbs (1839–1903, American) applies the first and second laws of thermodynamics to heterogeneous substances, helping to lay the foundations for chemical thermodynamics and, more broadly, for modern physical chemistry. He develops the concept of chemical potential and derives the following phase rule: $F = C + 2 - P$, where C is the number of separate substances in the system, P is the number of phases (gas, liquid, solid), and F is the number of degrees of freedom.

1876–1880

METEOROLOGY (Atmospheric Circulation)

Henrik Mohn (1835–1916, Norwegian) and Cato Guldberg (1836–1902, Norwegian) publish the two-volume *Études sur les mouvements de l'atmosphère* [Studies on the circulation of the atmosphere], providing a theoretical foundation for the study of dynamical meteorology. They incorporate such aspects as the Coriolis deflection (METEOROLOGY entries for 1835 and 1855) and the friction between earth and atmosphere.

1876–1896

BIOLOGY (Social Evolution)

In the three-volume *Principles of Sociology,* Herbert Spencer (1820–1903, English) applies the concept of evolution to human society. He analyzes the process leading to differentiation of the individual from the group and the coincident increase in individual freedom.

1877

ASTRONOMY (Mars; Extraterrestrial Life)

Giovanni Schiaparelli (1835–1910, Italian) detects a network of almost linear markings on Mars. He refers to them as *Canali,* or "channels," but this is later translated as "canals." Much speculation is generated on the probable level of intelligence of the Martian inhabitants able to construct these canals, presumably to carry polar water into the arid lower latitudes.

ASTRONOMY (Phobos and Deimos)

Asaph Hall (1829–1907, American) discovers two satellites of Mars, with diameters of 11 km and 6 km respectively. Astounded by the apparent prediction of the satellites by Jonathan Swift in his 1726 *Journey to Laputa,* Hall names them Phobos and Deimos, meaning Fear and Terror.

BIOLOGY (Evolution)

Othniel Marsh (1831–1899, American) helps substantiate the theory of evolution in his *Introduction and Succession of Vertebrate Life in America.*

CHEMISTRY (Low-Temperature Chemistry)

Oxygen, carbon monoxide, and nitrogen are all liquefied, requiring temperatures of 90 K, 83 K, and 77 K respectively (see 1869—CHEMISTRY entry on critical temperatures). This is achieved independently by Louis Cailletet (1832–1913, French) and Raoul Pictet (1846–1929, Swiss), as both respond to the need for easily transportable forms of the substances.

CHEMISTRY (Reactions)

Jacobus Van't Hoff (1852–1911, Dutch) classifies reactions and defines orders of reactions in terms of the number of molecules actively involved in the reaction.

CHEMISTRY (Refrigeration)

Carl von Linde (1842–1934, German) develops a commercially viable refrigerator, doing so by utilizing the Joule-Thomson cooling of ammonia gas as it is expanded without the addition of external work (see 1852—CHEMISTRY).

EARTH SCIENCES (Observational Oceanography; Historical Geography)

Alexander Agassiz (1835–1910, American) begins his oceanographic explorations in the Pacific and the Caribbean. In view of the similarity of the deep-sea animals he postulates that the Pacific and Caribbean had at one period been connected and that it was probably in the Cretaceous that the Panama isthmus rose to divide them.

HEALTH SCIENCES (Culturing)

Robert Koch (1843–1910, German) develops a method of obtaining pure cultures of individual bacterium.

MATHEMATICS (Differential Equations)

George Hill (1838–1914, American) invents methods for analyzing the trajectory of the moon's perigee, inserting, probably for the first time, infinite determinants into the study of differential equations.

MATHEMATICS (Hyperelliptic Theta Functions)

Arthur Cayley (1821–1895, English) discovers a parametric representation by hyperelliptic theta functions of the singular surface of the quadratic line complex.

MATHEMATICS (Logical Duality)

In the 37-page *Der Operationskreis des logikkalkuls,* Ernst Schröder (1841–1902, German) introduces the concept of logical duality, where two schemata are duals of each other if and only if their truth value structure is identical once T and F have been completely interchanged in one of them. Schröder also enunciates a principle of duality in Boolean algebra: if in a valid proposition, logical addition and multiplication are interchanged, the resulting proposition remains valid.

MATHEMATICS (Topology)

William Clifford (1845–1879, English) determines the topological equivalence of (1) the "many-sheeted" surfaces of Georg Riemann (1851—MATHEMATICS) and (2) boxes with holes, the number of holes in the boxes of (2) being the genus of the corresponding surface of (1).

PHYSICS (Applied Physics; Microphone)

Emile Berliner (1851–1929, German) invents the microphone.

PHYSICS (Applied Physics; Phonograph)

Thomas Edison (1847–1931, American) invents the phonograph. After developing his method for recording Morse code (1877—PHYSICS), Edison turns to the problem of recording sound and does so by positioning a needle to respond to a vibrating membrane and to create a track on a piece of tinfoil rolled round a cylinder, which revolves and advances to create a spiral path. After the recording, when the needle is positioned at the start and the cylinder is rotated at the same speed as for the recording, the needle causes the membrane to vibrate as it had earlier and thereby re-create the sound. The invention is acclaimed immediately.

PHYSICS (Applied Physics; Telegraphy)

Thomas Edison (1847–1931, American) develops a means of storing Morse code telegraph messages: the incoming message causes a stylus to move vertically, and this movement creates indentations on a revolving waxed paper disc. The message can then be obtained by turning the disc over and letting the stylus track across the bumps on the reverse side, making and breaking contact with a switch while doing so.

PHYSICS (Second Law of Thermodynamics; Ergodic Hypothesis)

Ludwig Boltzmann (1844–1906, Austrian) revises the second law of thermodynamics (1850—PHYSICS) to place it in a statistical framework. Each microscopic configuration possible for a gas with a given energy content is presumed to have an equal likelihood, and therefore the probability of any particular macroscopic condition is proportional to the number of microstates yielding it. Hence, in contrast to the original formulation of the second law, there is a finite, albeit small, probability for the entropy of a system to decrease. In line with the statistical viewpoint, Boltzmann and James Clerk Maxwell (1831–1879, Scottish) introduce an early version of the yet-to-be-named "ergodic hypothesis" that, given sufficient time, a system will pass through all possible microstates compatible with its total energy content.

1877–1878

PHYSICS (Sound)

Publication of John William Strutt's (Lord Rayleigh, 1842–1919, English) *Theory of Sound,* including original contributions to the analysis of resonance, vibration, and diffraction.

1878

ASTRONOMY (Moon)

George Darwin (1845–1912, English) suggests that before separating, the moon and earth had been a single spinning body. Extrapolating from the current rate of recession of the two bodies and the earth's slowly decreasing angular momentum, he calculates an earth-moon distance of 6,000 miles at 50 million years before the present, compared to the current 240,000 mile separation.

BIOLOGY (Physiology, Digestion)

Ivan Pavlov (1849–1936, Russian) describes the digestive effects in rabbits of preventing flow through the ducts of the pancreas.

CHEMISTRY (Dyes)

Adolf von Baeyer (1835–1917, German) synthesizes the dye indigo blue. Technical difficulties delay large-scale production until 1897.

CHEMISTRY (Ytterbia)

Jean de Marignac (1817–1894, Swiss) discovers the new earth ytterbia.

EARTH SCIENCES (Drift Currents)

K. Zöppritz presents a theoretical treatment of drift currents in "Hydrodynamische Probleme in Beziehung zur Theorie der Meeresströmungen." His neglect of the Coriolis force (1835—METEOROLOGY), which he attempts to justify on the basis of the slowness of movement of ocean currents, leads to the conclusion that the wind-driven flow of the ocean is in the direction of the wind independent of depth in the ocean, with a speed which decreases with depth. The model is revised by Vagn Ekman (1905—EARTH SCIENCES) after Fridtjof Nansen notes the deflection to the right of wind-induced sea ice drift and speculates that the cause derives from the earth's rotation (1902—EARTH SCIENCES).

EARTH SCIENCES (Experimental Geology: Folding and Faulting)

Gabriel Daubrée (1814–1896, French) describes the apparatus and results of experiments simulating the folding and faulting of strata through the exertion of horizontal and vertical pressures on beds of wax or metal. Daubrée finds that fairly uniform folds form when homogeneous beds under spatially uniform vertical pressures are subjected to sufficient horizontal pressure, the amount of pressure influencing the number and shape of the folds. Continued pressure in such a case results in a regular series of synclines and anticlines. By contrast, spatial nonuniformities in the bed thickness or in the vertical pressures can produce marked asymmetries in the resulting folds, with, for instance, fewer folds in regions of greater vertical pressure.

EARTH SCIENCES (Paleoclimatology)

Gerard DeGeer (1858–1943, Swedish) notes the regular laminations in the sediments of glacier-fed lakes, a pair of coarse and fine sediment layers occurring for each annual cycle due to the settling of glacial debris from the melt water each melt season. DeGeer proceeds to count varves in many Scandinavian lakes and to estimate from them that the retreat of the Pleistocene ice started about 12,000 years ago. By comparing the thicknesses of the varves throughout the 12,000 years, he is able to present a rough idea of the climatic sequence during that period.

MATHEMATICS (Dimension)

Georg Cantor (1845–1918, German) shows, in "Beitrag zur Mannigfaltigkeitslehre," that the points of a unit line segment can be placed in one-to-one correspondence with the points of a unit square and in fact with the points of a unit hypercube in *n*-dimensional space. This for some raises serious questions about the concept of dimension.

MATHEMATICS (Dimension)

Spurred by Georg Cantor's one-to-one mapping of the points of a unit square onto the points of a line segment (1878—MATHEMATICS), Jacob Lüroth (1844–1910, German) and Eugen Netto (1846–1919, German) independently prove that such a mapping cannot be continuous, and indeed that no continuous one-to-one mapping exists between a straight line and a space of dimension greater than 1.

MATHEMATICS (Group Theory)

Arthur Cayley (1821–1895, English) asserts that the properties of groups are defined by their multiplication tables, thereby asserting the essential equivalence of isomorphic groups.

PHYSICS (Applied Physics; Generators)

Thomas Edison (1847–1931, American) invents the bipolar dynamo.

PHYSICS (Bolometer)

Samuel Langley (1834–1906, American) invents the bolometer, an instrument for measuring radiation through focusing it on one of a pair of thermistors, the other being used as a reference.

PHYSICS (Velocity of Light)

Albert Michelson (1852–1931, American) publishes a paper entitled "On a Method of Measuring the Velocity of Light," a subject to become of lifelong interest for Michelson (for example, see PHYSICS entries for 1879 and 1887).

1879

ASTRONOMY (Moon)

George Darwin (1845–1912, English) publishes his theory that the moon originated by splitting from the earth (1878—ASTRONOMY), and suggests the cause of the splitting-off process to be an instability produced by resonant tides.

BIOLOGY (Experimental Psychology)

Wilhelm Wundt (1832–1920, German) founds the first experimental psychology laboratory.

CHEMISTRY (Incandescent Lamp)

Thomas Edison (1847–1931, American) and Joseph Swan (1828–1914, English) independently invent the incandescent lamp, spurred on by Humphry

Davy's earlier discovery that light is produced when a current is passed through a thin platinum wire. Edison tested 6,000 materials before deciding upon carbon for the filament. Carbon had been tried by others over the previous half century but unsuccessfully because they were unable to create an adequate vacuum to prevent the carbon from oxidizing. Edison obtains the necessary vacuum with the Sprengel pump newly arrived from Germany.

CHEMISTRY (Saccharin)

Constantine Fahlberg and Ira Remsen (1846–1927, American) produce orthobenzoyl sulfimide, or saccharin, and note its very sweet taste. Their production process is based on using potassium permanganate to oxidize orthotoluene sulfamide.

CHEMISTRY (Samarium)

Paul Lecoq de Boisbaudran (1838–1912, French) discovers samarium.

CHEMISTRY (Scandium)

Lars Fredrik Nilson (1840–1899, Swedish) discovers the element scandium, and Per Teodor Cleve (1840–1905, Swedish) shows it to be equivalent to the eka-boron predicted by Dmitry Mendeleev (1869—CHEMISTRY). As predicted, the element has atomic weight 45 (later adjusted to 44.956), forms an oxide of the form Sc_2O_3, has salts which are colorless and has a carbonate which is insoluble in water.

HEALTH SCIENCES (Puerperal Fever)

Louis Pasteur (1822–1895, French) identifies the bacteria (streptococcus) that causes puerperal, or childbed, fever.

MATHEMATICS (Infinite Point Sets)

In the first of a series of papers on infinite linear point sets, entitled "Ueber unendliche, lineare Punktmannigfaltigkeiten," Georg Cantor (1845–1918, German) sets out terminology and concepts, some of which had appeared earlier, which later assume major importance in his work: a set P is of the "second species" if no derived set $P^{(n)}$ (1872—MATHEMATICS), n finite, is empty; a set P is "everywhere dense" in an interval I if all subintervals of I contain points of P; two sets have the same "power" if the elements of the two sets can be placed in one-to-one correspondence. Cantor proceeds to develop his theory of point sets over the next several years and to complete a major work on transfinite set theory (1883—MATHEMATICS).

MATHEMATICS (Logic; Truth Functions)

Gottlob Frege (1848–1925, German) develops the concept of a truth function (a compound whose truth value is obtainable exclusively from the truth values of its component parts) and publishes *Concept Writing,* a major work leading toward the axiomatizing and mathematicizing of logic.

MATHEMATICS (Number Field; Group Characters)

In his *Eleventh Supplement* to the third edition of Gustav Peter Dirichlet's (1805–1859, German) *Vorlesungen über Zahlentheorie,* Richard Dedekind (1831–1916, German) provides perhaps the first explicit definition of a number field and formulates the concept of a group character. His definition of a number field F is that F is a collection of real or complex numbers such that for any a and b in F, $a+b$, $a-b$, and ab are also in F and if $b \neq 0$ then a/b is in F. The concept is later extended beyond the real and complex fields by Leopold Kronecker (1881—MATHEMATICS). Dedekind's concept of a group character is a function $g(x)$ defined on the elements x of an abelian group G such that $g(x) \neq 0$ for any element x of G and $g(xy) = g(x)\, g(y)$ for any two elements x and y of G. The concept is later extended from abelian groups to all groups.

MATHEMATICS (Power Series; Continued Fractions)

Edmond Laguerre (1834–1886, French) converts a power series which is divergent into a continued fraction which is convergent, increasing the importance of continued fractions to the study of the theory of functions.

PHYSICS (Hall Effect)

Edwin Hall (1855–1938, American) publishes a paper "On a New Action of the Magnet on Electric Currents" describing the thought processes and experiments leading to his discovery of the Hall effect, in which a difference of potential develops across a current-carrying conductor placed in a magnetic field.

PHYSICS (Stefan-Boltzmann Law)
ASTRONOMY (Solar Temperature)

Josef Stefan (1835–1893, Austrian) publishes in the paper "Ueber die Beziehung zwischen der Wärmestrahlung und der Temperatur" the empirical result that the total radiation from an object is proportional to the fourth power of its temperature. Among its applications, this law allows estimates of the sun's temperature. Ludwig Boltzmann later derives the law theoretically (1883—PHYSICS), as a result of which it is later generally referred to as the Stefan-Boltzmann Law.

PHYSICS (Velocity of Light)

Albert Michelson (1852–1931, American) carries out his first experiments to measure the velocity of light, preliminary to his better known experiments, both alone (1881—PHYSICS) and with Edward Morley (1887—PHYSICS).

SUPPLEMENTAL (Religion and Science)

Pope Leo XIII (Gioacchino Pecci, 1810–1903, Italian) encourages scientific and philosophical thought and particularly a revival of the philosophy of St. Thomas Aquinas in his encyclical *Aeterni Patris.*

1880

ASTRONOMY (Nebulae)

With his photograph of the Orion nebula, Henry Draper (1837–1882, American) becomes the first to successfully photograph a nebula.

ASTRONOMY (Satellite Orbits)

George Darwin (1845–1912, English) presents his research on satellite orbits in the paper "On the Secular Changes in the Elements of the Orbit of a Satellite Revolving About a Tidally Distorted Planet."

BIOLOGY (Chicken Cholera)

Louis Pasteur (1822–1895, French) discovers a procedure for immunizing chickens against chicken cholera. Chickens injected with an old culture of chicken cholera microbes become sick briefly but revive and are henceforth immune to new virulent cultures.

BIOLOGY (Extinct Birds)

Publication of Othniel Marsh's (1831–1899, American) *Odontornithes: A Monograph on the Extinct Toothed Birds of North America.*

BIOLOGY (Plant Motions)

In *Power of Movement in Plants,* Charles Darwin (1809–1882, English) endeavors to show that all plants, under such influences as light and gravity, have the capacity to move, and that the movements are all modifications of the plant's tendency to revolve.

EARTH SCIENCES (Mountain Formation)

Through laboratory experiments with a layer of soft clay and a band of thick rubber, Alphonse Favre (1815–1890, Swiss) demonstrates support for the theory that mountains such as the Alps, the Jura, and the Appalachians were likely formed under lateral compression.

EARTH SCIENCES (Seismology)

John Milne (1850–1913, English) invents the seismograph to record earthquakes.

HEALTH SCIENCES (Germ Theory)

Louis Pasteur (1822–1895, French) further develops the germ theory of disease in "On the Extension of the Germ Theory to the Etiology of Certain Common Diseases." He also suggests boric acid rather than carbolic acid (1867—HEALTH SCIENCES) as an antiseptic, boric acid having less odor and being less acidic.

HEALTH SCIENCES (Malaria)

Investigating the cause of the black pigment in the blood of malaria patients, Charles Laveran (1845–1922, French) discovers the parasite causing the disease and provides a detailed physical description of the parasite in the article "A New Parasite Found in the Blood of Malarial Patients: Parasitic Origin of Malarial Attacks."

HEALTH SCIENCES (Typhoid Fever)

Carl Eberth (1835–1926, German) discovers the germ causing typhoid fever, variously named typhoid bacillus, Eberth-Gaffky bacillus, or Eberthella typhosa.

MATHEMATICS (Computers)

Proceeding on a suggestion made by John Billings (1839–1913, American) while examining the laborious procedures involved in the compilation of the 1880 U.S. Census, Herman Hollerith (1860–1929, American) begins plans for the construction of a machine to partially automate future census tabulations (see 1890—MATHEMATICS).

MATHEMATICS (Fermat Numbers)

F. Landry factors the sixth Fermat number (1640—MATHEMATICS) $F_6 = 2^{64} + 1$ into the prime factors 274,117 and 67,280,421,310,721.

PHYSICS (Optics; Lorentz-Lorenz Relations)

Hendrik Lorentz (1853–1928, Dutch) and Ludwig Lorenz (1829–1891, Danish) discover the Lorentz-Lorenz relations connecting the refraction of light and the density of the refracting material.

1880–1882

HEALTH SCIENCES (Hysteria, Psychoanalysis)

Josef Breuer (1842–1925, Austrian) introduces use of catharsis and hypnosis in treating hysteria, finding the methods effective for the psychologically ill patient "Anna O." After finding Anna O's daily restless periods greatly eased on those days when she discussed her problems, he encourages such catharsis through hypnosis, eventually visiting her twice a day for intensive hypnosis sessions aimed at tracing memories back through time to reach crucial traumatic episodes. He concludes from the success of the treatment that Anna O's illness resulted from emotions which, not having been released in other manners, became embedded in the unconscious, and that the symptoms vanished after the unconscious causes were brought out through verbalization. Such conclusions prove fundamental to the later more general development of psychoanalysis by Sigmund Freud and Breuer (HEALTH SCIENCES entries for 1893 and 1895).

1881

BIOLOGY (Vegetable Mold; Worms)

Charles Darwin (1809–1882, English) publishes *The Formation of Vegetable Mould Through the Action of Worms, with Observations on their Habits,* in which he demonstrates once again the large effects which can occur from the accumulation of small changes acting over large spaces or long time intervals. Darwin argues that through loosening the soil and churning rock particles to ever smaller pieces, earthworms create the upper layer of the soil and shape the landscape.

CHEMISTRY/PHYSICS (Atomic Theory)

In a prize-winning essay, Joseph John Thomson (1856–1940, English) details the difficulties inherent in the current atomic theory, in which atoms are perceived as vortices in the ether.

HEALTH SCIENCES (Yellow Fever)

Carlos Finlay (1833–1915, Cuban) speculates that the mosquito is the carrier of yellow fever (see 1900—HEALTH SCIENCES).

MATHEMATICS (Calculus)

Vito Volterra (1860–1940, Italian) presents examples of functions that are bounded and differentiable but do not have Riemann integrals.

MATHEMATICS (Differential Equations)

Henri Poincaré (1854–1912, French) shifts the emphasis in the study of differential equations, suggesting a thoroughgoing qualitative or topologic analysis of a system before analysis on a quantitative level.

MATHEMATICS (Fields)

Leopold Kronecker (1823–1891, German) extends Richard Dedekind's fields of real and complex numbers (1879—MATHEMATICS) by including as fields all domains of rationality.

MATHEMATICS (Venn Diagrams)

After examining earlier uses of geometric diagrams in the study of syllogistic logic and finding them inadequate, John Venn (1834–1923, English) more systematically develops a method of geometric representation, including what are later termed Venn diagrams, in which, for instance, overlapping circles labeled *A* and *B* represent the proposition that "Some *A* are *B*" and a small *A* circle within a larger *B* circle represents the proposition that "All *A* are *B*." Venn describes his methods in *Symbolic Logic,* a book largely devoted to interpreting, revising, and extending the work of George Boole in the *Laws of Thought* (1854—MATHEMATICS).

METEOROLOGY (Condensation Nucleii)

John Aitken (1839–1919, Scottish) identifies ocean spray, volcanoes, condensed gases, and combustion as sources of atmospheric particulate matter facilitating condensation of water vapor.

PHYSICS (Interferometer)

Albert Michelson (1852–1931, American) invents and constructs an interferometer to measure distances precisely through dividing a beam of light in two and examining the interference patterns. He then uses this instrument to perform an early version of the later Michelson-Morley experiment done with Edward Morley (1887—PHYSICS).

1881–1884

MATHEMATICS (Vector Analysis)

J. Willard Gibbs (1839–1903, American) develops a general, three-dimensional vector analysis, independent of the then better-known quaternions (1843—MATHEMATICS), in the widely read pamphlet *Elements of Vector Analysis.*

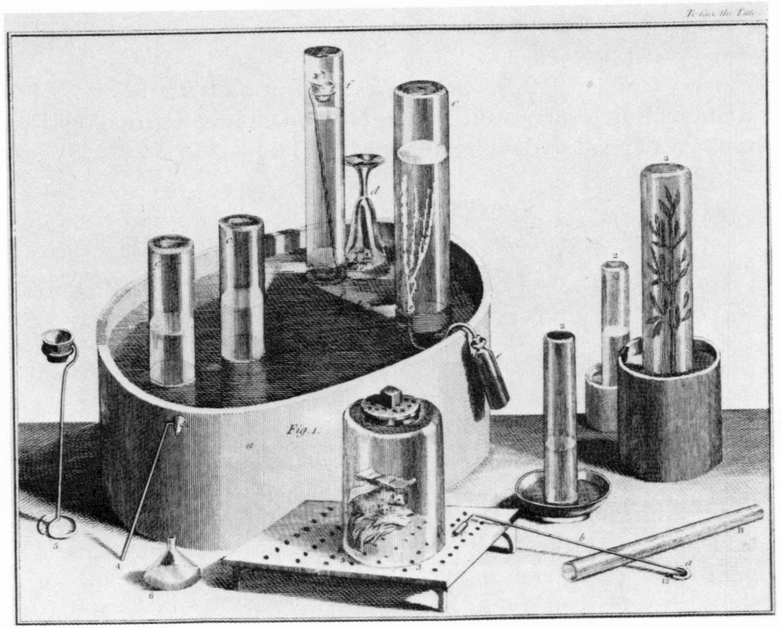

FIGURE 16. *Apparatus used by Joseph Priestley in investigating variously modified airs and their effects on plants, mice, and burning candles. The illustration is from Priestley's* Experiments and Observations on Different Kinds of Air *(see 1774–1777—*CHEMISTRY *entry). (By permission of the Houghton Library, Harvard University.)*

1882

ASTRONOMY (Solar Corona)

Arthur Schuster (1851–1934, German-English) photographs the spectrum of the solar corona.

BIOLOGY (Psychical Research)

The Society for Psychical Research is formed, with the intent of encouraging application of scientific methods to analysis of psychical phenomena.

CHEMISTRY (Liquefied Gas)

Jules Violle (1841–1923, French) designs a method of keeping liquefied gas sufficiently cool to remain in the liquid state. Using the fact that a vacuum does not transmit heat, Violle stores the liquid gas in glass containers with double walls separated by a vacuum and with silvering on the exterior wall to reduce radiation input.

EARTH SCIENCES (Pacific Ocean)
ASTRONOMY (Moon)

Osmond Fisher extends George Darwin's theory of the splitting off of the moon from the earth (1878—ASTRONOMY), suggesting that this splitting process formed the Pacific and other ocean basins.

HEALTH SCIENCES (Anthrax)

Louis Pasteur (1822–1895, French) demonstrates the success of an anthrax vaccine prepared from an inactive strain of the anthrax bacteria. In a widely publicized public demonstration, Pasteur vaccinates 24 sheep, 1 goat, and a few cattle, then, a few weeks later, injects these animals plus 24 more sheep, another goat, and a few more cattle with a heavy dose of active anthrax bacilli. The vaccinated animals all survive, while the unvaccinated ones die.

HEALTH SCIENCES (Tuberculosis)

Robert Koch (1843–1910, German) discovers *Mycobacterium tuberculosis,* the germ causing tuberculosis. This is supposedly the first definite discovery of a specific microbe causing a specific human disease.

MATHEMATICS (Divergent Series)

Otto Hölder (1859–1937, German) introduces new methods of treating divergent series through arithmetical summations.

MATHEMATICS (Geometry)

Guided partially by Benjamin Peirce's definition of mathematics as the science which draws necessary conclusions (1870—MATHEMATICS), Moritz Pasch (1843–1930, German) realigns the foundations of geometry, no longer seeking definitions for such terms as point, line, and plane but instead accepting those terms as undefined and proceeding from them along with unproved postulates providing relations among them. The work is published under the title *Vorlesungen über neuere Geometrie.*

MATHEMATICS (Klein Bottle)

Felix Klein (1849–1925, German) introduces the concept of a Klein bottle, a one-sided closed surface with no edge. (Such a bottle cannot be constructed in three-dimensional space.)

MATHEMATICS (π; Squaring the Circle)

Ferdinand Lindemann (1852–1939, German) proves that π is transcendental.

The basic outline of the proof follows: $e^{i\pi} + 1 = 0$; however, there are no a, b algebraic such that $e^a + b = 0$; thus $i\pi$ is not algebraic, implying—in conjunction with more elementary results—that π is not algebraic since i *is* algebraic. This leads to a proof that the ancient Greek problem of squaring a circle cannot be accomplished with straightedge and compass alone.

PHYSICS (Applied Physics; Electricity)

The age of widespread use of electricity begins as the first generating station for bringing electricity to private consumers opens in New York.

PHYSICS (Diffraction Gratings)

Henry Rowland (1848–1901, American) constructs a precision instrument for making diffraction gratings, successfully using it to produce 14,400 parallel, evenly spaced lines per inch. The machine becomes invaluable for future advances in spectroscopy.

1882–1883

METEOROLOGY/EARTH SCIENCES (International Polar Year)

Detailed meteorological and geophysical observations are made of the Arctic environment during the first International Polar Year.

1882–1891

EARTH SCIENCES (Geographic Impacts)

Friedrich Ratzel (1844–1904, German) traces various influences of geography on human history in the two-volume *Anthropogeographie.*

1883

BIOLOGY (Heart)

Ivan Pavlov (1849–1936, Russian) determines that augmenter nerves in the heart affect the force of the heart beat and the quantity of blood pumped outward. This supplements the result of previous researchers that the pacemaker controls the rate of the beat.

EARTH SCIENCES (Earthquakes)

James Ewing (1855–1935, Scottish) publishes a *Treatise on Earthquake Measurement.*

EARTH SCIENCES (Paleontology)

Melchior Neumayr (1845–1890, German) distinguishes five major zones in the distribution of animals during the Jurassic period—Boreal, Himalayan, Japanese, Mediterranean-Caucasian, and South Andean.

EARTH SCIENCES (Paleontology)

Edward Cope (1840–1897, American) publishes a survey report on *The Vertebrata of the Tertiary Formations of the West.*

HEALTH SCIENCES (Blood Transfusions)

In "Refusion in the Treatment of Carbonic Oxide Poisoning," William Halsted (1852–1922, American) describes a case of centripetal blood transfusion in which a patient subjected to carbonic oxide poisoning had his blood withdrawn, then reinserted after being defibrinated and reoxygenated.

HEALTH SCIENCES (Cholera)

Robert Koch (1843–1910, German) determines that the cause of Asiatic cholera rests in the comma bacillus.

HEALTH SCIENCES (Diphtheria)

Edwin Klebs (1834–1913, German-American) discovers the germ causing diphtheria (called the diphtheria bacillus or the Klebs-Loffler bacillus).

MATHEMATICS (Infinities; Continuum Hypothesis)

Georg Cantor states in his *Grundlagen einer allgemeinen Mannigfaltigkeitslehre* (1883—MATHEMATICS) the Continuum Hypothesis that there is no transfinite number greater than the number of integers and less than the number of reals. His many attempts over the next two decades to prove this conjecture all fall short.

MATHEMATICS (Transfinite Set Theory; Well-Ordered Sets)

Georg Cantor (1845–1918, German) develops transfinite set theory in the monograph *Grundlagen einer allgemeinen Mannigfaltigkeitslehre* [Foundations of a general theory of manifolds], introducing transfinite ordinal numbers (numbers as opposed to mere symbols) as a natural extension of the number concept. He emphasizes that the transfinites are completed, actual infinities and creates a consistent arithmetic for them, with both addition and multiplication associative but noncommutative and with two forms of prime numbers identified. The development includes the explicit introduction of the concept of a well-ordered set, that is, a well-defined set of elements in which there is a first element and every element (except the last if such exists) has a definite successor.

PHYSICS (Hydrodynamics; Reynolds' Number)

Osborne Reynolds (1842–1912, Irish) distinguishes between laminar and turbulent fluid flow in the article "An Experimental Investigation of the Circumstances which Determine Whether the Motion of Water Shall Be Direct or Sinuous, and of the Law of Resistance in Parallel Channels." The work follows experiments in which Reynolds inserted dye into liquids flowing through pipes with regulated velocities. Low fluid speeds result in a well-defined dye filament, signifying laminar flow, with the fluid particles following roughly parallel, roughly straightline paths, whereas high fluid speeds result in the dye moving irregularly throughout the fluid, signifying turbulent, irregular flow. The work leads to the subsequent naming of a nondimensional "Reynolds' number" to distinguish between laminar and turbulent flow.

PHYSICS (Stefan-Boltzmann Law)

Ludwig Boltzmann (1844–1906, Austrian) uses the second law of thermodynamics (1850—PHYSICS) and the electromagnetic theory of James Clerk Maxwell (1865—PHYSICS) to derive theoretically the proportionality of the radiation emitted from a body and the fourth power of the temperature of the body in Kelvin units. This result, earlier found experimentally by Josef Stefan (1879—PHYSICS), is later known as the Stefan-Boltzmann Law.

1883–1901

EARTH SCIENCES (Geology)

Publication of the five-volume *Das Antlitz der Erde* [The face of the earth] by Eduard Suess (1831–1914, Austrian).

1884

BIOLOGY (Dinosaurs)

Othniel Marsh (1831–1899, American) publishes *Dinocerata: A Monograph on an Extinct Order of Gigantic Mammals.*

CHEMISTRY (Affinity)

Jacobus Van't Hoff (1852–1911, Dutch) uses the maximum external work generated from a reversible, isothermal reaction as an index of chemical affinity.

EARTH SCIENCES (Overthrusting)

Marcel Alexandre Bertrand (1847–1907, French) describes large-scale overthrusting in the Alps and relates the geological structure of the Alps to that of the Pyrenees and Provence.

HEALTH SCIENCES (Anesthesiology)

Carl Koller (1857–1944, American) uses cocaine as a local anesthetic, to desensitize the region around the eye before eye surgery.

HEALTH SCIENCES (Tetanus)

Arthur Nicolaier (1862–1934, German) discovers the germ causing tetanus (the tetanus bacillus).

MATHEMATICS (Arithmetic; Cardinal Number)

Publication of Gottlob Frege's (1848–1925, German) *Grundlagen der Arithmetik* [The foundations of arithmetic], where he attempts to deduce arithmetic from logic. In this work Frege defines the cardinal number of a class to be the class of all classes similar to it, a definition later to gain significance but only after an independent restatement by Bertrand Russell in 1901. It has an advantage over some earlier definitions of number in that it applies to both finite and infinite classes. (Two classes are "similar" if they can be put into one-to-one correspondence with each other.)

MATHEMATICS (Differential Equations; Group Theory)

Publication of Marius Sophus Lie's (1842–1899, Norwegian) "Classification and integration of ordinary differential equations in X and Y with the aid of a group of P transformations." Lie also develops the notion and theory of a continuous group.

MATHEMATICS (Icosahedron; Algebraic Equations)

Felix Klein (1849–1925, German) develops his theory of the icosahedron and applies rotation groups of regular solids to solving algebraic equations.

MATHEMATICS (Integration; Theory of Content)

Georg Cantor (1845–1918, German) introduces a generalized integral for any point set, reducing to the standard Riemann integral (1854—MATHEMATICS) where such is defined, and develops a theory of content (or volume) for finite and transfinite point sets.

METEOROLOGY (Temperature)

Publication of William Ferrel's (1817–1891, American) "Temperature of the Atmosphere and the Earth's Surface."

1884–1887

CHEMISTRY (Ionization)

Svante Arrhenius (1859–1927, Swedish) develops a theory of electrolytic dissociation or ionization. He explains that acids, bases, and salts are broken into ions when dissolved in water and consequently conduct electricity, while other solutions do not.

1885

BIOLOGY (Fingerprints)

Francis Galton (1822–1911, English) establishes the individuality of human fingerprints.

CHEMISTRY (Gas Mantle)

Carl von Welsbach (1858–1929, Austrian) invents the incandescent gas mantle.

CHEMISTRY (Pasteurization)

Louis Pasteur (1822–1895, French) "pasteurizes" milk (see also 1856—BIOLOGY).

CHEMISTRY (Praseodymia; Neodymia)

Carl von Welsbach (1858–1929, Austrian) discovers the two earths praseodymia and neodymia by repeated fractionation of ammonium didymium nitrate.

EARTH SCIENCES (Continental Drift; Gondwanaland)

Eduard Suess (1831–1914, Austrian) postulates the prehistoric existence of a giant protocontinent termed Gondwanaland from which the current Southern Hemisphere land masses have all separated.

EARTH SCIENCES (Gulf Stream)

Prince Albert of Monaco (1848–1922, Monaco) sets out in the ship *Hirondelle* to study the Gulf Stream. Prior to this expedition it was unknown whether the Gulf Stream crosses the Atlantic after moving northward along the North American coastal zone. Crew members release 1,675 floats from various locations, 227 of which are later recovered and provide evidence for the transatlantic Gulf Stream hypothesis.

HEALTH SCIENCES (Rabies)

Louis Pasteur (1822–1895, French) develops and successfully administers a vaccination against rabies, then describes his procedures in "Method for Preventing Rabies after Bites." The inoculation derives from a virus taken from the spinal cord of a rabbit dead of rabies. Pasteur has successfully used the inoculation, in a sequence of increasingly virulent doses, on animals to develop in them an immunity from rabies when in July he is entreated to attempt the inoculation on a nine-year-old boy bitten by a mad dog. A sequence of 13 inoculations over 10 days proves successful.

HEALTH SCIENCES (Tuberculosis)

Edward Trudeau (1848–1915, American) opens the one-room Adirondack Cottage Sanatorium to provide rest, fresh air, and a cold dry climate for patients recovering from tuberculosis.

MATHEMATICS (Logic; Truth Values)

Charles Sanders Peirce (1839–1914, American) introduces the concept of truth values of a proposition, the forerunner to later truth tables. These truth values are binary, either true or false.

PHYSICS (Atomic Physics; Balmer Lines)

Johann Balmer (1825–1898, Swiss) creates a simple mathematical formula to list the particular wavelengths at which a hydrogen atom radiates light and publishes it in the paper "Notiz über die Spectrallinien des Wasserstoffs."

PHYSICS (Electromagnetism)

Galileo Ferraris (1847–1897, Italian) discovers the rotary magnetic field and contributes to developing alternating-current motors.

PHYSICS (Thermometry)

Henri LeChâtelier (1850–1936, French) improves the means of measuring high temperatures by demonstrating that a platinum wire in conjunction with a platinum-rhodium alloy wire enables a thermoelectric measurement of temperature.

1886

BIOLOGY (Evolution)

In *The Factors of Organic Evolution,* Herbert Spencer (1820–1903, English), although accepting large portions of Charles Darwin's theory of natural selection (1859—BIOLOGY), reiterates his support for the essentially Lamarckian position that acquired characteristics can be inherited, and that such inheritance is an important mechanism of evolution (1809—BIOLOGY).

CHEMISTRY (Aluminum; Hall-Héroult Process)

Charles Hall (1863–1914, American) and Paul Héroult (1863–1914, French) independently develop an electrolytic process for extracting aluminum from its ore, later called the Hall-Héroult process. Aluminum oxide is dissolved in molten cryolite, then electrolyzed. Molten aluminum forms at the cathode.

CHEMISTRY (Atomic Evolution)

William Crookes (1832–1919, English) hypothesizes that all existent elements evolved from a presumed primordial matter, termed "protyle." Crookes is influenced both by William Prout's hypothesis that all elements are built from hydrogen atoms (1816—CHEMISTRY) and by the uniformity among the elements revealed by the periodic table (1869—CHEMISTRY).

CHEMISTRY (Atomic Weights)

William Crookes (1832–1919, English) speculates that not all atoms of a given element have the same atomic weight. Using calcium as an example, he suggests that the atomic weight of 40 is actually an average over many atoms with weight 40, fewer with weight 39 or 41, fewer still with weight 38 or 42, and so on. (Compare against the later concept of isotopes, 1913—CHEMISTRY.)

CHEMISTRY (Fluorine)

After many unsuccessful and some fatal attempts by nineteenth-century chemists, fluorine is finally isolated by Ferdinand Henri Moissan (1852–1907, French). Moissan succeeds by dissolving potassium hydrogen fluoride in liquid hydrofluoric acid and electrolyzing the solution. Fluorine appears as a pale, yellow-green gas at the positive electrode.

CHEMISTRY (Germanium)

Clemens Winkler (1838–1904, German) discovers the eka-silicon predicted by Dmitry Mendeleev (1869—CHEMISTRY) and names it germanium. The properties are close to those predicted, the atomic weight, specific gravity, atomic volume, valence, and specific heat being 72.32, 5.47, 13.22, 4, and 0.076 respectively, compared to Mendeleev's predicted values of 72, 5.5, 13, 4, and 0.073.

CHEMISTRY (Holmium; Dysprosium)

By fractional precipitation, Paul Lecoq de Boisbaudran (1838–1912, French) separates holmium into two earths, naming one dysprosium and retaining the term holmium for the other.

MATHEMATICS (Analysis)

Karl Weierstrass (1815–1897, German) rigorously develops the theory of functions in *Abhandlungen aus der Funktionenlehre.*

MATHEMATICS (Asymptotic Series)

Henri Poincaré (1854–1912, French) develops a theory of asymptotic series representations of functions, proving such results as that the asymptotic representation of the product of two functions, if it exists, equals the product of the asymptotic representations of the two functions. He applies the theory to solving linear differential equations.

MATHEMATICS (Irrationals)

In *Vorlesungen über allgemeine Arithmetik,* Otto Stolz (1842–1905, Austrian) shows that all irrationals are nonperiodic decimals.

MATHEMATICS (Number)

Leopold Kronecker (1823–1891, German) utters his most famous statement: "Die ganzen Zahlen hat der liebe Gott gemacht, alles andere ist Menschenwerk" [God made the integers, all else is the work of man] at a meeting in Berlin.

METEOROLOGY (General)

Publication of William Ferrel's (1817–1891, American) *Recent Advances in Meteorology.*

PHYSICS (Canal Rays)

Eugen Goldstein (1850–1930, German) announces in "Ueber eine noch nicht untersuchte strahlungsform an der Kathode inducirter Entladungen" his discovery and analysis of Canal Rays appearing near the cathode in a high vacuum tube.

1886–1889

PHYSICS (Electromagnetism; Radio Waves)

Heinrich Hertz (1857–1894, German) confirms the electromagnetic theory of James Clerk Maxwell (1865—PHYSICS), producing, transmitting and receiving formerly unexhibited hertzian or radio waves in the laboratory, producing them with an unclosed circuit connected to an induction coil and detecting them with an unclosed loop of wire. He shows these waves to be transverse waves with wavelength longer than that of visible light but with velocity equal to that of visible light and with the same properties of reflection, refraction, and polarization. This establishes the basis for the later inventions of radio, television and radar. Hertz recognizes that the electromagnetic theory of Maxwell is embodied in the equations, not the physical model.

1887

ASTRONOMY (Solar Chemistry)

Publication of J. Norman Lockyer's (1836–1920, English) *The Chemistry of the Sun.*

CHEMISTRY (Calorimeter)

Robert Bunsen (1811–1899, German) invents the vapor calorimeter.

CHEMISTRY (Photographic Film)

Hannibal Goodwin (1822–1889, American) invents celluloid photographic film and develops a process for producing it.

EARTH SCIENCES (Climatology; Brückner Period)

Eduard Brückner (1862–1927, German-Austrian) describes a 35-year cycle in which periods of damp and cold weather alternate with periods of dry and warm weather.

EARTH SCIENCES (Mountain Formation)

Marcel Alexandre Bertrand (1847–1907, French) develops the theory that Europe has undergone three major periods of intense folding and orogeny, the first producing the Caledonian mountain chain, the second the Hercynian mountain chain, and the third the Alpine mountain chain.

MATHEMATICS (Arithmetic, Applied)

In *Counting and Measuring,* Hermann von Helmholtz (1821–1894, German) cautions against indiscriminant use of arithmetic for describing nature, providing some common examples of the inappropriateness of straightforward addition: for example, two volumes of hydrogen and one volume of oxygen yield two volumes of water rather than three.

MATHEMATICS (Functionals)

Vito Volterra (1860–1940, Italian) begins extensive research on functionals, or, in Volterra's terminology, "functions depending on other functions." Although individual functionals had long been familiar, because Volterra identifies and develops the concept, his work is later viewed as the start of functional analysis.

MATHEMATICS (Infinity)

Georg Cantor (1845–1918, German) proves that every infinite set has a subset which can be put in one-to-one correspondence with the set of positive integers.

MATHEMATICS (Reduction to Natural Numbers)

Leopold Kronecker (1823–1891, German) asserts that all geometry and mechanics can be expressed in terms of relations among the natural numbers. This is in the strict Pythagorean tradition, with the ideal of reducing all mathematics to the natural numbers. Kronecker banishes negatives through the use of congruences to the modulus $j + 1$ and fractions n/m through the use of congruences to the modulus $mk + nj$. He presents the details of his plan to base all of mathematics on the integers in the article "Ueber den Zahlbegriff."

MATHEMATICS (Topology; Jordan Curve Theorem)

Camille Jordan (1838–1921, French), in *Cours d'analyse,* states the Jordan Curve Theorem, that the Euclidean plane is divided into two parts by any simple closed curve.

PHYSICS (Michelson-Morley Experiment)

Albert Michelson (1852–1931, American) and Edward Morley (1838–1923, American) attempt to determine the velocity with which the earth moves through the surrounding ether. Using Michelson's interferometer (1881—PHYSICS) mounted on a stone floating in mercury, the experiment is constructed to direct two beams of light perpendicular to each other and have each reflected back to the starting point by mirrors at the same distance from the starting point. The aim is to detect a difference in travel times through interference patterns for the two returning beams. The unexpected result that the speed of light is independent of the direction of motion shows that the ether (if it exists) does not drift with respect to the earth. The experiment, an improved attempt at an earlier experiment by Michelson (1881—PHYSICS), provides the most precise determination yet achieved of the speed of light.

PHYSICS (Photoelectricity)

Through a series of experiments with spark gaps, Heinrich Hertz (1857–1894, German) determines that ultraviolet light falling on the electrodes increases the distance the spark jumps.

1887–1896

MATHEMATICS (Theory of Surfaces)

Publication of Jean Gaston Darboux's (1842–1917, French) four-volume *Le-*

çons sur la théorie générale des surfaces et les applications géométriques du calcul infinitésimal [Lessons on the general theory of surfaces and the geometrical applications of infinitesimal calculus], applying infinitesimal calculus to geometry in the study of the differential geometry of curves and surfaces.

1888

ASTRONOMY (Andromeda Nebula)

The spiral nature of the Andromeda nebula is revealed in a telescopic photograph.

ASTRONOMY (Nebulae)

Johann Dreyer (1852–1926, Danish-English) revises John Herschel's *A General Catalogue of Nebulae* (1864—ASTRONOMY) into *A New General Catalogue of Nebulae and Clusters of Stars.* This is the NGC standard reference source for nebulae and clusters.

ASTRONOMY (Rings of Saturn; Keeler Gap)

James Keeler (1857–1900, American) records observations of the Keeler Gap, the second most prominent gap in the ring system of Saturn, the first being the Cassini Division (1675—ASTRONOMY).

BIOLOGY (Central Nervous System)

Fridtjof Nansen (1861–1930, Norwegian) shows that the nerve fibers of the central nervous system divide into T formations in the spinal column.

EARTH SCIENCES (Paleovolcanology)

Publication of Archibald Geikie's (1835–1924, Scottish) *History of Volcanic Action During the Tertiary Period in Britain.*

MATHEMATICS (Algebraic Forms)

David Hilbert (1862–1943, German) extends Paul Gordan's result that every binary form has a finite complete system of invariants and covariants (1868—MATHEMATICS) to forms of any number of variables, not just binary forms.

MATHEMATICS (Calculators)

Leon Bollee (French) constructs a calculating machine that multiplies directly rather than by repeated additions.

MATHEMATICS (Geometrical Constructions)

Émile Lemoine (1840–1912, French) assigns a measure of simplicity to geometrical constructions, based on the number of times each of five elementary operations (such as placing one end of a compass on a given point) is employed.

MATHEMATICS (Mathematics as Logic)

Richard Dedekind (1831–1916, German), in *Was sind und was Sollen die Zahlen* [The nature and meaning of numbers], claims all of mathematics to be a branch of logic. He presents a theory of the integers using set-theoretic concepts and outlines a possible approach to placing the rationals on a logically well-founded axiomatic basis.

MATHEMATICS (Symbolism)

Giuseppe Peano (1858–1932, Italian) begins a long-term attempt to create a precise symbolism for all mathematics.

MATHEMATICS (Tensor Analysis)

Curbastro Ricci (1853–1925, Italian) develops tensor analysis through the algebra of quadratic differential forms and publishes it in *Della derivazione covariante e contra variante.*

PHYSICS (Electromagnetic Waves)

Oliver Lodge (1851–1940, English) finds that the electrical discharge from a Leyden jar (1745—PHYSICS) produces waves and standing waves along conducting wires and that these waves have wavelengths and other characteristics predicted by the electromagnetic theory of James Clerk Maxwell (1865—PHYSICS). This confirmation of Maxwell's theory, although substantial, is overshadowed by the independent results of Heinrich Hertz (1886–1889—PHYSICS).

PHYSICS (Theoretical Mechanics)
MATHEMATICS (Ultraelliptic Integrals)

Sonya Kovalevsky (1850–1891, Russian) receives the Prix Bordin of the French Academy of Sciences for her memoir "On the Problem of the Rotation of a Solid Body about a Fixed Point," in which she extends the work of Karl Weierstrass on ultraelliptic integrals and exhaustively solves a special case of the general problem of the rotation of a spinning top.

1888–1889

EARTH SCIENCES (Greenland)
METEOROLOGY (Observational)

An expedition led by Fridtjof Nansen (1861–1930, Norwegian) confirms that Greenland is nearly completely covered with ice. The detailed meteorological observations compiled during the winter lead to a better understanding of the weather conditions in northern Europe.

1888–1890

MATHEMATICS (Group Theory)

Publication of Marius Sophus Lie's (1842–1899, Norwegian) two-volume *Theories of Groups of Transformations.*

1888–1891

HEALTH SCIENCES (Diphtheria)

Émile Roux (1853–1933, French) Shibasaburo Kitasato (1852–1931, Japanese), and Emil von Behring (1854–1917, German) carry out extensive experimentation leading to an "antitoxin" (a term coined by Behring) for diphtheria. This is used on humans in 1891, with partial success.

1889

ASTRONOMY (Algol)

Hermann Vogel (1841–1907, German) determines that the periodic brightness variations in the star Algol are accompanied by shifts in the location of the spectral lines occurring in the spectrum of the star's light. These Doppler shifts suggest that there are two stars involved, which are orbiting and periodically obscuring each other.

BIOLOGY (Genetics)

Theodor Boveri (1862–1915, German) demonstrates through experiments with sea urchins that the genetic material is located in the nucleus. The crucial evidence comes from larvae derived through fertilization of nonnucleated egg fragments of one species with sperm from a different species. These larvae contain predominantly the characteristics of the latter species.

BIOLOGY (Genetics)
MATHEMATICS (Statistics; Standard Error)

Francis Galton (1822–1911, English) publishes *Natural Inheritance,* the outcome of his work over the past 15 years on statistical techniques related to genetics. This includes development of the concept of correlation coefficients and the formula for the standard error of estimate.

CHEMISTRY (Activation Energy)

Svante Arrhenius (1859–1927, Swedish) develops the concept of the "energy of activation," the amount of energy needed to initiate a chemical reaction. For example, when hydrogen and oxygen form water, energy is required not to form the H_2O molecule from hydrogen and oxygen, a process which is a "downward slide" along the "energy slope," but instead to break up the O_2 molecule in order to obtain the uncombined oxygen atoms. These atoms then readily join with the available hydrogen.

EARTH SCIENCES (Rivers; Cycle of Erosion)

William Morris Davis (1850–1934, American) presents in "The Rivers and Valleys of Pennsylvania" his first major published description of the cycle of erosion experienced by an idealized river, and with it the method of landscape analysis later known as the Davisian System. Davis describes the distinctive characteristics to be expected as a river progresses through youth, maturity, and old age, with the possibility at each stage of a disruption of the cycle through an uplift of the land. Periods without uplift are marked by erosion, with the end product of a full cycle being a nearly level plain termed a "peneplain."

MATHEMATICS (Arithmetic)

Proceeding from the work of Richard Dedekind (1888—MATHEMATICS) and that of Hermann Grassmann in 1861, Giuseppe Peano (1858–1932, Italian) presents a set of five postulates for the natural numbers and proceeds to develop the rational number system and deduce basic arithmetical statements. He thereby provides a logical structure for the number system, publishing his development in *Arithmetices principia nova methodo exposita* [Principles of arithmetic]. The five postulates are: 1 is a natural number, 1 is not the successor of any other natural number, each natural number has a successor, if successors are equal then the numbers are equal, and the axiom of mathematical induction. The concepts of set, natural number, successor, and "belong to" are left undefined.

METEOROLOGY (Instability; Clouds)
EARTH SCIENCES (Waves)

Hermann von Helmholtz (1821–1894, German) suggests that the hydrodynamic instability causing waves in the ocean is closely analogous to that causing billow clouds in the atmosphere.

1890

ASTRONOMY (Spectroheliograph)

George Hale (1868–1938, American) and Henri Deslandres (1853–1948, French) independently invent the spectroheliograph, an instrument to photograph the sun in a single wavelength.

BIOLOGY (Mind)

In *Principles of Psychology,* William James (1842–1910, American) rejects the idea that the mind is separate from the rest of the body, and declares that human thought, as well as action, results from the nervous system.

CHEMISTRY (Milk; Babcock Test)

By perfecting a test for determining the fractional contribution of butterfat in any given milk sample, Stephen Babcock (1843–1931, American) provides a means for the rapid grading of milk.

EARTH SCIENCES (Rivers and Valleys)

William Morris Davis (1850–1934, American) publishes "The Rivers and Valleys of Northern New Jersey, with Notes on the Classification of Rivers in General," a sequel to his famed paper of the previous year on "The Rivers and Valleys of Pennsylvania" (1889—EARTH SCIENCES).

HEALTH SCIENCES (Surgical Gloves)

Surgeons at Johns Hopkins Hospital in Baltimore begin using rubber gloves during surgery, a practice that soon spreads.

MATHEMATICS (Computers)

The 1890 U.S. Census is tabulated with the help of a machine developed and constructed by Herman Hollerith (1860–1929, American) following the initial suggestion of John Billings during the tabulation of the previous U.S. Census (1880—MATHEMATICS). The required statistics for each person, such as age and sex, are first recorded by punching holes in the appropriate locations on uniform cards the size of a dollar bill. The cards then run under a group of brushes which effect electrical circuits for each hole punched, causing the corresponding counter (for example, for five-person families) to advance one digit. The machine can also sort the cards on the basis of any particular characteristic, although part of the sorting process is manual, as the card is actually placed by hand into the bin opened by the machine. The savings in time and manual labor produced by using the Hollerith machine allow more questions to be asked in the 1890 census than in any previous census.

MATHEMATICS (Hilbert Basis Theorem; Algebraic Forms)

David Hilbert (1862–1943, German) presents the Hilbert Basis Theorem, asserting that every polynomial ideal has a finite basis, and extends the result of Paul Gordan that every binary form has a finite complete system of invariants and covariants (1868—MATHEMATICS) to algebraic forms in *n* variables.

MATHEMATICS (Space-Filling Curves)

Giuseppe Peano (1858–1932, Italian) constructs a continuous plane curve completely filling a square. This is the first "space-filling" curve.

PHYSICS (Ergodic Hypothesis)

Henri Poincaré (1854–1912, French) asserts that any mechanical system, given enough time, will return to its initial configuration if constrained within a finite volume and with a constant total energy content. He uses this to argue against the mechanistic viewpoint, as such a system would deny the second law of thermodynamics, in which a system's entropy can never decrease.

1890–1891

HEALTH SCIENCES (Mastectomy)

Within an article entitled "The Treatment of Wounds with Especial Reference to the Value of the Blood Clot in the Management of Dead Spaces," William Halsted (1852–1922, American) describes a method of radical mastectomy for treatment of breast cancer.

1890–1896

HEALTH SCIENCES (Beriberi; Vitamins)

Christiaan Eijkman (1858–1930, Dutch) publishes results establishing the importance of diet for preventing beriberi and the potential dangers in relying too heavily on processed foods. He had gone to the Dutch East Indies in 1886 as a junior member of a commission assigned to study the disease, which was widespread in the region, and remains an additional nine years after the senior members of the commission leave in 1887. While there, Eijkman is struck by the observation that chickens fed polished rice frequently contract polyneuritis, a disease similar to beriberi, and that they are cured upon a change of diet from polished to unpolished rice. He then carries out experiments which establish the preventive and curative effect of the unpolished rice. He attributes the cause of the disease to a toxin in the endosperm of the rice, this toxin being neutralized in unpolished rice by an antitoxin in the outer layers which are removed during polishing. This explanation is later overthrown as the refining process is discovered instead to deplete the unpolished rice of the essential element Thiamine (or vitamin B_1).

1890s

PHYSICS/CHEMISTRY (Atomic Theory)

The study of atoms begins to shift from the chemist to the physicist, as the atom is found not to be indivisible but rather composed of smaller particles, and different elements are found to be better distinguished by aspects of the electrical charges of these particles than by atomic weight.

1890–1905

MATHEMATICS (Symbolic Logic)

Ernst Schröder (1841–1902, German) publishes *Vorlesungen über die Algebra der Logik,* a thorough, four-volume exposition of the symbolic logic of the day.

1891

ASTRONOMY (Asteroids)

Maximilian Wolf (1863–1932, German) discovers the asteroid Brucia by identifying its path as a short line on a photographic plate taken through a long time exposure with a camera moving synchronously with the stars and hence showing stars as individual points. Others soon adopt the technique and with it discover numerous additional asteroids, adding to the approximately 300 asteroids discovered before 1891.

ASTRONOMY (Satellites of Jupiter)

Albert Michelson (1852–1931, American) measures the diameters of the known satellites of Jupiter, using a stellar interferometer.

ASTRONOMY (Sun)

Nils Dunér (1839–1914, Swedish) discovers that the sun does not have a uniform rotation rate, but rather its rotation rate varies from region to region.

EARTH SCIENCES (Deep-Sea Sediments)

John Murray (1841–1914, English) and Alphonse Renard (1842–1903, Belgian) present the first large-scale sediment map of the oceans, showing that sediment patterns basically follow latitude circles.

EARTH SCIENCES (Earth's Wobble)

Seth Chandler (1846–1913, American) announces discovery of a 14-month (428 day) periodic wobble in the earth's motions.

MATHEMATICS (Transfinite Set Theory)

Georg Cantor (1845–1918, German) uses a method of diagonalization in a new proof of the existence of nondenumerable sets (1874—MATHEMATICS), the new proof being independent of irrational numbers. Specifically, the set M of all elements $E = (x_1, \ldots, x_n, \ldots)$ where each $x_i = m$ or w, must be nondenumerable because for any attempted enumeration $E_i = (x_{1i}, \ldots, x_{ni}, \ldots)$, a new element $(y_1, \ldots, y_n, \ldots)$ of M can be formed by taking $y_j = w$ if $E_{jj} = m$ and y_j

$= m$ if $E_{jj} = w$. Cantor uses a similar method of diagonalization to establish the existence of a limitless hierarchy of transfinite cardinal numbers (or powers), showing that for any set L the set of all subsets of L has a cardinality greater than L. Thus there is no greatest cardinal number.

PHYSICS (Mass)

Roland Eötvös (1848–1919, Hungarian) establishes experimentally that the inertial mass and gravitational mass are proportional.

1891–1896

MATHEMATICS (Analysis)

Publication of Charles Picard's (1856–1941, French) three-volume *Traité d'analyse*.

1892

ASTRONOMY (Novae)

Hugo von Seeliger (1849–1924, German) revives an earlier hypothesis that novae form upon the collision of a dark star and a nebula.

ASTRONOMY (Satellites of Jupiter)

Edward Barnard (1857–1923, American) discovers a fifth satellite of Jupiter, the first four having been detected by Galileo Galilei (1610—ASTRONOMY).

ASTRONOMY (Sun)

Petr Lebedev (1866–1912, Russian) notes that if a small-enough particle is placed near the sun, the attractive force of the sun's gravitation could be more than compensated by the repulsive force from the light rays.

BIOLOGY (Genetics; Germ Plasm)

In *Das Keimplasma* [Germ-plasm, a theory of heredity], August Weismann (1834–1914, German) announces the germ-plasm theory of heredity, whereby the germ plasm, located in the sex cells, is the carrier of the hereditary endowment, half the germ plasm for an offspring coming from the mother and half from the father. In a doctrine of continuity, Weismann stresses that the germ plasm is transmitted unmodified to offspring and that changes in body cells are not transferred to the germ cells, hence that acquired characteristics are not inherited. He describes the process of meiosis, whereby the number of chromosomes is halved.

BIOLOGY (Filterable Viruses)
HEALTH SCIENCES (Tobacco-Mosaic Disease)

Dmitri Ivanovsky (1864–1920, Russian) determines that the agent in the sap of leaves causing tobacco-mosaic disease is not filtered out of the sap even with candle filters believed to remove all microbes and completely purify drinking water. The offending agent is a "filterable virus."

CHEMISTRY (Atomic Weights)

John William Strutt (Lord Rayleigh, 1842–1919, English) asserts that the ratio of the density of oxygen to that of hydrogen is 15.882 rather than 16.0.

CHEMISTRY (Dewar Flask; Thermos Bottles)

James Dewar (1842–1923, Scottish) increases the effectiveness of the Violle container (1882—CHEMISTRY) by silvering the interior as well as the exterior wall, thereby preventing heat from escaping as well as from entering. The Dewar flask (later marketed as Thermos Bottles) becomes common fare on polar and equatorial expeditions and spurs an increase in picnics, as well as being important scientifically for maintaining substances at unusually low or high temperatures.

CHEMISTRY (Electric Furnace)

Ferdinand Henri Moissan (1852–1907, French) invents the electric furnace, in which a strong electric current passes between two electrodes within a box of lime or other material highly resistant to heat. The extremely high temperatures obtainable from this furnace provide an impetus to further research in high-temperature chemistry, with Moissan himself using it to prepare and study various refractory oxides, borides, carbides, and silicides, to volatilize various metals, and to reduce metallic oxides.

CHEMISTRY (Nitrogen; Argon)
METEOROLOGY (Atmospheric Composition)

While trying to test the hypothesis of William Prout that each chemical element is a combination of hydrogen atoms (1816—CHEMISTRY) by accurate measurements of gas densities, John William Strutt (Lord Rayleigh, 1842–1919, English) determines that nitrogen obtained from the atmosphere is heavier by about 0.5% than that obtained from decomposing ammonia. This result is inexplicable until two years later, when Strutt and William Ramsay discover argon (1894—CHEMISTRY).

CHEMISTRY (Nomenclature)

The multinational Geneva Conference is convened to establish rules for standardization of terminology in organic chemistry and to agree upon official names for specific organic compounds. Among the agreed-upon rules for name endings are the following: saturated hydrocarbons should end in "ane," aldehydes in "al," ketones in "one," and alcohols in "ol."

HEALTH SCIENCES (Nervous Diseases)

John C. Shaw (1845–1900, American) publishes *Essentials of Nervous Diseases and Insanity: Their Symptoms and Treatment,* in which he describes a wide range of nerve-related injuries and diseases, including facial spasms, inflammatory and degenerative diseases of the spinal cord, muscular dystrophy, and epilepsy, plus a range of insanities, these being divided to simple and degenerative insanities, with the latter category including paranoia, imbecility, and so-called epileptic, hysterical, and alcoholic insanities.

HEALTH SCIENCES (Tetanus)

Emil von Behring (1854–1917, German) develops an antitoxin for tetanus.

PHYSICS (Applied Physics; Carburetor)

Making use of the Venturi principle (1797—PHYSICS), Wilhelm Maybach (1846–1929, German) creates a carburetor in which a jet of petroleum is inserted at a constricted area of air flow and there is transformed into a spray suitable for igniting the engine.

PHYSICS (Applied Physics; Converters)

David Salomons (1851–1925, American) and Pyke invent the direct current and alternating current converter.

PHYSICS (Electromagnetic Waves)

Publication of Heinrich Hertz's (1857–1894, German) *Electric Waves.*

PHYSICS (Fitzgerald Contractions and Lorentz Transformations)

George Fitzgerald (1851–1901, Irish) suggests that the failure of the Michelson-Morley experiment (1887—PHYSICS) to reveal the earth's motion through the ether derives from a contraction experienced by all moving bodies. Hendrik Lorentz (1853–1928, Dutch) buttresses the suggestion with a mathematical support, presenting the "Lorentz transformations" for the contractions of length and time. Lengths are transformed to their former value times the square root of $(1 - v^2/c^2)$, where v is the speed of the source and c is the speed of light. In contrast to the contraction later contained in the Einstein theory of relativity (1905—PHYSICS), the Fitzgerald contraction involves an actual physical change in the body. The contraction derives from the fact that matter is composed of charged particles.

PHYSICS (Pyrometer)

Henri LeChâtelier (1850–1936, French) creates a practical optical pyrometer for measuring the temperature of hot bodies from the light intensity in a narrow wavelength band by optical comparison of an image of the body in question and a glowing filament. He uses an oil lamp, an iris diaphragm, and a red glass filter for narrowing the wavelength range.

1892–1893

PHYSICS (Meter)

Albert Michelson (1852–1931, American) determines the length of a meter based on the wavelength of a particular line in the cadmium spectrum, thereby providing a reproducible standard.

1892–1899

ASTRONOMY (Celestial Mechanics)
MATHEMATICS (Differential Equations)

Publication of the three-volume *Les Méthodes nouvelles de la mécanique celeste* by Henri Poincaré (1854–1912, French). Poincaré's development of his methods further advances the mathematical analysis of systems of linear differential equations of the first order.

1893

EARTH SCIENCES (San Andreas Fault)

Andrew Lawson (1861–1952, Scottish) discovers the San Andreas Fault in California, which, with the development of plate tectonic theory, is later determined to be along the dividing line of the Pacific and North American Plates.

HEALTH SCIENCES (Cattle Fever)

Theobald Smith (1859–1934, American) publishes the monograph *Investigations Into the Nature, Causation, and Prevention of Texas or Southern Cattle Fever* in which he establishes that the disease results from destruction of red blood cells by a microparasite which is transmitted by ticks from the blood of healthy southern cattle to that of susceptible northern cattle. The careful verification of the tick-borne mechanism establishes that disease can be transmitted by an insect.

HEALTH SCIENCES (Heart Surgery)

Daniel Williams (1856–1931, American) performs open heart surgery on a patient with his pericardium punctured by a knife wound. Williams makes a six-inch incision, detaches the fifth rib, repairs the pericardium, and sutures the opening.

HEALTH SCIENCES (Hysteria; Psychoanalysis)

The publication of the paper "On the Psychical Mechanism of Hysterical Phenomena" by Sigmund Freud (1856–1939, Austrian) and Josef Breuer (1842–1925, Austrian) is often taken as marking the start of psychoanalysis. Building on the earlier efforts of Breuer (1880–1882—HEALTH SCIENCES), Freud and Breuer contend that the problems of hysterical patients can be directly traced to childhood traumas resulting in an undischarged emotional energy. The therapy suggested involves having the patient recall past events while hypnotized. (Freud later rejects the use of hypnotism.)

MATHEMATICS (Differential Equations)

Charles Picard (1856–1941, French) develops an existence theorem for differential equations based on successive approximations. This theorem in general is considered more practical than the Cauchy-Lipschitz Theorem (1876—MATHEMATICS).

MATHEMATICS (Field Theory)

Eliakim Moore (1862–1932, American) proves that any finite commutative field contains p^n elements, where p is prime and n is a natural number.

MATHEMATICS (Group Theory)

Eliakim Moore (1862–1932, American) develops the theory of the group of automorphisms of any finite group.

MATHEMATICS (Spherical Trigonometry)

Eduard Study (1862–1922, German) revises the subject of spherical trigonometry in light of the algebra and analysis of the nineteenth century.

MATHEMATICS (Standard Deviation)

Karl Pearson (1857–1936, English) defines the standard deviation of a set of measurements, namely, the square root of the sum of the squared deviations from the mean.

PHYSICS (Electromagnetism)
MATHEMATICS (Vector Analysis)

Oliver Heaviside (1850–1925, English) elaborates the theories of both electromagnetism and vector analysis in *Electromagnetic Theory*.

PHYSICS (Ether)

By demonstrating that moving massive steel discs fail to drag the ether with them, Oliver Lodge (1851–1940, English) contributes to the decline of the ether theory.

1893–1896

EARTH SCIENCES (*Fram* Expedition)

Fridtjof Nansen (1861–1930, Norwegian) leads the *Fram* expedition to the Arctic, a major purpose being to determine the upper water circulation and particularly to confirm the existence of a current from Siberia to Greenland, hypothesized by Nansen from the driftwood found arriving in Greenland, apparently from Siberian trees, and from the discovery in Greenland of the remains of the *Jeanette* expedition which had become stuck in the ice north of the Chukchi Sea in the winter of 1879–1880. Nansen's speculations regarding the Arctic Ocean circulation are largely confirmed with the drift of the *Fram* toward Svalbard after becoming (purposely) ice-bound in the Central Arctic. Results from the expedition are later published in a six-volume work (1900–1906—EARTH SCIENCES).

1893–1903

MATHEMATICS (Number; Logic; Implication)

Gottlob Frege (1848–1925, German), in the two-volume *Grundgesetze der Arithmetik* [The fundamental laws of arithmetic; published in 1893 and 1903], more fully develops his idea of number and the axiomatization of logic, begun by him in earlier works (see MATHEMATICS entries for 1879 and 1884). He formalizes the concept of material implication and distinguishes an object from the set containing only that object. Much to his dismay, Frege receives a letter from Bertrand Russell in 1902 shortly before publication of volume 2 of the *Grundgesetze* indicating a contradiction in Frege's system which upsets Frege's entire program.

1894

ASTRONOMY (Mars; Extraterrestrial Life)

Percival Lowell (1855–1916, American) undertakes a study of Mars during which he concludes that the channels recorded by Giovanni Schiaparelli (1877—ASTRONOMY) are canals constructed by Martians for the purpose of irrigation. He postulates the existence of a race of intelligent Martians in his 1895 volume, *Mars*.

BIOLOGY (Evolution)

William Bateson (1861–1926, English) presents, in *Materials for the Study of*

Variation, substantial evidence favoring evolution by discontinuous variations rather than by gradual adaptation to the environment. This is an unpopular stance, although it gains more currency six years later with the work of Hugo de Vries (1900—BIOLOGY).

CHEMISTRY (Argon)
METEOROLOGY (Atmospheric Composition)

John William Strutt (Lord Rayleigh, 1842–1919, English) and William Ramsay (1852–1916, Scottish) announce the discovery of argon following experimentation to explain the troublesomely high atomic weight of atmospheric nitrogen (see 1892—CHEMISTRY). Ramsay repeats eighteenth-century experiments of Henry Cavendish, removing the oxygen and nitrogen from the air, and analyzes the residual gas spectroscopically. He determines that the heavy "nitrogen" from the atmosphere in fact contains traces of a separate element, an inert gas which he names argon after the Greek word for "idle." Argon is denser than nitrogen and composes 1% of the atmosphere. Paul Lecoq de Boisbaudran (1838–1912, French) hypothesizes the existence of a full family of inert gases, of which argon is simply the first to be discovered. He lists the expected atomic weights of the inert gases at 20.0945, 36.40, 84.01, and 132.71.

CHEMISTRY (Catalysts; Reactions)

Wilhelm Ostwald (1853–1932, Russian-German) determines that a catalyst can only quicken a reaction, not create one. Without an external energy input, for instance through an electric current, a chemical reaction requires a downward movement along an energy slope. Ostwald establishes that the catalyst quickens the reaction by decreasing the energy of activation (1889—CHEMISTRY).

EARTH SCIENCES (Ice Ages)

Thomas Chamberlin (1843–1928, American) maps the North American ice coverage during the last ice age.

EARTH SCIENCES (Ice Ages)

James Dana (1813–1895, American) rejects James Croll's astronomical theory of the ice ages (1875—EARTH SCIENCES), basing his rejection on evidence from the retreat rates of Niagara Falls and the Falls of St. Anthony on the Mississippi that the last Ice Age ended 10,000 to 15,000 years ago, rather than the 80,000 years suggested by Croll's theory. Dana speculates instead that the Ice Age was caused by a worldwide uplifting of land resulting in a strip of dry land or shallow water blocking the warm water input to the Arctic from the North Atlantic.

EARTH SCIENCES (Sedimentary Cycle)

Marcel Alexandre Bertrand (1847–1907, French) postulates a full sedimen-

OF CHEMISTRY. 175

TABLE OF SIMPLE SUBSTANCES.

Simple ſubſtances belonging to all the kingdoms of nature, which may be conſidered as the elements of bodies.

New Names.	*Correſpondent old Names.*
Light	Light.
Caloric	Heat. Principle or element of heat. Fire. Igneous fluid. Matter of fire and of heat.
Oxygen	Dephlogiſticated air. Empyreal air. Vital air, or Baſe of vital air.
Azote	Phlogiſticated air or gas. Mephitis, or its baſe.
Hydrogen	Inflammable air or gas, or the baſe of inflammable air.

Oxydable and Acidifiable ſimple Subſtances not Metallic.

New Names.	*Correſpondent old names.*
Sulphur Phoſphorus Charcoal	The ſame names.
Muriatic radical Fluoric radical Boracic radical	Still unknown.

Oxydable and Acidifiable ſimple Metallic Bodies.

New Names.	*Correſpondent Old Names.*
Antimony	Regulus of Antimony.
Arſenic	Arſenic.
Biſmuth	Biſmuth.
Cobalt	Cobalt.
Copper	Copper.
Gold	Gold.
Iron	Iron.
Lead	Lead.
Manganeſe	Manganeſe.
Mercury	Mercury.
Molybdena	Molybdena.
Nickel	Nickel.
Platina	Platina.
Silver	Silver.
Tin	Tin.
Tungſtein	Tungſtein.
Zinc	Zinc.

Salifiable

FIGURE 17. *Antoine Lavoisier's table of chemical elements, as translated from his* Traité élémentaire de chimie [*Elementary treatise on chemistry*] *(see 1789—*CHEMISTRY *entry). (By permission of the Harvard College Library.)*

tary cycle with four recurring facies in various mountain chains: gneiss, schistous flysch, coarse flysch, and coarse sandstone.

HEALTH SCIENCES (Anger)

In one of the first modern scientific studies of anger, G. Stanley Hall (1846–1924, American) distributes detailed questionnaires on why people become angry and receives 2,184 filled-out responses. Among the most frequently mentioned causes are injustice, insults, condescension, annoying habits, and improper performance of inanimate objects, such as the failure of a pen to work.

HEALTH SCIENCES (Bubonic Plague)

Shibasaburo Kitasato (1852–1931, Japanese) and Alexandre Yersin (1863–1943, French) identify, independently, the germ causing bubonic plague. Yersin also develops a serum against the plague.

MATHEMATICS (Divergent Series)

Thomas Stieltjes (1856–1894, French) obtains a correspondence between divergent series and convergent continued fractions, making possible a definition for the integral of a divergent series, based on the integral of the corresponding fraction.

METEOROLOGY (General)

William Morris Davis (1850–1934, American) synthesizes the current state of the study of meteorology in his *Elementary Meteorology,* a popular college textbook for the next three decades.

PHYSICS/EARTH SCIENCES/METEOROLOGY (Reynolds' Stresses)

Osborne Reynolds (1842–1912, Irish) introduces turbulent shearing stresses, or Reynolds' stresses, into hydrodynamics in the paper "On the Dynamical Theory of Incompressible Viscous Fluids and the Determination of the Criterion."

PHYSICS (Mechanics)

Publication of Heinrich Hertz's (1857–1894, German) *Die Prinzipien der Mechanik* [Principles of mechanics].

1894–1908

MATHEMATICS (Arithmetic; Geometry; Symbolic Logic)

Giuseppe Peano (1858–1932, Italian) restates much of elementary mathematics, including arithmetic and geometry, with a revised symbolism (1888—MATHEMATICS) in his five-volume *Formulario mathematico* [Formulary of mathematics]. Like Moritz Pasch before him (1882—MATHEMATICS) and David Hilbert after him (1899—MATHEMATICS), Peano bases proofs on explicit sets of postulates and tries to show geometry to be an abstract hypothetico-deductive system. He presents axioms for the whole numbers and their arithmetic, and symbolizes such logical connectives and quantifiers as "and," "or," and "not." The work becomes a strong stimulus to twentieth-century mathematical logic.

1895

ASTRONOMY (Saturn's Rings)

James Keeler (1857–1900, American) and William Campbell (1862–1938, American) obtain observational evidence that Saturn's rings are composed of small individual masses, thereby confirming the hypothesis of James Clerk Maxwell (1857—ASTRONOMY). Keeler and Campbell base their conclusions on the Doppler shifts of sunlight reflected from the rings, these shifts indicating that for an individual ring the inner portions rotate faster than the outer portions, as would be expected from the theory of gravitation for rings composed of separate particles. The computed rotation rates are approximately 20 kilometers per second at the inner edge of the B ring and approximately 16 kilometers per second at the outer edge of the A ring.

BIOLOGY (Adrenaline)

George Oliver (1841–1915, English) and Edward Sharpey-Schäfer (1850–1935, English) identify a physiologically active substance secreted by the adrenal gland and capable of increasing blood pressure and accelerating other bodily functions, especially in the face of strong emotions. The substance is later termed "adrenaline" or "epinephrine."

CHEMISTRY (Calcium Carbide; Acetylene)

Ferdinand Henri Moissan (1852–1907, French) uses his electric furnace (1892—CHEMISTRY) to combine lime and carbon at 2,000°C, obtaining calcium carbide. The calcium carbide in conjunction with water produces the gas acetylene, which, although known and studied previously (1860—CHEMISTRY), now gains temporary popularity for lighting, because of the bright white light it produces and the new ability to prepare it in quantity.

CHEMISTRY (Helium)

Helium, recognized as an element in the sun (see 1868—ASTRONOMY), is discovered on earth by William Ramsay (1852–1916, Scottish) and independently by Per Cleve (1840–1905, Swedish). Ramsay isolates it from the mineral cleveite following the 1890 research of William Hillebrand (1853–1925, American), who had analyzed a nearly inert gas given off by the mineral and incorrectly identified it as nitrogen. When Ramsay obtains the gas, while actually seeking a nonatmospheric source of argon, he tests it spectroscopically and identifies it as helium upon finding the same spectral lines as those reported by Pierre Janssen and J. Norman Lockyer from the helium in the sun's spectrum (1868—ASTRONOMY). Helium gas is colorless, odorless, tasteless, insoluble, and incombustible.

CHEMISTRY (Helium)
METEOROLOGY (Atmospheric Composition)

Heinrich Kayser (1853–1940, German) discovers that helium exists in the earth's atmosphere.

EARTH SCIENCES (Deep Ocean Temperatures and Circulation)

Alexander Buchan (1829–1907, Scottish) maps worldwide water temperature distributions at various ocean depths. The data derive largely from the 1872–1876 voyage of the *Challenger* (1872–1876—EARTH SCIENCES). The deep-sea, 2,200-fathom charts suggest the spreading of Antarctic bottom water into the Atlantic, Pacific and Indian Oceans and the net flow of deep waters across the equator from the North to the South Atlantic.

EARTH SCIENCES (Oceanography)

A scientific team led by John Murray (1841–1914, English) completes publication of the 50-volume report of the data and analysis from the *Challenger* expedition (1872–1876—EARTH SCIENCES).

HEALTH SCIENCES (Psychoanalysis; Hysteria)

Josef Breuer (1842–1925, Austrian) and Sigmund Freud (1856–1939, Austrian) publish *Studien über Hysterie* [Studies in hysteria], a further development of the thesis contained in their 1893 paper that hysterical patients can be treated effectively by having them recall past events under hypnotism (1893—HEALTH SCIENCES).

HEALTH SCIENCES (X-Rays)

Within weeks of Wilhelm Röntgen's discovery of X-rays (1895—PHYSICS), physicians begin using them to examine bone fractures.

MATHEMATICS (Computers)

Otto Steiger mass produces a mechanical calculator called the Millionaire and sells over one thousand models in the first three years of production. Capable of direct multiplication, the machine becomes the standard for mechanical scientific calculation for the next 30 years.

MATHEMATICS (Topology)

In the 123-page *Analysis Situs,* Henri Poincaré (1854–1912, French) founds combinatorial topology for *n*-dimensional space and introduces the fundamental group of a complex, later known as the Poincaré group or the first homotopy group.

PHYSICS (Applied Physics; Diesel Engine)

Rudolf Karl Diesel (1858–1913, German) invents the Diesel engine.

PHYSICS (Applied Physics; Wireless Telegraphy)

Guglielmo Marconi (1874–1937, Italian) invents wireless telegraphy.

PHYSICS (Cathode Rays)

Jean Perrin (1870–1942, French) publishes a paper, "Nouvelles propriétés des rayons cathodiques" [New properties of cathode rays], in which he describes an experiment supporting the speculations of William Crookes, Joseph John Thomson, and others that cathode rays consist of negatively charged matter moving with considerable velocity.

PHYSICS (X-Rays)

Wilhelm Röntgen (1845–1923, German) discovers X-rays while experimenting with a Crookes cathode ray tube and electrical discharges through rarefied gases. He notices that upon inserting his hand between the tube and a screen of barium platinocyanide, the hand bones become pictured on the screen. Hence the rays emitted pass through skin and muscles but not bones. Röntgen announces the discovery and describes some of the properties of X-rays (sometimes known as Röntgen rays) in the paper "Über eine neue Art von Strahlen."

1895–1897

MATHEMATICS (Transfinite Numbers; Ordered Sets)

Georg Cantor (1845–1918, German) details his theory of transfinite ordinal and cardinal numbers in the two-part memoir "Beiträge zur Begründung der transfiniten Mengenlehre," concentrating on simply ordered sets in part 1 and on well-ordered sets in part 2. He defines multiplication, exponentiation, and other arithmetic operations for transfinite numbers and develops transfinite arithmetic, but remains unable to prove the continuum hypothesis (1883—MATHEMATICS) or to establish either the comparability of all transfinite numbers or the issue of whether every set can be well-ordered (see 1904—MATHEMATICS).

1895–1935

ASTRONOMY (Binary Stars)

Robert Aitken (1864–1951, American) discovers roughly 3,000 binary star systems with the Lick Observatory's refractor telescope.

1896

ASTRONOMY (Solar Spectrum)

Henry Rowland (1848–1901, American) lists 14,000 lines in the solar spectrum.

CHEMISTRY (Radioactivity)

A. Henri Becquerel's (1852–1908, French) discovery of radioactivity (1896—

PHYSICS) adds substance to the suspicion that there exists a genetic connection among the elements.

EARTH SCIENCES (Climatology; Carbon Dioxide Impact)

Svante Arrhenius (1859–1927, Swedish) examines the importance of carbon dioxide to the earth's heat balance and concludes that a doubling of the concentration of CO_2 in the atmosphere would result in an average global temperature increase of about 6 Kelvins, or 6°C. He publishes his results in the paper "On the Influence of Carbonic Acid in the Air upon the Temperature of the Ground."

HEALTH SCIENCES (Typhoid Fever)

Almroth Wright (1861–1947, English) introduces an inoculation against typhoid fever.

MATHEMATICS (Geometry of Numbers)

Hermann Minkowski (1864–1909, Russian) publishes *Geometrie der Zahlen* [Geometry of numbers], the first major work on the geometrical theory of numbers. Arithmetical problems are restated geometrically and solved through geometrical methods.

MATHEMATICS (Prime Number Theorem)

Charles de la Vallée-Poussin (1866–1962, Belgian) and Jacques Hadamard (1865–1963, French) prove the prime number theorem, establishing that the number of primes less than or equal to n is approximately $n/\log n$ for large values of n.

PHYSICS (Cloud Chamber)

Charles Wilson (1869–1959, Scottish) invents the Wilson cloud chamber, an instrument for making visible and photographing the paths of charged particles.

PHYSICS (Conductivity)

Joseph John Thomson (1856–1940, English) discovers that a gas at ordinary pressures can be made into an electrical conductor with the use of X-rays. He determines that the conductivity is due to charged particles.

PHYSICS (Kinetic Theory of Gases)

John William Strutt (Lord Rayleigh, 1842–1919, English) demonstrates that a porous sheet can be used to partially separate a mixture of gases by atomic weight, the lighter constituents preferentially passing through the filter.

PHYSICS (Radioactivity)

A. Henri Becquerel (1852–1908, French) discovers radioactivity while examining uranium ores during a sequence of experiments on the phosphorescent properties of various substances, and describes the discovery in the paper "Sur les radiations émises par phosphorescence" [On the radiations emitted by phosphorescence]. Wilhelm Röntgen's X-ray photographs (1895—PHYSICS) had stimulated Becquerel to investigate a full range of phosphorescent substances to determine the extent of X-ray emission. He finds that a photographic plate becomes fogged by a phosphorescent salt even when separated from the salt by black paper. He later determines that all uranium compounds, even nonphosphorescent ones, affect photographic plates and increase the conductivity of the surrounding air. The explanation is that uranium emits penetrating rays, soon distinguished as alpha rays, beta rays, and gamma rays. (The term "radioactivity" is introduced later, by Marie Curie.)

PHYSICS (Wien's Law; Wien's Displacement Law)

Wilhelm Wien (1864–1928, German) derives Wien's Law, an equation for the intensity of radiation J emitted at each wavelength L by a black body at an absolute temperature T: $J = a/(L^5 e^{b/LT})$, where a and b are constants. This equation is later revised by Max Planck, who subtracts a 1 in the denominator to obtain Planck's Law (1900—PHYSICS), but it serves in the meantime to yield, through differentiation, Wien's Displacement Law that the absolute temperature of the radiating black body is inversely proportional to the wavelength of maximum emission.

PHYSICS (Zeeman Effect)

Pieter Zeeman (1865–1943, Dutch) discovers that a magnetic field widens—and occasionally doubles—atomic spectral lines (the Zeeman Effect), and publishes an account of the discovery under the title "Ueber einen Einfluss der Magnetisirung auf die Natur des von einer Substanz emittirten Lichtes."

1896–1903

MATHEMATICS (Group Representations)

Georg Frobenius (1849–1917, German) expands the study of group representations to all finite abstract groups, and introduces and develops the concepts of reducible and completely reducible representations.

1897

ASTRONOMY (Telescopes)

The world's largest refracting telescope, with a 40-inch lens, is constructed at Yerkes Observatory in Wisconsin.

BIOLOGY (Digestion)

In *The Work of the Digestive Glands,* Ivan Pavlov (1849–1936, Russian) presents results of experiments during which he sham fed a dog, with the food leaving through an opening in the neck after having been swallowed. In spite of the failure of the food to pass beyond the neck, gastric juice was nonetheless secreted into the stomach. Pavlov's hypothesis that the secretion of the gastric juice derives from nerve impulses and not from the fullness of the stomach is further supported when the juice is not secreted upon additional sham feeding with the vagus nerve cut.

BIOLOGY (Fermentation)

Eduard Buchner (1860–1917, German) determines the cause of alcoholic fermentation to be the chemical zymase.

HEALTH SCIENCES (Bubonic Plague)

Robert Koch (1843–1910, German) demonstrates that fleas from rats transmit the germ causing bubonic plague.

MATHEMATICS (Algebraic Number Theory)

David Hilbert (1862–1943, German) reorganizes algebraic number theory based on unifying principles and greater generality in his *Der Zahlbericht.*

MATHEMATICS (International Congresses)

Georg Cantor (1845–1918, German) assumes a lead role in bringing about the First International Congress of Mathematicians, held in 1897 in Zurich.

MATHEMATICS (Lattice Theory)

Richard Dedekind (1831–1916, German) makes the first investigation of what are later termed Dedekind lattices. These are sets of classes A, B, C, . . . such that (1) if A contains B and B contains C then A contains C; (2) if A, B are in the lattice, then the intersection $[A,B]$ and the union (A,B) are in the lattice; (3) if the union (A,B) contains C and C contains A then $C = (A,[B,C])$.

MATHEMATICS (Transfinite Numbers)

Cesare Burali-Forti (1861–1931, Italian) announces a paradox on the ordinal of the set of ordinals which calls into question the logical structure of the current theory of transfinite numbers. The paradox states that if the set S of ordinals is well-ordered then it must have an ordinal, but then this ordinal would have to be within the set S and yet exceed any ordinal within S. He concludes that the ordinal numbers are not always comparable and that indeed two ordinals α and β can exist and yet satisfy none of the relations $\alpha < \beta$, $\alpha = \beta$, or $\alpha > \beta$.

PHYSICS (Cathode Rays)

John Townsend (1868–1957, Irish-English), Joseph John Thomson (1856–1940, English), and Harold Wilson (1874–1964, English-American) independently determine the approximate charge on cathode ray particles, finding it to equal the charge on an ion. In conjunction with Thomson's m/e value (1897—PHYSICS) they conclude that m, the mass of the cathode ray particle, is much less than the mass of a hydrogen atom.

PHYSICS (Electrons)
CHEMISTRY (Atomic Structure)

Analysis of cathode ray experiments—particularly the penetration of gold leaf and the magnitude of cathode ray deflection—leads Joseph John Thomson (1856–1940, English) to assert, in the paper "Cathode Rays," that there exist particles less massive than the hydrogen atom. These particles, termed corpuscles by Thomson, are later known as electrons. Thomson shows that they are negatively charged and can be extracted from atoms, the remains of which are positively charged ions. This establishes that atoms are not, as previously believed, indivisible. Thomson calculates the ratio m/e of the mass to the charge of the cathode ray to be 10^{-7}, three orders of magnitude smaller than the corresponding ratio for the hydrogen ion.

PHYSICS (Radioactivity)

Marie Curie (1867–1934, Polish-French) shows that the radioactive emission of a sample of uranium is proportional to the amount of uranium. She concludes that radioactivity is an atomic phenomenon. Curie also finds thorium to be radioactive.

1897–1898

MATHEMATICS (Fourier Analysis)

Arthur Schuster (1851–1934, German-English) develops a method of statistical confidence for Fourier analyses. By assuming the initial data to be independent and normally distributed, Schuster derives an exponential distribution for the standardized variance estimate.

1897–1906

ASTRONOMY/EARTH SCIENCES/METEOROLOGY (Periodicities)

Arthur Schuster (1851–1934, German-English) applies Fourier analysis to determine the major periodicities in various geological, meteorological and astronomical data.

1897–1928

BIOLOGY (Sex)

The seven-volume *Studies in the Psychology of Sex* by H. Havelock Ellis (1859–1939, English) helps modify public attitudes toward sexual difficulties.

1898

ASTRONOMY (Saturn; Phoebe)

William Pickering (1858–1938, American) discovers Phoebe, a satellite of Saturn. At 13 million kilometers, Phoebe is further from Saturn than any of the eight previously discovered satellites.

ASTRONOMY (Solar Corona)

Annie Maunder (1868–1947, English) photographs from India a coronal streamer extending outward from the sun to a distance of 6 solar radii.

BIOLOGY (Adrenaline)

John Jacob Abel (1857–1938, American) isolates epinephrine (adrenaline). This is the first hormone to be isolated, although the term "hormone" itself is not yet in use (see BIOLOGY entries for 1895 and 1905).

BIOLOGY (Viruses)
HEALTH SCIENCES (Foot and Mouth Disease)

Friedrich Loeffler (1852–1915, German) and Paul Frosch (1860–1928, German) publish a "Report of the Commission for Research on Foot-and-Mouth Disease," providing evidence from experiments with calves and heifers that the disease is caused by living organisms minute enough to pass through bacteriological filters. Lymph fluid from sick animals and without observable bacteria proves effective in transmitting the disease to healthy animals, and does so with equal effectiveness whether filtered or unfiltered.

BIOLOGY (Viruses)
HEALTH SCIENCES (Tobacco Mosaic Disease)

Martinus Beijerinck (1851–1931, Dutch), independently of Ivanovsky (1892—BIOLOGY), determines that tobacco mosaic disease is caused by a living liquid virus capable of passing through a fine filter. Through further experiments with the virus he determines that boiling destroys it but drying does not.

CHEMISTRY (Krypton)

William Ramsay (1852–1916, Scottish) and Morris Travers (1872–1961, English) discover krypton after boiling liquid air and examining the spectrum of the residual inert gas once oxygen and nitrogen are removed. One brilliant green line and another bright yellow line with a green tint confirm a new element. Atomic weight determinations place the element between bromine and rubidium in the periodic table.

CHEMISTRY (Low Temperature Chemistry)

James Dewar (1842–1923, Scottish), experienced in low temperature chemistry, liquefies hydrogen, having demonstrated that hydrogen, the one gas which at normal temperatures tends to warm upon expansion, exhibits the normal Joule-Thomson cooling (1852—CHEMISTRY) at temperatures below −80°C. Hence below −80°C the Joule-Thomson effect allows a mechanism for further cooling of hydrogen to below its critical temperature for liquefaction (see 1869—CHEMISTRY).

CHEMISTRY (Neon)

William Ramsay (1852–1916, Scottish) and Morris Travers (1872–1961, English) discover the inert gas neon after solidifying argon and allowing it to volatilize. The enthusiasm toward the brilliant crimson glow leads quickly to widespread commercial use of neon lights.

CHEMISTRY (Polonium)

After observing the high radioactivity of pitchblende and performing multiple fractionations, Marie Curie (1867–1934, Polish-French) discovers the element polonium and names it for her native country (Poland), publishing an account of the discovery in the paper "Sur une substance nouvelle radio-active, contenue dans la pechblende" [On a new radioactive substance contained in pitchblende]. Polonium is hundreds of times more radioactive than uranium.

CHEMISTRY (Radium)

Marie Curie (1867–1934, Polish-French) and Pierre Curie (1859–1906, French) discover radium, an element even more radioactive than polonium (1898—CHEMISTRY), and publish an account in the paper "Sur une nouvelle substance fortement radio-active, contenue dans la pechblende" [On a new strongly radioactive substance contained in pitchblende]. Eugène Demarçay (1852–1904, French) spectroscopically examines the substance for the Curies and confirms the existence of a previously unknown element. The Curies succeed in isolating radium from uranium, after considerable effort, in 1902.

CHEMISTRY (Thorium)

Independent work by Marie Curie (1867–1934, Polish-French) and Gerhard Schmidt (1865–1949, German) demonstrates that thorium is radioactive.

CHEMISTRY (Xenon)

William Ramsay (1852–1916, Scottish) and Morris Travers (1872–1961, English) discover the inert gas xenon, isolating it—after neon and krypton—from liquid air.

EARTH SCIENCES (Biogeography)

Andreas Schimper (1856–1901, German) classifies world vegetation in the two-volume *Plant Geography on a Physiological Basis.*

EARTH SCIENCES (Triassic Formations)

William Morris Davis (1850–1934, American) completes "The Triassic Formation of Connecticut," his fifteenth and final report in a study for the U.S. Geological Survey on the Triassic basins of New England and New Jersey. He presents a history of the Triassic volcanic sequence, criteria for distinguishing extrusive versus intrusive rocks, and methods of determining subsurface geologic structures from analysis of surface topographical features.

HEALTH SCIENCES (Malaria)

Working in India as a member of the Indian Medical Service, Ronald Ross (1857–1932, English) shows that mosquitoes, specifically the Anopheles mosquito, can transmit malaria to birds.

HEALTH SCIENCES (Nutrition)

Charles Langworthy (1864–1932, American) states ten laws of nutrition, based on the belief that foods have two primary purposes: to build and repair the body, and to provide energy.

MATHEMATICS (Computers)

An important step in the development of analog computers is made with the "analyzer" constructed by Albert Michelson (1852–1931, American) and Samuel Stratton (1861–1931, American). The analyzer can graph the sum of a Fourier series, given the coefficients, and can determine the coefficients if given the function. It is capable of handling Fourier series with up to 80 terms.

METEOROLOGY (Isohyets)

Although hindered by limited data, Alexander Supan (1847–1920, Austrian) plots isohyets (lines of equal precipitation) over the Atlantic and Indian oceans.

PHYSICS (Photoelectricity)

Joseph John Thomson (1856–1940, English) and Philipp Lenard (1862–1947,

German) independently determine that shining a light on a metallic surface can cause the surface to emit negatively charged particles, later found to be electrons and hence, from the light-induced origin, termed "photoelectrons."

PHYSICS (Rocket Propulsion)

Konstantin Tsiolkovsky (1857–1935, Russian) begins theoretical work on the fundamentals of rocket propulsion.

1898–1899

MATHEMATICS (Algebra)

Heinrich Weber (1842–1913, German) presents an up-to-date overview of algebraic topics in the three-volume *Algebra*.

1899

BIOLOGY (Philosophy)

In *Die Weltrathsel* [The riddle of the universe], Ernst Haeckel (1834–1919, German) declares that matter alone is real, that the mind depends upon the body and hence does not survive after it, and that all animals with a central nervous system possess consciousness.

CHEMISTRY (Actinium)

André Debierne (1874–1949, French) discovers actinium, a radioactive element found in uranium residues.

CHEMISTRY (Low-Temperature Chemistry)

James Dewar (1842–1923, Scottish) solidifies hydrogen.

CHEMISTRY/PHYSICS (Radioactivity)

Julius Elster (1854–1920, German) and Hans Geitel (1855–1923, German) recognize that radioactive elements are composed of unstable atoms which are gradually converting themselves to more stable atoms without pronounced radioactivity.

CHEMISTRY (Reaction Rates)

Svante Arrhenius (1859–1927, Swedish) determines an exponential relationship between the rate k of a chemical reaction, the temperature T, and the activation energy E: $k = Ae^{-E/RT}$.

EARTH SCIENCES (Climatology; Carbon Dioxide Impact)

Like Svante Arrhenius (1896—EARTH SCIENCES), Thomas Chamberlin (1843–1928, American) concludes that carbon dioxide contributes in essential ways to maintaining the heat balance of the earth and that changes in the CO_2 concentration of the atmosphere would affect the atmospheric temperature.

EARTH SCIENCES (Climatology; Glacial Epochs)

Thomas Chamberlin (1843–1928, American) publishes "An Attempt to Frame a Working Hypothesis of the Cause of Glacial Periods on an Atmospheric Basis," in which he includes ocean/atmosphere interactions as important mechanisms contributing to the initiation and/or perpetuation of glacial and interglacial stages. For instance, during periods of unusual warmth, there is an increase in low-latitude evaporation from the oceans, with a consequent increase in ocean salinity, density, and downwelling. The warm downwelled bottom water travels poleward, surfacing in high latitudes and mollifying the polar climates.

HEALTH SCIENCES (Aspirin)

The German chemical firm Bayer introduces acetylsalicyclic acid, tradenamed "aspirin."

HEALTH SCIENCES (Malaria)

Building on the work of Ronald Ross in India (1898—HEALTH SCIENCES), Giovanni Grassi (1854–1925, Italian) determines that malaria is spread to humans by the zanzarone or Anopheles mosquito, and only by those mosquitoes that have bitten malaria sufferers.

MATHEMATICS (Geometry)

In *Grundlagen der Geometrie* [Foundations of geometry], David Hilbert (1862–1943, German) presents the first widely recognized purely deductive development of Euclidean concepts, proceeding from undefined terms and unproved axioms to carefully derived propositions, with the possibility existing that such undefined terms as "point" and "line" can have representations other than those normally considered. Hilbert reconstructs Euclidean geometry in the algebraic framework of Cartesian coordinates and proceeds to show that if algebra is consistent then so are the Euclidean postulates. The work sparks renewed interest in Euclidean geometry, convincing a large number of mathematicians that geometry can be treated essentially as an abstract, purely formal system. Forerunners include the works of Moritz Pasch (1882—MATHEMATICS) and of Giuseppe Peano (1894–1908—MATHEMATICS).

MATHEMATICS (Geometry)

Frank Morley (1860–1937, American) proves that the angle trisectors of the three vertices of a triangle intersect by adjacent pairs so as to form the vertices of an equilateral triangle. This is one of the earliest instances of a mathematician proving a result in Euclidean geometry concerned with entities (such as angle trisectors) which cannot be constructed with straightedge and compass alone.

MATHEMATICS (Topology)

Henri Poincaré (1854–1912, French) introduces torsion coefficients, generalizes the Euler number, and derives the Euler-Poincaré formula.

PHYSICS (Light Pressure)

Petr Lebedev (1866–1912, Russian) presents a paper at an international conference describing his detection and measurement of the pressure of light, a pressure predicted by James Clerk Maxwell's theory (1865—PHYSICS) but not previously detected. The paper, entitled "An Experimental Investigation of the Pressure of Light," is published in 1901.

1900

BIOLOGY (Genetics)

Gregor Mendel's 1865 paper on pea plants (1865—BIOLOGY) is independently rediscovered by Carl Correns (1864–1933, German), Hugo de Vries (1848–1935, Dutch), and Erich Tschermak (1871–1962, Austrian), who begin reexamining Charles Darwin's concept of evolution (1859—BIOLOGY) in the context of Mendel's results. The units of heredity, or genes, analyzed by Mendel are believed to provide a mechanism through which natural selection can operate.

BIOLOGY (Genetics; Mutations)

As a result of several years of experimentation with plants in the 1880s, Hugo de Vries (1848–1935, Dutch) asserts that there exist sudden, noncontinuous changes in an organism which are transmitted to offspring. He terms these large, discontinuous variations "mutations" and claims that they, rather than the very slight variations described by Charles Darwin (1859—BIOLOGY), account for the key evolutionary steps in the creation of a new species. In his cultivation of 53,000 plants, covering eight generations, de Vries had found eight new varieties, each appearing abruptly, without transitional steps.

CHEMISTRY (Carbon)

Moses Gomberg (1866–1947, Russian-American) demolishes the belief that carbon is always quadrivalent by preparing compounds containing trivalent carbon.

CHEMISTRY (Radon)

Friedrich Dorn (1848–1916, German) discovers radon, an emanation from radium, and in the process explains why air becomes radioactive upon contact with radium compounds. Radon is a member of the family of inert gases, joining argon (1894—CHEMISTRY), krypton, neon, and xenon (1898—CHEMISTRY).

CHEMISTRY (Uranium X_1)
PHYSICS (Radioactivity)

William Crookes (1832–1919, English) discovers a highly radioactive form of uranium, uranium X_1.

HEALTH SCIENCES (Blood Types)

Karl Landsteiner (1868–1943, Austrian) divides human blood into three groups according to agglutinative properties and further notes the relevance to human blood transfusions, failed transfusions being expected when mixing incompatible blood types.

HEALTH SCIENCES (Dysentery)

Kiyoshi Shiga (1870–1957, Japanese) develops an immune serum for bacillary dysentery, following his earlier discovery of the bacillus causing the disease.

HEALTH SCIENCES (Yellow Fever)

By experimenting on American soldiers and Spanish immigrants in Cuba, a commission led by Walter Reed (1851–1902, American) confirms the speculations of Carlos Finlay that yellow fever is carried by mosquitoes (1881—HEALTH SCIENCES). Soldiers sleeping on the beds of those dying from yellow fever do not contract the disease, while those bitten by mosquitoes which had bitten the fever victims do. More specifically, Reed, with the assistance of Jesse Lazear (1866–1900, American) and Aristides Agramonte (1869–1931, Cuban), suggests that yellow fever is transmitted by a virus carried by the mosquito Aedes aegypti (or Stegormyia fasciata).

MATHEMATICS (Chi-Square Test)

Karl Pearson (1857–1936, English) publishes a formula yielding a measure of how well a set of observations fits a theoretical hypothesis, the chi-square test of goodness of fit.

MATHEMATICS (Geometry)

Max Dehn (1878–1952, German) establishes that there is no finite proof of Euclid's theorem that if two triangular pyramids have the same height the ratio of their volumes will equal the ratio of their bases. The proof given by Euclid relies on the method of exhaustion, which implies continuity. Carl Friedrich Gauss had urged in 1844 that a finite proof be sought.

MATHEMATICS (Unsolved Problems)

David Hilbert (1862–1943, German) presents a list of 23 unsolved problems in an address to the International Congress of Mathematicians in Paris. These problems, including Georg Cantor's Continuum Hypothesis (1883—MATHEMATICS), Riemann's zeta-function hypothesis (1859—MATHEMATICS), and the issue of which problems in the calculus of variations will have analytic solutions, were selected by Hilbert as likely to lead to productive research. They stimulate much work over the subsequent decades.

PHYSICS (Acoustics)

Wallace Sabine (1868–1919, American) examines acoustical issues related to the design of auditoriums and formulates a law relating reverberation time to a room's volume and its amount of absorbent material.

PHYSICS (Quantum Theory; Planck's Constant; Planck's Law)

In order to explain difficulties in the current theory of the emission and absorption of electromagnetic radiation, Max Planck (1858–1947, German) asserts that radiated energy exists as individual units, or quanta, and cannot be divided infinitesimally. He is led to this denial of a continuum of radiative energy while attempting to derive a formula to fit the experimental facts regarding radiation from hot bodies, in particular, to fit the observed shifting of the wavelength of maximum emission to lesser values as temperature increases. His suggested quantum theory asserts that the amount of energy contained in each quantum is directly proportional to the frequency of the radiation, the proportionality factor being a universal constant *(h)*, termed the "elementary quantum of action" by Planck and later known as Planck's constant. With the quantum postulate, Planck derives his famed blackbody radiation formula for the amount of energy J radiated at each wavelength L by a black body at a given absolute temperature T: $J = a/(L^5[e^{b/LT} - 1])$ where a and b are constants. Planck's Law contains as corollaries the results that with rising temperature the total amount radiated increases (the Stefan-Boltzmann Law; see PHYSICS entries for 1879 and 1883) and the peak emission shifts to shorter wavelengths (Wien's Displacement Law; 1896—PHYSICS). Soon after Planck's theoretical results become known, other researchers provide experimental confirmation.

PHYSICS (Rayleigh's Radiation Formula)

John William Strutt (Lord Rayleigh, 1842–1919, English) obtains Rayleigh's Radiation Formula for the amount of radiation emitted by a black body as a function of wavelength. Rayleigh's formula provides a close fit to observations at long wavelengths but fails noticeably at short wavelengths. Wien's earlier law (1896—PHYSICS), by contrast, provided a closer fit at short than at long wavelengths. Both Rayleigh's formula and Wien's law are soon surpassed by the work of Max Planck, who by assuming that energy is emitted only in finite quanta is able to derive a law which closely matches the observations at all wavelengths (Planck's Law; 1900—PHYSICS).

PHYSICS (Resistance in Metals; Ohm's Law)

Paul Drude (1863–1906, German) publishes a theory of metallic resistance based on electromagnetism and a theory of electrons. In the process, he provides a theoretical explanation for Ohm's Law (1827—PHYSICS).

1900–1906

EARTH SCIENCES (Arctic Oceanography)

Fridtjof Nansen (1861–1930, Norwegian) publishes results of the *Fram* expedition to the Arctic (1893–1896—EARTH SCIENCES) in the six-volume *Scientific Results of the Norwegian North Polar Expedition, 1893–1896*. Among the discoveries made from the extensive data collected are the unexpected great depth of the Arctic Ocean and the patterns of drift of Arctic sea ice (see 1902—EARTH SCIENCES). Also, Nansen correctly hypothesizes from the temperature and salinity distributions in the Arctic Ocean and adjacent seas that there exists an underwater ridge—later termed the Nansen Ridge—between Greenland and Svalbard.

1900–1910

MATHEMATICS (Discrete)

As the quantum theory develops in physics and genetics becomes important in biology, an awareness arises among applied mathematicians that certain natural phenomena formerly felt describable in terms of the continuous actually require a discrete mathematics instead. This spurs research, for instance, in difference (in contrast to differential) equations.

1901

ASTRONOMY (Stellar Spectra)

Annie Jump Cannon (1863–1941, American) describes the spectra of 1,122 of the brighter stars, a prelude to her later more massive compilations (ASTRONOMY entries for 1918–1924 and 1925–1936).

BIOLOGY (Adrenaline)

Jokichi Takamine (1854–1922, Japanese) isolates adrenaline, doing so independently of John Abel (1898—BIOLOGY).

CHEMISTRY (Atomic Theory; Prout's Hypothesis)

John William Strutt (Lord Rayleigh, 1842–1919, English) encourages further consideration of William Prout's hypothesis that each element is a combination of hydrogen atoms (1816—CHEMISTRY) by noting that atomic weights approximate whole numbers more closely than a random listing of weights by a factor exceeding 1,000.

CHEMISTRY (Europium)

Through extended fractionations of samarium magnesium nitrate, Eugène Demarçay (1852–1904, French) discovers the element europium.

HEALTH SCIENCES (Beriberi)

Gerrit Grijns (1865–1944, Dutch) asserts the existence of an essential nutrient in the rice polishings removed from unpolished rice during the polishing process. He explains the removal of this nutrient as the cause of beriberi in groups consuming only polished rice, thereby revising the earlier explanation of Christiaan Eijkman (1890–1896—HEALTH SCIENCES). Grijns supports his claim experimentally by showing that the protective nature of the rice polishings disappears after extended heating.

HEALTH SCIENCES (Psychiatry)

Sigmund Freud (1856–1939, Austrian) analyzes the meanings behind slips of the tongue, forgetfulness, and other such errors in his *Psychopathology of Everyday Life*.

MATHEMATICS (Symbolic Logic)

Giuseppe Peano (1858–1932, Italian) greatly increases the power of George Boole's symbolic logic (1854—MATHEMATICS) by introducing symbols for such notions as "there exists," "is contained in," and "the aggregate of all *x*'s such that."

1901–1903

BIOLOGY (Genetics)

Publication of Hugo de Vries's (1848–1935, Dutch) *Mutation Theory*, where he further develops the theory of mutations (see 1900—BIOLOGY) and its relationship to evolution.

1901–1906

EARTH SCIENCES (Radioactivity)

Julius Elster (1854–1920, German) and Hans Geitel (1855–1923, German) find radioactivity in various rocks and soils and show that radium and thorium are both found widely—for example, in the atmosphere, in the earth's crust, and in ocean water—though in very small quantities.

1902

BIOLOGY (Genetics)

Publication of William Bateson's (1861–1926, English) *Mendel's Principles of Heredity: A Defence.* Bateson is the first to apply Mendel's laws (1865—BIOLOGY) to animals as well as plants.

BIOLOGY (Digestion; Secretin)

Disputing Ivan Pavlov's assumption that the nervous system controls all digestive secretions, William Bayliss (1860–1924, English) and Ernest Starling (1866–1927, English) develop a theory of chemical or hormonal control, presenting evidence that a substance "secretin" (isolated by them at this time) is produced in the small intestine with the help of hydrochloric acid from the stomach, and that this substance causes the flow of pancreatic juice. To confirm that the flow is controlled chemically rather than through the nerves, they experiment with a dog and verify flow of the juice even with all nerves to the pancreas cut. (The term "hormone" is not introduced until 1905.)

CHEMISTRY (Electrovalency)

G. N. Lewis (1875–1946, American) explains valency (1852—CHEMISTRY) in terms of transfers of electrons between atoms.

CHEMISTRY (Thorium X)
PHYSICS (Radioactivity)

The discovery of thorium X is announced by Ernest Rutherford (1871–1937, New Zealander-English) and Frederick Soddy (1877–1956, English). Analysis of the emanation from thorium salt leads Rutherford and Soddy to conclude that this emanation is an inert gas given off by an intermediate substance which they label "thorium X." Upon analyzing thorium X, they find its half life to be four days. More importantly, they determine that both thorium and thorium X emit alpha radiation and conclude that thorium becomes thorium X by the spontaneous disintegration of its atoms into thorium X and alpha particles and that similarly thorium X becomes thorium emanation by spontaneous disintegration.

EARTH SCIENCES (Ocean Salinity)

Among the data on polar waters reported in Fridtjof Nansen's (1861–1930, Norwegian) *Scientific Results of the Norwegian North Polar Expedition 1893–1896* (1900–1906—EARTH SCIENCES) are the first sea water salinities determined by conductivity measures.

EARTH SCIENCES (Sea Ice Drift; Ocean Dynamics)

Based on data from the Norwegian North Polar Expedition on the *Fram* (EARTH SCIENCES entries for 1893–1896 and 1900–1906), Fridtjof Nansen (1861–1930, Norwegian) notes the tendency of sea ice to drift not in the direction of the wind but at an angle of 20° to 40° to the right of the wind. He speculates that the deviation results from the earth's rotation and that the motions in the water layers beneath the ice deviate even more from the wind direction, the motion of each layer deviating to the right of the layer above. He encourages Vagn Ekman to examine the dynamics mathematically (1905—EARTH SCIENCES).

HEALTH SCIENCES (Hookworms)

Charles Stiles (1867–1941, American) announces the discovery of the hookworm and claims that whites in the U.S. South are suffering from hookworm disease on a large scale.

MATHEMATICS (Lebesgue Integration; Lebesgue Measure)

Henri Lebesgue (1875–1941, French) introduces Lebesgue integration, in which the range of a function is subdivided into intervals instead of the domain as in Riemann integration (1854—MATHEMATICS). He thereby avoids the problems arising in Riemann integration near points of discontinuity, where the approach of x_{i+1} to x_i does not imply the approach of $f(x_{i+1})$ to $f(x_i)$. Lebesgue selects a value n_i in each interval Δy_i of the range, determines the "Lebesgue measure" [a nontrivial concept not defined in this entry] $m(E_i)$ of the set of points in the domain with functional values in the corresponding interval in the range, and replaces the Riemann sum $\Sigma\, f(x_i)\, \Delta\, x_i$ and the limit as Δx_i approaches 0 by the sum $\Sigma\, n_i m(E_i)$ and the limit as Δy_i approaches 0. The Lebesgue measure and Lebesgue integral greatly extend the power of the theory of functions, allowing integration of many functions over intervals for which the Riemann integral does not exist.

MATHEMATICS (Logic; Set Theory; Paradoxes)

Bertrand Russell (1872–1970, English) applies to the universal class Georg Cantor's proof that there is no greatest cardinal number (1891—MATHEMATICS) and thereby arrives at his first of several serious mathematical paradoxes: if *N* is the class of all classes not belonging to themselves, then does *N* belong to itself or not?

METEOROLOGY (Kennelly-Heaviside Layer)

Oliver Heaviside (1850–1925, English) and Arthur Kennelly (1861–1939, Indian) determine the existence of a layer in the atmosphere, the Kennelly-Heaviside layer, which reflects radiation at radio wavelengths.

METEOROLOGY (Troposphere, Stratosphere)

Teisserenc DeBort (1855–1913, French) summarizes results of 236 balloon soundings taken over three years to altitudes up to nine miles. DeBort shows that the decrease of temperature with height does not extend throughout the depth of the atmosphere, as generally believed, but levels off at an altitude of about seven miles. He terms the atmosphere below this level the "troposphere" and the atmosphere above it the "stratosphere," the latter term deriving from DeBort's misconception that the air above the troposphere has a uniform temperature, no motion, and a systematic stratification of gases according to density. Though the idea of such a stratification is later rejected, the term remains.

PHYSICS (Arc Transmitter)

Valdemar Poulsen (1869–1942, Danish) invents the arc transmitter, important for wireless telegraphy.

PHYSICS (Photoelectric Effect)

In experimenting on the photoelectric effect, Philipp Lenard (1862–1947, German) discovers that the energy of the electrons emitted increases as the frequency of the light increases but is not affected by the intensity of the light. The latter affects not the energy but the number of electrons emitted.

PHYSICS (Radioactivity)
CHEMISTRY (Radium)

Culminating four years spent in large part analyzing six tons of pitchblende, Marie Curie (1867–1934, Polish-French) and Pierre Curie (1859–1906, French) isolate a decigram of radium salts. They find radium to be far more radioactive than uranium and to have an atomic weight of 225.

PHYSICS (Radioactivity)

In the paper "The Cause and Nature of Radioactivity," Ernest Rutherford (1871–1937, New Zealander-English) and Frederick Soddy (1877–1956, English) offer a theory of disintegration of atoms as an explanation for radioactivity. Based on their work with thorium X (1902—CHEMISTRY), Rutherford and Soddy suggest that atoms spontaneously break up and thereby transform into other elements. The suggestion is strongly ridiculed, although later becomes generally accepted.

SUPPLEMENTAL (Truth of Scientific Theories)

Henri Poincaré (1854–1912, French) asserts that scientific theories, like Euclidean and non-Euclidean geometries, are neither true nor false but are creations of the mind whose conclusions must be validated or invalidated by experience.

1903

BIOLOGY (Genetics; Chromosomes)

Walter Sutton (1877–1916, American) stresses the importance of chromosomes in the transmission of traits to offspring in his paper "The Chromosome Theory of Heredity." His clear statement of the hypothesis that the hereditary factors are located on the chromosomes, along with further work by Theodor Boveri (1904—BIOLOGY), lead to the later labeling of this as the Sutton-Boveri Hypothesis, or the Sutton-Boveri Chromosome Theory of Inheritance.

HEALTH SCIENCES (Electrocardiograph)

Willem Einthoven (1860–1927, Dutch) develops a crude electrocardiograph (ECG), in the form of a string galvanometer, to record the electrical activity of the heart. He relates the development in the article "The String Galvanometer and the Human Electro-cardiogram."

MATHEMATICS (Logic)

In *The Principles of Mathematics,* Bertrand Russell (1872–1970, English) outlines a program—carried out later with Alfred North Whitehead (1910–1913—MATHEMATICS)—to construct pure mathematics from a small number of logical concepts and principles.

PHYSICS (Flight, Airplane)

Wilbur Wright (1867–1912, American) and Orville Wright (1871–1948, American) complete the first successful airplane flight, doing so at Kitty Hawk, North Carolina with a flight lasting 59 seconds.

PHYSICS (Atomic Model)

Philipp Lenard (1862–1947, German) proposes an atomic model in which the mass of an atom is contained in hypothesized electric doublets called "dynamides" which are distributed throughout the much greater volume of the atom. The model never gains widespread acceptance, as no convincing evidence is found for the dynamides.

PHYSICS (Conductivity)

Publication of *Conduction of Electricity through Gases* by Joseph John Thomson (1856–1940, English).

PHYSICS (General)

Albert Michelson (1852–1931, American) asserts that the basic laws of physics have all been discovered, and that the primary remaining need is for increased accuracy in measurements.

PHYSICS (Philosophy; Absolutism)

Along with many other British philosophers, Bertrand Russell (1872–1970, English) emerges firmly in the absolutist faction of the current debate over whether space and time are absolutes or relatives.

c.1903

BIOLOGY (Genetics)

Wilhelm Johannsen (1857–1927, Dutch) conducts experiments on the breeding of beans and determines that variations due to environment are not inherited.

1903–1909

EARTH SCIENCES (Pleistocene Glaciation)

In their three-volume *Die Alpen im Eiszeitalter* [The Alps in the glacial period], Albrecht Penck (1858–1945, German) and Eduard Brückner (1862–1927, German-Austrian) describe their extensive observations and analyses over the previous two decades on the land and glacial features of the Alps. They conclude, largely from river terraces, that there were four major glacial periods, named by them, in chronological order, the Günz, Mindel, Riss, and Würm Ice Periods. They estimate the length of each interglacial, determining the longest, at 240,000 years, to be between the Mindel and Riss glacial stages. The Würm ice age they estimate as having ended 20,000 years ago.

1904

ASTRONOMY (Earth's Orbit)

Ludwig Pilgrim (German) calculates the cycles of the earth's major orbital changes: the tilt angle, the eccentricity, and the precession of the equinoxes, and from them determines the variations in the earth's orbit over the past million years.

ASTRONOMY (Star Streams)

Jacobus Kapteyn (1851–1922, Dutch) analyzes the proper motions of the stars and concludes that stellar motions are not random but instead the stars appear to belong to two groups, or star streams, which are intermingled but have distinctly different mean motions with respect to the sun.

BIOLOGY (Chromosomes)

Through experiments with sea urchins, Theodor Boveri (1862–1915, German) determines the importance of the full set of chromosomes for the normal development of an embryo. Through various manipulations, he is able to obtain from the original 36-chromosome species offspring with a wide range of chromosome numbers. Only those with all 36 chromosomes develop normally. Boveri concludes from the work that each individual chromosome has a separate quality which it transmits.

CHEMISTRY/PHYSICS (Atomic Structure; Periodic Table)

Joseph John Thomson (1856–1940, English) formulates an atomic model in which the atom's weight is obtained from the electrons, which are arranged in concentric shells about the center of a sphere of positive electricity, in which they are embedded in a manner analogous to raisins in a bun. The number of electrons, along with the requirement of mechanical stability, determines their distribution within the shells. Thomson suggests that the cause of the periodicity in chemical properties apparent from the periodic table of elements lies in the distribution of electrons, atoms with the same number of electrons in their outermost shells having similar properties. Since the electrons are regarded as the only massive part of the atom, the hydrogen atom is perceived as having 1,836 electrons, its weight being 1,836 times the weight of a single electron.

CHEMISTRY (Radium)

Bertram Boltwood (1870–1927, American), Herbert McCoy (1870–1945, American), and Robert Strutt (1875–1947, English) independently determine radium to be produced from uranium through spontaneous transmutation.

EARTH SCIENCES (Paleoclimatology)

In an effort to examine a possible connection between the earth's orbital changes and its large-scale climatology, Ludwig Pilgrim (German) plots the current estimated timing of the earth's ice ages together with the changing values of the earth's eccentricity (1904—ASTRONOMY).

MATHEMATICS (Formalism)

David Hilbert (1862–1943, German), in an effort to approach mathematics from a formalist perspective, with meaning existent only in formal definitions and formally defined properties, outlines his suggestions for proving mathematics to be consistent and, in general, for firming the foundations of the subject.

MATHEMATICS (Lebesgue Integration)

In *Leçons sur l'intégration et la recherche des fonctions primitives,* Henri Lebesgue (1875–1941, French) further develops the concept of Lebesgue integration, first introduced in his doctoral thesis (1902—MATHEMATICS). The book introduces Lebesgue's ideas to a much wider audience.

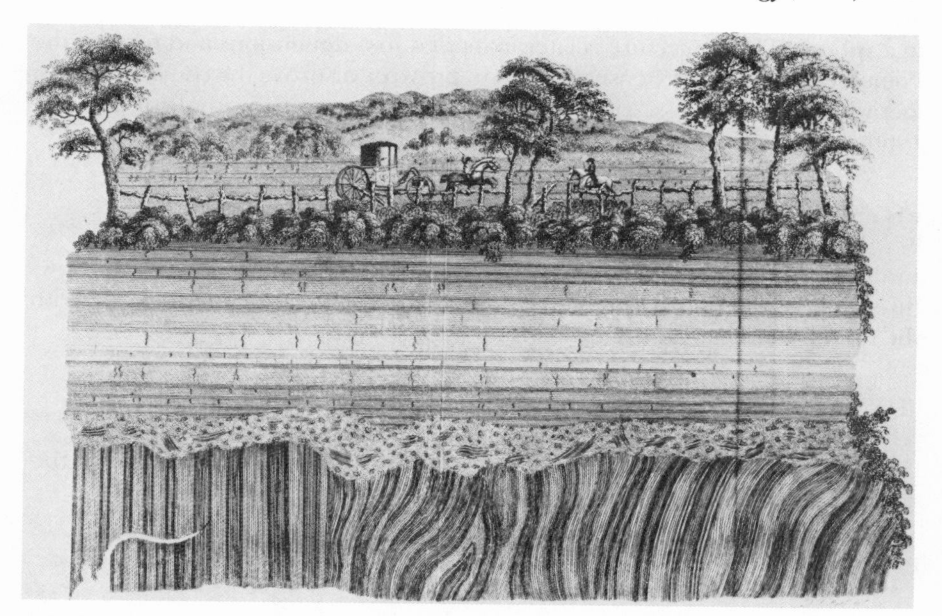

FIGURE 18. *A cross-sectional view of the earth's crust near Jedburgh, Scotland, from James Hutton's* Theory of the Earth, With Proofs and Illustrations *(see 1795—*EARTH SCIENCES *entry). In line with the book's central Uniformitarian theme that the basic geological processes occurring today are the same processes which occurred in the past, Hutton believes that the near-vertical strata at the bottom of the illustration were originally horizontal, deposited in the same sequential manner as strata continue to be deposited, then disrupted into their present positions by convulsions of the earth. The plant and animal life depicted above the geological structures help symbolize Hutton's belief in the continuing cycles of earth history. (By permission of the Houghton Library, Harvard University.)*

MATHEMATICS (Set Theory; Well-Ordering Theorem; Axiom of Choice)

Ernst Zermelo (1871–1953, German) proves that every set can be well-ordered, meaning that for every set S a relation $<$ can be introduced such that (1) for every pair of elements a, b in S either $a = b$, $a < b$, or $b < a$; (2) if a, b, c are in S, $a < b$, and $b < c$, then $a < c$; and (3) there is a first element in each nonempty subset of S. Zermelo employs in his proof the "axiom of choice," or "Zermelo's postulate": if M is a set of nonempty, nonoverlapping sets Q, there exists a set N containing exactly one element from each of the sets Q of M. This proof provokes heated controversy concerning the legitimacy of using the axiom of choice without providing an explicit rule for determining how the elements are to be selected from each set.

MATHEMATICS (Topology; Poincaré Conjecture)

Henri Poincaré (1854–1912, French) conjectures that every closed, orientable three-dimensional manifold which is simply connected is also homeomorphic to a sphere. The conjecture is later extended to n dimensions and termed the Poincaré conjecture. Extensive efforts to prove or disprove the conjecture have met with partial success, with Stephen Smale announcing a proof for $n > 4$ in 1960 and with Michael Freedman announcing a proof for $n = 4$ in 1982.

PHYSICS (Hydrodynamics; Viscous Boundary Layer)

Ludwig Prandtl (1875–1953, German) shows that as an inviscid fluid flows through a pipe a film layer immediately adjacent to the wall does not flow with the fluid but instead forms a viscous boundary layer.

PHYSICS (Relativity)

Henri Poincaré (1854–1912, French) puts forward the hypothesis that the same laws of nature are valid for all nonaccelerating systems.

PHYSICS (Relativity; Lorentz Transformations)

Hendrik Lorentz (1853–1928, Dutch) publishes the paper "Electromagnetic Phenomena in a System Moving with any Velocity Less than that of Light." He includes the "Lorentz transformations" (1892—PHYSICS) independently derived the following year by Albert Einstein and included in Einstein's landmark paper introducing the special theory of relativity (1905—PHYSICS).

1904–1905

CHEMISTRY (Actinium X)

Friedrich Giesel (1852–1927, German) discovers actinium X, an isotope of radium.

1905

BIOLOGY (Hormones)

Ernest Starling (1866–1927, English) introduces the term "hormone" in the paper "On the Chemical Correlation of the Functions of the Body" and clarifies the hormone concept, using the term for substances secreted into the blood stream and then transported with the blood to other locations where they produce decided effects. One such substance, secretin, was discovered earlier by Starling in collaboration with William Bayliss (1902—BIOLOGY). Starling now indicates that hormones are also secreted by the thyroid, gonads, and other endocrine glands.

CHEMISTRY (Dewar Flask)

James Dewar (1842–1923, Scottish) improves the Dewar flask (1892—CHEMISTRY) by using the adsorptive qualities of charcoal at low temperatures to approach more closely a vacuum state between the inner container and the outer covering. He also switches from glass to metal containers, allowing larger and stronger flasks.

CHEMISTRY (Radiothorium)

Otto Hahn (1879–1968, German) discovers radiothorium.

EARTH SCIENCES (Sea Ice Drift; Ekman Spiral)

Vagn Ekman (1874–1954, Swedish) publishes the paper "On the Influence of the Earth's Rotation on Ocean Currents," in which, following the encouragement of Fridtjof Nansen (1902—EARTH SCIENCES), he mathematically examines the systematic pattern of Arctic sea ice drift to the right of the wind direction, the further rightward drift of the ocean layers under the ice, and the similar rightward deflection of a wind-induced current in the absence of ice, all under various idealized assumptions regarding the ocean basin. Ekman considers the friction between the wind and ocean, the friction between layers of ocean, and the Coriolis force deflecting motions from a straight line path due to the earth's rotation (1835—METEOROLOGY). He determines the ocean's theoretical response both to a steady wind and to an impulsive but horizontally uniform wind, examining particularly the influence of the Coriolis force on the dynamical behavior of the ice and the upper layers of the ocean. Ekman shows not only that there should be a spiraling effect, waters moving slower and further to the right with depth (later termed the Ekman spiral), but also that the wind-driven currents should extend only to a depth of about 200 meters. Near the sea-bed, as the flow becomes influenced by friction with the ocean bottom, the spiraling is such that the direction of flow moves leftward with depth and the speed decreases with depth.

HEALTH SCIENCES (Cornea Transplant)

Ophthalmologist Eduard Zirm (1863–1944, Austrian) performs the first known modern cornea transplant, taking sections from the cornea of an injured boy and attaching them to the eyes of a blinded workman. The operation is successful for one eye, with sight regained for the remaining three years of the workman's life, although sight in the other eye is lost due to glaucoma. Zirm reports on the operation in the 1906 article "Eine erfolgreiche totale Keratoplastik."

HEALTH SCIENCES (Nutrition; Milk)

Cornelius Pekelharing (1848–1922, Dutch) determines that there exists a still unknown substance in milk that is essential to proper nutrition. Pekelharing feeds two groups of experimental mice identical diets with the single exception that the diet of one group is supplemented by milk. The mice in the group receiving milk remain healthy, while those in the other group all die within four weeks. The results remain unchanged when a very small amount of the whey portion of milk replaces the full milk supplement.

MATHEMATICS (Calculators)

Christel Hamann (1870–1948, German) creates a calculating machine capable of automatic division.

MATHEMATICS (Fermat Numbers)

Morehead and Western independently prove that Fermat number F_7 (see MATHEMATICS entries for 1640 and 1880) is composite, although neither actually factors it.

MATHEMATICS (Linear Associative Division Algebras)

Joseph Wedderburn (1882–1948, American) proves that any linear associative division algebra whose coordinates are elements of a finite field must be multiplicatively commutative.

MATHEMATICS (Logical Paradoxes)

Bertrand Russell (1872–1970, English) points out that a large class of paradoxes all derive from allowing an object to be defined in terms of a set containing it and hence can be eliminated by disallowing such definitions. Examples of such paradoxes include the paradox on the class of all classes not belonging to themselves (1902—MATHEMATICS) and the paradox of the barber who is presumed to shave each man in his town who does not shave himself but not to shave anyone else (paradoxical question: does the barber shave himself?).

METEOROLOGY (Planetary Boundary Layer: Ekman Spirals)

Vagn Ekman (1874–1954, Swedish) suggests applying his theory of the spiraling of the velocity vectors of upper ocean waters (1905—EARTH SCIENCES) to the atmosphere. The theoretical wind vectors are directed further to the right (in the Northern Hemisphere) with increasing height, with the flow in the layer above the frictional influence of the earth's surface being nearly geostrophic, at which point winds are directed parallel to the isobars, with high pressure on the right and low pressure on the left (or, in the Southern Hemisphere, with low pressure on the right and high pressure on the left). The geostrophic flow is produced by a balance between the pressure gradient force, directing winds from high to low pressure, and the Coriolis effect (1835—METEOROLOGY) resulting from the rotation of the earth-atmosphere system.

PHYSICS (Brownian Motion; Atomic Theory)

In the paper "Die von der molekularkinetischen Theorie der Wärme geforderte Bewegung von in ruhenden Flüssigkeiten suspendierten Teilchen" [On the movement of small particles suspended in a stationary liquid demanded by the molecular kinetic theory of heat], Albert Einstein (1879–1955, German-

Swiss-American) develops a mathematical analysis of Brownian motion and uses it to support the atomic hypothesis and the molecular kinetic theory of heat. Some consider the work, along with later experimental confirmation by Jean Perrin (1908—PHYSICS) to provide the most convincing evidence to date for the existence of molecules and atoms.

PHYSICS (Photoelectric Effect; Particle-Wave Duality of Light)

In the paper "Über einen die Erzeugung und Verwandlung des Lichtes betreffenden heuristischen Gesichtspunkt" [On a heuristic viewpoint concerning the production and transformation of light], Albert Einstein (1879–1955, German-Swiss-American) develops a quantum theory of light and uses it to explain the photoelectric effect, that is, the emission of electrons from certain metals when struck by light. Einstein asserts that light consists of individual quanta which exhibit particle as well as wavelike properties, thus introducing the particle-wave duality. The photoelectric effect arises because the direct impact of a light quantum (or photon) on the metal can cause the emission of an electron. Einstein's success in explaining the photoelectric effect helps establish the quantum hypothesis, first introduced by Max Planck (1900—PHYSICS).

PHYSICS (Relativity)

In the paper "Zur Elektrodynamik bewegter Körpen" [On the electrodynamics of moving bodies], Albert Einstein (1879–1955, German-Swiss-American) introduces the special theory of relativity, appropriate for inertial systems moving in uniform motion relative to each other. He bases the theory on two fundamental hypotheses: (1) the speed of light is invariant in any inertial reference frame, that is, it does not depend in any way on the motion of the source or the observer (an hypothesis consistent with the results of the Michelson-Morley experiment, 1887—PHYSICS); and (2) the laws of nature have the same form in all nonaccelerating systems (from Henri Poincaré, 1904—PHYSICS). Einstein's derivations from these two hypotheses are startling: two events which are simultaneous to one observer are not necessarily simultaneous to another observer; two individuals moving relative to each other will determine a different pair of values for the length and mass of a measured object (with ideal measuring instruments); no body can move faster than the speed of light. More specifically, in the case of two observers measuring the length of a moving object (for instance a train), one from within the object and one from outside, the observer from outside will measure the length as shorter by a factor of the square root of $(1 - v^2/c^2)$, where c is the speed of light and v is the speed of the object relative to the outside observer. Similarly, the rates of clocks within the object will appear to the outside observer as slowed by this same factor; *but,* to the observer within the object, the clock on the outside is the one which is slowed, again by the same factor. Key conclusions are that velocity is relative and that no observer's point of reference is any more fundamental than any other observer's. Einstein has no need for the ether hypothesis assumed in the work of Hendrik Lorentz, Poincaré, and others.

PHYSICS (Relativity, $E = mc^2$)

Albert Einstein (1879–1955, German-Swiss-American) publishes a short paper supplementary to his introduction of special relativity (1905—PHYSICS) in which he concludes that the mass (m) of a body is a measure of its energy content (E), with the two being related according to the formula $E = mc^2$, where c is the speed of light. The formula is not presented as $E = mc^2$ in the paper, entitled "Ist die Trägheit eines Körpers von seinem Energiegehalt abhängig?" [Does the inertia of a body depend upon its energy content?], but the loss of mass of a body giving off radiative energy L is determined to be L/c^2. In such a process matter is converted to energy, with the reverse process of energy conversion to matter being also possible. Hence the separate laws of the conservation of matter and the conservation of energy must be rejected.

SUPPLEMENTAL (Intelligence Testing)

Alfred Binet (1857–1911, French) and Théodore Simon (1873–1961, French) publish a "Metrical Scale of Intelligence" to distinguish the intelligence levels of tested children. By determining from supposedly "average" children the types of questions that they can answer correctly at each age level, Binet derives the concept of "mental age." The mental age of a random child is then determined according to Binet and Simon's Metrical Scale based on the child's responses to a set of questions selected for the apparent age separation it produces among the group of average children.

1905–1907

ASTRONOMY (Hertzsprung-Russell Diagram; Giant Stars)

In a pair of papers published in 1905 and 1907, each titled "Zur Strahlung der Sterne," Ejnar Hertzsprung (1873–1967, Danish) identifies two major luminosity sequences of stars, the one including the fainter stars having brightness decrease as spectral class increases, and the other showing the reverse trend. Later plotting of such relationships leads to the Hertzsprung-Russell diagram (1908—ASTRONOMY). The 1905 and 1907 papers also announce Hertzsprung's discovery of giant stars, that is, stars significantly more voluminous than the mean. These stars fall in the high-luminosity sequence, with brightness and spectral class increasing together. Hertzsprung also shows that these stars have particularly sharp absorption lines.

1906

ASTRONOMY (Mars; Extraterrestrial Life)

Percival Lowell (1855–1916, American) publishes *Mars and Its Canals,* where he elaborates his belief (1894—ASTRONOMY) that Mars is inhabited and that the linear markings first recorded by Giovanni Schiaparelli (1877—ASTRONOMY) are artificial waterways constructed by intelligent beings.

ASTRONOMY (Milky Way)

Jacobus Kapteyn (1851–1922, Dutch) constructs an ellipsoidal model of the Milky Way Galaxy similar to that of William Herschel (1785—ASTRONOMY) but with a distance scale added, indicating a diameter of about 3,000 parsecs. He determines the shape of the galaxy by analyzing star counts in different regions of the sky.

ASTRONOMY (Pluto)

Percival Lowell (1855–1916, American) initiates a systematic search for a planet beyond Neptune to account for those peculiarities in the orbit of Uranus unexplained by Neptune itself. Such efforts lead eventually to the discovery of Pluto (1930—ASTRONOMY).

ASTRONOMY (Rotation of Sun)

Walter Adams (1876–1956, American) and Carrington examine the rotation of the sun, noting in particular that sunspots rotate around the solar globe proportionately faster as the latitude decreases.

BIOLOGY (Plant Dispersal)

In *Plant Dispersal,* the second volume of *Observations of a Naturalist in the Pacific Between 1896 and 1899,* Henry Guppy (1854–1926, English) provides empirically based arguments for the phenomenon of plant dispersal in the Pacific region. He includes experiments and data particularly on the method of dispersal by flotation in sea water, indicating which seeds and fruits float and how long they can do so. In trying to account for the current geographical distribution of plants, he also hypothesizes former large-scale plant dispersal by a hypothetical, now-extinct race of pigeons.

CHEMISTRY (Nernst Heat Theorem)

Walther Nernst (1864–1941, German) formulates the Nernst heat theorem, or third law of thermodynamics, that if a system could be cooled to absolute zero, then any further chemical changes would leave the entropy unaffected. This is later strengthened to: the entropy of any system at a temperature of absolute zero is zero.

EARTH SCIENCES (Magnetic Reversals)

Bernard Brunhes (French), interpreting evidence from ancient French lava flows, discovers the phenomenon of geomagnetic reversal, noting evidence of the latest such reversal, occurring 700,000 years ago, in iron-rich particles in an ancient lava flow in France. The period since this latest reversal is later labeled the Brunhes Epoch.

EARTH SCIENCES (Volcanoes)

Clarence Dutton (1841–1912, American) suggests that radioactivity is the cause of volcanic activity.

HEALTH SCIENCES (Spotted Fever)

Howard Ricketts (1871–1910, American), through experiments with guinea pigs, determines that Rocky Mountain spotted fever is transmitted by tick bites.

HEALTH SCIENCES (Syphilis)

August von Wassermann (1866–1925, German) creates a blood test for syphilis.

HEALTH SCIENCES (Vitamins)

By determining that food contains other essential ingredients in addition to carbohydrates, minerals, fats, proteins, and water, Frederick Hopkins (1861–1947, English) approaches the discovery of the vitamin concept (1912—HEALTH SCIENCES).

MATHEMATICS (Finite Projective Geometries)

Oswald Veblen (1880–1960, American) and William Bussey (b.1879, American) construct projective geometries with only a finite number of points and a finite number of lines on any plane.

MATHEMATICS (Logical Paradoxes; Impredicative Definitions)

Henri Poincaré (1854–1912, French), agreeing with Bertrand Russell's determination that the current logical paradoxes result from allowing an object to be defined in terms of a set containing it (1905—MATHEMATICS), labels all such definitions "impredicative" and declares them illegitimate. He thereby eliminates the troublesome paradoxes, but a wide range of nonparadoxical statements are eliminated as well.

MATHEMATICS (Projective Differential Geometry)

Ernest Wilczynski (1876–1932, German-American) presents his theory of projective differential geometry, based on invariants and covariants of systems of linear homogeneous differential equations, in *Projective Differential Geometry of Curves and Ruled Surfaces*.

MATHEMATICS (Topology)

Stimulated by a desire to unify Georg Cantor's set theory (MATHEMATICS entries for 1874, 1879, 1883, 1895–1897) and the use of functions as space points,

Maurice Fréchet (1878–1973, French) creates a geometry of abstract spaces and introduces the concepts of separability and completeness for topological spaces.

METEOROLOGY (Polar Front)

William Shaw (1854–1945, English) approaches the concept of the polar front in his *Life History of Surface Air Currents.*

1906–1909

PHYSICS (Alpha Particles)

Ernest Rutherford (1871–1937, New Zealander-English) makes intensive studies of alpha particles and shows them to be the nuclei of helium atoms.

1906–1923

MATHEMATICS (History of Determinants)

Thomas Muir (1844–1934, Scottish) publishes a five-volume history of determinants: *Theory of Determinants in Their Historical Order of Development.*

1907

BIOLOGY (Conditioned Reflexes)

Ivan Pavlov (1849–1936, Russian) publishes details of his extensive experimentation with dogs in *Conditioned Reflexes.* Among his described experiments is the successful conditioning of a dog to secrete saliva upon the sound of a bell, the conditioning having been accomplished by ringing the bell during all meals. Pavlov also maps a dog's cerebral cortex according to the locations of the centers of hearing, vision, barking, and memory, doing so after noting responses when various parts of the brain are cut off.

BIOLOGY (Genetics)

Publication of Hugo de Vries's (1848–1935, Dutch) *Plant Breeding,* providing further evidence for the importance of mutations in plant evolution (see BIOLOGY entries for 1900 and 1901–1903).

CHEMISTRY (Ionium)

Bertram Boltwood (1870–1927, American) discovers ionium, later recognized as an isotope of thorium of atomic mass 230, and shows it to be radioactive, naturally transforming into radium.

CHEMISTRY (Immunochemistry)

Publication of Svante Arrhenius's (1859–1927, Swedish) *Immunochemistry.*

CHEMISTRY (Isotopes)

Herbert McCoy (1870–1945, American) and William Ross (1875–1947, American) present an explicit description of isotopes and solidify the evidence that isotopes are chemically inseparable.

CHEMISTRY (Lutetium)

Georges Urbain (1872–1938, French) discovers lutetium.

CHEMISTRY (Mesothorium)

Otto Hahn (1879–1968, German) discovers mesothorium I and mesothorium II.

HEALTH SCIENCES (Nutrition; Palatability)

Elmer McCollum (1879–1967, American) increases the palatability of purified foods in an attempt to overcome the unexplained apparent necessity of natural foods in animal diets. Only later, with the establishment of the vitamin concept (1912—HEALTH SCIENCES), is it recognized that the problem is not simply one of palatability.

HEALTH SCIENCES (Trypanosomes)

Paul Ehrlich (1854–1915, German) discovers trypan red, a dye which will destroy trypanosomes when injected into the bloodstream of animals infested with them.

MATHEMATICS (Algebraic Invariants)

Emmy Noether (1882–1935, German) advances the study of algebraic invariants with her doctoral thesis "On Complete Systems of Invariants for Ternary Biquadratic Forms."

MATHEMATICS (Intuitionism)

The intuitionist interpretation of mathematics is forcefully asserted in Luitzen Brouwer's (1881–1966, Dutch) dissertation *On the Foundations of Mathematics.* Very much opposed to the logistic interpretation of mathematics as a subset of logic, the intuitionists hold mathematics to originate from within the human mind, which inherently contains basic intuitions of such concepts as the integers, multiplication, and, for some, the principle of mathematical induction. Brouwer rejects Georg Cantor's theory of transfinite numbers (MATHEMATICS entries for 1883 and 1895–1897) and Ernst Zermelo's axiom of choice (1904—MATHEMATICS).

MATHEMATICS (Projective Geometry)

Oswald Veblen (1880–1960, American) and John Young (1879–1932, American) present a set of postulates for projective geometry.

PHYSICS (Space-Time)

Hermann Minkowski (1864–1909, Russian) suggests treating all events in a four-dimensional, space-time context.

1907–1909

EARTH SCIENCES (Antarctica; South Magnetic Pole; Paleoclimatology)

Ernest Shackleton (1874–1922, English) leads one of the first expeditions to attempt to reach the South Pole. Although he fails in that objective, he arrives within 160 kilometers of the pole, locates the south magnetic pole, and returns with important information on the continent of Antarctica, including fossils suggesting that Antarctica some time in the past had a temperate climate.

1908

ASTRONOMY (Hertzsprung-Russell Diagram)

Following his earlier work on the luminosity of stars (1905–1907—ASTRONOMY), Ejnar Hertzsprung (1873–1967, Danish) plots a scatter diagram of a star's absolute magnitude versus its spectral class, creating what is later to be termed a Hertzsprung-Russell diagram due to its independent development by Henry Russell (1914—ASTRONOMY).

ASTRONOMY (Solar Storms)

George Hale (1868–1938, American) examines solar magnetic storms and determines that the Zeeman effect (1896—PHYSICS) is apparent in the spectra of sunspots.

ASTRONOMY (Variable Stars)

Henrietta Leavitt (1868–1921, American) concludes from a study of 16 Cepheid variable stars in the Small Magellanic Cloud that the periods of the variables are correlated with their absolute magnitudes, brighter stars having longer periods. She presents the results in the memoir "1777 Variables in the Magellanic Clouds."

BIOLOGY (Enzymes)

Publication of William Bayliss's (1860–1924, English) *The Nature of Enzyme Action.*

CHEMISTRY (Ammonia)

Fritz Haber (1868–1934, German) synthesizes ammonia.

CHEMISTRY (Avogadro's Number)

Jean Perrin (1870–1942, French) calculates a value of 6×10^{23} particles/mole for Avogadro's number (1811—CHEMISTRY).

CHEMISTRY (Low-Temperature Chemistry; Helium)

Heike Kamerlingh-Onnes (1853–1926, Dutch) liquefies helium, a task requiring a temperature of 4.2 K.

MATHEMATICS (Differential Equations)

Eliakim Moore (1862–1932, American) examines differential equations with an infinity of variables. This is apparently the first instance in which the number of variables is not constrained to be finite.

MATHEMATICS (Logic)

Luitzen Brouwer (1881–1966, Dutch) publishes a short but powerful paper asserting the essential "unreliability of the principles of logic" for deriving meaningful statements for the real world.

MATHEMATICS (Set Theory)

Ernst Zermelo (1871–1953, German) establishes set theory on an axiomatic basis, providing a set of seven axioms, including the controversial axiom of choice (1904—MATHEMATICS). To avoid such paradoxes as those of Bertrand Russell (1902—MATHEMATICS), Zermelo limits the concept of set by requiring that the defining property E of any subset be such that for each element of the set it can be determined whether or not property E holds.

PHYSICS (Particle Counters; Alpha Particles)

Ernest Rutherford (1871–1937, New Zealander-English) and Hans Geiger (1882–1945, German) invent an electrical alpha-particle counter in which alpha rays are sent through a gas in a cylindrical ionization chamber with a thin wire stretched through it and a strong electric field such that an alpha particle generates a voltage step measurable by an attached voltmeter. Rutherford and Geiger use the counter to determine that one gram of radium emits alpha particles at a rate of 3.4×10^{10} per second. Combining this with the total emitted charge, it is seen that the magnitude of the charge on an individual alpha particle is twice that on the electron (with reversed sign). This leads Rutherford to confirm that alpha particles are helium atoms stripped of two electrons. Over the next several years Geiger continues working on improving the instrument, which later becomes known as the Geiger counter.

PHYSICS (Relativity)

Hermann Minkowski (1864–1909, Russian) shows the square root of $S^2 - ct^2$ to be invariant even for observers moving relative to each other, provided that the relative motion is uniform, that is, nonaccelerated. (*S* is the spatial distance between two events separated by time *t*, while c is the speed of light.) Thus, though neither space nor time itself is an invariant, there does exist a form of absolute interval in a space-time context. Minkowski emphasizes the four-dimensional nature of space-time in the paper "Space and Time" presented at a conference at Cologne.

PHYSICS (Suspended Distributions)

Jean Perrin (1870–1942, French) quantifies the distribution of suspended particles of gum resin in a liquid and confirms that the density drops exponentially with height. Brownian motion (1827—PHYSICS) is assumed, the suspended particles being continually bombarded by the molecules of the liquid. Perrin's experiments help confirm Albert Einstein's theoretical analysis of Brownian motion (1905—PHYSICS).

c.1908

METEOROLOGY (Greenhouse Effect)

Svante Arrhenius (1859–1927, Swedish) describes the "greenhouse effect" brought about by carbon dioxide in the atmosphere: since carbon dioxide absorbs terrestrial radiation more strongly than it reflects solar radiation, it allows solar radiation to enter the earth-atmosphere system more easily than it allows terrestrial radiation to escape, thereby contributing an overall warming effect.

1909

ASTRONOMY (Moon)

Thomas See (1866–1962, American) suggests that the moon formed in the outer solar system, lost energy, attained an orbit closer to the sun, and was finally captured by the earth. He explains the moon's craters as having formed from meteorite bombardment, and hypothesizes that the moon's surface consists of rock fragments and fine dust.

BIOLOGY (Genetics)

Wilhelm Johannsen (1857–1927, Dutch) introduces the term "gene," replacing the Mendelian "factor."

EARTH SCIENCES (Moho Discontinuity)

The research of Andrija Mohorovičič (1857–1936, Serbian) reveals a sharp change in the velocity of earthquake waves occurring at about 20 miles below the earth's surface, and as a result he deduces the existence of a sharp discontinuity (the Moho discontinuity) separating the earth's mantle from its more-rigid outer crust. The depth of this separation varies with location from about 8 to about 40 miles.

HEALTH SCIENCES (Nutrition; Amino Acids)

Thomas Osborne (1859–1929, American) and Lafayette Mendel (1872–1935, American) begin extensive experimentation with rats to determine the nutritive value of individual amino acids, by deleting them one at a time from a controlled diet.

HEALTH SCIENCES (Nutrition; Feces)

Thomas Osborne (1859–1929, American) and Lafayette Mendel (1872–1935, American) find that experimental rats on restricted, purified diets are healthier when allowed to eat their own feces than when prevented from doing so. They also find that the health improves markedly when the rats are allowed to eat the feces of rats fed natural rather than purified foods.

HEALTH SCIENCES (Nutrition; Phosphorus)

Elmer McCollum (1879–1967, American) shows that organic phosphorus is not explicitly necessary in animal diets, as calcium phosphate in the diet will allow animals to synthesize the needed organic phosphates.

HEALTH SCIENCES (Spotted Fever)

Howard Ricketts (1871–1910, American), in the paper "A Micro-Organism which Apparently has a Specific Relationship to Rocky Mountain Spotted Fever," describes a bacillus found in the blood of guinea pigs and monkeys infected with the disease. He suggests that the bacillus may be the causative agent in spotted fever.

MATHEMATICS (Fermat Numbers)

Morehead and Western show the Fermat number F_8 to be composite (see MATHEMATICS entries for 1640, 1880, and 1905).

MATHEMATICS (Number Theory; Waring's Problem)

David Hilbert (1862–1943, German) proves that for every integer n there exists an integer m such that every integer is a sum of m n^{th} powers. This expands upon the hypothesis of Edward Waring (1770—MATHEMATICS) that each positive integer is a sum of 9 cubes ($n = 3, m = 9$) and of 19 fourth powers ($n = 4, m = 19$).

1910

BIOLOGY (Genetics; Fruit Flies)

While breeding fruit flies *(Drosophila),* Thomas Morgan (1866–1945, American) obtains a mutant white-eyed male amongst an otherwise all red-eyed collection of flies. Upon breeding the mutant with the red-eyed females, he finds an expected mixed distribution in the first generation of offspring but a second generation consisting of red-eyed females and white-eyed males. From this he postulates his principle of sex-linkage, whereby some hereditary factors are linked to either the male or female chromosome. Extensive further work on *Drosophila* leads later to Morgan's *Theory of the Gene* (1926—BIOLOGY).

CHEMISTRY (Liquid Fuels)

Friedrich Bergius (1884–1949, German) develops a process for producing liquid fuels based on hydrogenation of coal.

CHEMISTRY (Radium)

Marie Curie (1867–1934, Polish-French) and André Debierne (1874–1949, French) use electrolysis to isolate metallic radium.

CHEMISTRY (Titanium)

Matthew Hunter (b.1878) prepares a sample of 99.9% pure titanium.

HEALTH SCIENCES (Syphilis)

Paul Ehrlich (1854–1915, German) and his assistant Sahachiro Hata (1873–1938, Japanese) find a cure for syphilis in Ehrlich's 606th tested compound, an organic arsenical named Salvarsan or arsphenamine, and publish their results in *Die experimentelle Chemotherapie die Spirillosen (Syphilis, Rückfallfieber, Hühnerspirillose, Frambösie).*

MATHEMATICS (Binary Forms)

Emmy Noether (1882–1935, German) extends from two to *n* variables the earlier result of Paul Gordan (1868—MATHEMATICS) that all binary forms have a finite complete system of invariants and covariants.

MATHEMATICS (Calculus)

Henri Lebesgue (1875–1941, French) publishes the memoir "Sur l'intégration des fonctions discontinues," in which he extends integration and differentiation to *n* dimensions and introduces and develops the notion of a countably additive set function.

MATHEMATICS (Fields)

Ernst Steinitz (1871–1928, German) generalizes Leopold Kronecker's theory of algebraic magnitudes (1881—MATHEMATICS) and presents an algebraic theory of fields including an attempt at enumerating all possible fields by extending the simplest ones through algebraic and transcendental adjunctions.

MATHEMATICS (Projective Geometry)

In the first volume of their *Projective Geometry,* Oswald Veblen (1880–1960, American) and John Young (1879–1932, American) axiomatize projective geometry, following the formalistic approach that David Hilbert suggested for all mathematics (1904—MATHEMATICS).

PHYSICS (Electric Charge; Millikan Oil-Drop Experiment)

Robert Millikan (1868–1953, American) determines the charge of an electron by adjusting an upward electric force on an oil drop that has acquired an electron charge until the oil drop comes to a suspended rest. At this point, the electric force just balances the downward gravitational force, from which balance the electron's charge can be computed.

1910–1911

CHEMISTRY (Atomic Weights; Isotopes)

By measuring deflections of positive rays in a cathode ray tube, Joseph John Thomson (1856–1940, English) calculates masses of various atoms. He shows neon, with atomic weight 20.183, to consist of appropriate proportions of isotopes of mass 20 and 22.

1910–1912

EARTH SCIENCES (Antarctica)

Scientific information is amassed during the first two expeditions to reach the South Pole, the first headed by Roald Amundsen (1872–1928, Norwegian) and reaching the Pole on December 14, 1911, the second headed by Robert Scott (1868–1912, English) and reaching the Pole on January 17, 1912. The Scott expedition, which ends tragically, with all five members of the Pole group dying on the return trip over the ice, conducts geological, meteorological, oceanographic, geomagnetic, auroral, glaciological, and biological studies, the latter being done at an emperor penguin rookery near Cape Crozier. The Amundsen expedition, which concentrates more exclusively on reaching the Pole and does so more easily than the Scott expedition largely because of using sled dogs, also collects rock samples near the coast and carries out oceanographic experiments.

1910–1913

MATHEMATICS (Mathematics as a Subset of Logic)

Alfred North Whitehead (1861–1947, English) and Bertrand Russell (1872–1970, English) publish a classic three-volume work, *Principia Mathematica,* propounding that all pure mathematics is a subset of formal logic and proceeding to derive mathematics from a small number of undefined concepts and principles of logic. The work further develops and systematizes the symbolism and methods of Giuseppe Peano (1901—MATHEMATICS).

1911

ASTRONOMY (Parallax)

Publication of Henry Russell's (1877–1957, American) *Determinations of Stellar Parallax.*

CHEMISTRY (Low-Temperature Chemistry; Superconductivity)

While examining the electrical resistance of mercury at extremely low temperatures, Heike Kamerlingh-Onnes (1853–1926, Dutch) finds that at 4.12 K this resistance vanishes altogether, the metal having become "superconductive." Along with superconductivity comes diamagnetism, preventing magnetic lines of force of normal strength from penetrating. (Upon increasing the strength of the magnetic field, both the diamagnetism and the superconductivity can be destroyed.)

CHEMISTRY (Uranium Y)

Uranium Y is discovered by G. N. Antonoff.

EARTH SCIENCES (Ocean Layering)

Brennecke determines a layering in the structure of the waters of the South Atlantic, with southward moving waters from the North Atlantic sandwiched between northward moving waters from the Antarctic in the bottom and intermediate ocean layers. Brennecke determines the temperature inversion between the intermediate waters and the underlying North Atlantic waters through examination of data from the *Deutschland* east of South America.

HEALTH SCIENCES (Depression)

Karl Abraham (1877–1925, Austrian) suggests that depression following the death of a loved one can arise from directing hostility toward oneself.

HEALTH SCIENCES (Schizophrenia)

The term "schizophrenia" is introduced by Paul Bleuler (1857–1939, Swiss) in *Dementia Praecox,* where he concludes that schizophrenia does not derive from a reduced mental state but rather from a disharmony within the mind.

MATHEMATICS (Topology)

Luitzen Brouwer (1881–1966, Dutch) proves that the dimension of a cartesian space is topologically invariant.

MATHEMATICS (Topology)

Luitzen Brouwer (1881–1966, Dutch) proves that every vector field on an even-dimensional sphere must have at least one singular point and that every continuous transformation of the n-dimensional simplex must have at least one fixed point. He also extends the Jordan Curve Theorem (1887—MATHEMATICS) to: the n-dimensional Euclidean space is separated into two regions by any $(n-1)$-dimensional manifold.

METEOROLOGY/PHYSICS (Cosmic Radiation)

Victor Hess (1883–1964, Austrian) discovers cosmic radiation following measurements during balloon ascents to an altitude of 5,350 meters. He finds that radiation decreases with height up to an altitude of about 150 meters above sea level, then begins to increase with height, with the increase leading to a radiation level at 5,000 meters several times that at sea level. Since only the low level decrease can be explained by known radiation sources and since the high-level radiation remains constant day or night, he concludes a cosmic origin for the high-level radiation.

METEOROLOGY (Rain)

Alfred Wegener (1880–1930, German) suggests that rain generally begins as ice, high in the atmosphere, and subsequently melts as it descends to warmer air layers. Although speculative at first—but based somewhat on the principle that ice, in view of its relatively low surrounding vapor pressure, attracts water vapor—observational evidence accumulates with the later work of Tor Bergeron in 1922 and 1933 (1922—METEOROLOGY).

PHYSICS (Ergodicity)

In an encyclopedia article on the foundations of statistical mechanics, Paul Ehrenfest (1880–1933, Austrian) and Tatyana Ehrenfest (Russian) suggest that there are probably no strictly ergodic systems (see PHYSICS entries for 1877 and 1890) but that likely there exist quasi-ergodic systems which, although not passing through every possible microstate, pass within any predetermined $\delta > 0$ distance of each microstate.

PHYSICS/CHEMISTRY (Nuclear Atom)

Ernest Rutherford (1871–1937, New Zealand-English) proposes the concept of a nuclear atom, suggesting that the atom consists of a very small central nucleus surrounded by electrons. His colleagues Hans Geiger (1882–1945, German) and Ernest Marsden had found that most alpha particles shot through a thin sheet of gold foil are scattered only a few degrees but about 1 in 10,000 bounces back from the foil. Rutherford concludes that the atom must have a small, massive nucleus with a positive charge and that it is upon striking this nucleus that the 1 in 10,000 alpha particles bounces back. The nucleus is surrounded by negatively charged electrons which revolve around it in set orbits. The total positive charge in the nucleus balances the total negative charge of the electrons. This nuclear model of the atom revises Joseph John Thomson's earlier model (1904—CHEMISTRY), in which negative electrons are embedded in a sphere of positive electricity.

PHYSICS (Number of Electrons)

Charles Barkla (1877–1944, English) shows that the number of electrons in an atom is approximately half the atomic weight, doing so by analyzing the amount of scattering of X-rays by various light elements.

PHYSICS (Relativity; Gravitation; Mass)
ASTRONOMY (Gravitation)

Albert Einstein (1879–1955, German-Swiss-American) calculates the deflection of starlight passing close to the sun, the deflection being caused by the action of the sun's gravitational field on the light photons. He presents a calculated value of deflection of 0.83 seconds of arc in the paper "On the influence of gravitation on the propagation of light," where he also shows that the gravitational mass of a body increases identically with the inertial mass when the energy is increased, both masses increasing according to $m = E/c^2$. (Johann von Soldner had calculated starlight deflection using principles of Newtonian physics [1801—PHYSICS].)

1912

ASTRONOMY (Nebular Radial Velocities)

Vesto Slipher (1875–1969, American) measures the radial velocity of the Andromeda nebula, from Lowell Observatory in Arizona. Basing his calculations on the Doppler-Fizeau Effect (1848—ASTRONOMY), he finds the nebula to be approaching the solar system at about 125 miles per second. Later he performs similar calculations on other nebulae.

ASTRONOMY (Stellar Distances)

Henrietta Leavitt (1868–1921, American) determines a method—termed the

Cepheid variable method—for estimating distances to stars far beyond the range of measurable stellar parallaxes (1839—ASTRONOMY). Extending her earlier study of Cepheid variable stars in the Small Magellanic Cloud (1908—ASTRONOMY) from 16 to 25 stars, Leavitt finds a direct proportionality between the apparent brightnesses and the periods of the variables, although admitting considerable scatter. Since the distance to the earth is approximately equal for all of the Magellanic Cloud stars, this proportionality translates to a direct proportion between period and intrinsic brightness. Leavitt calculates the proportionality constant from Cepheids relatively close to the solar system, with distances obtainable by other means. Extrapolation then allows calculation of the intrinsic brightness of any Cepheid once the period is determined; and the intrinsic brightness, when compared with the apparent brightness, allows calculation of the distance. Harlow Shapley (1885–1972, American) soon applies the method to calculating the distances to other Cepheid variable stars.

BIOLOGY (Unconscious)

Publication of Carl Jung's (1875–1961, Swiss) *Wandlungen und Symbols der Libido* [Psychology of the unconscious], in which he divides the unconscious into the personal and collective unconscious.

CHEMISTRY (Molar Heat Capacities)

Peter Debye (1884–1966, Dutch-American) revises Pierre Dulong and Alexis Petit's value of 6 cal/mole°C for the molar heat capacity of most substances (1819—CHEMISTRY) by correcting at very low temperatures for vibrational frequency of the atoms. The resulting theoretical curve, which descends to 0 cal/mole°C at 0°K, is found to agree well with experiment and is seen as an early success for quantum physics.

EARTH SCIENCES (Continental Drift)

Alfred Wegener (1880–1930, German) suggests the idea of continental drift in an address entitled "The Geophysical Basis of the Evolution of the Large-Scale Features of the Earth's Crust (Continents and Oceans)" at a professional meeting in Frankfurt am Main. His fuller development of the theory appears later (1915—EARTH SCIENCES).

EARTH SCIENCES (Oceanography)

Gerhard Schott (1866–1961, German) presents probably the most comprehensive treatment to date of the oceanography of the Atlantic Ocean in his *Geographie des Atlantischen Ozean* [Geography of the Atlantic Ocean]. In addition to discussing hydrological conditions, such as water temperature, salinity, and velocity, he describes basin geology, marine life, ocean climates, the history of exploration, the fishing industry, and shipping routes.

HEALTH SCIENCES (Antineuritic Substance; Thiamin)

Edward Vedder (1878–1952, American) and Robert Williams (1886–1965, American) seek to isolate the antineuritic dietary substance able to cure pigeons of neuritis, a disease similar to beriberi in humans and one which develops upon restriction of the diet to polished rice. They verify earlier work indicating that the antineuritic substance is an organic base soluble in both water and alcohol and further determine that it is adsorbable on charcoal. Continued work by Williams and other investigators leads to the synthesis of vitamin B_1 (thiamin) in the 1930s.

HEALTH SCIENCES (Syphilis)

Paul Ehrlich (1854–1915, German) introduces Neosalvarsan as a cure for syphilis. An advantage of the new compound over Salvarsan (1910—HEALTH SCIENCES) is that Neosalvarsan is readily soluble in water.

HEALTH SCIENCES (Vitamin-Deficiency Diseases)

Casimir Funk (1884–1967, Polish-American), who, along with Frederick Hopkins (1906—HEALTH SCIENCES), is variously referred to as the discoverer of the vitamin concept, publishes a paper hypothesizing that certain diseases, particularly beriberi, scurvy, pellagra, and perhaps rickets, are caused by deficiencies in specific trace components and introduces the term "vitamine." The name is later changed to "vitamin" when it is determined that the labeled substances are not all amines.

MATHEMATICS (Intuitionism)

Luitzen Brouwer (1881–1966, Dutch) encourages intuitionism in mathematics (1907—MATHEMATICS) and attacks formalism (1904—MATHEMATICS) and logicalism (1910–1913—MATHEMATICS). Brouwer suggests that humans are born with an innate intuition essentially of the natural numbers (specifically, of an unending sequence of objects formed by repeated additions of 1) and that logic is properly a subset of mathematics rather than vice versa. He denies that the law of the excluded middle is universally valid, asserts that proofs of existence should always be accompanied by methods of construction, and declares that theorems about infinite classes must be supported by methods of proof that require a strictly finite number of steps.

PHYSICS (X-Rays)

Max von Laue (1879–1960, German) obtains a diffraction pattern for X-rays through a crystal and thereby has the sought-after evidence that X-rays are waves. Previously, X-rays had resisted being diffracted even by the most precise gratings. Von Laue's success is a result of his correct speculation that the regular spacing of the atoms in crystals can serve as a grating of the desired precision. Von Laue proceeds to measure the wavelengths of the X-rays.

1912–1913

EARTH SCIENCES (Paleoclimatology)

Milutin Milankovitch (1879–1958, Yugoslavian) publishes several papers attributing past glaciations to changes in solar radiation due to changes in the earth's orbital parameters (precession, eccentricity, and tilt) and presenting his ongoing progress in calculating the distribution of solar radiation as a function of earth latitude and time. His more complete statement is published later (1920—EARTH SCIENCES).

1912–1917

ASTRONOMY (Nebular Radial Velocities)

Vesto Slipher (1875–1969, American) determines that, in contrast to the Andromeda nebula (1912—ASTRONOMY), most other near nebulae are receding from the solar system and at enormous speeds, a typical recession velocity being 400 miles per second.

1913

ASTRONOMY (Small Magellanic Cloud)

Ejnar Hertzsprung (1873–1967, Danish) calculates the distance of the Small Magellanic Cloud from the solar system.

CHEMISTRY (Isotopes)

Frederick Soddy (1877–1956, English) coins the term "isotope" for those elements, recently identified, which seem to belong in the same location in the periodic table, having similar chemical properties, and yet have different atomic weights and different physical properties. ("Isotope" is Greek for "same place.")

CHEMISTRY (Isotopes; Lead)

Theodore Richards (1868–1928, American) determines that lead from radioactive minerals has a lower atomic weight than ordinary lead, thereby reaffirming the existence of isotopes.

CHEMISTRY (Radioactivity)

Frederick Soddy (1877–1956, English), Kasimir Fajans (1887–1975, Polish-American), and Alexander Russell (b.1888) independently discover the law of radioactive displacement of elements, showing a connection between a radioactive element and its products resulting from emission of alpha and beta rays.

CHEMISTRY (Uranium X_2)

Kasimir Fajans (1887–1975, Polish-American) and O. H. Göhring (German) discover uranium X_2, a short-lived product of the disintegration of uranium X_1 (1900—CHEMISTRY).

HEALTH SCIENCES (Syphilis)

Hideyo Noguchi (1876–1928, Japanese) isolates the spirochete *Treponema pallidum* as the cause of syphilis. Noguchi also develops a skin test for syphilis.

HEALTH SCIENCES (Vitamin A)

Elmer McCollum (1879–1967, American) and Marguerite Davis conclude from various dietary studies on rats that certain fats contain a previously unidentified but indispensable nutrient. This fat-soluble substance is later named vitamin A.

MATHEMATICS (Algebra; Absolute Value)

József Kürschák (1864–1933, Hungarian) extends the notion of "absolute value" from the complex numbers to any abstract field F, assigning to each element p of F a unique real number $|p|$ constrained by the following postulates: $|p| \geqslant 0$, with $|p| = 0$ if and only if p is the 0 element of the field; $|pq| = |p| \cdot |q|$ for any elements p and q in F; and $|p + q| \leqslant |p| + |q|$ for any elements p and q in F.

MATHEMATICS (Logical Connectives)

Henry Sheffer (1883–1964, American) reduces the number of connectives necessary for truth-functional analysis from two—negation and conjunction—to one, the neither-nor connective.

PHYSICS (Atomic Model)

(See first the entry under 1913—PHYSICS [Hydrogen Model; Quantum Theory].) In accounting for the spectra of elements other than hydrogen, Niels Bohr (1885–1962, Danish) expands his quantized model of the hydrogen atom (1913—PHYSICS) to atoms with more than one electron, and concludes that not all electrons are involved equally in radiation emission. In the case of the sodium atom, with 11 electrons, Bohr explains the simplicity of the observed spectrum by assuming that only the outermost electron jumps from one orbit to another and thereby emits and absorbs radiation.

PHYSICS (Atomic Number)

Henry Moseley (1887–1915, English) examines the X-ray spectra of over 50 elements and notes a regularity in the shifting of spectral lines when the elements are arranged according to atomic weight. He finds that bombardment of the various elements with cathode rays yields a systematic sequence of vibration frequencies, and from this work derives the concept of atomic number, which he recognizes as equal to the nuclear charge. Upon a reordering of the periodic table according to atomic numbers rather than atomic weights, hydrogen remains the first element, with atomic number 1, but a few elements do change positions (e.g., argon now precedes potassium), improving the periodicity. Moseley predicts the discovery of three missing elements, those with atomic numbers 43, 61, and 75, and describes part of his research in the paper "The High Frequency Spectra of the Elements."

PHYSICS (Constancy of the Speed of Light)
ASTRONOMY (Binary Stars)

Through analysis of the movements of a binary star system, Willem de Sitter (1872–1934, Dutch) provides support for Albert Einstein's assumption of the constancy of the speed of light (1905—PHYSICS). He shows that if the velocity of light were to depend systematically upon the velocity of the light source, then the observed movements of binary stars would show anomalous behavior which is not in fact observed.

FIGURE 19. *Apparatus used by William Herschel to examine heating properties of light of different colors and to discover infrared radiation (see 1800—PHYSICS entry). (By permission of the Harvard College Library).*

PHYSICS (Hydrogen Model; Quantum Theory)

Concerned with the issue of the mechanism by which an atom radiates light, Niels Bohr (1885–1962, Danish) incorporates Max Planck's quantum postulate (1900—PHYSICS) into a theoretical model for the hydrogen atom fundamentally different from any previous conception. By inserting the quantum postulate he is able to retain Ernst Rutherford's postulate of the atomic nucleus (1911—PHYSICS) and furthermore eliminate the major perceived problem with the Rutherford model, namely the suspicion that the atom would not be stable since the electron should radiate and spin into the nucleus. The hydrogen model Bohr develops includes a limited number of possible orbits for the one hydrogen electron and specifies that emission and absorption of energy take place only as the electron shifts from one orbit to another and only in specific discrete amounts (quanta), the amount of energy involved in any individual case equaling the difference between the energy levels of the two stationary states. The success in calculating the wavelengths of hydrogen emission enables Bohr to explain the regularity of Balmer lines (1885—PHYSICS) through quantum characteristics. The Bohr atom leads to wider acceptance of the quantum theory.

PHYSICS (Stark Effect)

Johannes Stark (1874–1957, German) finds that an electric field causes a splitting of the spectral lines of hydrogen.

c.1913

ASTRONOMY (Cepheid Variables)

Harlow Shapley (1885–1972, American) shows the Cepheid variables to be pulsating stars, rejecting the other major alternative that the Cepheids are binaries periodically eclipsing each other.

1914

ASTRONOMY (Hertzsprung-Russell Diagram)

Henry Russell (1877–1957, American) publishes the paper "Relations Between the Spectra and other Characteristics of the Stars," including a "Hertzsprung-Russell diagram" of the relationship between a star's spectral class and its absolute magnitude. Russell developed the diagram independently of Ejnar Hertzsprung (1908—ASTRONOMY).

ASTRONOMY (Spectroscopic Parallax)

Walter Adams (1876–1956, American) compares stars with similar spectral types but very dissimilar luminosities and determines that certain spectral lines are strong for the very luminous stars and others for the faint stars. This allows estimates of luminosity, and thereby of distance, from the relative strengths of the individual spectral lines.

ASTRONOMY (White Dwarfs)

Walter Adams (1876–1956, American) calculates a surface temperature of approximately 8,000 K for the star Sirius B. This temperature, along with the low apparent brightness of the star, suggests—to the astronomers of the time—a stellar radius less than 1% of the radius of the sun and a density exceeding the sun's density by a factor of a million. It is regarded as the first known "white dwarf."

BIOLOGY (Evolution)

William Bateson (1861–1926, English) announces that the Darwinian theory of evolution by natural selection (1859—BIOLOGY) has been effectively reduced to a topic of purely historical interest. The argument is that the process of evolution has been shown, through extensions of the work of Gregor Mendel (1865—BIOLOGY), to proceed by large rather than small variations. (Darwin regains favor later with the findings that large variations are often harmful and that the fossil record does reveal gradual transitions. Aspects of the Darwinian theory are also supported in the works of Ronald Fisher [1930—BIOLOGY].)

BIOLOGY (Thyroxine)

Edward Kendall (1886–1972, American) isolates the hormone thyroxine, secreted by the thyroid gland.

CHEMISTRY (Lead)

A radioactive isotope of lead is discovered by Theodore Richards (1868–1928, American).

EARTH SCIENCES (Gutenberg Discontinuity)

Beno Gutenberg (1889–1960, German-American) determines the existence of a sharp discontinuity, the Gutenberg discontinuity, in the earth at a distance of about 2,150 miles from the surface. He speculates that this discontinuity divides a liquid core, of radius 1,800 miles, from the solid mantle.

HEALTH SCIENCES (Pellagra)

Carl Voegtlin (b.1879, Swiss) establishes, after dietary experiments on pellagra sufferers in a South Carolina hospital, that human pellagra is a dietary deficiency disease.

HEALTH SCIENCES (Vitamin A)

Elmer McCollum (1879–1967, American) and Marguerite Davis experiment with vitamin A, the fat-soluble nutrient earlier discovered by them (1913—HEALTH SCIENCES), and identify both locations where it exists, such as in the kidney and in ether extracts from dried leaves, and locations where it does not exist, such as in several specific vegetable oils and in tissue fats. Thomas Osborne (1859–1929, American) and Lafayette Mendel (1872–1935, American) determine that vitamin A is contained in cod liver oil, an animal fat, but not in almond oil, a vegetable fat.

MATHEMATICS (Topology)

In *Grundzüge der Mengenlehre* [Essentials of set theory], Felix Hausdorff (1868–1942, German) constructs a theory of abstract spaces from the neighborhood concept, defining a topological space as a set of elements along with a family of subsets (called neighborhoods) belonging to each element and satisfying three "natural" postulates: (1) any two neighborhoods of x have a common subset which is also a neighborhood of x; (2) if N_1 is a neighborhood of x and y is contained in N_1, then there exists a neighborhood N_2 of y which is also a subset of N_1; (3) if $x \neq y$ then there exist neighborhoods N_1 of x and N_2 of y such that the intersection of N_1 and N_2 is the null set. Hausdorff proves that each metric space can be extended to a complete metric space in essentially one way only.

PHYSICS (Atom)

James Franck (1882–1964, German) and Gustav Hertz (1887–1950, German) provide experimental confirmation of the Bohr model of the atom (1913—PHYSICS). By bombarding mercury vapor with a beam of electrons, they cause the vapor to emit radiation and calculate that the radiation emitted can be attributed to the transition of electrons from one stationary, quantized state to another.

PHYSICS (Beta Particles)

James Chadwick (1891–1974, English) discovers that the energy spectrum of beta particles is continuous, in contrast to the specific energy levels of alpha particles and gamma rays.

SUPPLEMENTAL (Intelligence Testing; IQ)

Expanding upon the intelligence testing methods of Alfred Binet and Théodore Simon (1905—SUPPLEMENTAL) and others, William Stern (1871–1938, German-American) suggests that the supposed "mental age" obtained by the tests be divided by the individual's chronological age to obtain an intelligence quotient, or IQ.

1915

BIOLOGY (Bacteriophage)

In the paper "An Investigation on the Nature of Ultra-Microscopic Viruses," Frederick Twort (1877–1950, English) reports the discovery of an ultramicroscopic virus, later named bacteriophage, which lyses (or causes the disintegration of) bacteria, thus indicating that even bacteria are susceptible to diseases caused by invisible viruses.

BIOLOGY (Genetics)

Publication of *The Mechanism of Mendelian Heredity* by Thomas Morgan (1866–1945, American), Calvin Bridges (1889–1938, American), Alfred Sturtevant (1891–1970, American), and Hermann Muller (1890–1967, American). The basic theme of the work is that invisible genes contained within the chromosomes of the cell nucleus determine the hereditary traits of the offspring.

BIOLOGY (Quantification in Biochemistry)

Publication of *Quantitative Laws in Biological Chemistry* by Svante Arrhenius (1859–1927, Swedish).

EARTH SCIENCES (Continental Drift)

In *Die Entstehung der Kontinente und Ozeane* [The origin of continents and oceans], Alfred Wegener (1880–1930, German) develops the theory of continental drift, including the thesis of a super primordial continent of Pangeae which broke into fragments, these to drift slowly apart and form the current, still-drifting continents. The 1915 edition is little read, but a revised 1922 third edition sparks vigorous discussion. Although Wegener exhibits such geological evidence as similar rock formations in India, Madagascar, and East Africa, plus geographical, geodetic, geophysical, paleontological, biological, and paleoclimatic evidence, the theory fails to gain widespread support until the early 1960s.

HEALTH SCIENCES (Nutrition; Vitamin A; Vitamin B)

Elmer McCollum (1879–1967, American) and Marguerite Davis initiate what they term the "biological method of analysis"in nutrition studies with rats by feeding the rats one base food (wheat in one set of studies and unpolished rice in another set) plus 0, 1, 2, or 3 supplements from among the following: purified protein, a salt mixture, or the fat soluble nutrient they identified earlier (1913—HEALTH SCIENCES), later named vitamin A. They find two supplements to be significantly better nutritionally than one and three to be markedly better still. They conclude that rats require both the previously-identified fat-soluble nutrient and a water-soluble, anti-beriberi nutrient. By labeling these "fat-soluble A" and "water-soluble B," they initiate the labeling of vitamins by letters.

HEALTH SCIENCES (Pellagra)

Unconvinced by earlier results of Carl Voegtlin (1914—HEALTH SCIENCES), Joseph Goldberger (1874–1929, Austria-Hungarian-American) and colleagues examine the impact of diet on pellagra victims, establishing from studies at state asylums in South Carolina, Georgia, and Mississippi that milk, meat, and eggs in the diet of the hospital staffs explains their continued health in the face of the pellagra afflicting the patients. Goldberger and colleagues then successfully use this knowledge to eradicate pellagra in a Mississippi orphanage by adding liberal helpings of milk, meat, and eggs to the children's diet. Their overwhelming conclusion is that pellagra results from dietary deficiency, not poisoning or infection.

PHYSICS (Electron Paths)

Modifying the Bohr model of the atom (1913—PHYSICS), Arnold Sommerfeld (1868–1951, German) suggests that the electron paths are elliptical and that the main electron orbits have subsidiary orbits with slightly different energy values. The number of subsidiary orbits is governed by a new quantum number K. A second subsidiary quantum number M is needed to account for the splitting of atomic spectral lines, as in the Zeeman effect (1896—PHYSICS).

PHYSICS (Photoelectric Effect)

Robert Millikan (1868–1953, American) experimentally verifies the mathematical theory of the photoelectric effect presented by Albert Einstein (1905—PHYSICS).

1915–1916

PHYSICS (Atomic Model)

William Wilson (1875–1965) in 1915 and, independently, Arnold Sommerfeld (1868–1951, German) in 1916 generalize Bohr's rule for determining the orbits in stationary states of atoms (1913—PHYSICS).

1916

ASTRONOMY (Cosmology; Cosmological Constant)

In applying his general theory of relativity (1916—PHYSICS) to the nature of the universe as a whole, Albert Einstein (1879–1955, German-Swiss-American) concludes that space is non-Euclidean and curved, that the amount of curvature depends on the amount of matter, and that the universe is unbounded but finite. He inserts a cosmological constant into his equations to maintain a static, nonexpanding universe (1917—ASTRONOMY).

ASTRONOMY (Schwarzschild Singularities; Black Holes)

Using relativistic equations of Albert Einstein, Karl Schwarzschild (1873–1916, German) shows that through the forces of gravitation, very massive stars might condense to such an extent that the theoretical escape velocity would exceed the velocity of light. No signal or material body could then escape this condensed star, later termed a "black hole" or "Schwarzschild singularity."

BIOLOGY (Genetics)

Publication of Calvin Bridges (1889–1938, American) and Thomas Morgan's (1866–1945, American) *Sex-Linked Inheritance in Drosophila.*

CHEMISTRY (Shared Electrons)

G. N. Lewis (1875–1946, American) hypothesizes the phenomenon of shared electrons in nonionic chemical compounds.

EARTH SCIENCES (Earth's Core)

Albert Michelson (1852–1931, American) determines that the earth has a molten core.

HEALTH SCIENCES (Depression)

Extending the concepts of Karl Abraham (1911—HEALTH SCIENCES), Sigmund Freud (1856–1939, Austrian) in *Mourning and Melancholia* suggests that the inward feeling of guilt over a major loss results in lowered self-confidence, an inward-directed hostility, and a need to suffer.

HEALTH SCIENCES (Pellagra)

To further confirm that pellagra is a dietary deficiency disease and not infective, an earlier result (1915—HEALTH SCIENCES) not universally accepted, Joseph Goldberger (1874–1929, Austria-Hungarian-American) and colleagues report the failure of attempts to infect themselves with pellagra by inoculations of blood, feces, and urine from pellagra victims.

MATHEMATICS (Logic)

P. J. Daniell (b.1889, English) creates the symmetric difference, the first inverse to the logical addition of Boolean algebra. This allows Boolean algebra to be incorporated into the algebra of rings and ideals.

PHYSICS (General Relativity)
ASTRONOMY (Planetary Orbits)

Albert Einstein (1879–1955, German-Swiss-American) publishes the paper

"Die Grundlagen der allgemeinen Relativitätstheorie" [The foundation of the general theory of relativity], where he expands from his special theory of 1905 to consider accelerated motions. The work is based on the postulate that the basic laws of physics assume the same form in all systems, whatever the relative motions. To develop his equations, Einstein abandons the standard Euclidean conception of space and instead turns to the non-Euclidean, four-dimensional geometry of Georg Riemann (1854—MATHEMATICS) and Hermann Minkowski (1907—PHYSICS), obtaining a theoretical space with a decided curvature. Einstein postulates that gravitational motions proceed along the shortest path in space-time, due not to a *force* of gravity but to the curvature of space-time resulting from the presence of mass and the associated gravitational fields. He shows that in the case of a planet revolving around a sun this path will be an ellipse that rotates over time. His calculations of the amount of rotation for the elliptical orbit of Mercury, although not agreeing exactly with observations, agree enough to be considered the first successful application of the general relativity theory. (The unusual precession of Mercury's orbit had been explained by Urbain LeVerrier from a Newtonian context by assuming an unseen planet between Mercury and the sun [1859—ASTRONOMY].) Another prediction is that the frequency of light emitted by a source should be lower the higher the density of the source. This expected red-shifting of stellar light is not yet measurable for the solar spectrum but is later confirmed for the white dwarf companion of Sirius (1925—ASTRONOMY). Whereas Einstein portrays Newton's theory of gravitation as a first approximation to the current gravitational results, he rejects Euclidean geometry as appropriate for the geometry of space even to a first approximation. Another important conceptual change is that the velocity of light is no longer perceived as constant, as light rays must follow curvilinear paths when bent in the presence of gravitational fields. Einstein suggests testing the hypothesis by measuring the deflection of starlight as it passes the vicinity of the sun (1919—PHYSICS/ASTRONOMY). The deflection predicted by Einstein is twice that which would be accountable from Newtonian theory.

PHYSICS (Spectral Lines)

Arnold Sommerfeld (1868–1951, German) creates a general theory of spectral lines based on quantum concepts.

1917

ASTRONOMY (Cosmology; Cosmological Constant)

Albert Einstein (1879–1955, German-Swiss-American) publishes "Cosmological Considerations on the General Theory of Relativity," a paper in which he introduces a so-called universal constant λ to his field equations of gravitation (1916—PHYSICS). This cosmological constant complicates the equations but allows a quasi-static distribution of matter in space, a requirement Einstein later regrets having made. The resulting universe is spatially finite, with a positive curvature and a finite mass calculable from the mean density distribution.

ASTRONOMY (Red Shifts)

After solving Albert Einstein's general relativity equations for a static universe, Willem de Sitter (1872–1934, Dutch) indicates that light from great distances will be red shifted, the greatest red shifts occurring for the most distant objects. Red shifts later come to be more directly associated with velocities of recession than with distance.

BIOLOGY (Bacteriophage)
HEALTH SCIENCES (Dysentery)

In the paper "An Invisible Microbe that is Antagonistic to the Dysentery Bacillus," Félix d'Hérelle (1873–1949, English) reports experiments on an antidysentery microbe found in the stools of patients recovering from the disease. He concludes that the microbe is a bacteriophage, destroying the dysentery bacteria, and speculates that other bacteriophages will be discovered for countering other specific bacteria.

CHEMISTRY (Protactinium)

Three groups independently discover protactinium: Otto Hahn (1879–1968, German) and Lise Meitner (1878–1968, Austrian-Swedish); Frederick Soddy (1877–1956, English), John Cranston, and Alexander Fleck (b.1889, English); and Kasimir Fajans (1887–1975, Polish-American).

HEALTH SCIENCES (Measles)

John Brownlee (1868–1927, English) examines the historical records of measles epidemics in London over the period 1703–1917, using Fourier analysis to determine the periodicities.

MATHEMATICS (Parallel Displacement)

Tullio Levi-Civita (1873–1941, Italian) generalizes the concept of parallelism with his introduction of parallel displacement in geometry.

1918

ASTRONOMY (Milky Way)

In the paper "Remarks on the Arrangement of the Sidereal Universe," Harlow Shapley (1885–1972, American) places the solar system at the outer edge of the Milky Way Galaxy, thereby further distancing the modern perceptions from the ancient concept of the earth's being at or near the center of the universe.

EARTH SCIENCES (Climatic Classification)

Wladimir Köppen (1846–1940, Russian-German) develops an empirical climatic classification based on temperature and precipitation, yielding five major climate types: tropical, dry, warm temperate, snow, and ice. Tropical climates, for an example, are those with the average monthly temperature of each month exceeding 18°C (64.4°F). Köppen publishes his classification scheme in the article "Klassification der Klimate nach Temperatur, Niederschlag, und Jahreslauf."

HEALTH SCIENCES (Nutrition; Phosphorus)

Thomas Osborne (1859–1929, American) and Lafayette Mendel (1872–1935, American) add further evidence on the importance of phosphorus in the diet by presenting experimental results showing severely restricted growth in young rats fed an otherwise healthy but phosphorus-deficient diet.

HEALTH SCIENCES (Yellow Fever)

Hideyo Noguchi (1876–1928, Japanese) isolates the microbe *Leptospira ictoroides* and believes it to be the causative agent of yellow fever. Over the next several years yellow fever vaccines and antiserums are prepared from the microbe, but the experimental data prove inconclusive. (A decade later Noguchi and a co-worker die of yellow fever while engaged in further research in Africa.)

MATHEMATICS (Analysis)

In *Das Kontinuum,* Hermann Weyl (1885–1955, German-American) identifies difficulties in the foundation supporting the use of formalism in analysis. This contributes to broadening the movement from a more formalistic to a more intuitionistic approach in mathematical studies (see MATHEMATICS entries for 1904, 1907, and 1912).

MATHEMATICS (Non-Riemannian Geometry)
PHYSICS (Unified Field Theory)

Hermann Weyl (1885–1955, German-American) creates the first non-Riemannian geometry while attempting to unify the gravitational field theory of Albert Einstein (1916—PHYSICS) and the electromagnetic field theory of James Clerk Maxwell (1865—PHYSICS).

1918–1924

ASTRONOMY (Stellar Spectra)

Annie Jump Cannon (1863–1941, American) publishes the spectral classifications of 225,300 stars in the nine-volume *Henry Draper Catalogue.* Including virtually all stars down to the ninth magnitude, the classifications were done by Cannon over the period 1911–1915 from low-dispersion objective prism plates at a rate of approximately three stars per minute.

1918–1925

EARTH SCIENCES/METEOROLOGY (*Maud* Expedition)

Detailed oceanographic and meteorologic observations are made during the *Maud* expedition to the north polar region. Chief scientist Harald Sverdrup (1889–1957, Norwegian), who replaces Roald Amundsen (1872–1928, Norwegian) as leader of the expedition in 1922, uses the data to determine the dynamics of Arctic tides (1927—EARTH SCIENCES) and the general features of Arctic oceanography and meteorology, published in 1933 in the three-volume *Scientific Results: The Norwegian North Polar Expedition with the "Maud."*

1919

ASTRONOMY (Cosmology)

Willem de Sitter (1872–1934, Dutch) solves Albert Einstein's field equations (1916—PHYSICS) for a universe without any material bodies. He determines that two particles inserted into this hypothetical massless universe would move away from each other and do so at a steadily increasing rate. (Aleksandr Friedmann later considers a universe with matter [1922—ASTRONOMY].)

ASTRONOMY (Stellar Evolution)

James Jeans (1887–1946, English) discusses stellar evolution from the perspective of the dynamics of rotating gases in *Problems of Cosmogony and Stellar Dynamics*.

CHEMISTRY (Mass Spectrograph; Isotope Percentages)

Francis Aston (1877–1945, English) uses the mass spectrograph, with which the mass of an ion can be determined by the curve it follows through a magnetic field, to determine the relative proportions of various isotopes of an element, doing so through analysis of the distribution of darkening on a photographic plate placed at the end of the path through the magnetic field.

HEALTH SCIENCES (Vitamin A)

Harry Steenbock (1886–1967, American) and E. G. Gross determine from dietary experiments with animals that yellow foods such as carrots and sweet potatoes are strong sources of vitamin A whereas white and red foods are not.

HEALTH SCIENCES (Water Purification)

Small fish are placed into key water reservoirs to eat mosquito larvae and thereby prevent such diseases as yellow fever.

HEALTH SCIENCES (Yellow Fever)

Usage begins of a preventative serum against yellow fever (see 1918—HEALTH SCIENCES).

METEOROLOGY (Air Masses; Fronts)

Jakob Bjerknes (1897–1975, Norwegian-American) and Vilhelm Bjerknes (1862–1951, Norwegian) show that the atmosphere, far from being homogeneous, can be viewed as divided into relatively distinct "air masses." Drawing an analogy with battlefield fronts, they appropriate the term "front" for the boundary formed when warm and cold air masses meet, and suggest that cyclones originate as waves along the front. They specifically develop a theory of the polar front and its importance to large-scale atmospheric motions.

PHYSICS (Atomic Physics; Artificial Transmutation)

Ernest Rutherford (1871–1937, New Zealander-English) succeeds, through atomic bombardment with subatomic particles, to alter an atomic nuclei and thereby transform one element into another, an achievement sought for centuries by alchemists. By bombarding nitrogen with alpha particles he succeeds in creating a stable oxygen isotope (plus a proton):

$$_2He^4 + {}_7N^{14} = {}_8O^{17} + {}_1H^1.$$

This is the first demonstrated artificial transmutation of one element into another.

PHYSICS (Atomic Theory)

Albert Einstein (1879–1955, German-Swiss-American), in the paper "Do Gravitational Fields Play an Essential Part in the Structure of the Elementary Particles of Matter?" argues that gravitational fields contribute to holding an atom together. He ascribes 75% of the energy constituting matter to electromagnetic forces and the remaining 25% to gravitational forces. Although he is able to forego use of the cosmological constant λ he introduced earlier (1917—ASTRONOMY), this is at the expense of allowing determination of a unique distribution of electricity, the current results indicating that an equilibrium state could be conceivable with any spherically symmetric distribution.

PHYSICS (Proton)

Ernest Rutherford (1871–1937, New Zealander-English) discovers the proton, as it appears as a product of the bombardment of nitrogen with alpha particles (see above, 1919—PHYSICS).

PHYSICS (Relativity)
ASTRONOMY (Eclipse; Starlight Deflection)

Observations during a solar eclipse confirm the bending of light rays passing a strong gravitational field (the sun) as predicted by Albert Einstein's general theory of relativity (1916—PHYSICS). British expeditions to Sobral, Brazil, and Principe, West Africa, the latter led by Arthur Eddington (1882–1944, English), photograph stars during an eclipse, in an attempt to test the Einstein theory. Upon comparison with photographs of the same stars taken several months earlier, the new photographs reveal the starlight to be deflected by amounts comparable with the values predicted by Einstein. The measurements reveal shifts in the apparent star positions of 1.64 seconds of arc, only 0.11 seconds of arc off the predicted value of 1.75 seconds. This is the second major verification of general relativity, the first being Einstein's explanation of the motions of Mercury (1916—PHYSICS/ASTRONOMY).

SUPPLEMENTAL (Acausality)

Paul Kammerer (1881–1926, Austrian) analyzes events that are meaningfully related but have no known causal connection. In the work *Das Gesetz der Serie,* he presents 100 examples, then develops the thesis that there exists a universal principle of acausality analogous to the principle of universal gravitation.

1919–1921

CHEMISTRY (Chemical Compounds)

Irving Langmuir (1881–1957, American) establishes that the electronic structure of a chemical compound tends to be more stable than the electronic structures of the separate combining elements, in terms of the compound's having completed electronic shells and subshells or at least shells and subshells which are more-nearly completed than are those of the individual elements. This is accomplished in some cases by the sharing of electrons between the combining elements and in other cases by the transference of an electron from one to the other element. For instance, in sodium fluoride the fluorine atoms gain an electron from the sodium atoms, with the result that both sets of atoms end with an electronic structure similar to the inert gas neon.

1919–1936

EARTH SCIENCES (Climatology; Tree Rings)

Publication of Andrew Douglass's (1867–1962, American) three-volume *Climatic Cycles and Tree Growth* (in 1919, 1928, and 1936). Douglass includes extensive use of tree rings for dating.

1920

ASTRONOMY (Star Diameters)

Albert Michelson (1852–1931, American), using a stellar interferometer, makes the first measurement of the diameter of a star other than the sun, determining Alpha Orionis (Betelgeuse) to have a diameter of 260 million miles.

BIOLOGY (Blood; Lungs; Kidneys)

A. Damiens discovers that bromine is a normal constituent of the blood, lungs and kidneys of humans, dogs, and other animals.

BIOLOGY (Territorial Drive)

In *Territory in Bird Life*, H. Eliot Howard (1873–1940, English) relates pioneering research revealing that male birds quarrel over territory far more frequently than they quarrel over females. He describes the tendency of the male bird to isolate itself and remain carefully spaced from other birds, hypothesizing that the biological function of such behavior is to increase the chance of an adequate distribution of food for the newborn.

EARTH SCIENCES (Climatology; Orbital Hypothesis for Climatic Change)

In *Théorie mathématique des phénomènes thermiques produits par la radiation solaire* [Mathematical theory of the thermal phenomena produced by solar radiation], Milutin Milankovitch (1879–1958, Yugoslavian) publishes formulae for determining incoming solar radiation and surface air temperature as a function of location and time, and presents results of his calculations for the present climates of earth, Mars, and Venus. He furthermore proposes that the oscillations in the earth's orbit and the tilt of its axis cause differences in the reception of solar radiation which are sufficient to cause the earth's ice ages.

METEOROLOGY (Richardson Number)

In a paper on "Some Measurements of Atmospheric Turbulence," Lewis Fry Richardson (1881–1953, English) includes a dimensionless quantity reflecting the turbulent structure of the atmosphere. Later termed the gradient Richardson number, its value is calculated as $(g/T)(\partial T/\partial z)/(\partial u/\partial z)^2$, where g is the acceleration due to gravity, T is temperature, u is wind speed, and z is height.

PHYSICS (Rocket Propulsion)

Robert Goddard (1882–1945, American) develops the mathematics of rocket propulsion in the paper "A Method of Reaching Extreme Altitudes."

c.1920

BIOLOGY (Evolution)

Paul Kammerer (1881–1926, Austrian) offers support for the Lamarckian evolutionary concept of the transmission of acquired characteristics through his experiments with salamanders, toads, and other amphibians. He claims that by breeding the animals in unnatural, controlled environments, he has induced adaptive modifications which have become hereditary, sometimes within one generation, other times not for several generations. Major questions later arise regarding the validity of the experiments.

1920–1921

MATHEMATICS (Logic; Truth Tables)

Jan Lukasiewicz (1878–1956, Polish), Emil Post (1897–1954, American), and Ludwig Wittgenstein (1889–1951, Austrian) independently introduce truth tables as a method for truth-value analysis. In a truth table, truth values of a schema are tabulated for each combination of truth values of the individual letters.

1921

ASTRONOMY (Stellar Spectra)

Theoretical work by Meghnad Saha (1894–1956, Indian) establishes that a star's spectrum depends at least as much on its temperature as on its chemical composition, since the temperature determines which atoms can produce absorption lines. He formulates the thermal ionization equation and uses it in interpreting stellar spectra.

CHEMISTRY (Atomic Theory)

The systematic evidence for the existence of atoms and molecules presented in Jean Perrin's (1870–1942, French) *Les Atomes* helps convince many of the remaining doubters.

CHEMISTRY (Uranium Z)

Otto Hahn (1879–1968, German) discovers uranium Z, a product of the disintegration of uranium X_1. Uranium Z constitutes only 0.35% of the composition resulting from the uranium X_1 disintegration, uranium X_2 (discovered in 1913) accounting for the remaining 99.65%.

HEALTH SCIENCES (Insulin; Diabetes)

Frederick Banting (1891–1941, Canadian), Charles Best (1899–1978, Canadian), John MacLeod (1876–1935, Scottish), and James Collip (1892–1965, Canadian) develop a method to extract insulin from the human pancreas and to purify it to combat diabetes by injecting it into the blood of diabetics. After first finding that the extract lowers blood glucose in dogs, they demonstrate its safety for humans through injecting Banting and Best, and finally in January of 1922 they begin actual use of the procedure with insulin injections given to the diabetic boy Leonard Thompson (1908–1935, Canadian).

HEALTH SCIENCES (Nutrition; Vitamins A and D)

Edward Mellanby (1884–1955, English) destroys the vitamin A in butter and cod liver oil by oxidation and finds that the oxidized butter can no longer protect against rickets but that the oxidized cod liver oil still can (see 1922—HEALTH SCIENCES for Elmer McCollum's further studies).

HEALTH SCIENCES (Rickets)

P. G. Shipley and colleagues determine that exposure to sunlight has a positive effect in preventing rickets. Experimenting with two groups of rats maintained on the same diet, they expose one group to sunlight for four hours a day while keeping the other group continually indoors under laboratory conditions. Within three weeks the rats in the latter group suffer from severe rickets, whereas those in the former group exhibit no symptoms for the full two months of the experiment.

MATHEMATICS (Game Theory)

Émile Borel (1871–1956, French) initiates the modern mathematical theory of games by subjecting several familiar games to precise analysis.

MATHEMATICS (Metaphysics)

In his *Logisch-philosophische Abhandlung,* Ludwig Wittgenstein (1889–1951, Austrian) asserts that all mathematics is a tautology.

METEOROLOGY (Dynamic Meteorology)

Publication of Vilhelm Bjerknes's (1862–1951, Norwegian) classic work in dynamic meteorology: *On the Dynamics of the Circular Vortex with Applications to the Atmosphere and to Atmospheric Vortex and Wave Motion.* Bjerknes applies thermodynamics and hydrodynamics to the atmosphere and oceans with the long-term goal of accurate weather prediction.

c.1921

MATHEMATICS (*N*-Valued Logical Systems)

Emil Post (1897–1954, American) generalizes the two-valued (true-false) logic of Aristotle to *n*-valued truth systems.

1921–1924

PHYSICS (Artificial Transmutation)

Following the first artificial transmutation of an element by Rutherford (1919—PHYSICS), Ernest Rutherford (1871–1937, New Zealander-English) and James Chadwick (1891–1974, English) succeed in ejecting protons from all elements with atomic number under 20 except helium, carbon, and oxygen. In each case, bombardment of the nucleus is performed with α particles.

1922

ASTRONOMY(Cosmology; Expanding Universe)

The mathematician Aleksandr Friedmann (1888–1925, Russian) solves Albert Einstein's general relativity equations (1916—PHYSICS) for a homogeneous, isotropic universe with matter, advancing on Willem de Sitter's solution for a universe without matter (1919—ASTRONOMY). Friedmann obtains three possible cosmological models, all based on a universe which expands after eruption from an initial singularity. The universe after the eruption could be open, flat, or closed, expanding without limit if open, and eventually recollapsing if closed. Friedmann's is the first cosmology clearly suggesting the possibility of an expanding universe.

ASTRONOMY (Star Colors; Hertzsprung Gap)

A catalog of star colors and luminosities compiled by Ejnar Hertzsprung (1873–1967, Danish) reveals a curious lack of bright stars in a color range later referred to as the Hertzsprung gap.

EARTH SCIENCES (Ocean Dynamics: Atlantic Circulation)

Upon analysis of temperature and salinity data in the Atlantic, Alfred Merz (1880–1925, German) and Georg Wüst (b.1890, German) describe the basic features of Atlantic circulation. They emphasize cross-equatorial exchange, claiming for the top kilometer a surface northward flow across the equator, for depths between one and four kilometers a southward flow across the equator, and for bottom water, northward flows across the equator in the west and southward flows in the east.

HEALTH SCIENCES (Infection)
BIOLOGY (White Blood Corpuscles)

Alexis Carrel (1873–1944, French-American) discovers leucocytes (white blood corpuscles) and records their role in preventing the spread of infection.

HEALTH SCIENCES/BIOLOGY (Lysozyme)

Alexander Fleming (1881–1955, Scottish) discovers lysozyme, an antibacterial protein found in saliva.

HEALTH SCIENCES (Vitamin D)

Elmer McCollum (1879–1967, American) and colleagues extend the experiments of Edward Mellanby on removing vitamin A from cod liver oil (1921—HEALTH SCIENCES) and determine that oxidized cod liver oil can cure rickets but is not effective as a preventative against xerophthalmia. They conclude that the antiricketic substance is not vitamin A but is a previously unidentified nutrient which they name vitamin D. They proceed to test, in experiments with rats, the relative levels of vitamin D in various sources, basing their judgments on the amount of calcification of the rat bones after a ricket-producing diet is relieved for six-to-eight days by the vitamin D source.

HEALTH SCIENCES (Vitamin E)

Herbert Evans (1882–1971, American) and K. J. Scott suggest the existence of a fertility substance contained in select natural foods but not among the already identified nutrients. In experiments with rats reared on growth-inducing diets, they find low fertility in the first generation of rats and total sterility in the second generation, unless supplementing the diets with fresh lettuce, wheat germ, or dried alfalfa leaves, in which case normal fertility is attained. The fertility substance in the supplemental foods is later named vitamin E (1923—HEALTH SCIENCES).

MATHEMATICS (Geometry of Paths)

Oswald Veblen (1880–1960, American) and Luther Eisenhart (1876–1965, American) generalize the Riemannian theory of geodesics into the geometry of paths. After the initial work in 1922, Veblen, Eisenhart and Tracy Thomas (b.1899, American) develop the subject in depth over the following two decades.

MATHEMATICS (Set Theory)

Abraham Fraenkel (1891–1965, German) modifies Ernst Zermelo's axiomatization of set theory (1908—MATHEMATICS), obtaining the Zermelo-Fraenkel system. From the Zermelo-Fraenkel axioms, the natural numbers—and hence all their direct derivatives—can be developed.

METEOROLOGY (Air Temperatures; Stratosphere)

Frederick Lindemann (1886–1957, English) and G. M. B. Dobson determine atmospheric temperatures at levels above those reached by balloon soundings by examining meteor trail records in the context of the laws of thermodynamics, obtaining approximate air densities from the height, brightness, and length of the trails, and then calculating air temperature from the density and other variables. They obtain the surprising result that the air is warmer at an altitude of about 30 miles than at lower levels in the stratosphere, the higher value being about 70°F and the lower one being under −60°F.

METEOROLOGY (Rain)

Tor Bergeron (Norwegian) notes that fog extends to the ground when the air temperature is above freezing but not when it is below freezing. Reasoning that in the latter case ice in the trees attracts the fog droplets, he uses this as evidence that ice particles high in the atmosphere would attract water vapor, and hence as evidence supporting Albert Wegener's speculation that rain generally begins as ice (1911—METEOROLOGY).

METEOROLOGY (Weather Prediction)

Lewis Fry Richardson (1881–1953, English), in his publication *Weather Prediction by Numerical Process,* details the equations, physical processes, and computational methods required to predict weather numerically. His results in a test six-hour forecast for a portion of Europe are highly unrealistic (including a pressure change of 140 millibars) and the calculations required for solving the equations are so impractical with current computing capabilities that he estimates 64,000 as the number of human calculators needed to keep pace with the weather itself. Nonetheless, the basic scheme of his pioneering effort is soundly based and contributes to developments later in the century after computer advances make numerical weather forecasting feasible.

PHYSICS (Atomic Structure)

Publication of Niels Bohr's (1885–1962, Danish) *The Theory of Spectra and Atomic Constitution.*

1922–1924

MATHEMATICS (Calculus of Variations)

Leonida Tonelli (1885–1946, Italian) publishes the two-volume *Fondamenti di Calcolo delle Variazioni* on the calculus of variations. The work is noted particularly for the emphasis on the theory of functionals and a variety of existence theorems.

1923

ASTRONOMY (Andromeda)

Edwin Hubble (1889–1953, American) examines 12 Cepheid variables in the Andromeda nebula and estimates their distance from the solar system at approximately 750,000 light years, thereby placing the stars well outside the Milky Way and establishing that the Andromeda nebula is in actuality a galaxy. The distance estimate in turn allows an estimate of the diameter of the Andromeda galaxy, which is calculated to be approximately that of the Milky Way. Hubble's method utilizes the relationship between the period of a Cepheid variable and its intrinsic brightness, first formulated by Henrietta Leavitt (1912—ASTRONOMY).

ASTRONOMY (Galaxies)

Using the Mount Wilson telescope, Edwin Hubble (1889–1953, American) distinguishes individual stars composing galaxies other than the Milky Way, especially the Andromeda galaxy (see above, 1923—ASTRONOMY), thereby essentially ending the ongoing debate on whether such galaxies exist. He estimates distances from the solar system by comparing star brightnesses to those in the Milky Way and finds some distances to exceed a billion trillion miles. As he continues his examination of galaxies through the decade, he concludes that most are either spiral or elliptical in shape. His continuing efforts are among the first to indicate the structure of the entire universe.

CHEMISTRY (Hafnium)

Dirk Coster (1889–1950, Dutch) and György Hevesy (1885–1966, Hungarian) discover hafnium, Hevesy doing so after a suggestion by his supervisor Niels Bohr to seek a new element in zirconium ores.

EARTH SCIENCES (Ocean Circulation)

Vagn Ekman (1874–1954, Swedish) expands upon his earlier mathematical theory of wind-driven ocean circulation (1905—EARTH SCIENCES) in the paper "Über Horizontal zirkulation bei winderzeugten Meeresströmungen," here examining effects of nonuniform winds, variable ocean bottom topography, and variable latitude.

HEALTH SCIENCES (Rickets; Irradiation)

Harry Goldblatt (b.1891, American) and K. M. Soames experiment with irradiation on rats fed a diet designed to produce rickets. Antiricketic features are found in the livers of the irradiated rats though not in the livers of the nonirradiated rats.

HEALTH SCIENCES (Vitamin E)

B. Sure recommends the label "vitamin E" for the fertility substance suggested by the experiments of Herbert Evans and K. J. Scott (1922—HEALTH SCIENCES) to exist in lettuce, wheat germ, and dried alfalfa leaves. He also extends the list of foods presumed to contain vitamin E, adding polished rice, yellow maize, and rolled oats.

MATHEMATICS (Analysis)

Luitzen Brouwer (1881–1966, Dutch) rejects use of the Law of the Excluded Middle for proofs dealing with infinite sets, and lists several theorems whose current proofs he rejects on that basis. Among the rejected theorems are the Weierstrass-Bolzano Theorem (c.1865—MATHEMATICS) and the Heine-Borel Theorem (1872—MATHEMATICS).

MATHEMATICS (Topology; Curves)

Karl Menger (b.1902) and Pavel Uryson (1898–1924, Russian) define "curve" as a one-dimensional set of points which is both closed and connected. By doing so they make the property of being a curve invariant under any homeomorphism and they eliminate space-filling curves.

PHYSICS (Compton Effect)

Through his discovery of the Compton effect, whereby the wavelength of X-rays and gamma rays increases following collisions with electrons, Arthur Compton (1892–1962, American) helps verify the quantum postulate for radiation (1900—PHYSICS). Compton explains the effect by attributing particle-like characteristics to the beam of X-rays: When an X-ray quantum collides with an electron, the energy (frequency) of the X-ray is decreased, and its wavelength correspondingly increased, by the amount needed to accelerate the electron, as though two particles had collided and energy and momentum were transferred between them but conserved within the system as a whole. The effect cannot be as readily explained when treating the rays exclusively as waves.

PHYSICS (Quantum Theory; Correspondence Principle)

Niels Bohr (1885–1962, Danish) introduces the Correspondence Principle, requiring the predictions of quantum theory to correspond to those of classical physics when applied to systems which are successfully described by the classical approach. Bohr specifically requires that if a quantal transition occurs between states with very large quantum numbers then the frequency of the photon emitted or absorbed as determined from quantum theory must match the frequency determined from classical electromagnetics.

1924

EARTH SCIENCES (Paleoclimatology)

Wladimir Köppen (1846–1940, Russian-German) and Alfred Wegener (1880–1930, German) publish *Die klimate der geologischen Vorzeit* [The climates of the remote geological past], in which they use Wegener's theory of continental drift (1915—EARTH SCIENCES) and varying astronomical conditions to explain the Quaternary Ice Age and the succession of interglacial and glacial periods. Köppen and Wegener include plots by Milutin Milankovitch of summer radiation at 55°N, 60°N, and 65°N as a function of time for the past 600,000 years. Milankovitch had constructed these plots to verify his thesis on the impact of the earth's orbital parameters on the earth's climate and particularly on the occurrence of ice ages (1920—EARTH SCIENCES), and had selected the season and latitudes on the basis of advice from Köppen that the important quantity for widespread glaciation is the amount of summer heat in temperate latitudes. Upon comparing Milankovitch's plots of mid-latitude summer radiation over the past 600,000 years with ice-age sequences established by Albrecht Penck and Eduard Brückner (1903–1909—EARTH SCIENCES), Köppen becomes convinced of the validity of Milankovitch's theory.

HEALTH SCIENCES (Rickets; Irradiation; Vitamin D)

After confirming studies of Harry Goldblatt and K. M. Soames on the antiricketic properties induced in the livers of irradiated rats (1923—HEALTH SCIENCES), Harry Steenbock (1886–1967, American) and A. Black further the experiments by directly irradiating the rat liver and the rickets-producing diet. Both the liver and the diet develop antiricketic properties. These studies lead to the determination that irradiation contributes to the formation of vitamin D.

MATHEMATICS (Computers)

Thomas J. Watson (1874–1956, American) reorganizes the Computer-Tabulating-Recording Company, formed in 1911 from Herman Hollerith's Tabulating Machine Company, into the International Business Machines Corporation (IBM). Hollerith's company had been formed in 1896 to make tabulating machines and punched cards, as an outgrowth of his machine constructed for the U.S. Census (1890—MATHEMATICS).

MATHEMATICS (Topology; Manifolds)

Tibor Radó (1895–1965, Hungarian) shows that every manifold of dimension 2 can be triangulated.

METEOROLOGY (Appleton Layer)

Edward Appleton (1892–1965, English) discovers that radio emissions are reflected by an ionized layer of the atmosphere, now known as the Appleton layer. He measures the distance to the layer by timing radio signals sent to it and returned.

PHYSICS (Atomic Physics; Electrons; Particle/Wave Duality)

Louis de Broglie (1892–1960, French) conjectures that electrons (previously regarded as particles) have also wave characteristics, so that, like electromagnetic radiation, matter has a dual particle/wave nature. De Broglie suggests that, as with electromagnetic radiation, the wavelength of the electron is the ratio of Planck's constant h and the momentum p, that is, $\lambda = h/p$. Hence, since p is the product of the relativistic mass and the velocity, the so-called de Broglie wavelength is inversely proportional to the electron velocity. The de Broglie postulate is consistent with Niels Bohr's postulate of discrete orbits in the hydrogen atom (1913—PHYSICS), as orbits become constrained to having a circumference which is an integral number of de Broglie wavelengths.

PHYSICS (Radon)

Using an ionization chamber, Irene Curie (1897–1956, French) and C. Chamié determine the half life of radon to be 3.823 days.

SUPPLEMENTAL (Religion and Science: Evolution)

The Presbyterian Church of San Antonio (Texas) declares that the theory of evolution is an untenable hypothesis.

1924–1935

ASTRONOMY (Galaxies)

Edwin Hubble (1889–1953, American) discovers thousands of galaxies beyond the Milky Way and classifies them into three types: spiral, elliptical, and irregular. For the near galaxies he uses Cepheid variables to calculate their distances, whereas for the distant ones he uses spectral red shifts.

1925

ASTRONOMY (Andromeda Nebula)

Edwin Hubble (1889–1953, American) determines the distance from earth to the Andromeda nebula to be two million light years.

ASTRONOMY (Gravitation)

Walter Adams (1876–1956, American) finds a red shift in the spectral lines of the light from the very intense gravitational field of the white dwarf companion of Sirius. This helps confirm Albert Einstein's general theory of relativity (1916—PHYSICS), although the 19 kilometers per second displacement differs somewhat from the predicted value of 17 kilometers per second.

CHEMISTRY (Rhenium; Masurium)

Ida Noddack (b.1896, German) and Walter Noddack (1893–1960, German) discover the element rhenium, with atomic number 75, and believe also that they discover an element masurium with atomic number 43. The latter was presumably spurious since element 43, finally found in 1940 among the products of uranium fission and named "technetium," is radioactive and not found in nature under normal conditions.

EARTH SCIENCES (Oceanography: Mixing Length)
METEOROLOGY (Mixing Length)

In the paper "Bericht über Untersuchungen zer ausgebildeten Turbulenz," Ludwig Prandtl (1875–1953, German) introduces the "mixing length" concept for turbulent motion, doing so through analogy with the concept of the mean free path length for molecular motion (1858—CHEMISTRY).

HEALTH SCIENCES (Chemical Warfare)

Edward Vedder (1878–1952, American) publishes *The Medical Aspects of Chemical Warfare.*

HEALTH SCIENCES (Pellagra)

Joseph Goldberger (1874–1929, Austria-Hungarian-American) and W. F. Tanner, following a sequence of dietary experiments, classify foods according to their usefulness as pellagra preventives and curatives. Dried yeast ranks alone as highly effective, with milk and selected other substances showing definite but lesser effectiveness. Goldberger concludes that there exists an unknown pellagra preventative that occurs in large quantities in yeast and in lesser quantities in specified other foods.

HEALTH SCIENCES (Scarlet Fever)

George Dick (1881–1967, American) and Gladys Dick (1881–1963, American) find an antitoxin for scarlet fever.

MATHEMATICS (Formalism)

In the paper "On the Infinite," David Hilbert (1862–1943, German) further develops the formalist approach to mathematics, viewing the subject as a formal system based on abstract symbols whose meanings extend only to their formal definitions and formally defined properties, not to any external physical reality. Hilbert opposes as circular the aim of Bertrand Russell and Alfred North Whitehead (MATHEMATICS entries for 1903 and 1910–1913) to derive mathematics from logic, since logic itself depends on number; and he opposes the intuitionism of Luitzen Brouwer (MATHEMATICS entries for 1907 and 1912), since too much valuable mathematics must be rejected from an intuitionist framework (see, for example, 1923—MATHEMATICS).

MATHEMATICS (Topology)

Pavel Uryson (1898–1924, Russian) proves that for every normal topological space there can be defined a metric space which preserves limit points. He also proves that the subsets of the Hilbert cube provide homeomorphic images for every separable metric space.

METEOROLOGY (Air Pollution)

William Shaw (1854–1945, English) addresses the issue of air pollution in *The Smoke Problem of Great Cities.*

PHYSICS (Electron Spin)

On the basis of spectral experiments, George Uhlenbeck (b.1900, Dutch-American) and Samuel Goudsmit (b.1902, Dutch-American) postulate that electrons spin and introduce a quantum spin number denoting the direction of the spin.

PHYSICS (Matrix Mechanics)

Werner Heisenberg (1901–1976, German) recommends replacing concrete atomic models based on unobservable quantities such as electron orbits by analysis based solely on observables such as spectral line frequencies and intensities. He, Max Born (1882–1970, German), and Pascual Jordan (1902–1980, German) then develop a form of quantum theory called "matrix mechanics" in which such observables are manipulated through the well-developed noncommutative algebra of matrices (1857–1858—MATHEMATICS), first recognized by Born as a proper formalism for the subject matter.

PHYSICS (Pauli Exclusion Principle)

Wolfgang Pauli (1900–1958, Austrian-American) postulates that in a given atom no two electrons can have the identical set of four quantum numbers. This postulate provides a criterion for the electronic structure of atoms, and helps to explain the periodic table and the chemical combining properties of the elements.

SUPPLEMENTAL (Science and Politics: Evolution)

The Texas State Text Book Board forbids the use of any textbook which discusses the theory of evolution. Elsewhere, in the highly publicized Scopes trial, John Scopes (b.1901, American) is accused of teaching Charles Darwin's theory (1859—BIOLOGY) in a high school in Tennessee, where, earlier in the year, a law had been passed forbidding the teaching in public schools of any theory contrary to the biblical account of creation. Although Scopes is convicted at first and fined $100, the state supreme court later reverses the decision on a technicality.

1925–1927

EARTH SCIENCES (Observational Oceanography)

Publication of Georg Wüst's (b.1890, German) four-volume *Results of German Atlantic Expedition METEOR*, describing the temperature, salinity, and currents of the Atlantic Ocean.

1925–1936

ASTRONOMY (Stellar Spectra)

Annie Jump Cannon (1863–1941, American) extends the compilation of

225,300 stellar spectral classifications of the *Henry Draper Catalogue* (1918–1924—ASTRONOMY) with approximately 47,000 additional classifications published in the *Henry Draper Extension.*

1926

ASTRONOMY (Galactic Nebulae)

Edwin Hubble (1889–1953, American) publishes *A General Study of Diffuse Galactic Nebulae.*

BIOLOGY (Genetics)

Through his studies of the prolific fruit fly *Drosophila,* Thomas Morgan (1866–1945, American) establishes the importance of the gene and the chromosome in transmitting inherited characteristics. Morgan concludes from his examination of hundreds of mutant characters in *Drosophila* that inherited characters are connected to paired Mendelian factors or genes and that genes are linearly joined into chromosomes. Although the inheritance of characteristics from genes positioned along different chromosomes occurs independently, as in Gregor Mendel's second law (1865—BIOLOGY), this is not true for characteristics from genes along the same chromosome. These latter characteristics tend to be inherited as a group; however, since linkages along the chromosome may be broken, the further apart two genes are in the chromosome chain, the less certain it is that their characteristics will be inherited together. Morgan describes the work in *The Theory of the Gene.*

CHEMISTRY (Enzymes)

James Sumner (1887–1955, American) carries out the first crystallization of an enzyme, doing so for jackbean urease.

EARTH SCIENCES (Ekman Current Meter)

Vagn Ekman (1874–1954, Swedish) in "On a New Repeating Current Meter" describes a new current meter for mechanically determining the speed and direction of ocean currents.

EARTH SCIENCES (Mediterranean Outflow)

Björn Helland-Hansen (1877–1957, Norwegian) and Fridtjof Nansen (1861–1930, Norwegian) determine that outflow from the Mediterranean causes the observed high salinities at the middle levels of the subtropical North Atlantic.

HEALTH SCIENCES (Pernicious Anemia)

George Minot (1885–1950, American) and William Murphy (b.1892, American) successfully treat pernicious anemia through recommendation of a diet heavy in liver.

FIGURE 20. *Polished, striated bedrock near Neuchâtel, Switzerland, as depicted in Louis Agassiz's* Études sur les glaciers [*Studies on glaciers*] *(see 1840—*EARTH SCIENCES *entry). Agassiz notes how such distinctive rock features are known to develop at present in glaciated regions through glacial actions and presents the existence of similar surfaces miles from existing glaciers as evidence of the greater extent of glacial coverage in the past. More broadly, he hypothesizes a former Ice Age. (Courtesy of Albert V. Carozzi, University of Illinois at Urbana-Champaign and Jean Guinand, rector, University of Neuchâtel, Switzerland.)*

PHYSICS (Electrons)

Max Born (1882–1970, German) rejects the concept that an individual electron may be spread out over space, and instead refers to "statistical probabilities" of the electron's being in any particular spatial volume.

PHYSICS (Hydrogen Atom)

Paul Dirac (1902–1984, English-American) derives the Balmer-spectrum energy levels of the hydrogen atom using an algebraic version of quantum mechanics based on Poisson brackets, a version Dirac shows to obtain the fundamental results of the matrix mechanics of Werner Heisenberg, Max Born, and Pascual Jordan (1925—PHYSICS).

PHYSICS (Rockets)

Robert Goddard (1882–1945, American) develops and successfully launches a liquid fuel rocket, following 17 years of theoretical work. The rocket reaches a height of 41 feet and a speed of approximately 60 miles per hour.

PHYSICS (Wave Mechanics)

Extending the work of Louis de Broglie (1924—PHYSICS), Erwin Schrödinger

(1887–1961, Austrian) mathematically develops wave mechanics, including formulation of the Schrödinger wave equation. He seeks an explanation of atomic radiation exclusively through wave theory, eliminating the wave/particle duality and the electron's supposed quantum jumps. He is able to explain the stationary states of Niels Bohr (1913—PHYSICS) by applying de Broglie's concept of electron waves to the hydrogen atom, showing in the process that the orbit of an electron is restricted to insure an integral number of electron wavelengths.

PHYSICS (Wave Mechanics; Quantum Mechanics)

Erwin Schrödinger (1887–1961, Austrian) and Carl Eckart (1902–1971, American) independently establish that Schrödinger's wave mechanics (1926—PHYSICS) is mathematically equivalent to the matrix mechanics of Werner Heisenberg, Max Born, and Pascual Jordan (1925—PHYSICS) as regards atomic structure and spectra.

1926–1931

METEOROLOGY (General)

Publication of William Shaw's (1854–1945, English) four-volume *Manual of Meteorology*.

1927

ASTRONOMY (Comets)

Fernand Baldet indicates that comets are extremely light bodies and are noticeably influenced by radiation pressure.

ASTRONOMY (Composition of Extraterrestrial Matter)

Ira Bowen (1898–1973, American) hypothesizes that the elements composing the rest of the universe are the same as those composing the earth.

ASTRONOMY (Cosmology; Big Bang Hypothesis)

Georges Lemaitre (1894–1966, Belgian) proposes that the universe erupted into being from a "primeval atom" and may still be expanding. The hypothesis is later known as the hypothesis of the Big Bang and elaborated by numerous researchers.

ASTRONOMY (Interstellar Space)

Ira Bowen (1898–1973, American) examines the polarization of distant starlight and its origin in interstellar matter.

ASTRONOMY (Milky Way)

Jan Hendrik Oort (b.1900, Dutch) shows, from his discovery of systematic effects in the radial velocities of numerous individual stars, that the Milky Way system is rotating and that the motions of the galaxy satisfy the basic principles of Kepler's Laws for the solar system (ASTRONOMY entries for 1609 and 1619).

BIOLOGY (Embryology)

Publication of Thomas Morgan's (1866–1945, American) *Experimental Embryology*.

BIOLOGY (Genetics; Mutations)

Hermann Muller (1890–1967, American) discovers means of artificially inducing mutations through ionizing radiations. By irradiating with X-rays, he succeeds in increasing the mutation rate in the fruit fly *Drosophila* by a factor of 150.

CHEMISTRY (Chemical Bond)

Walter Heitler (b.1904, German) and Fritz London (1900–1954, German-American) examine the nature of the chemical bond, using wave mechanics to develop a theory of the forces between atoms as they bond together through the exchange of electrons. The Heitler and London studies center on the bonding between the two atoms of a hydrogen molecule.

CHEMISTRY (Prout's Hypothesis)

Francis Aston (1877–1945, English) calls into question William Prout's hypothesis that each element is a combination of hydrogen atoms (1816—CHEMISTRY) as he shows that even when individual isotopes are considered, the atomic mass of an element is generally not an exact multiple of the mass of a hydrogen atom but rather is somewhat more than an exact multiple for the lightest and heaviest elements and somewhat less than an exact multiple for the elements of intermediate mass.

EARTH SCIENCES (Arctic Oceanography)

Harald Sverdrup (1889–1957, Norwegian) publishes a description of the tidal data collected on the *Maud* expedition (1918–1925—EARTH SCIENCES) in the paper "Dynamics of Tides on the North Siberian Shelf, Results from the *Maud* Expedition."

HEALTH SCIENCES (Agglutinogens)

Karl Landsteiner (1868–1943, Austrian) and Philip Levine (b.1900, Russian-American) report their discovery of the M and N agglutinogens in the paper "A New Agglutinable Factor Differentiating Individual Human Bloods."

PHYSICS (Electrons)

Clinton Davisson (1881–1958, American) and Lester Germer (1896–1971, American) obtain diffraction patterns from beams of electrons, thereby offering evidence for Louis de Broglie's postulate (1924—PHYSICS) of a wave nature for electrons.

PHYSICS (Quantum Theory; Metallic Structure)

Arnold Sommerfeld (1868–1951, German) analyzes the structure of metals from the point of view of quantum mechanics.

PHYSICS (Uncertainty Principle)

Werner Heisenberg (1901–1976, German) postulates the "uncertainty principle" that it is impossible to determine accurately and simultaneously both members of specific pairs of atomic variables. For instance, it is impossible to determine both the momentary position and velocity of an electron, as the high energy of the radiation needed to determine the position would necessarily alter the velocity in the process of determining the position. Heisenberg quantifies the minimum value of the product of the uncertainties in the two variables (for example, position and velocity), this minimum being proportional to Planck's constant h (1900—PHYSICS).

PHYSICS (Wave Equation; Antimatter)

Paul Dirac (1902–1984, English-American) derives a relativistically invariant form of the Schrödinger wave equation (1926—PHYSICS), thereby uniting quantum mechanics and relativity theory. The derivation produces energy states due to electron spin, so that the electron spin enters more naturally in the Dirac wave mechanics than in the previous nonrelativistic wave mechanics. The new wave mechanics also predicts the existence of positively charged electrons, termed "antielectrons" or "positrons" and first observed experimentally in 1932. Dirac correctly predicts that antimatter, such as antielectrons, will annihilate when brought in contact with ordinary matter, such as electrons, both particles being converted into radiant energy.

1927–1928

MATHEMATICS (Topology; Knot Theory)

James Alexander (1888–1971, American) advances the theory of knots by identifying specific invariants which distinguish various knot types.

1928

ASTRONOMY (Expanding Universe)

Howard Robertson (American) demonstrates mathematically that Willem de

Sitter's cosmology with increased red shifts for more distant objects (1917—ASTRONOMY) can be interpreted as an expanding, although empty, universe.

HEALTH SCIENCES (Child Psychology)

Publication of John Watson's (1878–1958, American) *Psychological Care of Infant and Child.*

HEALTH SCIENCES (Pellagra-Preventing Factor; Niacin)

Joseph Goldberger (1874–1929, Austria-Hungarian-American) and George Wheeler (b.1885, American) find that either autoclaved yeast or extracts of heated yeast can cure black tongue, a canine disease similar to pellagra in humans. They designate the curative factor in the yeast the "P-P factor," for Pellagra-Preventing. This curative factor is later determined to be the vitamin nicotinic acid, or niacin.

HEALTH SCIENCES (Penicillin)

Alexander Fleming (1881–1955, Scottish) discovers penicillin from a mold, *Penicillium notatum,* in a culture dish and describes its chemical and antibacterial properties. Although its potential medical value is surmised early on by Fleming, it cannot be realized until the substance is stabilized and purified. This is accomplished by Howard Florey, Ernst Chain, and co-workers, who publish encouraging results of animal experiments in 1940 in the article "Penicillin as a Chemotherapeutic Agent." Results of trial experiments on humans are published in 1941, and large-scale production of penicillin begins in 1943.

MATHEMATICS (Game Theory)

John von Neumann (1903–1957, Hungarian-American) develops a mathematical theory of games.

MATHEMATICS (Lattice Algebra)

Karl Menger (b.1902) restructures projective and affine geometries into lattice algebra.

MATHEMATICS (Topology)

Karl Menger (b.1902) and A. Georg Nöbeling (b.1907) prove that any compact metric space of dimension n is homeomorphic to a subset of the Euclidean space of dimension $2n + 1$.

METEOROLOGY (Air Masses)

Tor Bergeron (Norwegian) classifies air masses into maritime tropical, continental tropical, maritime polar, and continental polar.

PHYSICS (Complementarity Principle)

Niels Bohr (1885–1962, Danish) asserts the complementarity principle, stating the necessity of including certain seemingly mutually exclusive complementary aspects, for instance the simultaneous wave and particle behavior of an electron, for the complete description of atomic phenomena.

PHYSICS (Electrons)

George Thomson (1892–1975, English) independently of Clinton Davisson and Lester Germer (1927—PHYSICS) publishes experimental confirmation of Louis de Broglie's speculation on the wave nature of electrons (1924—PHYSICS), having found diffraction patterns from the scattering of cathode rays by a thin metallic film. From the rings in the diffraction pattern he determines the wavelength, consistently finding the de Broglie wavelength $\lambda = h/p$ (1924—PHYSICS).

PHYSICS (Geiger-Müller Counter)

Hans Geiger (1882–1945, German) and Walther Müller develop the Geiger-Müller counter, improving the Geiger counter earlier developed by Ernest Rutherford and Geiger (1908—PHYSICS). The instrument contains a galvanometer attached to a high-voltage battery which in turn is connected to a short copper tube with a stretched wire through it, the wire also being connected to the galvanometer. The Geiger-Müller counter has a higher sensitivity and a longer operational lifetime than the 1908 instrument and can count particles from many forms of ionizing radiation, not just alpha particles.

PHYSICS (Iconoscope)

Vladimir Zworykin (1889–1982, Russian-American) creates the iconoscope, a forerunner of the television camera.

PHYSICS (Radioactivity)

George Gamow (1904–1968, Russian-American) formulates a theory of alpha radioactive decay, in which the atomic nucleus is surrounded by a threshold potential through which the alpha particle has a finite probability of passing.

1929

ASTRONOMY (Andromeda Galaxy)

Edwin Hubble (1889–1953, American) publishes a study of the Andromeda galaxy, including distance estimates from analysis of 40 Cepheid variables. Hubble estimates the galaxy at 0.9 million light years from the solar system. The work helps lead to the general acceptance that most stars are grouped into galaxies.

ASTRONOMY (Hubble's Law)

After comparing the radial velocities of two dozen galaxies as determined by Vesto Slipher (1912–1917—ASTRONOMY) with their distances as determined by himself, Edwin Hubble (1889–1953, American) publishes an empirical relationship between the velocity and distance of a galaxy. Known as Hubble's Law, it asserts that the redshift of a galaxy, or, under the normal interpretation of redshifts, its velocity of recession, is directly proportional to its distance from the solar system: $V = Hr$, where H is Hubble's constant. Hubble does not join others in proclaiming that the evidence implies an expanding universe.

ASTRONOMY (Sun)

George Gamow (1904–1968, Russian-American), Robert Atkinson, and Fritz Houtermans hypothesize that thermonuclear processes produce the heat and light from the sun.

CHEMISTRY (Oxygen)

William Giauque (b.1895, American) and Herrick Johnston (b.1898, American) determine that there exist at least three isotopes of oxygen, with atomic weights 16, 17, and 18. These three isotopes generally occur in the ratio of 99,759 to 37 to 204 respectively. In light of the discovery of the isotopes, it is suggested that the atomic weight scale be adjusted to set 0^{16} precisely at atomic weight 16 instead of setting common oxygen, with its small contributions from the two heavier isotopes, at 16.

EARTH SCIENCES (Dynamic Oceanography: Pacific Circulation)

Complementing his earlier (1922—EARTH SCIENCES) analysis of Atlantic circulation, Georg Wüst (b.1890, German) sketches the ocean-wide circulation in the Pacific, concluding that there exist major differences between the two oceans. He determines that in the Pacific there is no significant interchange between the two hemispheres, there is sinking near the equator, the deep water moves generally poleward (in both hemispheres), and the intermediate water above the deep water and the bottom water below it both move generally equatorward.

EARTH SCIENCES (Dynamic Oceanography: Indian Ocean Circulation)

In *Die Zirkulation des Indischen Ozeans* [The circulation of the Indian Ocean], L. Moller examines data from the Indian Ocean, determining for the deep water a southward flow and for the intermediate water above and the bottom water below a northward flow. She finds a significant interchange between the Northern and Southern Hemispheres.

HEALTH SCIENCES (Child Psychology)

Jean Piaget (1896–1980, Swiss) presents a theory of the intellectual development of the child in *The Child's Conception of the World.* Piaget indicates that the child's development proceeds in set stages and that almost all children proceed through these stages in the same order.

HEALTH SCIENCES (EEG)

Hans Berger (1873–1941, Austrian) invents the electroencephalogram after discovering that electrodes placed on the scalp allow recording of the electrical activity of the brain. He publishes a description in the paper "Über das Elektrenkephalogramm des Menschen."

MATHEMATICS (Numerical Stability)

Richard Courant (1888–1972, German-American), Kurt Friedrichs (b.1901, German-American), and Lewy show that to ensure numerical stability in solving partial differential equations by finite approximations, the ratio of the grid size to the timestep must be greater than the speed of the fastest moving relevant waves.

PHYSICS (Unified Field Theory)

Albert Einstein (1879–1955, German-Swiss-American) publishes a preliminary mathematical attempt to unify the studies of electromagnetism and gravitation. He continues working on this unified field theory for years, although many of his colleagues are convinced that all such attempts are destined to fail because of the inherent uncertainty factor introduced by quantum theory.

1930

ASTRONOMY (Dark Clouds)

Through analysis of star clusters, R. J. Trumpler (1886–1956, Swiss-American) determines that dark regions in the sky, whose existence has been known for over a century, are regions of absorption rather than of no stars. Obscuring dark clouds block the light from the more-distant stars.

ASTRONOMY (Milky Way)

Harlow Shapley (1885–1972, American) calculates the Milky Way Galaxy to be 250,000 light years in diameter.

ASTRONOMY (Pluto)

Clyde Tombaugh (b.1906, American) discovers the planet Pluto, locating it as an object slowly shifting against the field of fixed stars on photographs from Lowell Observatory (Arizona). The discovery confirms the calculations of Percival Lowell (1906—ASTRONOMY) and thereby increases the enthusiasm for mathematical astronomy.

BIOLOGY (Evolution; Genetics)

With *The Genetic Theory of Natural Selection,* Ronald Fisher (1890–1962, English) helps to reestablish Charles Darwin's theory of evolution (1859—BIOLOGY). Fisher revises the more recent mutation theory (1900—BIOLOGY) to posit that the mutations occur by chance and that it is natural selection which controls the direction of evolution by weeding out the harmful mutations and perpetuating the favorable ones. Fisher is able mathematically to link data on mutation rates with data on evolution rates through the assumption that the rate at which a population's fitness increases is proportional to its genetic variance.

BIOLOGY (Evolution; Plant Dispersal)

Henry Ridley (1855–1956, English) includes, in *The Dispersal of Plants Throughout the World,* numerous instances of long-distance dispersal of plants through such means as marine birds carrying seeds and fruits.

BIOLOGY (Genetics; Evolution)

Trofim Lysenko (1898–1976, Russian) claims that winter wheats can be changed so that not only they themselves can be sown as spring wheats but so that their seeds and all future offspring will be spring wheats as well. The practice of "vernalizing" winter wheats (moistening and refrigerating the seed) to allow spring sowing had long been successfully carried out in many regions and was generally accepted by the scientific community; but the additional claim of Lysenko meets with much skepticism, especially in the West. If valid, the Lysenko claim would be a serious blow to those theories of evolution insisting that acquired characteristics are not inherited.

EARTH SCIENCES (Paleoclimatology)

Milutin Milankovitch (1879–1958, Yugoslavian) publishes *Mathematical Climatology and the Astronomical Theory of Climatic Change,* making further advances to his earlier theory of climatic history based on changes in the earth's orbital parameters (see EARTH SCIENCES entries for 1920 and 1924). Having supplemented his 1924 plots for the latitude band 55–65°N by corresponding plots for other latitude bands, he now shows that the 41,000 year cycle for the tilt of the earth's axis is most important for high latitudes, whereas the 22,000 year cycle for the precession of the equinoxes is most important for low latitudes.

MATHEMATICS (Fourier Analysis)

Norbert Wiener (1894–1964, American) shows that for the continuous spectrum of nonperiodic data, the variance and autocovariance functions contain the same information.

MATHEMATICS (Logic; Completeness)

Kurt Gödel (1906–1978, Austrian-American), in his dissertation and a published paper prepared from it, proves that every valid formula in the predicate calculus of the first order is provable, that is, this calculus is "complete." He also presents an abstract of the paper "On Formally Undecidable Propositions of *Principia Mathematica* and Related Systems," published in 1931, which startles the mathematics community by proving the incompleteness of a formal system consisting of the Peano axioms (1889—MATHEMATICS), the axiom of choice (1904—MATHEMATICS), and the logic of Alfred North Whitehead and Bertrand Russell's *Principia Mathematica* (1910–1913—MATHEMATICS). This essentially establishes that any sufficiently complex mathematical system will have problems that can be expressed within the system but cannot be solved within it.

MATHEMATICS (Logic; Nicod's Criterion)

Jean Nicod (1893–1924, French) introduces, in *Foundations of Geometry and Induction,* Nicod's Criterion that the proposition that all *A*'s are *B*'s is (1) supported by evidence of an *A* that is also a *B*, (2) weakened by evidence of an *A* that is not a *B*, and (3) unaffected by evidence of a *B* that is not an *A* or by evidence of a *C* that is neither an *A* nor a *B*.

PHYSICS (Cyclotron)
CHEMISTRY (Elements)

Nils Edlefsen (1893–1971, American) constructs a cyclotron, an instrument used to produce directed beams of charged particles. The construction is based largely on the ideas of Ernest Lawrence (1901–1958, American), who then further develops the machine over the next eight years. The cyclotron facilitates the discovery of new chemical elements since such elements can be created by bombarding known elements with positive particles or neutrons.

PHYSICS (Nuclear Structure)

George Gamow (1904–1968, Russian-American) formulates a model—termed the liquid drop model—for the nuclear structure of heavy elements.

SUPPLEMENTAL (Philosophy of Science)

James Jeans (1887–1946, English) warns, in *The Mysterious Universe,* that extrapolation beyond mathematical formulae to attempted explanations of the essence of nature are always uncertain.

Sources

*Key historic work (or collection of works of one person).
**Collection of key historic works or excerpts thereof.

Science Prior to 1400

Butterfield, H. *The Origins of Modern Science.* Rev. ed. 1957. Reprint. New York: Free Press, 1965, 255 pp.

Crombie, A. C. *Medieval and Early Modern Science.* 2 vols. Cambridge: Harvard University Press, 1963, 296, 380 pp.

Durant, W. *The Story of Civilization.* Vol. 4. *The Age of Faith.* New York: Simon & Schuster, 1950, 1196 pp.

Goldstein, T. *Dawn of Modern Science from the Arabs to Leonardo da Vinci.* Boston: Houghton Mifflin Co., 1980, 297 pp.

**Grant, E., ed. *A Source Book in Medieval Science.* Cambridge: Harvard University Press, 1974, 864 pp.

Grant, E. *Physical Science in the Middle Ages.* New York: John Wiley & Sons, 1971, 128 pp.

Lindberg, D. C., ed. *Science in the Middle Ages.* Chicago: University of Chicago Press, 1978, 549 pp.

Pedersen, O., and Pihl, M. *Early Physics and Astronomy: An Historical Introduction.* New York: American Elsevier, 1974, 413 pp.

*Polo, M *The Travels of Marco Polo.* Translated, with introduction, by R. Latham. New York: Penguin Books, 1958, 378 pp.

Sarton, G. *Introduction to the History of Science.* Vol. 2. *Introduction to the History of Science from Rabbi Ben Ezra to Roger Bacon.* Baltimore: Williams & Wilkins Co., 1931, 1251 pp.

Sarton, G. *Introduction to the History of Science.* Vol. 3. *Science and Learning in the Fourteenth Century.* Baltimore: Williams and Wilkins Co., 1948, 2155 pp.

Wiet, G., Elisseeff, V., Wolff, P., and Naudou, J. *History of Mankind.* Vol. 3. *The Great Medieval Civilizations.* New York: Harper & Row, 1975, 1082 pp.

Astronomy

Birney, D. S. *Modern Astronomy.* Boston: Allyn & Bacon, 1969, 338 pp.

Brush, S. G. "Nickel for Your Thoughts: Urey and the Origin of the Moon." *Science* 217, no. 4563 (1982):891–98.

Clayton, D. D. *The Dark Night Sky.* New York: Quadrangle, 1975, 206 pp.

Drake, S. *Galileo At Work. His Scientific Biography.* Chicago: University of Chicago Press, 1978, 536 pp.

Ferris, T. *The Red Limit.* New York: William Morrow & Co., 1977, 287 pp.

Gingerich, O. "Unlocking the Chemical Secrets of the Cosmos." *Sky and Telescope* 62, no. 1 (1981):13–15.

Gregory, S. A., and Thompson, L. A. "Superclusters and Voids in the Distribution of Galaxies." *Scientific American* 246, no. 3 (1982):106–14.

Koestler, A. *The Sleepwalkers.* New York: Macmillan Co., 1959, 624 pp.

*Laplace, P. S. *Oeuvres.* 14 vols. Paris: Imprimerie Royale, 1843.

Lessing, E. *Discoverers of Space*. Freiburg, West Germany: Herder & Herder, 1969, 176 pp.
Morrison, D. *Voyages to Saturn*. NASA Special Publication SP–451. Washington, D.C.: U.S. Government Printing Office, 1982, 227 pp.
Morrison, D., and Samz, J. *Voyage to Jupiter*. NASA Special Publication SP–439. Washington, D.C.: U.S. Government Printing Office, 1980, 199 pp.
Moyer, G. "The Gregorian Calendar." *Scientific American* 246, no. 5 (1982):144–52.
Nicolson, I. *Gravity, Black Holes and the Universe*. New York: John Wiley & Sons, 1981, 264 pp.
Pollack, J. B., and Cuzzi, J. N. "Rings in the Solar System." *Scientific American* 245, no. 5 (1981):104–29.
**Shapley, H., ed. *A Source Book in Astronomy 1900–1950*. Cambridge: Harvard University Press, 1960, 423 pp.
**Shapley, H., and H. E. Howarth, eds. *A Source Book in Astronomy*. New York: McGraw-Hill Book Co., 1929, 412 pp.
Shklovskii, I. S., and Sagan, C. *Intelligent Life in the Universe*. San Francisco: Holden Day, 1966, 509 pp.
Silk, J. *The Big Bang: The Creation and Evolution of the Universe*. San Francisco: W. H. Freeman & Co., 1980, 394 pp.
Stephenson, F. R. "Historical Eclipses." *Scientific American* 247, no. 4 (1982):170–83.
Van de Kamp, P. *Basic Astronomy*. New York: Random House, 1952, 400 pp.
Whitney, C. A. *The Discovery of Our Galaxy*. New York: Alfred A. Knopf, 1971, 316 pp.

Biology and Health Sciences

Ackerknecht, E. H. *A Short History of Medicine*. New York: Ronald Press Co., 1968, 275 pp.
Allen, G. *Life Science in the Twentieth Century*. New York: John Wiley & Sons, 1975, 258 pp.
**Brock, T. D., ed. *Milestones in Microbiology*. Englewood Cliffs, N. J.: Prentice-Hall, 1961, 273 pp.
Bulloch, W. *The History of Bacteriology*. London: Oxford University Press, 1938, 422 pp.
Carlquist, S. "Chance Dispersal." *American Scientist* 69, no. 5 (1981):509–16.
Castiglioni, A. *A History of Medicine*. 1936. Translated by E. B. Krumbhaar. 2d ed. New York: Alfred A. Knopf, 1958, 1192 pp.
**Clendening, L., ed. *Source Book of Medical History*. 1942. Reprint. New York: Dover Publications, 1960, 685 pp.
Coleman, W. *Biology in the Nineteenth Century: Problems of Form, Function, and Transformation*. New York: John Wiley & Sons, 1971, 187 pp.
*Darwin, C. *The Origin of Species By Means of Natural Selection*. 1859. Reprint. New York: Avenel Books, 1979, 476 pp.
Darwin, F., ed. *The Autobiography of Charles Darwin and Selected Letters*. 1892. Reprint. New York: Dover Publications, 1958, 365 pp.
Davis, A. B. "The Development of Anesthesia." *American Scientist* 70, no. 5 (1982):522–28.
De Kruif, P. *Microbe Hunters*. New York: Harcourt, Brace, & Co., 1926, 363 pp.
*Dobell, C., ed. *Antony van Leeuwenhoek and His "Little Animals."* 1932. Reprint. New York: Dover Publications, 1960, 435 pp.
Dubos, R. J. *Louis Pasteur: Free Lance of Science*. 1950. Reprint. New York: Charles Scribner's Sons, 1976, 421 pp.
Fishbein, J., ed. *Illustrated Medical and Health Encyclopedia*. 4 vols. Westport, Conn.: H. S. Stuttman Co., 1978, 1524 pp.
*Freke, H. *On the Origin of Species by Means of Organic Affinity*. London: Longman & Co., 1861, 135 pp.
Freud, S. *A General Introduction to Psychoanalysis*. 1924. Reprint. New York: Pocket Books, 1969, 480 pp.
Fried, J. J. *The Mystery of Heredity*. New York: John Day Co., 1971, 180 pp.
*Galton, F. "Classification of Men According to Their Natural Gifts." 1869. Reprint. In *The World of Mathematics*, edited by J. R. Newman. New York: Simon & Schuster, 1956, 1173–88.
Garrison, F. H. *Contributions to the History of Medicine*. New York: Hafner Publishing Co., 1966, 989 pp.

Garrison, F. H. *An Introduction to the History of Medicine.* 4th ed. Philadelphia: W. B. Saunders Co., 1929, 996 pp.

Gould, S. J. "A Visit to Dayton." *Natural History* 90, no. 10 (1981):8–22.

*Harvey, W. "An Anatomical Disquisition on the Motion of the Heart and Blood in Animals." 1628. Reprint. In *Great Books of the Western World,* edited by R. M. Hutchins, translated by S. P. Thompson. Chicago: William Benton, 1952, 28:265–304.

*Harvey, W. "Anatomical Exercises on the Generation of Animals." 1651. Reprint. In *Great Books of the Western World,* edited by R. M. Hutchins. Chicago: William Benton, 1952, 28:329–496.

Ihde, A. J., and Becker, S. L. "Conflict of Concepts in Early Vitamin Studies." *Journal of the History of Biology* 4, no. 1 (1971):1–33.

Koestler, A. *Janus: A Summing Up.* New York: Random House, 1978, 354 pp.

Lechevalier, H. A., and Solotorovsky, M. *Three Centuries of Microbiology.* 1965. Reprint. New York: Dover Publications, 1974, 536 pp.

Leicester, H. M. *Development of Biochemical Concepts from Ancient to Modern Times.* Cambridge: Harvard University Press, 1974, 286 pp.

*Lister, J. "On the Antiseptic Principle in the Practice of Surgery." 1867. Reprint. In *The Founders of Modern Medicine,* edited by E. Metchnikoff. Freeport, N.Y.: Books for Libraries Press, 1971, 289–302.

Luce, G. G. *Biological Rhythms in Psychiatry and Medicine.* Chevy Chase, Md.: National Institute of Mental Health, 1970, 183 pp.

McCollum, E. V. *A History of Nutrition: The Sequence of Ideas in Nutrition Investigations.* Boston: Houghton Mifflin Company, 1957, 451 pp.

Magner, L. N. *A History of the Life Sciences.* New York: Marcel Dekker, 1979, 489 pp.

*Malthus, T. R. "Mathematics of Population and Food." 1798. Reprint. In *The World of Mathematics,* edited by J. R. Newman. New York: Simon & Schuster, 1956, 1192–99.

*Matthew, P. *On Naval Timber and Arboriculture.* London: Longman, Rees, Orme, Brown, & Green, 1831, 391 pp.

Mayr, E. *The Growth of Biological Thought: Diversity, Evolution, and Inheritance.* Cambridge: Harvard University Press, 1982, 974 pp.

*Mendel, G. "Mathematics of Heredity." 1866. Reprint. In *The World of Mathematics,* edited by J. R. Newman. New York: Simon & Schuster, 1956, 937–49.

Morton, L. T. *Garrison and Morton's Medical Bibliography.* 2d ed. New York: Argosy Book Stores, 1954, 655 pp.

Rothschuh, K. E. *History of Physiology.* Huntington, New York: Robert E. Krieger Publishing, 1973, 379 pp.

Sands, H., and Minters, F. C. *The Epilepsy Fact Book.* New York: Charles Scribner's Sons, 1977, 116 pp.

*Schwann, T. *Microscopical Researches into the Accordance in the Structure and Growth of Animals and Plants.* 1838. Translated by H. Smith. London: C. & J. Adlard, 1847, 268 pp.

Singer, C. *A History of Biology. A General Introduction to the Study of Living Things.* Rev. ed. New York: Henry Schuman, 1950, 579 pp.

Singer, C., and Underwood, E. A., *A Short History of Medicine.* Oxford: Clarendon Press, 1962, 854 pp.

Slaughter, F. G. *Semmelweis: The Conqueror of Childbed Fever.* New York: Collier Books, 1961, 159 pp.

*Snow, J. *On the Mode of Communication of Cholera.* 2d ed. London: John Churchill, 1855, 162 pp.

Springer, S. P., and Deutsch, G. *Left Brain, Right Brain.* San Francisco: W. H. Freeman & Co., 1981, 243 pp.

Talbott, J. H. *A Biographical History of Medicine: Excerpts and Essays on the Men and Their Work.* New York: Grune & Stratton, 1970, 1211 pp.

*Wallace, A. R. *The Malay Archipelago: The Land of the Orangutan and the Bird of Paradise.* 1869. Reprint. New York: Dover Publications, 1962, 518 pp.

Chemistry

Asimov, I. *Asimov on Chemistry.* Garden City, N.Y.: Doubleday & Co., 1974, 267 pp.

Crosland, M. P. *Historical Studies in the Language of Chemistry.* 1962. Reprint. New York: Dover Publications, 1978, 406 pp.

Farber, E. *The Evolution of Chemistry. A History of Its Ideas, Methods, and Materials.* New York: Ronald Press Co., 1952, 349 pp.

Garard, I. D. *Invitation to Chemistry.* Garden City, N.Y.: Doubleday & Co., 1969, 420 pp.

Gelender, M. *Review Text in Chemistry.* New York: Amsco School Publications, 1959, 500 pp.

*Hales, S. *Statical Essays: Containing Haemastaticks.* 1733. Reprint. New York: Hafner Publishing Co., 1964, 384 pp.

Ihde, A. J. *The Development of Modern Chemistry.* New York: Harper & Row, 1964, 851 pp.

**Knight, D. M. *Classical Scientific Papers: Chemistry.* London: Mills & Boon Limited, 1968, 391 pp.

*Lavoisier, A. L. *Elements of Chemistry, in a New Systematic Order, Containing all the Modern Discoveries.* 1790. Translated by R. Kerr. Reprint. New York: Dover Publications, 1965, 539 pp.

Leicester, H. M. *The Historical Background of Chemistry.* 1956. Reprint. New York: Dover Publications, 1971, 260 pp.

**Leicester, H. M., and Klickstein, H. S., eds. *A Source Book in Chemistry 1400–1900.* Cambridge: Harvard University Press, 1963, 554 pp.

*Mendeleeff, D. "Periodic Law of the Chemical Elements." 1889. Reprint. In *The World of Mathematics,* edited by J. R. Newman. New York: Simon & Schuster, 1956, 913–18.

Moore, F. J. *A History of Chemistry.* New York: McGraw-Hill, 1939, 447 pp.

Partington, J. R. *A History of Chemistry.* 4 vols. New York: Macmillan & Co., 1961.

Weeks, M. E., and Leicester, H. M. *Discovery of the Elements.* 7th ed. Easton, Pa.: Journal of Chemical Education, 1968, 896 pp.

Woolley, A. E. *Photography: A Practical and Creative Introduction.* New York: McGraw-Hill, 1974, 383 pp.

Earth Sciences

Adams, A. B. *Eternal Quest: The Story of the Great Naturalists.* New York: G. P. Putnam's Sons, 1969, 509 pp.

Adams, F. D. *The Birth and Development of the Geological Sciences.* 1938. Reprint. New York: Dover Publications, 1954, 506 pp.

Arx, W. S. von. *An Introduction to Physical Oceanography.* Reading, Mass.: Addison-Wesley Publishing Co., 1962, 422 pp.

Brooks, C. E. P. *Climate through the Ages.* 2d rev. ed. New York: Dover Publications, 1970, 395 pp.

Carozzi, A. V. "Glaciology and the Ice Age." *Journal of Geological Education* 32 (1984):158–70.

Chorlton, W. et al. *Planet Earth: Ice Ages.* Alexandria, Va.: Time-Life Books, 1983, 176 pp.

Cooke, D. W. *Variations in the Seasonal Extent of Sea Ice in the Antarctic during the Last 140,000 Years.* New York: Columbia University, 1978, 287 pp.

Darwin, F., ed. *The Autobiography of Charles Darwin and Selected Letters.* 1892. New York: Dover Publications, 1958, 365 pp.

Deacon, M. *Scientists and the Sea 1650–1900. A Study of Marine Science.* London: Academic Press, 1971, 445 pp.

Edwards, W. N., *The Early History of Palaeontology.* London: British Museum, 1967, 58 pp.

Ehlen, T. "A Tall Ship Tracks Magellan, Elcano around the World." *Smithsonian* 11, no. 3 (1980):116–25.

Faul, H., and Faul, C. *It Began with a Stone: A History of Geology from the Stone Age to the Age of Plate Tectonics.* New York: John Wiley & Sons, 1983, 270 pp.

Gabler, R. E., Sager, R. J., Brazier, S., and Pourciau, J. *Essentials of Physical Geography.* New York: Holt, Rinehart, & Winston, 1977, 513 pp.

Geikie, A. *The Founders of Geology.* 2d ed. 1905. Reprint. New York: Dover Publications, 1962, 486 pp.

*Gilbert, W. "On the Lodestone and Magnetic Bodies and On the Great Magnet the Earth." 1600. Reprint. In *Great Books of the Western World,* edited by R. M. Hutchins, translated by P. F. Mottelay. Chicago: William Benton, 1952, 28:1–121.

Gillispie, C. C. *Genesis and Geology: A Study in the Relations of Scientific Thought, Natural Theology, and Social Opinion in Great Britain, 1790–1850.* Cambridge: Harvard University Press, 1951, 315 pp.

Gould, S. J. "Hutton's Purposeful View." *Natural History* 91, no. 5 (1982):6–12.

Hallam, A. *Great Geological Controversies.* Oxford: Oxford University Press, 1984, 182 pp.

Imbrie, J., and Imbrie, K. P. *Ice Ages, Solving the Mystery.* Hillside, N.J.: Enslow Publishers, 1979, 224 pp.

James, P. E. *All Possible Worlds: A History of Geographical Ideas.* Indianapolis: Odyssey Press, 1972, 622 pp.

** Mather, K. F., and Mason, S. L. *A Source Book in Geology.* New York: Hafner Publishing, 1964, 702 pp.

Milankovitch, M. *Canon of Insolation and the Ice-Age Problem.* 1941. Jerusalem: Israel Program of Scientific Translations, 1969, 484 pp.

Neumann, G., and Pierson, W. J., Jr. *Principles of Physical Oceanography.* Englewood Cliffs, N.J.: Prentice-Hall, 1966, 545 pp.

Perry, A. H., and Walker, J. M. *The Ocean-Atmosphere System.* London: Longman, 1977, 160 pp.

Rapp, G., Jr., and Gifford, J. A. "Archaeological Geology." *American Scientist* 70, no. 1 (1982):45–53.

Richardson, P. L. "Benjamin Franklin and Timothy Folger's First Printed Chart of the Gulf Stream." *Science* 207, no. 4431 (1980):643–45.

Schlee, S. *The Edge of an Unfamiliar World: A History of Oceanography.* New York: E. P. Dutton & Co., 1973, 398 pp.

Simon, C. "Chandler Wobble." *Science News* 120, no. 17 (1981):268–69.

Strahler, A. N. *Introduction to Physical Geography.* 3d ed. New York: John Wiley & Sons, 1973, 468 pp.

Tarling, D., and Tarling, M. *Continental Drift: A Study of the Earth's Moving Surface.* Garden City, N.Y.: Doubleday & Co., 1971, 140 pp.

Tilling, R. I. "Volcanic Cloud May Alter Earth's Climate." *National Geographic* 162, no. 5 (1982):672–75.

*Wallace, A. R. *The Malay Archipelago: The Land of the Orangutan and the Bird of Paradise.* 1869. Reprint. New York: Dover Publications, 1962, 518 pp.

Warren, B. A., and Wunsch, C., eds. *Evolution of Physical Oceanography.* Cambridge: Massachusetts Institute of Technology Press, 1981, 623 pp.

*Wegener, A. *The Origin of Continents and Oceans.* 3d ed. 1922. Translated by J. G. A. Skerl. London: Methuen & Co., 1924, 212 pp.

Mathematics, Logic

Aleksandrov, A. D., Kolmogorov, A. N., and Lavrent'ev, M. A., eds. *Mathematics: Its Content, Methods and Meaning.* 1956. 3 vols. Translated by S. H. Gould and T. Bartha. Cambridge: Massachusetts Institute of Technology Press, 1963, 1092 pp.

Barker, S. F. *Philosophy of Mathematics.* Englewood Cliffs, N.J.: Prentice-Hall, 1964, 111 pp.

Bell, E. T. *Development of Mathematics.* 2d ed. New York: McGraw-Hill Book Co., 1945, 637 pp.

Bell, E. T. *Men of Mathematics.* New York: Simon & Schuster, 1937, 592 pp.

*Boole, G. *An Investigation of the Laws of Thought on Which Are Founded the Mathematical Theories of Logic and Probabilities.* 1854. Reprint. New York: Dover Publications, 1958, 424 pp.

Boyer, C. B. *A History of Mathematics.* New York: John Wiley & Sons, 1968, 717 pp.

Bunt, L. N. H., Jones, P. S., and Bedient, J. D. *The Historical Roots of Elementary Mathematics.* Englewood Cliffs, N.J.: Prentice-Hall, 1976, 299 pp.

Cajori, F. *A History of Mathematical Notations.* 2 vols. LaSalle, Ill.: Open Court Publishing Co., 1928–1929, 818 pp.

*Cardano, G. *The Book on Games of Chance.* c.1545. Translated by S. H. Gould, with notes by O. Ore. In *Cardano the Gambling Scholar,* by O. Ore. New York: Dover Publications, 1965, 182–241.

Dauben, J. W. *Georg Cantor: His Mathematics and Philosophy of the Infinite.* Cambridge: Harvard University Press, 1979, 404 pp.

*Descartes, R. *La Géométrie.* 1637. Reprint. In *The Geometry of René Descartes.* Translated by E. Smith and M. L. Latham. New York: Dover Publications, 1954, 244 pp.

Gardner, M. "Euclid's Parallel Postulate and Its Modern Offspring." *Scientific American* 245, no. 4 (1981):23–34.
*Gauss, C. F. *Disquisitiones Arithmeticae.* 1801. Translated by A. A. Clarke. New Haven: Yale University Press, 1966, 472 pp.
Goldstine, H. H. *The Computer from Pascal to von Neumann.* Princeton: Princeton University Press, 1972, 378 pp.
Groza, V. S. *A Survey of Mathematics: Elementary Concepts and Their Historical Development.* New York: Holt, Rinehart & Winston, 1968, 327 pp.
Hall, T. *Carl Friedrich Gauss.* Translated by Albert Froderberg. Cambridge: Massachusetts Institute of Technology Press, 1970, 176 pp.
Heath, T. L. *Greek Mathematics.* 1931. Reprint. New York: Dover Publications, 1963, 552 pp.
**Heijenoort, J. van. *From Frege to Gödel: A Source Book in Mathematical Logic, 1879–1931.* Cambridge: Harvard University Press, 1967, 665 pp.
Herstein, I. N., and Kaplansky, I. *Matters Mathematical.* 2d ed. New York: Chelsea Publishing Co., 1978, 246 pp.
Hofstadter, D. R. *Gödel, Escher, Bach: An Eternal Golden Braid.* 1979. Reprint. New York: Vintage Books, 1980, 777 pp.
Ivins, W. M., Jr. *Art & Geometry: A Study in Space Intuitions.* 1946. Reprint. New York: Dover Publications, 1964, 113 pp.
Klein, J. *Greek Mathematical Thought and the Origin of Algebra.* Cambridge: Massachusetts Institute of Technology Press, 1968, 360 pp.
Kline, M. *Mathematical Thought from Ancient to Modern Times.* New York: Oxford University Press, 1972, 1238 pp.
Kline, M. *Mathematics: The Loss of Certainty.* New York: Oxford University Press, 1980, 366 pp.
**Midonick, H. O., ed. *The Treasury of Mathematics.* New York: Philosophical Library, 1965, 820 pp.
Moreau, R. *The Computer Comes of Age: The People, the Hardware, and the Software.* Cambridge: Massachusetts Institute of Technology Press, 1984, 227 pp.
**Newman, J. R., ed. *The World of Mathematics.* 4 vols. New York: Simon & Schuster, 1956, 2535 pp.
*Newton, I. *The Mathematical Works of Isaac Newton.* 1669–1710. Reprints. Edited by D. T. Whiteside. 2 vols. New York: Johnson Reprint Corporation, 1964, 333 pp.
Ore, O. *Cardano the Gambling Scholar.* 1953. Reprint. New York: Dover Publications, 1965, 249 pp.
Osen, L. M. *Women in Mathematics.* Cambridge: Massachusetts Institute of Technology Press, 1974, 185 pp.
*Pascal, B. *Oeuvres Complètes.* c.1650. Reprints. Edited by J. Chevalier. Brussels: Editions Gallimard, 1954, 1529 pp.
Pomerance, C. "The Search for Prime Numbers." *Scientific American* 247, no. 6 (1982):136–47.
Quine, W. van O. *Methods of Logic.* Rev. ed. New York: Holt, Rinehart & Winston, 1967, 272 pp.
Rayner, J. N. *An Introduction to Spectral Analysis.* London: Pion Limited, 1971, 174 pp.
Reid, C. *Hilbert.* New York: Springer-Verlag, 1970, 290 pp.
Rothman, T. "The Short Life of Evariste Galois." *Scientific American* 246, no. 4 (1982):136–49.
Russell, B. *The Autobiography of Bertrand Russell.* Vol. 1. *1872–1914.* Boston: Little, Brown & Co., 1951, 356 pp.
Shaw, J. B. *Lectures on the Philosophy of Mathematics.* Chicago: Open Court Publishing Co., 1918, 206 pp.
**Smith, D. E. *A Source Book in Mathematics.* New York: McGraw-Hill Book Co., 1929, 701 pp.
Steinmann, J. *Pascal.* Translated by M. Turnell. New York: Harcourt, Brace, & World, 1966, 304 pp.
Struik, D. J. *A Concise History of Mathematics.* 3d ed. New York: Dover Publications, 1967, 195 pp.
**Struik, D. J., ed. *A Source Book in Mathematics, 1200–1800.* Cambridge, Harvard University Press, 1969, 427 pp.

Weyl, H. "David Hilbert and His Mathematical Work." In *Hilbert,* by C. Reid. New York: Springer-Verlag, 1970, 245–283.
*Whitehead, A. N., and Russell, B. *Principia Mathematica.* 1913. Reprint. Cambridge: Cambridge University Press, 1962, 410 pp.
Wilder, R. *Evolution of Mathematical Concepts: An Elementary Study.* New York: John Wiley & Sons, 1968, 224 pp.

Meteorology

Allen, O. E. et al. *Planet Earth: Atmosphere.* Alexandria, Va.: Time-Life Books, 1983, 176 pp.
Benedict, R. P. *Fundamentals of Temperature, Pressure and Flow Measurement.* 2d ed. New York: John Wiley & Sons, 1977, 517 pp.
*Franklin, B. *Benjamin Franklin: The Autobiography and Other Writings.* c.1760. Reprints. Edited by L. J. Lemisch. New York: New American Library, 1961, 350 pp.
Frisinger, H. H. *The History of Meteorology to 1800.* New York: Science History Publications, 1977, 148 pp.
Middleton, W. E. K. *A History of the Thermometer and its Uses in Meteorology.* Baltimore: Johns Hopkins Press, 1966, 249 pp.
Middleton, W. E. K. *Invention of the Meteorological Instruments.* Baltimore: Johns Hopkins Press, 1969, 362 pp.
*Milankovitch, M. *Théorie Mathématique des Phénomènes Thermiques Produits par la Radiation Solaire.* Paris: Gauthier-Villars, 1920, 334 pp.
*Pascal, B. *The Physical Treatises of Pascal.* c.1650. Translated by I. H. B. Spiers and A. G. H. Spiers, introduction and notes by F. Barry. New York: Octagon Books, 1973, 181 pp.
*Richardson, L. F. *Weather Prediction by Numerical Process.* Cambridge, England: University Press, 1922, 240 pp.
Steinmann, J. *Pascal.* 1962. Translated by M. Turnell. New York: Harcourt, Brace, & World, 1966, 304 pp.
*Torricelli, E. "Letters on the Pressure of the Atmosphere." 1644. In *The Physical Treatises of Pascal.* New York: Octagon Books, 1973, 163–70.

Physics

Bennett, C. E. *Physics.* New York: Barnes & Noble, 1964, 208 pp.
Clark, R. W. *Einstein: The Life and Times.* New York: World Publishing Co., 1971, 718 pp.
Cohen, I. B. *The Birth of a New Physics.* Garden City, N.Y.: Doubleday & Co., 1960, 200 pp.
*Descartes, R. *The Method, Meditations and Philosophy.* 1637, 1641, 1644. Translated by J. Vietch, introduction by F. Sewall. Washington, D.C.: M. Walter Dunne, 1901, 371 pp.
Dirac, P. A. M. "Address at the Einstein Session of the Pontifical Academy of Sciences." *Science* 207, no. 4436 (1980):1161–62.
Drake, S. *Galileo At Work. His Scientific Biography.* Chicago: University of Chicago Press, 1978, 536 pp.
Drake, S. "Newton's Apple and Galileo's Dialogue." *Scientific American* 243, no. 2 (1980):150–56.
Einstein, A., and Infeld, L. *The Evolution of Physics.* New York: Simon & Schuster, 1938, 302 pp.
*Franklin, B. *Benjamin Franklin: The Autobiography and Other Writings.* c.1760. Edited by L. J. Lemisch. New York: New American Library, 1961, 350 pp.
French, A. P. *Special Relativity.* New York: W. W. Norton & Co., 1968, 286 pp.
*Galilei, G. *Dialogues Concerning Two New Sciences.* 1638. New York: Dover Publications, 1954, 300 pp.
Gamow, G. *Thirty Years That Shook Physics.* Garden City, N.Y.: Doubleday & Co., 1966, 224 pp.
Gingerich, O. "The Galileo Affair." *Scientific American* 247, no. 2 (1982):132–42.
Halliday, D. *Introductory Nuclear Physics.* New York: John Wiley & Sons, 1955, 493 pp.
Hart, I. B. *The World of Leonardo da Vinci: Man of Science, Engineer and Dreamer of Flight.* New York: Viking Press, 1961, 374 pp.
Heisenberg, W. *Physics and Beyond.* New York: Harper & Row, 1971, 247 pp.

*Huygens, C. "Treatise on Light." 1690. In *Great Books of the Western World,* edited by R. M. Hutchins, translated by S. P. Thompson. Chicago: William Benton, 1952, 34:549–619.

Jungk, R. *Brighter Than a Thousand Suns: A Personal History of the Atomic Scientists.* 1956. Translated by J. Cleugh. New York: Harcourt, Brace, & World, 1958, 360 pp.

**Lorentz, H. A., Einstein, A., Minkowski, H., and Weyl, H. *The Principle of Relativity.* 1895–1919. Reprints. New York: Dover Publications, 1923, 216 pp.

McCloskey, M. "Intuitive Physics." *Scientific American* 248, no. 4 (1983):122–30.

**Magie, W. F. *A Source Book in Physics.* New York: McGraw-Hill, 1935, 620 pp.

Meyer, H. W. *A History of Electricity and Magnetism.* Cambridge: Massachusetts Institute of Technology Press, 1971, 325 pp.

*Moseley, H. G. J. "Atomic Numbers." 1914. Reprint. In *The World of Mathematics,* edited by J. R. Newman. New York: Simon & Schuster, 1956, 842–50.

*Newton, I. *Mathematical Principles of Natural Philosophy.* 1687. Translated by A. Motte. Berkeley: University of California Press, 1960, 680 pp.

*Newton, I. *Opticks.* 4th ed. 1730. Reprint. New York: Dover Publications, 1979, 406 pp.

Resnick, R., and Halliday, D. *Physics.* New York: John Wiley & Sons, 1966, 646 pp.

Rozental, S., ed. *Niels Bohr: His Life and Work as Seen by Friends and Colleagues.* New York: Interscience Publishers, 1967, 355 pp.

Schiff, L. I. *Quantum Mechanics.* 3d ed. New York: McGraw-Hill, 1968, 544 pp.

Scott, J. F. *The Scientific Work of René Descartes.* London: Taylor & Francis, 1952, 211 pp.

*Stevin, S. "The Elements of Hydrostatics" and "Commencing the Practice of Hydrostatics." 1583. In *The Physical Treatises of Pascal.* New York: Octagon Books, 1973, 135–62.

Trefil, J. S. "Einstein's Theory of General Relativity Is Put to the Test." *Smithsonian* 11, no. 1 (1980):74–83.

Von Laue, M. *History of Physics.* New York: Academic Press, 1950, 150 pp.

Weisskopf, V. F. "Address at the Einstein Session of the Pontifical Academy of Sciences." *Science* 207, no. 4436 (1980):1163–65.

Whittaker, E. T. *A History of the Theories of Aether and Electricity from the Age of Descartes to the Close of the Nineteenth Century.* London: Longmans, Green, & Co., 1910, 475 pp.

Whittaker, E. T. *A History of the Theories of Aether and Electricity: The Modern Theories 1900–1926.* New York: Philosophical Library, 1954, 319 pp.

Williams, L. P. *Michael Faraday, A Biography.* New York: Basic Books, 1965, 531 pp.

Williams, L. P. *The Origins of Field Theory.* New York: Random House, 1966, 148 pp.

General

Andrade, E. N. *Sir Isaac Newton.* London: Collins, 1954, 140 pp.

Asimov, I. *Biographical Encyclopedia of Science and Technology.* Garden City, N.Y.: Doubleday & Co., 1972, 662 pp.

*Bacon, F. *Advancement of Learning* (1605), *Novum Organum* (1620), and *New Atlantis* (1614–1617). In *Great Books of the Western World,* edited by R. M. Hutchins. Vol 30. Chicago: Encyclopaedia Britannica, 1952, 214 pp.

Berkeley, E. B., and Berkeley, D. S. *The Reverend John Clayton: A Parson with a Scientific Mind. His Scientific Writings and Other Related Papers.* Charlottesville: University Press of Virginia, 1965, 170 pp.

Burnam, T. *The Dictionary of Misinformation.* New York: Thomas Y. Crowell, 1975, 302 pp.

**Burnham, J. C., ed. *Science in America: Historical Selections.* New York: Holt, Rinehart, & Winston, 1971, 495 pp.

Butler, A., ed. *Everyman's Dictionary of Dates.* 4th ed. New York: E. P. Dutton & Co., 1964, 455 pp.

Bynum, W. F., Browne, E. J., and Porter, R., eds. *Dictionary of the History of Science.* Princeton: Princeton University Press, 1981, 494 pp.

Carruth, G. et al., eds. *The Encyclopedia of American Facts and Dates.* 6th ed. New York: Thomas Y. Crowell Co., 1972, 922 pp.

Cohen, I. B. *Album of Science: From Leonardo to Lavoisier, 1450–1800.* New York: Charles Scribner's Sons, 1980, 306 pp.

Debus, A. G., ed. *World's Who's Who in Science: A Biographical Dictionary of Notable Scientists from Antiquity to the Present.* Hannibal, Mo.: Western Publishing Co., 1968, 1855 pp.

De Solla Price, D. *Science Since Babylon.* New Haven: Yale University Press, 1975, 215 pp.
Duncan, R., and Weston-Smith, M., eds. *The Encyclopedia of Ignorance.* New York: Pocket Books, 1978, 443 pp.
Durant, W. *The Story of Civilization.* Vol. 5. *The Renaissance.* New York: Simon & Schuster, 1953, 778 pp.
Durant, W. *The Story of Civilization.* Vol. 6. *The Reformation: A History of European Civilization from Wyclif to Calvin: 1300–1564.* New York: Simon & Schuster, 1957, 1025 pp.
Durant, W. J. *The Story of Philosophy.* 1927. Reprint. New York: Simon & Schuster, 1949, 592 pp.
Feuer, L. *The Scientific Intellectual.* New York: Basic Books, 1963, 441 pp.
Forbes, R. J., and Dijksterhuis, E. J. *A History of Science and Technology.* Vol. 1. Baltimore: Penguin Books, 1963, 294 pp.
Gillispie, C. C., ed. *Dictionary of Scientific Biography.* 15 vols. New York: Charles Scribner's Sons, 1970.
Greene, J. E., ed. *100 Great Scientists.* New York: Washington Square Press, 1964, 498 pp.
Grun, B. *The Timetables of History.* New York: Simon & Schuster, 1982, 676 pp.
Guerlac, H. *Essays and Papers in the History of Modern Science.* Baltimore: Johns Hopkins University Press, 1977, 540 pp.
Hall, A. R., and Hall, M. B. *A Brief History of Science.* New York: Signet Science Library Books, 1964, 352 pp.
Harré, R. *Great Scientific Experiments: 20 Experiments That Changed Our View of the World.* Oxford: Phaidon Press, 1981, 224 pp.
Harris, W. H., and Levey, J. S., eds. *The New Columbia Encyclopedia.* 4th ed. New York: Columbia University Press, 1975, 3052 pp.
Howard, A. V. *Chambers's Dictionary of Scientists.* New York: Dutton & Co., 1951, 499 pp.
Kuhn, T. S. *The Structure of Scientific Revolutions.* 2d ed. Chicago: University of Chicago Press, 1970, 210 pp.
Langer, W. L., ed. *An Encyclopedia of World History.* Boston: Houghton Mifflin Co., 1958, 1243 pp.
**McKenzie, A. E. E. *The Major Achievements of Science.* New York: Simon & Schuster, 1960, 575 pp.
Marias, J. *History of Philosophy.* New York: Dover Publications, 1967, 505 pp.
Mason, S. F. *A History of the Sciences.* New York: Collier Books, 1971, 638 pp.
More, L. T. *Isaac Newton.* 1934. Reprint. New York: Dover Publications, 1962, 675 pp.
**Moulton, F. R., and J. J. Schifferes, eds. *The Autobiography of Science.* 2d ed. Garden City, N.Y.: Doubleday & Co., 1960, 748 pp.
Newman, J. R. *Science and Sensibility.* New York: Simon & Schuster, 1961, 689 pp.
Neyman, J., ed. *The Heritage of Copernicus.* Cambridge: Massachusetts Institute of Technology Press, 1974, 542 pp.
Ploski, H. A., and Marr, W., II, eds. *The Negro Almanac: A Reference Work on the Afro American.* New York: Bellwether Co., 1976, 1206 pp.
Quigley, C. *The Evolution of Civilizations: An Introduction to Historical Analysis.* Indianapolis: Liberty Press, 1979, 442 pp.
Sarton, G. *The Life of Science.* Bloomington: Indiana University Press, 1948, 197 pp.
Struik, D. J. *Yankee Science in the Making.* Boston: Little, Brown & Co., 1948, 430 pp.
Taton, R., ed. *The Beginnings of Modern Science from 1450 to 1800.* New York: Basic Books, 1965, 667 pp.
Webster's Biographical Dictionary. Springfield, Mass.: G. & C. Merriam Co., 1963, 1697 pp.
West, S. "The Roots of Science." *Science News* 117, no. 21 (1980):332–33.
Westfall, R. S. *Never at Rest: A Biography of Isaac Newton.* Cambridge: Cambridge University Press, 1980, 908 pp.
Who Was Who in American History—Science and Technology. Chicago: Marquis Who's Who, 1976, 688 pp.
Williams, T. I., ed. *A Biographical Dictionary of Scientists.* 2d ed. New York: John Wiley & Sons, 1974, 641 pp.
Yule, J., ed. *Concise Encyclopedia of the Sciences.* New York: Van Nostrand Reinhold Co., 1982, 590 pp.

Many additional standard reference works were also used, including most prominently the *Encyclopaedia Britannica* (Chicago: William Benton, 1970); the *New Encyclopaedia Britannica,* 15th ed. *Macropaedia* (Chicago: Encyclopaedia Britannica, 1975); and the Oxford English Dictionary. Also, individual random dates were obtained from various articles in *American Scientist, Science, Science News, Scientific American, Smithsonian,* and other journals over the period 1970–1984.

Name Index

The indexing is done according to date and major subject heading, not page number. Middle names are italicized in the index if the individual is referred to in the text by his middle name rather than his first name. Major subject headings are identified as follows: A—Astronomy, B—Biology, C—Chemistry, E—Earth Sciences, H—Health Sciences, Ma—Mathematics, Met—Meteorology, P—Physics, S—Supplemental.

A

Abbe, Cleveland (1838–1916, American)—1873 Ma

Abel, John Jacob (1857–1938, American)—1898 B, 1901 B

Abel, Niels Henrik (1802–1829, Norwegian)—1800 Ma, 1824 Ma, 1825 Ma, 1826 Ma, 1827 Ma, 1829 Ma, 1830–1832 Ma, 1832 Ma

Abraham, Karl (1877–1925, Austrian)—1911 H, 1916 H

Abreu, Aleixo de (1568–1630, Portuguese)—1623 H

Achenwall, Gottfried (1719–1772, German)—1749 Ma

Acosta, José de (1539–1600, Spanish)—1590 E

Adams, John Couch (1819–1892, English)—1845 A, 1846 A

Adams, Walter Sydney (1876–1956, American)—1906 A, 1914 A (2 entries), 1925 A

Addison, Thomas (1793–1860, English)—1855 H

Adelard of Bath (c.1075–1160, English)—c.1255 Ma

Adet, Pierre Auguste (1763–1834, French)—1787 C, 1813 C

Adhémar, Joseph Alphonse (1797–1862, French)—1842 E, 1852 E, 1864 E

Agassiz, Alexander (1835–1910, American)—1877 E

Agassiz, Jean *Louis* Rodolphe (1807–1873, Swiss-American)—1787 E, 1828 B, 1833–1844 E, 1836 E, 1837 E, 1839 E, 1840 E, 1841 E, 1846 E, 1852–1855 E, 1860 B

Agnesi, Maria Gaetana (1718–1799, Italian)—1748 Ma

Agramonte, Aristides (1869–1931, Cuban)—1900 H

Agricola, Georgius (Georg Bauer, 1494–1555, German)—1546 E (4 entries), 1556 E

Ailly, Pierre d' (1350–1420, French)—1410 E

Airy, George Biddell (1801–1892, English)—1825 H, 1845 E

Aitken, John (1839–1919, Scottish)—1881 Met

Aitken, Robert Grant (1864–1951, American)—1895–1935 A

Albert, Prince of Monaco (1848–1922, Monaco)—1885 E

Alberti, Leone Battista (1404–1472, Italian)—c. 1450 Met, 1667 Met

Albertus Magnus, Saint (Albert the Great, c.1200–1280, German)—c.1220 B, c.1250 B (2 entries), c.1260 E (2 entries)

Alcmaeon of Crotona (born c.535 B.C., Italian-Greek)—1564 B

Aldrovandi, Ulisse (1522–1605, Italian)—1554 B

Alembert, Jean le Rond d' (1717–1783, French)—1743 P, 1744 P, 1747 Ma, 1747 Met, 1751–1772 S, 1754 A, 1754 Ma, 1788 P

Alexander, James Waddell (1888–1971, American)—1927–1928 Ma

Alfonso El Sabio (Alfonso X of Castile, 1221–1284, Spanish)—1272 A, 1483 A
Al-Haytham, Ibn—*see* Ibn Al-Haytham, alphabetized under H
Alhazen—*see* Ibn Al-Haytham, alphabetized under H
Al-Khwārizmī—see Al-Khwārizmī, alphabetized under K
Al-Kindi—see Al-Kindī, alphabetized under K
Al-Nafis, Ibn (1210–1288, Egyptian)—c.1260 B
Alpetragius (Abū Ishāq al-Bitrūjī al-Ishbīlī, fl. c.1190, Spanish)—1217 A
Alter, David (1807–1881, American)—1854 C
Amici, Giovan Battista (1786–1868, Italian)—1806 B, 1823–1830 B, 1827 B
Amodio, Andrea (Italian)—1410 B
Amontons, Guillaume (1663–1705, French)—1687 Met, 1699 C, 1702 Met, 1704 Met
Ampère, André-Marie (1775–1836, French)—1806 Ma, 1814 C, 1820 P, 1827 P
Amundsen, Roald (1872–1928, Norwegian)—1910–1912 E, 1918–1925 E
Andrews, Thomas (1847–1907, English)—1869 C
Anglicus, Bartholomew (fl. c.1250, English)—c.1235 S, 1250 S
Angström, Anders Jonas (1814–1874, Swedish)—1853 C, 1862 A, 1867 Met, 1868 A
Antonoff, G. N.—1911 C
Apian, Peter (Peter Bennewitz, Peter Bienewitz, Petrus Apianus, 1495–1552, German)—1524 E, 1534 Ma, 1540 A, 1753 E/A
Apianus, Petrus—*see* Apian, Peter
Appert, Nicholas (1750–1841, French)—1810 C
Appleton, Edward Victor (1892–1965, English)—1924 Met
Aquinas, Saint Thomas (c.1225–1274, Italian)—c.1270 S, 1879 S
Arago, Dominique François Jean (1786–1853, French)—1819 P
Arbuthnot, John (1667–1735, Scottish)—1731 H, 1733 B
Archimedes (c.287–212 B.C., Greek)—1269 S, 1543 S, 1558 Ma, 1565 Ma, 1586 P, 1589–1592 P, 1596 Ma, 1612 P, 1615 Ma, 1647 Ma, 1651 Ma, 1667 Ma.
Arduino, Giovanni (1714–1795, Italian)—1759 E, 1779 E/C
Arfwedson, Johan August (1792–1841, Swedish)—1817 C
Argand, Jean Robert (1768–1822, French)—1797 Ma, 1806 Ma
Argelander, Friedrich Wilhelm August (1799–1875, German)—1862 A
Aristotle (384–322 B.C., Greek)—1215 S, 1217 A, c.1220 B, c.1220 P, c.1230 S, c.1250 B (2 entries), 1268 B, 1269 S, c.1270 S, 1277 S, 1277–1279 P, c.1280 P, c.1290 A, c.1300 S, c.1320 Ma, 1328 P, 1330 C, c.1330 P, c.1350 P, 1498 S, 1531 S, 1537 P, 1543 P (2 entries), 1557 E, 1565–1586 B, 1572 A, 1577 A, 1585 P, 1586 P, 1589–1592 P, 1604 B, 1617 C, 1624 S, 1627 S, 1628 B, 1632 A, 1638 P, 1648 C, 1651 B, 1661 C, 1684 Met, 1686–1704 B, 1847 Ma.
Arrhenius, Carl Axel (1757–1824, Swedish)—1794 C
Arrhenius, Svante August (1859–1927, Swedish)—1884–1887 C, 1889 C, 1896 E, 1899 C, 1899 E, 1907 C, c.1908 Met, 1915 B
Aston, Francis William (1877–1945, English)—1919 C, 1927 C
Atkinson, Robert—1929 A
Atwood, George (1745–1807, English)—1784 P
Audubon, John James (1785–1851, American)—1827–1839 B
Auenbrugger, Joseph *Leopold* (1722–1809, Austrian)—1761 H
Augustine of Hippo, Saint (354–430)—c.1220 P
Auzout, Adrien (1622–1691, French)—1641 A, 1666 A, 1667 A
Averroës (Abū'l-Walīd Muhammad Ibn Ahmad Ibn Muhammad Ibn Rushd, 1126–1198, Arabic)—c.1280 P, c.1300 P
Avicebron (Spanish)—c.1300 C
Aviso, Urbano d' (Italian)—1666 Met
Avogadro, Amedeo (1776–1856, Italian)—1811 C, 1814 C, 1860 C, 1864 C, 1865 C, 1908 C

B

Babbage, Charles (1792–1871, English)—1812 Ma, 1822 Ma, 1830 S, 1833 Ma, 1834 Ma, 1853 Ma
Babcock, Stephen Moulton (1843–1931, American)—1890 C
Baccius, Andreas—1603 E
Bachet de Méziriac, Claude-Gaspar (1581–1638, French)—1621 Ma, 1670 Ma
Bacon, Francis (1561–1626, English)—1531 S, 1605 S, 1620 A, 1620 S, 1627 S, 1644 S

Bacon, Roger (c.1219–c.1292, English)—1267–1268 E, 1267–1268 S, 1268 B, c.1330 B, 1604 P
Baer, Karl Ernst von (1792–1876, Russian)—1827 B, 1828–1837 B
Baeyer, Adolf Johann Friedrich Wilhelm von (1835–1917, German)—1878 C
Baffin, William (1584–1622, English)—1616 E
Bagehot, Walter (1826–1877, English)—1873 B
Bakewell, Robert (1768–1843, English)—1829 E, 1841 E
Balard, Antoine Jérome (1802–1876, French)—1825 C, 1826 C
Baldet, Fernand—1927 A
Ballot, Christoph Buys—*see* Buys Ballot, Christoph
Balmer, Johann Jakob (1825–1898, Swiss)—1885 P, 1913 P, 1926 P
Banister, John (1650–1692, American)—1680 B
Banks, Joseph (1743–1820, English)—1768–1771 B
Banting, Frederick Grant (1891–1941, Canadian)—1921 H
Barkla, Charles Glover (1877–1944, English)—1911 P
Barnard, Edward Emerson (1857–1923, American)—1892 A
Barrow, Isaac (1630–1677, English)—1669 Ma, 1670 Ma
Bartholin, Erasmus (1625–1692, Danish)—1669 P
Bartram, William (1739–1823, American)—1791 B
Basedow, Karl Adolf von (1799–1854, German)—1840 H
Bateson, William (1861–1926, English)—1894 B, 1902 B, 1914 B
Bauer, Georg—*see* Agricola, Georgius
Bauhin, Gaspard (1560–1624, Swiss)—1588 B, 1623 B, 1737 B
Bayer, Johann (1572–1625, German)—1603 A
Bayes, Thomas (1702–1761, English)—1736 Ma, 1763 Ma
Bayliss, William Maddock (1860–1924, English)—1902 B, 1905 B, 1908 B
Beaufort, Francis (1774–1857, Irish)—1805 Met
Beaumont, Jean Baptiste Armand Louis Leonce Elie de (1798–1874, French)—1830 E, 1840 E, 1852 E
Beaumont, William (1785–1853, American)—1833 B
Beccaria, Giambatista (1716–1781, Italian)—1753 Met/P
Becher, Johann Joachim (1635–1682, German)—1669 C (2 entries), 1680 B, 1700 C
Becquerel, Alexandre-Edmond (1820–1891, French)—1839 P
Becquerel, *Antoine Henri* (1852–1908, French)—1896 C, 1896 P
Beguin, Jean (c.1550–c.1620, French)—1611 C
Béguyer de Chancourtois, Alexandre-Émile (1820–1886, French)—1862 C
Behaim, Martin (1459–1507, German)—1492 E
Behring, Emil Adolph von (1854–1917, German)—1888–1891 H, 1892 H
Beijerinck, Martinus Willem (1851–1931, Dutch)—1898 B/H
Beke, Charles Tilstone (1800–1874, English)—1835 E
Bell, Alexander Graham (1847–1922, American)—1876 P
Bell, Charles (1774–1842, Scottish)—1811 B, 1822 B
Belon, Pierre (1517–1564, French)—1555 B
Beltrami, Eugenio (1835–1899, Italian)—1868 Ma
Benedetti, Giovanni Battista (1530–1590, Italian)—1585 P
Bennewitz, Peter—*see* Apian, Peter
Berger, Hans (1873–1941, Austrian)—1929 H
Bergeron, Tor (Norwegian)—1911 Met, 1922 Met, 1928 Met
Berghaus, Heinrich (1797–1884, German)—1837–1848 E
Bergius, Friedrich (1884–1949, German)—1910 C
Bergman, Torbern Olof (1735–1784, Swedish)—1775 C, 1779 C, 1782 C (2 entries), 1784 C/E
Berkeley, George (1685–1753, Irish)—1709 S, 1734 Ma, 1736 Ma, 1742 Ma
Berliner, Emile (1851–1929, German)—1877 P
Bernard, Claude (1813–1878, French)—1843 B, 1843 H, 1865 H
Bernard of Gordon (c.1285–c.1320, French)—c. 1310 B, c.1310 H
Bernard of Verdun (fl. late 13th century, French)—c.1290 A
Bernhardi, Reinhard (German)—1832 E
Bernoulli, Daniel (1700–1782, Swiss)—1728 Ma, 1738 C, 1738 P/E
Bernoulli, Jakob (Jacques) I (1654–1705, Swiss)—1690 Ma, 1691 Ma, 1694 Ma,

1696 Ma (2 entries), 1697 Ma, 1701 Ma, 1713 Ma (2 entries), 1812 Ma
Bernoulli, Johann (Jean) I (1667–1748, Swiss)—1696 Ma, 1697 Ma, 1701 Ma
Bernoulli, Nikolaus (1687–1759, Swiss)—1720 Ma, 1755 Ma
Berthelot, Pierre Eugène *Marcellin* (1827–1907, French)—1860 C, 1860 C/B
Berthollet, Claude Louis (1748–1822, French)—1785 C, 1787 C, 1789 C, 1790 A, 1790 C, 1794 C, 1799 C, 1801 C, 1803 C
Bertrand, Joseph Louis François (1822–1900, French)—1845 Ma, 1850 Ma
Bertrand, Marcel Alexandre (1847–1907, French)—1884 E, 1887 E, 1894 E
Berzelius, Jöns Jacob (1779–1848, Swedish)—1789 C, 1803 C (2 entries), 1808 C, 1810–1820 C, 1813 C, 1814 C (2 entries), 1814 E, 1818 C (2 entries), 1820 C, 1823 C, 1824 C, 1825 C, 1826 C (2 entries), 1829 C, 1830 C, 1833 C, 1834 C, 1836 C
Bessel, Friedrich Wilhelm (1784–1846, German)—1818 A, 1824 Ma, 1838 A, 1839 A, 1841 E, 1844 A, 1862 A
Bessemer, Henry (1813–1898, English)—1856 C
Bessy, Bernard Frénicle de (1605–1675, French)—1640 Ma
Best, Charles Herbert (1899–1978, Canadian)—1921 H
Betti, Enrico (1823–1892, Italian)—1871 Ma
Bianco, Andrea (Italian)—1436 E
Bichat, Marie-François-Xavier (1771–1802, French)—1800 B
Bienewitz, Peter—*see* Apian, Peter
Billings, John Shaw (1839–1913, American)—1880 Ma, 1890 Ma
Billingsley, Henry (d.1606, English)—1570 Ma
Binet, Alfred (1857–1911, French)—1905 S, 1914 S
Biot, Jean Baptiste (1774–1862, French)—1803 A, 1804 Met, 1820 P
Biringuccio, Vannoccio (1480–c.1539, Italian)—1540 C
al-Bitrūjī—*see* Alpetragius
Bjerknes, Jakob Aall Bonnevie (1897–1975, Norwegian-American)–1919 Met
Bjerknes, Vilhelm Frimann Koren (1862–1951, Norwegian)—1919 Met, 1921 Met
Black, A.—1924 H
Black, Joseph (1728–1799, Scottish)—1754 C, 1756 C, 1760 C, 1762 C, 1770 C, 1772 C, 1776 P, 1778 C, 1789 C
Bleuler, Paul Engen (1857–1939, Swiss)—1911 H
Blumenbach, Johann Friedrich (1752–1840, German)—1776 B
Boccaccio, Giovanni (1313–1375, Italian)—1348 H, 1353 H
Bode, Johann Elert (1747–1826, German)—1766 A, 1772 A, 1781 A, 1801 A (2 entries)
Boethius, Anicius Manlius Severinus (c.480–524/525, Roman)—c.1240 Ma
Bohr, Niels Henrik David (1885–1962, Danish)—1913 P (2 entries), 1914 P, 1915 P, 1915–1916 P, 1922 P, 1923 C, 1923 P, 1924 P, 1926 P, 1928 P
Boisbaudran, Paul Émile Lecoq de (1838–1912, French)—1874 C, 1879 C, 1886 C, 1894 C/Met
Bollee, Leon (French)—1888 Ma
Boltwood, Bertram Borden (1870–1927, American)—1904 C, 1907 C
Boltzmann, Ludwig Eduard (1844–1906, Austrian)—1860s C, 1877 P, 1883 P, 1900 P
Bolyai, Farkas (or Wolfgang) (1775–1856, Hungarian)—1832 Ma, 1832–1833 Ma
Bolyai, János (or Johann) (1802–1860, Hungarian)—1816 Ma, 1823 Ma, 1832 Ma, 1854 Ma, 1868 Ma, 1871 Ma
Bolzano, Bernard (1781–1848, Czechoslovakian)—1817 Ma, 1834 Ma, c.1840 Ma, 1850 Ma, 1861 Ma, c.1865 Ma, 1923 Ma
Bombelli, Rafael (1526–1572, Italian)—1572 Ma
Bond, George Phillips (1825–1865, American)—1848 A, 1850 A
Bond, William Cranch (1789–1859, American)—1850 A
Bonnet, Charles (1720–1793, Swiss)—1740 B, 1770 B
Bonpland, A. J. A.—1805–1834 Met/E
Bonus, Petrus, of Ferrara (fl.1325, Italian)—1330 C
Boole, George (1815–1864, English)—1844 Ma, 1847 Ma, 1854 Ma, 1881 Ma, 1901 Ma
Borel, Émile (Félix-Édouard-Justin) (1871–1956, French)—1872 Ma, 1921 Ma, 1923 Ma
Borelli, Giovanni Alfonso (Alphonse) (1608–1679, Italian)—1666 A, 1680 B
Borgognoni of Lucca, Theodoric (1205–1298, Italian)—1266 H
Born, Max (1882–1970, German)—1925 P, 1926 P (3 entries)
Borough, William (1536–1599, English)—1581 E

Bošković, Rudjer (1711–1787, Yugoslavian-Italian)—1758 C, 1844 C
Bosse, Abraham (1602–1676, French)—1648 Ma
Bouguer, Pierre (1698–1758, French)—1735–1744 E
Boullay, Polydore (1806–1835, French)—1828 C
Boulliau, Ismael (1605–1694, French)—1667 A
Boulton, Mathew (1728–1809, English)—1776 P
Bourne, William (c.1535–1582, English)—1578 E
Boussingault, Jean Baptiste (1802–1887, French)—1844 B
Bouvard, Alexis (1767–1843, French)—1821 A
Boveri, Theodor (1862–1915, German)—1889 B, 1903 B, 1904 B
Bowditch, Henry Pickering (1840–1911, American)—1871 B
Bowditch, Nathaniel (1773–1838, American)—1802 A
Bowen, Ira Sprague (1898–1973, American)—1927 A (2 entries)
Boyle, Robert (1627–1691, Irish-English)—1647 A, 1660 Met, 1661 C, 1662 C (2 entries), 1663 B, 1665 Met, 1666 C, 1669 Met, 1671 C, 1673 E, 1676 C, 1687 C, 1738 C, 1766 C, 1774 C, 1834 C
Boylston, Zabdiel (1679–1766, American)—1718 H, 1721 H
Bradley, James (1693–1762, English)—1727–1747 A, 1728 A, 1748 A
Bradwardine, Thomas (c.1290–1349, English)—1328 P
Brahe, Tycho (1546–1601, Danish)—1572 A, 1573 A, 1576 A, 1577 A, c.1580 Ma, 1588 A, 1598 A, 1600 A, 1602 A, 1605 A, 1609 A, 1627 A, 1647 A, 1651 A
Braid, James (1795–1860, Scottish)—1841 H
Brand, Hennig (c.1630–c.1692, German)—1677 C
Brandt, Georg (1694–1768, Swedish)—1737–1738 C
Braun, Alexander Carl Heinrich (1805–1877, German)—1854 B
Bravais, Auguste (1811–1863, French)—1848 C/Ma, 1851 C, 1870 Ma
Bravais, Louis François (French)—1827 H, 1863 H
Brennecke—1911 E
Breuer, Josef (1842–1925, Austrian)—1880–1882 H, 1893 H, 1895 H
Brewster, David (1781–1868, Scottish)—1811 P, 1816 P, 1834 C
Bridges, Calvin Blackman (1889–1938, American)—1915 B, 1916 B
Briggs, Henry (1561–1630, English)—1624 Ma, 1627 Ma
Bring, Erland (1736–1798, Swedish)—1786 Ma
Broca, Pierre *Paul* (1824–1880, French)—1836 H/B, 1861 B
Broglie, Louis Victor de (1892–1960, French)—1924 P, 1926 P, 1927 P, 1928 P
Brongniart, Adolphe Théodore (1801–1876, French)—1823–1830 B, 1828 B
Brongniart, Alexandre (1770–1847, French)—1808 E, 1811 E, 1813 E
Brougham, Henry (English)—1801–1804 P
Brouncker, William (1620–1684, English)—1658 Ma, 1668 Ma
Brouwer, Luitzen Egbertus Jan (1881–1966, Dutch)—1907 Ma, 1908 Ma, 1911 Ma (2 entries), 1912 Ma, 1923 Ma, 1925 Ma
Brown, Alexander *Crum* (1838–1922, Scottish)—1864 C
Brown, Robert (1773–1858, Scottish)—1810 B, 1827 P, 1831 B, 1905 P, 1908 P
Brownlee, John (1868–1927, English)—1917 H
Brückner, Eduard (1862–1927, German-Austrian)—1887 E, 1903–1909 E, 1924 E
Brunfels, Otto (c.1489–1534, German)—1530 B
Brunhes, Bernard (French)—1906 E
Bruno, Giordano (1548–1600, Italian)—1584 A, 1600 S
Bruno of Longoburgo (fl. c.1250, Italian)—1252 H
Brunschwig (or Brunswyck or Brunschwygk), Hieronymus (c.1450–c.1512, German-French)—1500 C/H, 1512 C
Buache, Philippe (1700–1773, French)—1737 E
Buat, Louis Gabriel Comte du (1734–1809, French)—1786 E
Buch, Christian *Leopold* von (1774–1853, German)—1802–1803 E, 1809 E, 1810 E, 1825 E, 1826 E, 1830 B/E
Buchan, Alexander (1829–1907, Scottish)—1869 Met, 1895 E
Buchner, Eduard (1860–1917, German)—1897 B

Buckland, William (1784–1856, English)—1820 E, 1823 E, 1840 E
Buffon, Georges Louis Leclerc, Comte de (1707–1788, French)—1749 B, 1749 E, 1749–1788 B/E, 1753–1767 B, 1777 Ma, 1778 E, 1781–1786 B/E, 1785 A, 1788–1804 B
Bunsen, Robert Wilhelm Eberhard (1811–1899, German)—1843 C, 1855 C, 1859 A, 1859 C, 1860 C, 1861 C, 1887 C
Burali-Forti, Cesare (1861–1931, Italian)—1897 Ma
Burbank, Luther (1849–1926, American)—1875 B
Burdach, Karl Friedrich (1776–1847, German)—1800 B, 1802 B
Buridan, Jean (Johannus Buridanus, c.1295–c.1358, French)—c.1350 A, c.1350 P
Burnet, Thomas (1635–1715, English)—1681 E
Burton, John (1710–1771, English)—1751 H
Busch, August Ludwig (1804–1855, German)—1851 A
Bussey, William Henry (b.1879, American)—1906 Ma
Bussy, Antoine Alexandre Brutus (1794–1882, French)—1828 C, 1831 C
Butler, Charles (1559–1647, English)—1609 B
Butlerov, Aleksandr Mikhailovich (1828–1886, Russian)—1864 C
Buys Ballot, Christoph Hendrik Diederik (Didericus) (1817–1890, Dutch)—1857 Met, 1858 C/P, 1875 Met
Bylot, Robert (English)—1616 E
Byron, Augusta Ada (Countess of Lovelace, 1815–1852, English)—1833 Ma

C

Cadwalader, Thomas (1708–1799, American)—1745 H
Cagniard de la Tour, Charles (1777–1859, French)—1822 P, 1836 B
Cailletet, Louis Paul (1832–1913, French)—1877 C
Calandri, Filippo (fifteenth century, Italian)—1491 Ma
Calcar, Jan Stevenszoon van (1499–c.1546, Italian)—1543 B
Calvin, John (1509–1564, French)—1553 B
Camerarius, Rudolph Jakob (1665–1721, German)—1694 B
Camillus, Leonardus—1502 E
Campanus of Novara, Johannes (d.1296, Italian)—c.1255 Ma, 1482 Ma
Camparella, Tommaso (1568–1639, Italian)—1623 S
Campbell, William Wallace (1862–1938, American)—1857 A, 1895 A
Camper, Peter (1722–1789, Dutch)—1760–1762 B
Cannizzaro, Stanislao (1826–1910, Italian)—1860 C, 1864 C
Cannon, Annie Jump (1863–1941, American)—1901 A, 1918–1924 A, 1925–1936 A
Cantor, Georg Ferdinand Ludwig Philipp (1845–1918, German)—1872 Ma, 1874 Ma, 1878 Ma (2 entries), 1879 Ma, 1883 Ma (2 entries), 1884 Ma, 1887 Ma, 1891 Ma, 1895–1897 Ma, 1897 Ma, 1900 Ma, 1902 Ma, 1906 Ma, 1907 Ma
Carcavi, Pierre de (c.1600–1684, French)—1659 Ma
Cardano, Girolamo (1501–1576, Italian)—1539 P, 1545 Ma, c. 1545 Ma, 1546 Ma, 1550 B, 1550 Met, c.1550 C, 1683 Ma
Carlisle, Anthony (1768–1840, English)—1800 C
Carnot, Lazare-Nicolas-Marguerite (1753–1823, French)—1797 Ma, 1803 Ma, 1803 P
Carnot, Nicolas Léonard *Sadi* (1796–1832, French)—1824 P, 1849 P, 1850 P, 1851 P
Carpenter, William B. (1813–1885, Irish)—1868 E, 1869 E
Carrel, Alexis (1873–1944, French-American)—1922 H/B
Carrington—1906 A
Carrington, Richard Christopher (1826–1875, English)—1853–1861 A, 1859 A
Casseri, Giulio (c.1552–1616, Italian)—1600–1601 B, 1609 B, 1626 B
Cassini, Gian (Giovanni) Domenico (Jean-Dominique Cassini, 1625–1712, Italian-French)—1668 A, 1671 A, 1672 A, 1675 A, 1684 A, 1866 A, 1888 A
Castelli, Benedetto (1578–1643, Italian)—1639 Met
Catherine II, Empress (1729–1796, Russian)—1768–1774 S
Cauchy, Augustin Louis (1789–1857, French)—1754 Ma, 1776 Ma, 1815 Ma, 1820–1830 Ma, 1821 Ma, 1825 Ma, 1826 Ma, 1833 Ma, 1836 Ma, 1837 Ma,

1843 Ma, 1847 Ma, 1851 Ma, 1854 Ma, 1872 Ma, 1876 Ma, 1893 Ma
Cavalieri, Francesco *Bonaventura* (1598–1647, Italian)—1635 Ma (2 entries), 1647 Ma, 1655 Ma
Cavendish, Henry (1731–1810, English)—1766 C, 1772 C, 1781 C, 1783 C, 1785 C/Met, 1789 C, 1798 E, 1798 P, 1894 C/Met
Caventou, Joseph Bienaimé (1795–1877, French)—1817 B, 1818 C
Cayley, Arthur (1821–1895 English)—1839 Ma, 1843 Ma, 1844 Ma, 1845 Ma, 1849 Ma, 1854 Ma, 1857–1858 Ma, 1859 Ma, 1870 Ma, 1871 Ma, 1877 Ma, 1878 Ma
Cayley, George (1773–1857, English)—1809–1854 P
Caxton, William (c.1421–1491, English)—1476 S
Celsius, Anders (1701–1744, Swedish)—1694 P, 1742 P, 1743 P
Cesalpino, Andrea (1519–1603, Italian)—1583 B, 1694 B, 1737 B, 1749 B
Ceulen, Ludolph van (1540–1610, Dutch)—1596 Ma
Chadwick, James (1891–1974, English)—1914 P, 1921–1924 P
Chain, Ernst Boris (1906–1979, German)—1928 H
Chamberlen, Hugh (English)—1696 H
Chamberlen, Peter (1572–1626, English)—1696 H
Chamberlin, Thomas Chrowder (1843–1928, American)—1894 E, 1899 E (2 entries)
Chambers, Robert (1802–1871, Scottish)—1844 B, 1852–1855 E
Chamié, C.—1924 P
Chamisso, Adelbert von (1781–1838, German)—1818 E
Chancourtois, Alexandre Béguyer de—*see* Béguyer de Chancourtois
Chandler, Seth Carlo (1846–1913, American)—1752 E, 1891 E
Charles, Jacques-Alexandre-César (1746–1823, French)—1783 Met, 1787 C, 1801 C, 1802 C, 1834 C
Charles II (1630–1685, English)—1660 S, 1661 Met
Charpentier, Johann (Jean) de (1786–1855, German)—1834 E, 1835 E, 1836 E
Châtelet, Gabrielle-Émile le Tonnelier de Breteuil, Marquise du (1706–1749, French)—1759 A/P
Chaucer, Geoffrey (c.1343–1400, English)—1391 A, 1392 A
Chauliac, Guy de (c.1290–c. 1368, French)—1348 H, 1360 H
Chebyshev (Chebycheff, Tchebycheff), Pafnuty (Pafnuti) Lvovich (1821–1894, Russian)—1845 Ma, 1850 Ma (2 entries)
Cheseaux, Jean Philippe Loÿs de—*see* Loÿs de Cheseaux, Jean-Philippe
Chevreul, Michel Eugène (1786–1889, French)—1823 C
Chladni, Ernst Florenz Friedrich (1756–1827, German)—1787 P, 1794 A
Christin, Jean Pierre (French)—1743 P
Christoffel, Elwin Bruno (1829–1900, German)—1861 P/Ma, 1869 Ma
Clairaut, Alexis Claude (1713–1765, French)—1731 Ma, 1743 E/A, 1757 A, 1758 A
Clapeyron, Benoit Pierre (Paul) *Émile* (1799–1864, French)—1834 C
Clark, Alvan (1804–1887, American)—1862 A
Clark, Alvan Graham (1832–1897, American)—1862 A
Clausius, Rudolf Julius Emanuel (1822–1888, German)—1850 P, 1854 C, 1857 C/P, 1858 C/P
Clavius, Christoph (1537–1612, Italian)—1593 Ma
Clément, Nicholas (1779–1841, French)—1806 C, 1811 C
Cleve, Per Teodor (1840–1905, Swedish)—1879 C, 1895 C
Clifford, William Kingdon (1845–1879, English)—1870 P, 1873–1876 Ma, 1877 Ma
Clusius, Carolus—*see* L'Écluse, Charles de
Colding, Ludwig August (1815–1888, Danish)—1842 P, 1871 Met, 1872 E
Collins, John (1625–1683, English)—1669 Ma
Collip, James Bertram (1892–1965, Canadian)—1921 H
Collomb, Edouard (1801–1875, French)—1847 E
Colombo, Realdo (Realdus Columbus, c.1510–1559, Italian)—1555 B, 1556 B, 1559 B (2 entries)
Colonne, Marie-Pompée (Italian)—1734 E
Columbus, Christopher (1451–1506, Italian)—c.1450 E, 1492 E, 1500 E
Columbus, Realdus—*see* Colombo, Realdo
Commandino, Federico (1509–1575, Italian)—1558 Ma, 1565 Ma
Compton, Arthur Holly (1892–1962, American)—1923 P
Comte, Auguste (1798–1857, French)—1830–1842 S, 1840–1842 S

Condamine, Charles Marie de la (1701–1774, French)—1735–1744 E
Condillac, Étienne Bonnet, Abbé de (1714–1780, French)—1754 B, 1780 C
Conrad, Timothy Abbott (1803–1877, American)—1832 E, 1839 E
Conti, Niccolo de (c.1469, Italian)—1442 E
Cook, James (1728–1779, English)—1768–1771 B, 1769 A
Cooke, William Fothergill (1806–1879, English)—1837 P
Cope, Edward Drinker (1840–1897, American)—1870 B, 1883 E
Copernicus, Nicolaus (1473–1543, Polish)—1514 A, 1540 A, 1543 A (2 entries), 1543 P, 1551 A, 1576 A, 1584 A, 1588 A, 1600 A, 1605 A, 1609 A, 1613 A, 1616 S, 1627 A, 1632 A, 1633 S, 1638 S, 1640 A, 1647 A, 1651 A, 1675 P/A, 1690 A, 1781 A, 1835 S
Cordus, Valerius (1515–1544, German)—1535 H, 1540 C, 1544 B
Coriolis, Gaspard Gustave de (1792–1843, French)—1831 Met, 1835 Met, 1855 Met, 1857 Met, 1875 Met, 1878 E, 1902 E, 1905 E, 1905 Met
Correns, Carl Franz Joseph Erich (1864–1933, German)—1900 B
Corvisart, Jean Nicolas (1755–1821, French)—1761 H
Cosa, Juan de la (c.1460–1510, Spanish)—1500 E
Coster, Dirk (1889–1950, Dutch)—1923 C
Cotes, Roger (1682–1716, English)—1714 Ma, 1730 Ma
Cotton, John (1584–1652, American)—1654 S
Coulier, Paul Jean (French)—1875 Met
Coulomb, Charles Augustin de (1736–1806, French)—1777 P, 1784 P, 1832 P/E
Couper, Archibald Scott (1831–1892, Scottish)—1858 C
Courant, Richard (1888–1972, German-American)—1929 Ma
Courtois, Bernard (1777–1838, French)—1811 C
Cranston, John Arnold—1917 C
Crawford, Adair (1748–1795, English)—1779 B/C, 1783 C, 1786 C
Crelle, August Leopold (1780–1855, German)—1826 Ma, 1828 P/Ma
Cremona, Luigi (1830–1903, Italian)—1863 Ma
Crescenzi, Pietro (Peter of Crescenzi, 1230–c.1310, Italian)—1306 B
Croll, James (1821–1890, Scottish)—1864 E, 1867 E, 1874 E, 1875 E, 1876 E, 1894 E
Cronstedt, Axel Fredrik (1722–1765, Swedish)—1751 C
Crookes, William (1832–1919, English)—1861 C, 1873–1876 Met, 1886 C (2 entries), 1895 P (2 entries), 1900 C/P
Crosthwait, Joseph (English)—1725 A
Cruikshank, William Cumberland (1745–1800, English)—1800 H
Cullen, William (1710–1790, Scottish)—1755 Met
Curie, Irene (1897–1956, French)—1924 P
Curie, Marie (Maria Sklodowska, 1867–1934, Polish-French)—1896 P, 1897 P, 1898 C (3 entries), 1902 P/C, 1910 C
Curie, Pierre (1859–1906, French)—1898 C, 1902 P/C.
Cusa, Nicholas (Nicholas of Cusa, Nicolaus Cusanus, c.1401–1464, German)—1440 A, c.1450 H, c.1450 Met, c.1450 S, 1463 P
Cuvier, Georges Chrétien Leopold (1769–1832, French)—1726 E, 1798 B, 1808 E, 1811 E, 1812 E (2 entries), 1813 E, 1817 B, 1830 B/E

D

Daguerre, Louis Jacques Mande (1789–1851, French)—1824 C, 1839 C
D'Alembert, Jean le Rond—*see* Alembert, Jean le Rond d'
Dalin, O.—1745 H
Dalton, John (1766–1844, English)—1758 C, 1789 C, 1793 Met, 1794 H, 1801 C/Met, 1801 C, 1802 C, 1802 Met, 1803 C/P, 1803 C, 1804 C, 1807 C, 1808 C, 1809 C, 1810–1820 C, 1811 C, 1813 C, 1818 C, 1844 C
Damiens, A.—1920 B
Dampier, William (1652–1715, English)—1699 E/Met
Dana, James Dwight (1813–1895, American)—1846 B/E, 1846 E, 1847 E, 1862 E, 1894 E
Daniell, John Frederic (1790–1845, English)—1836 P, 1837 P
Daniell, P. J. (b.1889, English)—1916 Ma
Dante Alighieri (1265–1321, Italian)—1320 E
Darboux, Jean Gaston (1842–1917, French)—1887–1896 Ma
Darwin, Charles Robert (1809–1882, English)—1789 C, 1831 B, 1831–1836 E/B, 1836 E, 1842 E, 1844 B, 1844 E,

1846 B/E, 1852 B, 1858 B, 1859 B (2 entries), 1860 B, 1861 B, 1862 B, 1867 B, 1868 B, 1869 B, 1871 B, 1872 B, 1875 B (2 entries), 1876 B, 1880 B, 1881 B, 1886 B, 1900 B (2 entries), 1914 B, 1925 S, 1930 B
Darwin, Erasmus (1731–1802, English)—1794 B
Darwin, George Howard (1845–1912, English)—1878 A, 1879 A, 1880 A, 1882 E/A
Daubree, Gabriel Auguste (1814–1896, French)—1862 E, 1866 E, 1878 E
Davis, Marguerite—1913 H, 1914 H, 1915 H
Davis, William Morris (1850–1934, American)—1889 E, 1890 E, 1894 Met, 1898 E
Davisson, Clinton Joseph (1881–1958, American)—1927 P, 1928 P
Davy, Edmund William (1785–1851, English)—1836 C
Davy, Humphry (1778–1829, English)—1748 C/P, 1758 C, 1799 P, 1800 C/H, 1802 C, 1807 C, 1808 C (2 entries), 1810 C (2 entries), 1811 C, 1812 C, 1816 C, 1818 C, 1821 P, 1831 C, 1871 P, 1879 C
Dax, Marc (d.1837, French)—1836 H/B, 1861 B
Debierne, André Louis (1874–1949, French)—1899 C, 1910 C
DeBort, Teisserenc (1855–1913, French)—1902 Met
Debye, Peter Joseph William (Wilhelm) (1884–1966, Dutch-American)—1912 C
Decker, Ezechiel de (fl. c.1630, Dutch)—1627 Ma
Dedekind, Julius Wilhelm *Richard* (1831–1916, German)—1833–1835 Ma, 1872 Ma, 1879 Ma, 1881 Ma, 1888 Ma, 1889 Ma, 1897 Ma
Dee, John (1527–1608, English)—1570 Ma
DeGeer, Gerard Jakob (1858–1943, Swedish)—1878 E
Dehn, Max (1878–1952, German)—1900 Ma
Delamain, Richard (died c.1645, English)—1630 Ma
Delambre, Jean Baptiste Joseph (1749–1822, French)—1791 P, 1808 A
Del Ferro, Scipio—*see* Ferro, Scipione
Della Porta, Giambattista—*see* Porta, Giambattista Della
Del Río, Andrés Manuel—see Río, Andrés Manuel del
DeLuc, Jean Andre (1727–1817, Swiss)—1773 Met
De Mairan, Jean Jacques—*see* Mairan, Jean Jacques D'Ortous de
Demarçay, Eugène Anatole (1852–1904, French)—1898 C, 1901 C
Democritus (c.460–c.370 B.C., Greek)—1473 C, 1623 A/S, 1750 A, 1803 C/P
De Moivre, Abraham—*see* Moivre, Abraham de
De Morgan, Augustus (1806–1871, English)—1831 Ma, 1847 Ma, 1860 Ma, 1874 Ma
Desaguliers, John Theophilus (Jean Théophile) (1683–1744, English)—1716 P
Desargues, Girard (Gérard) (1591–1661, French)—1635–1648 S, 1636–1639 Ma, 1639 Ma, 1648 Ma, 1685 Ma, 1825–1826 Ma
DeSaussure, Horace Bénédict—*see* Saussure, Horace Bénédict de
Descartes, René du Perron (1596–1650, French)—1613 P, 1629 Ma, 1635–1648 S, 1637 Ma (3 entries), 1637 Met, 1637 P, 1637 S, 1639 Ma, 1644 A, 1644 E, 1644 P (2 entries), 1644 S, 1659 P, 1664 B, 1669 P, 1685 Ma, 1707 Ma, 1752 Ma, 1803 Met, 1847 Ma
Deslandres, Henri Alexandre (1853–1948, French)—1890 A
Desmarest, Nicholas (1725–1815, French)—1752 E, 1756 E, 1765 E, 1774 E, 1775 E, 1815 E
Désormes, Charles Bernard (1777–1862, French)—1806 C, 1811 C
Dettonville, A. (Blaise Pascal, 1623–1662, French)—1659 Ma
Deville, Henri Étienne Sainte-Claire (1818–1881, French)—1854 C
de Vries, Hugo Marie—*see* Vries, Hugo Marie de
Dewar, James (1842–1923, Scottish)—1892 C, 1898 C, 1899 C, 1905 C
Dick, George Frederick (1881–1967, American)—1925 H
Dick, Gladys Henry (1881–1963, American)—1925 H
Diderot, Denis (1713–1784, French)—1751–1772 S, 1754 B
Diemerbroeck, Isbrand van (1609–1674, Dutch)—1646 H
Diesel, Rudolf Christian Karl (1858–1913, German)—1895 P
Dietrich von Freiberg (Theodoric of Freiburg or Freiberg) (c.1250–1311, German)—c.1300 P, 1304 Met/P
Digges, Thomas (c.1546–1595, English)—1576 A

Diophantus of Alexandria (fl. c.250 A.D., Alexandrian Greek)—1489 Ma, 1621 Ma, 1670 Ma, 1769–1770 Ma
Dirac, Paul Adrien Maurice (1902–1984, English-American)—1926 P, 1927 P
Dirichlet, Gustav Peter Lejeune (1805–1859, German)—1837 Ma (2 entries), 1839 Ma, 1879 Ma
Dixon, Jeremiah (1733–1799, English)—1761 A
Döbereiner, Johann Wolfgang (1780–1849, German)—1823 C, 1829 C
Dobson, G. M. B.—1922 Met
Dolland, John (1706–1761, English)—1757 A
Dolomieu, Gratet de (1750–1801, French)—1776 E, 1784 E
Dondi, Giacomo (1298–1359, Italian)—c.1350 E, c.1350 H
Doppler, Johann Christian (1803–1853, Austrian)—1842 P, 1848 A, 1868 A, 1912 A
Dorn, Friedrich Ernst (1848–1916, German)—1900 C
Douglass, Andrew Ellicott (1867–1962, American)—1919–1936 E
Douglass, William (c.1691–1752, American)—1736 H
Draper, Henry (1837–1882, American)—1872 A, 1880 A, 1918–1924 A
Draper, John William (1811–1882, American)—1839–1840 A, 1844 A, 1876 A
Dreyer, Johann Louis (Ludwig) Emil (1852–1926, Danish-English)—1864 A, 1888 A
Drude, Paul Karl Ludwig (1863–1906, German)—1900 P
Drummond, Thomas (1797–1840, English)—1825 C
Dubois de Chémant, Nicolas (1753–1824, French)—1788 H
DuBois-Reymond, Paul David Gustav (1831–1889, German)—1873 Ma (2 entries)
Duchâteau (French)—1788 H
Ducos du Haroun, Louis (1837–1920, French)—1869 C
DuFay, Charles François de Cisternai (1698–1739, French)—1734 P
Dufrénoy, Ours *Pierre Armand* (1792–1857, French)—1840 E
Dulong, Pierre Louis (1785–1838, French)—1813 C, 1819 C, 1912 C
Dumas, Jean-Baptiste-André (1800–1884, French)—1824 B, 1828 C, 1834 C (2 entries), 1837 C, 1840 C, 1844 B
Duner, Nils Christofer (1839–1914, Swedish)—1891 A
Duns Scotus, Joannes (John) (c.1266–1308, Scottish)—c.1300 S, c.1330 S.
Dutrochet, Rene Joachim *Henri* (1776–1847, French)—1826 C, 1837 B
Dutton, Clarence Edward (1841–1912, American)—1906 E

E

Earl, George Windsor—1845 E
Eberth, Carl Joseph (1835–1926, German)—1880 H
Eckart, Carl (1902–1971, American)—1926 P
Eddington, Arthur Stanley (1882–1944, English)—1919 P/A
Edison, Thomas Alva (1847–1931, American)—1876 P, 1877 P (2 entries), 1878 P, 1879 C
Edlefsen, Nils (1893–1971, American)—1930 P/C
Ehrenfest, Paul (1880–1933, Austrian)—1911 P
Ehrenfest, Tatyana (Russian)—1911 P
Ehrlich, Paul (1854–1915, German)—1907 H, 1910 H, 1912 H
Eijkman, Christiaan (1858–1930, Dutch)—1890–1896 H, 1901 H
Einstein, Albert (1879–1955, German-Swiss-American)—1801 P, 1859 A, 1872 P, 1892 P, 1904 P, 1905 P (4 entries), 1908 P, 1911 P/A, 1913 P/A, 1915 P, 1916 A (2 entries), 1916 P/A, 1917 A (2 entries), 1918 Ma/P, 1919 A, 1919 P, 1919 P/A, 1922 A, 1925 A, 1929 P
Einthoven, Willem (1860–1927, Dutch)—1903 H
Eisenhart, Luther Pfahler (1876–1965, American)—1922 Ma
Eisenstein, Ferdinand Gotthold Max (1823–1852, German)—1844 Ma (2 entries), 1847 Ma, 1850 Ma
Ekeberg, Anders Gustaf (1767–1813, Swedish)—1802 C
Ekman, Vagn Walfrid (1874–1954, Swedish)—1878 E, 1902 E, 1905 E, 1905 Met, 1923 E, 1926 E
Elcano, Juan Sebastian de (d.1526, Spanish)—1522 E
Elhuyar (or Elhuyart), Fausto (or Don Fausto) d' (1755–1833, Spanish)—1783 C
Elhuyar (or Elhuyart), Juan José d' (1754–1796, Spanish)—1783 C
Eliot, Jared (1685–1763, American)—1748 B

Ellis, Henry (1721–1806, English)—1751 E, 1797 E
Ellis, *Henry Havelock* (1859–1939, English)—1897–1928 B
Elster, Johann Philipp Ludwig *Julius* (1854–1920, German)—1899 C/P, 1901–1906 E
Empedocles of Acragas (c.492–c.432 B.C., Greek)—1754 B
Encke, Johann Franz (1791–1865, German)—1818 A, 1824 A
Eötvös, Roland, Baron von (1848–1919, Hungarian)—1891 P
Epicurus (341–270 B.C., Greek)—1473 C, 1626 C, c.1640 C
Espy, James (1785–1860, American)—1841 Met
Euclid (c.330–c.275 B.C., Greek)—1220 Ma, c.1250 Ma, c.1255 Ma, 1482 Ma, 1570 Ma, 1637 Ma, 1637 S, 1687 Ma, 1733 Ma, 1766 Ma, 1781 S, 1786 Ma, 1795 Ma, 1796 Ma, 1816 Ma, 1823 Ma, 1826 Ma, 1854 Ma, 1871 Ma, 1873 Ma, 1899 Ma, 1900 Ma, 1911 Ma, 1916 P/A, 1928 Ma
Euler, Leonhard (1707–1783, Swiss)—1640 Ma (2 entries), 1706 Ma, 1720 Ma, 1727–1728 Ma, 1729 Ma, 1732 Ma, 1735 Ma, 1736 Ma (2 entries), 1736 P, 1737 Ma (2 entries), 1738 P, 1739 Ma, 1740 Ma, 1742 Ma, 1744 Ma (4 entries), 1744 P, 1746 Ma, 1748 Ma (2 entries), c.1750 Ma (2 entries), 1752 E, 1752 Ma, 1755 Ma, 1755 Met/P, 1761 Ma (2 entries), 1761 P/E, 1765 P, 1768–1770 Ma, 1770 Ma, 1774 Ma, 1775 P, 1788 P (2 entries), 1795 Ma, 1852 Ma, 1899 Ma
Eustachi (Eustachio), Bartolomeo (c. 1510–1574, Italian)—1552 B, 1563 B, 1564 B
Evans, Herbert McLean (1882–1971, American)—1922 H, 1923 H
Evelyn, John (1620–1706, English)—1661 Met
Ewing, James Alfred (1855–1935, Scottish)—1883 E

F

Fabrici, Girolamo (or Hieronymus Fabricius) (c.1533–1619, Italian)—1603 B, 1604 B, 1621 B, 1628 B
Fabricius, Hieronymus—*see* Fabrici, Girolamo
Fabricius, Johan Christian (1745–1808, Danish)—1775 B
Fagnano dei Toschi, Giulio Carlo (1682–1766, Italian)—1716 Ma
Fahlberg, Constantine—1879 C
Fahrenheit, Daniel Gabriel (1686–1736, German)—1714 P, 1760 C
Fajans, Kasimir (1887–1975, Polish-American)—1913 C (2 entries), 1917 C
Fallopio, Gabriele (1523–1562, Italian)—1557 E, c.1560 B
Faraday, Michael (1791–1867, English)—1748 C/P, 1758 C, 1820 C, 1821 P, 1823 C, 1825 C (2 entries), 1831 P, 1833 C, 1833 P, 1844 C, 1845 P (2 entries), 1846 P, 1855 P, 1873 P
Fauchard, Pierre (1678–1761, French)—1728 H
Faventies, Valerius—1561 E
Favre, Alphonse (1815–1890, Swiss)—1880 E
Ferdinand II, Grand Duke of Tuscany (1610–1670, Italian)—1657 S/Met
Fermat, Pierre de (1601–1665, French)—1621 Ma, 1629 Ma, 1630–1665 Ma, 1637 Ma, 1638 Ma, 1640 Ma (2 entries), 1654 Ma, 1657 Ma, 1657 P, 1658 Ma, 1659 Ma, 1670 Ma, 1679 Ma, 1732 Ma, 1736 Ma, 1770 Ma, 1819 Ma, 1830–1832 Ma, 1849 Ma, 1880 Ma, 1905 Ma, 1909 Ma
Fernel, Jean François (1497–1558, French)—1554 H
Ferrari, Ludovico (1522–1565, Italian)—1544 Ma, 1545 Ma, 1770 Ma
Ferraris, Galileo (1847–1897, Italian)—1885 P
Ferrel, William (1817–1891, American)—1835 Met, 1855 Met, 1856 Met/E, 1875 Met, 1884 Met, 1886 Met
Ferro, Scipione (Scipio Del Ferro, 1465–1526, Italian)—1515 Ma, 1535 Ma
Fibonacci, Leonardo (Leonardo of Pisa, c.1170–c.1250, Italian)—1202 Ma, 1220 Ma, c.1220 Ma, 1225 Ma
Finlay, Carlos Juan (1833–1915, Cuban)—1881 H, 1900 H
Fisher, Osmond—1882 E/A
Fisher, Ronald Alymer (1890–1962, English)—1914 B, 1930 B
Fitzgerald, George Francis (1851–1901, Irish)—1892 P
Fizeau, Armand Hippolyte Louis (1819–1896, French)—1845 A, 1848 A, 1849 P, 1851 P, 1868 A, 1912 A
Flamsteed, John (1646–1719, English)—1712 A, 1725 A, 1798 A
Fleck, Alexander (b.1889, English)—1917 C
Fleming, Alexander (1881–1955, Scottish)—1922 H/B, 1928 H

Florey, Howard Walter (b.1898, Australian)—1928 H
Floyer, John (1649–1734, English)—1697 H, 1707 H, 1724 H, 1726 H
Forbes, Edward, Jr. (1815–1854, English)—1843 E, 1846 E
Forchhammer, Johan Georg (1794–1865, Danish)—1865 E
Fothergill, John (1712–1780, English)—1748 H
Foucault, Jean Bernard Léon (1819–1868, French)—1845 A, 1851 E, 1852 P
Fourcroy, Antoine François de (1755–1809, French)—1787 C, 1790 C, 1792 C, 1794 C, 1806 C, 1808 B
Fourier, Jean Baptiste Joseph (1768–1830, French)—1728 Ma, 1807 Ma, 1822 Ma, 1822 P, 1822 P/Ma, 1827 P, 1854 Ma, 1873 Ma, 1897–1906 A/E/Met, 1917 H
Fracastoro (Fracastorius), Girolamo (c.1478–1553, Italian)—1530 H, 1546 H
Fraenkel, Adolf *Abraham* (1891–1965, German)—1922 Ma
Franck, James (1882–1964, German)—1914 P
Frankland, Edward (1825–1899, English)—1852 C
Franklin, Benjamin (1706–1790, American)—1734 P, 1746 P, 1750 P, 1752 Met/P, 1753 Met/P, 1770 E, 1784 E
Fraunhofer, Joseph (Joseph von Fraunhofer, 1787–1826, German)—1814 A, 1859 A, 1870 A
Fréchet, Maurice Rene (1878–1973, French)—1906 Ma
Frederick II, King of Denmark—1576 A
Frederick II of Hohenstaufen (1194–1250, Italian)—1238 B, 1240 H, c.1245 B
Freedman, Michael (American)—1904 Ma
Frege, Friedrich Ludwig *Gottlob* (1848–1925, German)—1879 Ma, 1884 Ma, 1893–1903 Ma
Freke, Henry—1851 B, 1861 B
Frend, William (1757–1841, English)—1796 Ma
Fresnel, Augustin Jean (1788–1827, French)—1818 P, 1819 P
Freud, Sigmund (1856–1939, Austrian)—1880–1882 H, 1893 H, 1895 H, 1901 H, 1916 H
Friedel, Charles (1832–1899, French)—1862 C
Friedmann, Aleksandr Aleksandrovich (1888–1925, Russian)—1919 A, 1922 A
Friedrichs, Kurt Otto (b.1901, German-American)—1929 Ma
Frisius, Reiner Gemma—*see* Gemma Frisius, Reiner
Frobenius, Georg Ferdinand (1849–1917, German)—1896–1903 Ma
Frosch, Paul (1860–1928, German)—1898 B/H
Fuchs, Leonhard (1501–1566, German)—1542 B
Füchsel, George Christian (1722–1773, German)—1762 E
Funk, Casimir (1884–1967, Polish-American)—1912 H

G

Gadolin, Johan (1760–1852, Finnish)—1794 C
Gahn, Johan Gottlieb (1745–1818, Swedish)—1769 C, 1774 C
Galen (129/130–199/200, Greek-Roman)—c.1260 B, 1537 H, 1543 B, 1553 B, 1555 B, 1628 B
Galilei, Galileo (1564–1642, Italian)—1328 P, 1537 P, c.1583 P, 1586 P, 1589–1592 P (2 entries), 1592 Met, 1603 H, 1603 S, 1604 P, 1609 A (2 entries), 1609 P, 1610 A (2 entries), c.1610 B, 1612 A, 1612 P, 1613 A, 1613 Met, 1613 P, 1616 S, 1620 S, 1623 A/S, 1632 A, 1633 S, 1637 A, 1638 Ma, 1638 Met, 1638 P, 1639 Met, 1641 P, 1644 P, 1644 S, 1655 A, 1656 P, 1657 P, 1738 P/E, 1750 A, 1835 S, 1892 A
Galileo—*see* Galilei, Galileo
Gall, Franz Joseph (1758–1828, Austrian)—1810–1819 B
Galle, Johann Gottfried (1812–1910, German)—1845 A, 1846 A
Galois, Evariste (1811–1832, French)—1824 Ma, 1830–1832 Ma, 1846 Ma
Galton, Francis (1822–1911, English)—1863 Met, 1869 B, 1869 Ma, 1875 B, 1885 B, 1889 B/Ma
Galvani, Luigi (1737–1798, Italian)—1780 B, 1791 B
Gama, Vasco da (c.1469–1524, Portuguese)—1500 E
Gamow, George (1904–1968, Russian-American)—1928 P, 1929 A, 1930 P
Gascoigne, William (c.1612–1644, English)—1641 A, 1666 A
Gassendi, Pierre (1592–1655, French)—1624 S, 1626 C, 1631 A, 1635–1648 S, c.1640 C, 1641 P
Gauss, Carl Friedrich (1777–1855, German)—1769–1770 Ma, 1794 Ma, 1795 Ma, 1796 Ma (2 entries), 1796–1814 Ma, 1799 Ma, 1800 Ma, 1801 A,

1801 Ma, 1802 A, 1806 Ma, 1809 A, 1809 Ma, 1812 Ma, 1816 Ma (2 entries), 1817 Ma, 1823 Ma, 1825 Ma, 1827 Ma, 1829 Ma, 1829 P, 1830 Ma, 1831 Ma (2 entries), 1832 Ma (2 entries), 1833 P/Ma, 1833 P, 1839 P, 1843 P, 1845–1850 Ma, 1849 Ma, 1854 Ma, 1871 Ma, 1900 Ma

Gay-Lussac, Joseph Louis (1778–1850, French)—1787 C, 1789 C, 1802 C, 1804 Met, 1808 C, 1809 C (3 entries), 1810 C, 1811 C, 1814 C (2 entries), 1815 C, 1823 C, 1828 C

Geber (Spanish)—c.1310 C

Geiger, Hans (Johannes) Wilhelm (1882–1945, German)—1908 P, 1911 P/C, 1928 P

Geikie, Archibald (1835–1924, Scottish)—1863 E, 1888 E

Geikie, James (1839–1915, Scottish)—1874 E

Geitel, F. K. *Hans* (or Hans Friedrich) (1855–1923, German)—1899 C/P, 1901–1906 E

Gellibrand, Henry (1597–1636, English)—1635 E

Gemma Frisius, Reiner (1508–1555, Flemish)—1533 Ma

Gentile da Foligno (d.1348, Italian)—1348 H

Geoffroy, Étienne Francois (1672–1731, French)—1718 C

George III (1738–1820, English)—1765 P/E

Gergonne, Joseph Diaz (1771–1859, French)—1810 Ma, 1825–1826 Ma

Gerhardt, Charles Frédéric (1816–1856, French)—1843 C

Germain, Sophie (1776–1831, French)—1816 P/Ma, 1821 P/Ma

Germer, Lester Halbert (1896–1971, American)—1927 P, 1928 P

Gerstner, Franz Joseph (1756–1832, German)—1802 E

Gesner, Konrad (1516–1565, German)—1551–1558 B, 1565 B

Giauque, William Francis (b.1895, American)—1929 C

Gibbes, Lewis Reeve (1810–1894, American)—1870–1875 C

Gibbs, *Josiah Willard* (1839–1903, American)—1870–1879 C, 1873 C, 1876–1878 C, 1881–1884 Ma

Giesel, Friedrich Oskar (Otto) (1852–1927, German)—1904–1905 C

Gilbert, Grove Karl (1843–1918, American)—1870–1879 E, 1875 E

Gilbert, William (1544–1603, English)—1600 A, 1600 E/P, 1609 E

Giles of Rome (Aegidus, 1247–1316, Italian)—c.1276 B, c.1290 A, c.1300 C, 1604 B

Girard, Albert (1595–1632, Dutch)—1629 Ma, 1799 Ma

Glisson, Francis (c.1597–1677, English)—1650 H

Goddard, Robert Hutchings (1882–1945, American)—1920 P, 1926 P

Gödel, Kurt (1906–1978, Austrian-American)—1930 Ma

Godin, Louis (1704–1760, French)—1735–1744 E

Goethe, Johann Wolfgang von (1749–1832, German)—1784 B, 1795 B/S, 1810 P

Gohring, O. H. (German)—1913 C

Goldbach, Christian (1690–1764, Prussian)—1742 Ma

Goldberger, Joseph (1874–1929, Austria-Hungarian-American)—1915 H, 1916 H, 1925 H, 1928 H

Goldblatt, Harry (b.1891, American)—1923 H, 1924 H

Goldstein, Eugen (1850–1930, German)—1876 P, 1886 P

Gomberg, Moses (1866–1947, Russian-American)—1900 C

Goodricke, John (1764–1786, English)—1782 A

Goodwin, Hannibal Williston (1822–1889, American)—1887 C

Goodyear, Charles (1800–1860, American)—1839 C

Gordan, Paul Albert (1837–1912, German)—1868 Ma, 1870 Ma, 1890 Ma, 1910 Ma

Goudsmit, Samuel Abraham (b.1902, Dutch-American)—1925 P

Graham, C. M.—1822 H

Graham, Thomas (1805–1869, Scottish)—1834 C

Grandi, Luigi *Guido* (1671–1742, Italian)—1703 Ma

Grant, Robert Edmond (1793–1874, Scottish)—1826 B

Grassi, Giovanni Battista (1854–1925, Italian)—1899 H

Grassmann, Hermann Günther (1809–1877, German)—1843 Ma, 1844 Ma, 1862 Ma, 1889 Ma

Graunt, John (1620–1674, English)—1662 Ma

Gray, Stephen (1666–1736, English)—1729 P, 1732 P

Greaves, John (1602–1652, English)—c.1646 E
Green, George (1793–1841, English)—1828 P/Ma
Greenough, George Bellas (1778–1855, English)—1819 E, 1840 E
Greenwood, Isaac (1702–1745, American)—1728 Met
Gregory (or Gregorie), James (1638–1675, Scottish)—1663 A, 1667 Ma, 1668 A, 1668 Ma, 1670 Ma, 1671 Ma
Gregory of Rimini (d.1358, Italian)—1344 Ma
Gregory of St. Vincent (1584–1667, French)—1668 Ma
Gregory IX, Pope (c.1143–1241, Italian)—1231 S
Gregory XIII, Pope (1502–1585, Italian)—1576 A, 1582 A
Grew, Nehemiah (1641–1712, English)—1682 B, 1695 H, 1704 S
Grijns, Gerrit (1865–1944, Dutch)—1901 H
Grimaldi, Francesco Maria (1618–1663, Italian)—1651 A, 1665 P, 1678 P
Gross, E.G.—1919 H
Grosseteste, Robert (c.1168–1253, English)—c.1220 P, c.1230 P, c.1230 S, c.1231–1235 P, c.1235 Met, 1277–1279 P, c.1330 S
Grove, William Robert (1811–1896, English)—1839 C
Gruner, Gottlieb Sigmund (1717–1778, Swiss)—1760 E
Guericke (Gericke), Otto von (1602–1686, German)—1645 P, c.1645 Met, 1654 Met/P, 1660 P, 1662 C, 1672 P (2 entries)
Guettard, Jean-Étienne (1715–1786, French)—1746 E, 1752 E, 1757 E, 1765 E, 1770 E, 1779 E
Guldberg, Cato Maximilian (1836–1902, Norwegian)—1803 C, 1867 C, 1876–1880 Met
Gunter, Edmund (1581–1626, English)—1620 Ma, 1622 P
Guppy, Henry Brougham (1854–1926, English)—1906 B
Gurney, Goldsworthy (1798–1875, English)—1825 C
Gutenberg, Beno (1889–1960, German-American)—1914 E
Gutenberg, Johann (c.1397–1468, German)—1289 S, 1440 S, 1456 S
Guthrie, Francis (d.1899)—1852 Ma
Guthrie, Samuel (1782-1848, American)—1831 C
Guyton de Morveau, Louis Bernard—*see* Morveau, Louis Bernard Guyton de

H

Haber, Fritz (1868–1934, German)—1908 C
Hadamard, Jacques Salomon (1865–1963, French)—1896 Ma
Hadley, George (1685–1768, English)—1686 Met, 1735 Met, 1747 Met, 1855 Met
Haeckel, Ernst Heinrich Philipp August (1834–1919, German)—1899 B
Hahn, Otto (1879–1968, German)—1905 C, 1907 C, 1917 C, 1921 C
Hahnemann, Christian Friedrich *Samuel* (1755–1843, German)—1810 H
Haldeman, Samuel (1812–1880, American)—1843–1844 B
Hale, George Ellery (1868–1938, American)—1890 A, 1908 A
Hales, Stephen (1677–1761, English)—c.1705 B, 1727 B, 1733 B
Hall, Asaph (1829–1907, American)—1877 A
Hall, Charles Martin (1863–1914, American)—1886 C
Hall, Chester Moor (1703–1771, English)—1733 A
Hall, Edwin Herbert (1855–1938, American)—1879 P
Hall, *Granville Stanley* (1846–1924, American)—1894 H
Hall, James (1761–1832, Scottish)—1798 E, 1805 E
Hall, James (1811–1898, American)—1847–1894 E
Haller, Albrecht von (1708–1777, Swiss)—1752 B, 1756 B
Halley, Edmond (Edmund) (1656–1743, English)—1443–1472 A, 1676–1678 A, 1677 A, 1678 A, 1679 A/P, 1685 A, 1686 Met, 1693 Ma, 1694 E, 1698–1700 E, 1701 E, 1704 S, 1705 A, 1712 A, 1715 A, 1715 E, 1716 A, 1718 A, 1734 Ma, 1735 Met, 1747 Met, 1758 A
Halloy, Jean Baptiste Julien d'Omalius d'—*see* Omalius d'Halloy, Jean Baptiste Julien d'
Halsted, William Stewart (1852–1922, American)—1883 H, 1890–1891 H
Ham, Johan—1677 H
Hamann, Christel Bernhard Julius (1870–1948, German)—1905 Ma
Hamilton, William Rowan (1805–1865, Irish)—1828 P, 1831 Ma, 1833 P, 1833–1835 Ma, 1837 Ma, 1843 Ma, 1844 Ma, 1853 Ma, 1866 Ma, 1873–1876 Ma
Hancock, Henry (1809–1880, English)—1848 H

Hankel, Hermann H. (1839–1873, German)—1870 Ma
Hansen, Gerhard Henrik Armauer (1841–1912, Norwegian)—1873 H
Hare, Robert (1781–1858, American)—1803 C
Haroun, Louis Ducos du—*see* Ducos du Haroun, Louis
Harriot, Thomas (1560–1621, English)—1588 E
Harris, Chapin Aaron (1806–1860, American)—1837 H
Harris, John (1667–1719, English)—1704 S
Harrison, John (1693–1776, English)—1765 P/E
Hartley, David (1705–1757, English)—1749 H
Harvey, William (1578–1657, English)—1603 B, 1616 B, 1628 B, 1651 B (2 entries), 1660 B
Harvey, William Henry (1811–1866, Irish)—1841 B
Hassenfratz, Jean Henri (1755–1827, French)—1787 C, 1813 C
Hata, Sahachiro (1873–1938, Japanese)—1910 H
Hatchett, Charles (1765–1847, English)—1801 C
Hauksbee, Francis (c.1666–1713, English)—1705 P, 1706 P, 1729 P
Hausdorff, Felix (1868–1942, German)—1914 Ma (2 entries)
Haüy, René Just (1743–1821, French)—1801 E, 1822 E
Ibn Al-Haytham, Abū 'Alī al-Hasan Ibn al-Hasan (*Alhazen*, 965–c.1040, Egyptian)—1268 B, 1277–1279 P, 1604 P
Heaviside, Oliver (1850–1925, English)—1893 P/Ma, 1902 Met
Hegel, Georg Wilhelm Friedrich (1770–1831, German)—1812–1816 Ma
Heine, Heinrich *Eduard* (1821–1881, German)—1872 Ma, 1923 Ma
Heisenberg, Werner Karl (1901–1976, German)—1925 P, 1926 P (2 entries), 1927 P
Heitler, Walter Heinrich (b.1904, German)—1927 C
Helland-Hansen, Björn (1877–1957, Norwegian)—1926 E
Helmholtz, Hermann Ludwig Ferdinand von (1821–1894, German)—1807 P, 1842 P, 1847 P, 1851 H, 1853 A, 1854 A, 1858 P, 1862 P, 1871 P, 1877 Ma, 1889 Met/E
Helmont, Johann (John) Baptista (Baptist) van (1579–1644, Belgian)—1624 C, 1648 B (2 entries), 1648 C, 1648 H, 1648 S
Hencke, Karl Ludwig (1793–1866, German)—1845 A
Henderson, Thomas (1798–1844, Scottish)—1839 A
Henle, Friedrich Gustav *Jacob* (1809–1885, German)—1840 H
Henlein, Peter (1480–1542, German)—1502 P
Henri de Mondeville (c.1260–c.1320, French)—1266 H, 1320 H
Henry, Joseph (1797–1878, American)—1825 P, 1832 P
Henry the Navigator (1394–1460, Portuguese)—1416–1434 E, c.1420 A/E
Henry, William (1774–1836, English)—1803 C, 1808
Hérelle, Félix d' (1873–1949, English)—1917 B/H
Hérigone, Pierre (died c.1643, French)—1637 Ma
Hermann, J.H.–1814 Ma
Hermite, Charles (1822–1901, French)—1844 Ma, 1849 Ma, 1854–1857 Ma, 1858 Ma, 1873 Ma
Héroult, Paul Louis Toussaint (1863–1914, French)—1886 C
Herschel, Caroline Lucretia (1750–1848, English)—1783 A, 1798 A, 1828 A
Herschel, John Frederick William (1792–1871, English)–1812 Ma, 1833 A, 1834–1837 A, 1864 A, 1888 A
Herschel, William (1738–1822, German-English)—1781 A (2 entries), 1783 A (2 entries), 1785 A (2 entries), 1789 A, 1800 P, 1801 P, 1828 A, 1833 A, 1906 A
Hertwig, Wilhelm August *Oscar* (1849–1922, German)—1875 B
Hertz, Gustav (1887–1950, German)—1914 P
Hertz, Heinrich Rudolf (1857–1894, German)—1886–1889 P, 1887 P, 1888 P, 1892 P, 1894 P
Hertzsprung, Ejnar (1873–1967, Danish)—1905–1907 A, 1908 A, 1913 A, 1914 A, 1922 A
Hess, Germain Henri (1802–1850, Russian)—1840 C
Hess, Victor Franz (1883–1964, Austrian)—1911 Met/P
Hevelius, Johannes (1611–1687, Polish)—1647 A, 1661 A, 1687 A
Hevesy, György (Georg von) (1885–1966, Hungarian)—1923 C

Hewson, William (1739–1774, English)—1771 B, 1774 B
Hickman, Henry Hill (1800–1830, English)—1824 H
Higgins, William (1762/1763–1825, Irish)—1789 C
Hilbert, David (1862–1943, German)—1888 Ma, 1890 Ma, 1894–1908 Ma, 1897 Ma, 1899 Ma, 1900 Ma, 1904 Ma, 1909 Ma, 1910 Ma, 1925 Ma (2 entries)
Hill, George William (1838–1914, American)—1877 Ma
Hillebrand, William Francis (1853–1925, American)—1895 C
Hire, Philippe de la (1640–1718, French)—1685 Ma
Hisinger, Wilhelm (1766–1852, Swedish)—1803 C (2 entries)
Hjelm, Peter Jacob (1746–1813, Swedish)—1781 C
Hodgkin, Thomas (1798–1866, English)—1832 H
Hofer, Hubert Franz (German)—1778 C
Hoff, Jacobus Hendricus Van't—*see* Van't Hoff, Jacobus Hendricus
Hofmann, August Wilhelm von (1818–1892, German)—1849 C
Holder, Otto Ludwig (1859–1937, German)—1882 Ma
Hollerith, Herman (1860–1929, American)—1880 Ma, 1890 Ma, 1924 Ma
Holmes, Oliver Wendell (1809–1894, American)—1843 H
Homberg, Wilhelm (or Guillaume or Willem)(1652–1715, German)—1700 C, 1702 C
Home, Francis (1719–1813, Scottish)—1765 H
Hooke, Robert (1635–1702, English)—1657 P, 1658 P, 1662–1666 P, 1663 Met, 1664 A (2 entries), 1665 B, 1665 C/Met, 1665 Met (2 entries), 1665 P, 1667 Met, 1674 A, 1677 B, 1678 P, 1679 A/P, 1684 A, 1685 Met, 1705 E, 1738 C
Hooker, Joseph Dalton (1817–1911, English)—1844–1847 B, 1859 B, 1875–1897 B
Hooker, William Jackson (1785–1865, English)—1838 B, 1846–1864 B
Hopkins, Frederick Gowland (1861–1947, English)—1906 H, 1912 H
Horrocks, Jeremiah (1618–1641, English)—1639 A
Houtermans, Fritz—1929 A
Howard, *Henry Eliot* (1873–1940, English)—1920 B
Howard, Luke (1772–1864, English)—1803 Met, 1832 Met
Hubble, Edwin Powell (1889–1953, American)—1923 A (2 entries), 1924–1935 A, 1925 A, 1926 A, 1929 A (2 entries)
Huddart, Joseph (1741–1811, English)—1777 H
Hudde, Jan (Johann; Johannes)(1628–1704, Dutch)—1657 Ma, 1658 Ma
Hudson, Henry (d.1611, English)—1609 E
Huggins, William (1824–1910, English)—1866 A, 1868 A
Hugh of Lucca (Italian)—1266 H
Humboldt, Friedrich Wilhelm Heinrich *Alexander von* (1769–1859, German)—1805–1834 Met/E, 1811 E, 1837–1848 E, 1845 Met, 1845–1862 S, 1852 E
Hume, David (1711–1776, Scottish)—1739 B
Hume, James—1637 Ma
Hunt, Thomas *Sterry* (1826–1892, American)—1855 E
Hunter, John (1728–1793, Scottish-English)—1771 H, 1774 B, 1778 H, 1785 H
Hunter, Matthew Albert (b.1878)—1910 C
Hutton, James (1726–1797, Scottish)—1785 E, 1794 E, 1795 E (5 entries), 1802 E (2 entries), 1830–1833 E, 1859 B
Huxley, Thomas Henry (1825–1895, English)—1859 B, 1869 B
Huygens (Huyghens), Christiaan (1629–1695, Dutch)—1612 A, 1655 A, 1656 P, 1656 P/Ma, 1657 Ma, 1659 A, 1659 P, 1661 P, 1669 P, 1672 P, 1673 Ma, 1673 P, 1675 A, 1678 P (2 entries), 1680 P, 1687 A/P, 1690 A, 1690 P, 1738 P/E, 1785 A
Hyatt, John Wesley (1837–1920, American)—1869 C
Hyatt, Isaiah Smith (American)—1869 C

I

Ibn al-Haytham—*see* alphabetized under H
Ibn al-Khwārizmī—*see* alphabetized under K
Ibn Rushd—*see* Averroës
Ingenhousz, Jan (1730–1799, Dutch)—1779 B
Ivanovsky (Ivanovski), *Dmitri* (Dmitrii) Iosifovich (1864–1920, Russian)—1892 B

J

Jackson, John Hughlings (1835–1911, English)—1827 H, 1863 H, 1864 H, 1868 H/B, 1876 H/B

Jacobi, Carl Gustav Jacob (1804–1851, German)—1786 Ma, 1829 Ma, 1832 Ma, 1834 Ma, 1837 Ma, 1839 C, 1841 Ma, 1850 Ma
Jacquard, Joseph Marie (1752–1834, French)—1805 Ma, 1833 Ma
James, William (1842–1910, American)—1890 B
Jameson, Robert (1774–1854, Scottish)—1808 E
Jamieson, Thomas (1829–1913, Scottish)—1865 E
Jansen, Zacharias (Joannides, 1588–c.1630, Dutch)—1608 A, 1609 B
Janssen, Pierre Jules César (1824–1907, French)—1868 A/C, 1868 A, 1895 C
Jeans, James Hopwood (1887–1946, English)—1919 A, 1930 S
Jeffreys, John *Gwyn* (1809–1885, Welsh)—1862–1869 B, 1869 E
Jenkin, Fleeming (1833–1885, Welsh)—1867 B
Jenner, Edward (1749–1823, English)—1796 H, 1798 H
Joannides—*see* Jansen, Zacharias
Johannes de Lineriis (French)—c.1340 Ma
Johannsen, Wilhelm Ludvig (Ludwig) (1857–1927, Dutch)—c.1903 B, 1909 B
John of Glogau (c.1305–1377, Silesian)—1348 H
John of Holywood (Johannes de Sacrobosco, c.1200–1256, Irish)—c.1240 Ma
John Philoponus (Johannes Philoponus, fl.c.500 A.D., Greek)—c.1280 P, c.1350 P
John XXII, Pope (1244–1334, French)—1317 S
Johnston, Herrick Lee (b.1898, American)—1929 C
Jones, William (1675–1749, English)—1706 Ma
Jordan, Camille (1838–1921, French)—1870 Ma, 1887 Ma, 1911 Ma
Jordan, Pascual (1902–1980, German)—1925 P, 1926 P (2 entries)
Jordanus de Nemore (Nemorarius, fl.c.1220, French)—c.1220 E/Ma, c.1220 Ma, c.1220 P
José de Acosta—*see* Acosta, José de
Joule, James Prescott (1818–1889, English)—1841 P, 1842 P, 1843 P, 1847 P, 1849 P, 1850 P, 1851 C/P, 1852 C, 1877 C, 1898 C
Jung, Carl Gustav (1875–1961, Swiss)—1912 B
Jungius, Joachim (1587–1657, German)—1608 S
Jurin, James (English)—1723 Met, 1728 Met
Jussieu, Antoine Laurent de (1748–1836, French)—1789 B
Jussieu, Bernard de (1699–1777, French)—1789 B

K

Kamerlingh-Onnes, Heike (1853–1926, Dutch)—1908 C, 1911 C
Kammerer, Paul (1881–1926, Austrian)—1919 S, c.1920 B
Kant, Immanuel (1724–1804, German)—1755 A, 1781 S, 1785 A, 1786 P
Kapteyn, Jacobus Cornelius (1851–1922, Dutch)—1904 A, 1906 A
Karl Theodor of Bavaria (German)—1780 Met
Kayser, Heinrich Johannes Gustav (1853–1940, German)—1895 C/Met
Keeler, James Edward (1857–1900, American)—1857 A, 1888 A, 1895 A
Kekule von Stradonitz, Friedrich *August* (1829–1896, German)—1858 C, 1861 C (2 entries), 1865 C
Kelly, William (1811–1888, American)—1856 C
Kelvin, Lord—*see* Thomson, William
Kendall, Edward Calvin (1886–1972, American)—1914 B
Kennelly, Arthur Edwin (1861–1939, Indian)—1902 Met
Kepler, Johannes (1571–1630, German)—1596 A, 1600 A, 1602 A, 1604 A, 1604 Ma, 1604 P, 1605 A, 1606 A, 1609 A (2 entries), 1609 A/P, 1609 Ma, 1611 A, 1615 Ma, 1619 A, 1620 A, 1621 A (2 entries), 1624 Ma, 1627 A, 1630 A, 1634 S, 1639 Ma, 1647 Ma, 1664–1666 A/P, 1666 A, 1672 A, 1679 A/P, 1687 A/P, 1927 A
Al-Khwārizmī, Abū Ja'far Muhammad Ibn Mūsā (c.800–c.850, Arabic)—c.1240 Ma
Al-Kindī, Abū Yūsuf Ya'Qūb Ibn Ishāq al-Sabbāh (c.801–c.873, Arabic)—1277–1279 P
Kircher, Athanasius (1602–1680, German)—1665 E
Kirchhoff, Gustav Robert (1824–1887, German)—1859 A, 1859 A/C, 1859 C, 1859 P, 1860 C, 1860 P, 1861 C, 1870 A
Kirchhoff, Konstantin Sigizmundovich (Gottlieb Sigismund Constantin) (1764–1833, Russian)—1812 C.

Kirkman, Thomas Penyngton (1806–1895, English)—1858 Ma
Kirkwood, Daniel (1814–1895, American)—1857 A, 1866 A
Kirwan, Richard (1733–1812, Irish)—1784 E, 1787 E, 1787 Met
Kitasato, Shibasaburo (1852–1931, Japanese)—1888–1891 H, 1894 H
Klaproth, Martin Heinrich (1743–1817, German)—1789 C, 1795 C, 1803 C
Klaus, Karl Karlovich (1796–1864, Estonian)—1844 C
Klebs, Edwin (1834–1913, German-American)—1883 H
Klein, Christian *Felix* (1849–1925, German)—1870 Ma, 1871 Ma, 1872 Ma, 1873 Ma, 1882 Ma, 1884 Ma
Kleist, Ewald Georg (Jürgens) von (c.1700–1748, German)—1745 P
Klügel, Georg Simon (1739–1812, German)—1763 Ma
Knight, Gowin (1713–1772, English)—1748 C/P, 1758 C
Koch, Heinrich Hermann *Robert* (1843–1910, German)—1876 H, 1877 H, 1882 H, 1883 H, 1897 H
Koller, Carl (1857–1944, American)—1884 H
Kölreuter, Joseph Gottlieb (1733–1806, German)—1763 B
Köppen, Wladimir Peter (1846–1940, Russian-German)—1918 E, 1924 E
Koster, Lauren Janszoon (c.1370–c.1440, Dutch)—1440 S
Kovalevsky, Sonya (Sofya Vasilyevna Kovalevskaya, 1850–1891, Russian)—1888 P/Ma
Kramer, J.G.H. (Austrian)—1720 H
Kronecker, Leopold (1823–1891, German)—1853 Ma, 1881 Ma, 1886 Ma, 1887 Ma, 1910 Ma
Kuhn, Bernard Friedrich (1762–1825, Swiss)—1787 E, 1794 E
Kummer, Ernst Eduard (1810–1893, German)—1846 Ma, 1849 Ma
Kürschák, József (1864–1933, Hungarian)—1913 Ma

L

LaCaille, Nicholas Louis de (1713–1762, French)—1750 A, 1751–1753 A, 1755 A
Lacépède, Bernard-Germain-Étienne de la ville-sur-illon, Comte de (1756–1825, French)—1788–1804 B
Laënnec, Théophile-*René*-Hyacinthe (1781–1826, French)—1816 H, 1819 H
Lagrange, Joseph Louis (1736–1813, French)—1759–1792 Ma, 1760 Ma, 1760 P, 1762 Ma, 1767 Ma, 1769–1770 Ma, 1770 Ma (3 entries), 1772 Ma/A, 1773 Ma, 1773 P, 1774 Ma, 1778 P, 1778 P/E, 1790 P, 1797 Ma, 1798 Ma (2 entries), 1799–1806 Ma, 1801 Ma, 1807 Ma, 1831 Ma
Laguerre, Edmond Nicolas (1834–1886, French)—1879 Ma
La Hire, Philippe de (1640–1718, French)—1685 Ma
Lalande, Joseph Jérôme Lefrançais de (1732–1807, French)—1751–1753 A, 1801 A
Lamarck, Jean Baptiste Pierre Antoine de Monet de (1744–1829, French)—1778 B, 1794 B, 1800 B, 1801 B, 1802 B (2 entries), 1802 E, 1809 B, 1816–1822 B, 1859 B, 1868 B, 1886 B
Lambert, Johann Heinrich (1728–1777, German-French)—1766 Ma, 1770 Ma, 1771 Met, 1774 Met, 1777 Met, 1786 Ma
LaMettrie, Julien Offray de (1709–1751, French)—1748 B
Lamy, Claude Auguste (1820–1878, French)—1862 C
Landen, John (1719–1790, English)—1758 Ma
Landry, F.—1880 Ma
Landsteiner, Karl (1868–1943, Austrian)—1900 H, 1927 H
Lanfranchi, Guido (d.1315, Italian)—c.1290 H, 1296 H
Langley, Samuel Pierpont (1834–1906, American)—1878 P
Langmuir, Irving (1881–1957, American)—1919–1921 C
Langworthy, Charles Ford (1864–1932, American)—1898 H
Laplace, Pierre Simon, Marquis de (1749–1827, French)—1760 P, 1776 E, 1777 Ma, 1783 C, 1785 A, 1785 M/A, 1786 C, 1787 A, 1788 A, 1790 P, 1796 A (2 entries), 1799–1825 A (2 entries), 1812 Ma, 1812 P, 1813 P, 1820 Ma, 1825 A, 1857 A, 1870 Ma
Lassell, William (1799–1880, English)—1848 A
Laue, Max von (1879–1960, German)—1912 P
Laugier, André (1770–1832, French)—1817 C
Laurent, Auguste (Augustin) (1807–1853, French)—1834 C
Laveran, Charles Louis Alphonse (1845–1922, French)—1880 H

Lavoisier, Antoine Laurent (1743–1794, French)—1700 C, 1769 C, 1772 A (2 entries), 1774 C (2 entries), 1775 C, 1777 C/Met, 1778 C, c.1778 B, 1780 C, 1781 C, 1782 C, 1783 C (2 entries), 1786 C, 1787 C, 1789 C (3 entries), 1790 C, 1790 P, 1794 C, 1794 S, 1797–1812 C, 1800 C, 1810 C
Lawes, John Bennet (1814–1900, English)—1842 C
Lawrence, Ernest Orlando (1901–1958, American)—1930 P/C
Lawson, Andrew Cowper (1861–1952, Scottish)—1893 E
Lawson, John (d.1711, English)—1709 B
Lazear, Jesse William (1866–1900, American)—1900 H
Leavitt, Henrietta Swan (1868–1921, American)—1908 A, 1912 A, 1923 A
Lebedev, Petr (Pyotr)Nikolaevich (1866–1912, Russian)—1892 A, 1899 P
LeBel, Joseph Achille (1847–1930, French)—1874 C
Lebesgue, Henri Leon (1875–1941, French)—1902 Ma, 1904 Ma, 1910 Ma
LeChâtelier, Henri Louis (1850–1936, French)—1885 P, 1892 P
L'Écluse, Charles de (Carolus Clusius, 1526–1609, French)—1576 B
Leeuwenhoek, Antony van (1632–1723, Dutch)—1668 B, 1673 B, 1676 B, 1677 B, 1680 B, 1683 B, 1692 B
Legendre, Adrien-Marie (1752–1833, French)—1784 Ma, 1786 Ma, 1794 Ma, 1795 Ma, 1798 Ma, 1806 Ma, 1812 Ma, 1825–1832 Ma, 1830 Ma, 1832 Ma, 1837 Ma
Lehmann, Johann Gottlob (1719–1767, German)—1753 E, 1756 E, 1762 E
Leibniz, Gottfried Wilhelm (1646–1716, German)—1658 Ma, 1666 Ma, 1668 Ma, 1670 Ma, 1673 Ma (2 entries), 1673–1676 Ma, 1677 Ma, 1679 Ma, 1682 P, 1684 Ma, 1686 Ma, 1687 A/P, 1690 Ma, 1692 Ma, c.1693 E, 1694 Ma, 1695 Ma, 1696 Ma, 1698 Ma, 1700 S, c.1700 Ma, 1702 Met, 1705 Ma, 1712 Ma, 1734 Ma, 1748 Ma, 1812 Ma
Lemaitre, Georges Edouard (1894–1966, Belgian)—1927 A
Lémery, Nicolas (1645–1715, French)—1675 C, 1697 C
Lemoine, Émile Michel Hyacinthe (1840–1912, French)—1888 Ma
Lenard, Philipp Eduard Anton (1862–1947, German)—1898 P, 1902 P, 1903 P
Lenz, Heinrich Friedrich Emil (1804–1865, Russian)—1834 P
Leo XIII, Pope (Gioacchino Pecci, 1810–1903, Italian)—1879 S
Leonardo da Vinci (1452–1519, Italian)—c.1500 S, c.1513 B
Leonardo of Pisa—*see* Fibonacci, Leonardo
LeRoy, Charles (1726–1779, French)—1751 Met (2 entries)
LeRoy, Pierre (1717–1785, French)—1765 P/E
Leslie, John (1766–1832, Scottish)—1800 Met, 1804 P
Leucippus (5th century B.C., Greek)—1803 C/P
Leverrier, M. (French)—1855 Met
LeVerrier, Urbain Jean Joseph (1811–1877, French)—1843 A, 1845 A, 1846 A, 1859 A, 1864 E, 1916 P/A
Levi ben Gerson (1288–1344, French)—1321 Ma
Levi-Civita, Tullio (1873–1941, Italian)—1917 Ma
Levine, Philip (b.1900, Russian-American)—1927 H
Lewis, G.N. (1875–1946, American)—1902 C, 1916 C
Lewy—1929 Ma
L'Hospital (L'Hôpital), Guillaume-François-Antoine de (1661–1704, French)—1696 Ma
Libau, Andreas (Libavius, 1540–1616, German)—1597 C/H, 1611–1613 C
Lie, Marius Sophus (1842–1899, Norwegian)—1884 Ma, 1888–1890 Ma
Liebig, Justus von (1803–1873, German)—1823 C, 1831 C, 1832 C, 1834 C (2 entries), 1837 C, 1843 C, 1855 C
Lightfoot, John (English)—1654 E
Lilio, Luigi (d.1576, Italian)—1576 A, 1582 A
Lind, James (1716–1794, Scottish)—1753 H
Linde, Carl von (1842–1934, German)—1877 C
Lindemann, Carl Louis *Ferdinand* (1852–1939, German)—1882 Ma
Lindemann, Frederick Alexander (1886–1957, English)—1922 Met
Linnaeus (or von Linné), Carolus (Carl) (1707–1778, Swedish)—1735 B, 1737 B (2 entries), 1749 B (2 entries), 1753 B, 1758 B, 1763 H, 1776 B (2 entries), 1779 C, 1784 C/E
Liouville, Joseph (1809–1882, French)—1836 Ma, 1844 Ma, 1846 Ma

Lippershey, Hans (1587–1619, Dutch)—1608 A, 1609 A, 1609 B
Lipschitz, Rudolph Otto Sigismund (1832–1903, German)—1820–1830 Ma, 1876 Ma, 1893 Ma
Lister, Joseph (1827–1912, English)—1865 H, 1867 H
Lister, Martin (1638–1712, English)—1671 E, 1678 E, 1683 E, 1684 Met
Listing, Johann Benedict (1806–1882, German)—1847 Ma, 1858 Ma
Lobachevsky, Nikolai Ivanovich (1792–1856, Russian)—1816 Ma, 1823 Ma, 1826 Ma, 1829 Ma, 1854 Ma, 1868 Ma, 1871 Ma
Locke, John (1632–1704, English)—1690 B, 1781 S
Lockyer, *Joseph Norman* (1836–1920, English)—1868 A/C, 1868 A, 1887 A, 1895 C
Lodge, Oliver Joseph (1851–1940, English)—1888 P, 1893 P
Loeffler (or Löffler), Friedrich August Johannes (1852–1915, German)—1883 H, 1898 B/H
Logan, William Edmond (1798–1875, Canadian)—1855 E
Lombroso, Cesare (1836–1909, Italian)—1876 B
Lomonosov, Mikhail Vasilievich (or Vasilyevich) (1711–1765, Russian)—1750 C, 1751 C, 1761 A
London, Fritz (1900–1954, German-American)—1927 C
Long, Crawford Williamson (1815–1878, American)—1842 H
Lonsdale, William (1794–1871, English)—1837 E, 1839 E
Lorentz, Hendrik Antoon (1853–1928, Dutch)—1880 P, 1892 P, 1904 P, 1905 P
Lorenz, Ludwig (Ludvig)Valentin (1829–1891, Danish)—1880 P
Lorenzo de Medici (1449–1492, Italian)—1469 S
Loschmidt, Johann *Joseph* (1821–1895, Austrian)—1861 C, 1865 C
Louis XIV (1638–1715, French)—1666 S
Lovits (Lowitz), Johann *Tobias* (1757–1804, German)—1785 C
Lowell, Percival (1855–1916, American)—1894 A, 1906 A (2 entries), 1930 A
Lower, Richard (1631–1691, English)—1665 H, 1669 B, 1670 B
Lowig, Carl (1803–1890, German)—1825 C
Lowitz, Tobias—*see* Lovits, Johann Tobias
Loÿs de Cheseaux, Jean-Philippe (1718–1751, Swiss)—1744 A (2 entries), 1826 A
Lucretius (c.95–c.55 B.C., Roman)—1473 C
Lukasiewicz, Jan (1878–1956, Polish)—1920–1921 Ma
Lüroth, Jacob (1844–1910, German)—1878 Ma
Lyell, Charles (1797–1875, English)—1830 E, 1830–1833 E, 1833 E, 1840 E, 1841 E, 1846 E, 1859 B, 1863 E, 1865 E
Lysenko, Trofim Denisovich (1898–1976, Russian)—1930 B

M

MacGillivray, William (1796–1852, Scottish)—1827–1839 B
Mach, Ernst (1838–1916, Czechoslovakian)—1872 P
Macintosh, Charles (1766–1842, Scottish)—1823 C
Maclaren, Charles (1782–1866, Scottish)—1839 E, 1841 E
Maclaurin, Colin (1698–1746, Scottish)—1715–1717 Ma, 1720 Ma, 1742 Ma
MacLeod, John James Rickard (1876–1935, Scottish)—1921 H
Maclure, William (1763–1840, American)—1817 E
Macquer, Pierre Joseph (1718–1784, French)—1749 C, 1751 C, 1766 C
Magellan, Ferdinand (c.1480–1521, Portuguese)—1521 E, 1522 E
Magendie, François (1783–1855, French)—1811 B, 1822 B
Magnus, Albertus—*see* Albertus Magnus
Magnus, Heinrich Gustav (1802–1870, German)—1835 P, 1837 B
Magnus, Ludwig Imanuel (1790–1861, German)—1833 Ma
Maillet, Benoit de (1656–1738, French)—1748 B, 1748 E (2 entries)
Mairan, Jean Jacques (or Jean Baptiste) D'Ortous (or Dortous) de (1678–1771, French)—1716 E, 1733 Met
Mallet, Robert (1810–1881, Irish)—1846 E
Malpighi, Marcello (1628–1694, Italian)—1660 B (2 entries), 1669 B, 1673 B, 1675–1679 B
Malthus, Thomas Robert (1766–1834, English)—1798 B, 1852 B, 1859 B
Malus, Étienne Louis (1775–1812, French)—1808 P
Manutius, Aldus (1450–1515, Italian)—1482–1515 S, 1498 S
Marcet, Alexander (1770–1822, Swiss-English)—1820 E

Marconi, Guglielmo (1874–1937, Italian)—1895 P
Marggraf, Andreas Sigismund (1709–1782, German)—1746 C, 1747 C, 1758 C
Maricourt, Marquis de—*see* Peter of Maricourt
Marignac, Jean Charles Galissard de (1817–1894, Swiss)—1878 C
Mariotte, Edme (1620–1684, French)—1660 B, 1676 C
Marius, Simon (Simon Mayr, 1573–1624, German)—1612 A
Marsden, Ernest—1911 P/C
Marsh, Othniel Charles (1831–1899, American)—1871 B, 1877 B, 1880 B, 1884 B
Marsili (or Marsigli), Luigi Fernando (1658–1730, Italian)—1725 E
Martius, Karl Friedrich Philipp von (1794–1868, German)—1828 B
Mascagni, Paolo (1755–1815, Italian)—1779 C, 1787 B
Maskelyne, Nevil (1732–1811, English)—1761 A, 1774 E, 1798 E
Mason, Charles (1728–1786, English)—1761 A, 1769 A
Mästlin, Michael (1550–1631, German)—1577 A, 1578 A, 1582 A
Mather, Cotton (1663–1728, American)—1721 H
Mathieu, Émile Léonard (1835–1890, French)—1868 Ma
Matthew of Paris—*see* Paris, Matthew
Matthew, Patrick—1831 B
Matthiessen, Augustus (1831–1870, English)—1855 C
Maunder, Annie Scott Dill Russell (1868–1947, English)—1898 A
Maunder, Edward *Walter* (1851–1928, English)—1874 A
Maupertuis, Pierre Louis Moreau de (1698–1759, French)—1736–1737 E, 1744 P, 1750 A, 1751 B, 1833 P
Maury, Matthew Fontaine (1806–1873, American)—1847 E, 1850 Met, 1853 E, 1854 E, 1855 E/Met, 1855 Met
Maxwell, James Clerk (1831–1879, Scottish)—1855 P, 1857 A, 1859 A, 1859 C/P, 1860s C, 1865 P, 1873 Ma, 1873 P, 1877 P, 1883 P, 1886–1889 P, 1888 P, 1895 A, 1899 P, 1918 Ma/P
Maybach, Wilhelm (1846–1929, German)—1892 P
Mayer, Johann *Tobias* (1723–1762, German)—1750 A, 1753 E/A
Mayer, Julius *Robert* (von) (1814–1878, German)—1840 B/P, 1842 P, 1848 A
Mayow, John (1641–1679, English)—1674 C/Met
Mayr, Simon—*see* Marius, Simon
McCollum, Elmer Verner (1879–1967, American)—1907 H, 1909 H, 1913 H, 1914 H, 1915 H, 1921 H, 1922 H
McCoy, Herbert Newby (1870–1945, American)—1904 C, 1907 C
Méchain, Pierre-François-André (1744–1804, French)—1784 A, 1791 P
Megenburg, Conrad von (1309–1374, German)—c.1360 E
Meitner, Lise (1878–1968, Austrian-Swedish)—1917 C
Mellanby, Edward (1884–1955, English)—1921 H, 1922 H
Mendel, Johann *Gregor* (1822–1884, Austrian)—1865 B, 1900 B, 1902 B, 1914 B, 1915 B, 1926 B
Mendel, Lafayette Benedict (1872–1935, American)—1909 H (2 entries), 1914 H, 1918 H
Mendeleev, Dmitry (Dmitri) Ivanovich (1834–1907, Russian)—1869 C (2 entries), 1870–1875 C, 1875 C, 1879 C, 1886 C
Menger, Karl (b.1902)—1923 Ma, 1928 Ma (2 entries)
Menghini, Vincenzo Antonio (1704–1759, Italian)—1745 B
Mercator, Gerardus (or Gerhard Kremer) (1512–1594, Flemish)—1541 E, 1568 E, 1594 E
Mercator, Nicolaus (Nicholas) (c.1619–1687, Flemish)—1668 Ma
Mercuriale, Girolamo (Italian)—1573 B
Mercuriali, Geronimo (1530–1606, Italian)—1572 H
Méré, Chevalier de (1610–1685, French)—1654 Ma
Merle, William (d.1347, English)—1337–1344 Met
Mersenne, Marin (1588–1648, French)—1635–1648 S, 1636 P, 1640 Ma
Merz, Alfred (1880–1925, German)—1922 E
Mesmer, Franz (Friedrich) Anton (1734–1815, Austrian)—1774 H
Messier, Charles Joseph (1730–1817, French)—1771 A, 1784 A
Meyer, *Julius Lothar* (1830–1895, German)—1864 C, 1869 C (2 entries), 1870–1875 C
Michael Scot (c.1175–c.1234, Scottish)—1217 A, c.1220 B

Michell, John (1724–1793, English)—1760 E, 1777 P, 1784 A, 1796 A, 1798 E
Michelson, Albert Abraham (1852–1931, American)—1878 P, 1879 P, 1881 P, 1887 P, 1891 A, 1892 P, 1892–1893 P, 1898 Ma, 1903 P, 1905 P, 1916 E, 1920 A
Milankovitch, Milutin (1879–1958, Yugoslavian)—1912–1913 E, 1920 E, 1924 E, 1930 E
Millikan, Robert Andrews (1868–1953, American)—1910 P, 1915 P
Milne, John (1850–1913, English)—1760 E, 1880 E
Minkowski, Hermann (1864–1909, Russian)—1896 Ma, 1907 P, 1908 P, 1916 P/A
Minot, George Richards (1885–1950, American)—1926 H
Mitchill, Samuel Latham (1764–1831, American)—1794 C
Mitscherlich, Eilhard (1794–1863, German)—1820 C
Möbius, August (Augustus) Ferdinand (1790–1868, German)—1827 Ma, 1830 Ma, 1858 Ma, 1863 Ma
Moerbeke, William of (c.1220–c.1286, Flemish)—1269 S
Mohn, Henrik (1835–1916, Norwegian)—1876–1880 Met
Mohorovičič, Andrija (1857–1936, Serbian)—1909 E
Mohr, Carl *Friedrich* (1806–1879, German)—1837 P
Moissan, Ferdinand-Frédéric-Henri (1852–1907, French)—1886 C, 1892 C, 1895 C
Moivre, Abraham de (1667–1754, French-English)—1718 Ma, 1730 Ma, 1733 Ma, 1748 Ma
Moller, L.—1929 E
Monceau, Henri-Louis Duhamel du (1700–1782, French)—1736 C (2 entries)
Mondeville, Henri—*see* Henri de Mondeville
Mondino De'Luzzi (c.1275–1326, Italian)—1316 B
Monge, Gaspard (1746–1818, French)—1781 Ma, 1790 P, 1795 Ma, 1799 Ma, 1807 Ma
Montagu, Edward (Earl of Sandwich, 1625–1672, English)—1661 E
Montagu, Mary Wortley (1689–1762, English)—1717 H
Montgolfier, Étienne Jacques de (1745–1799, French)—1783 Met
Montgolfier, Michel *Joseph* de (1740–1810, French)—1783 Met
Moore, Eliakim Hastings (1862–1932, American)—1893 Ma (2 entries), 1908 Ma
Morehead—1905 Ma, 1909 Ma
Morgagni, Giovanni Battista (1682–1771, Italian)—1761 H
Morgan, Thomas Hunt (1866–1945, American)—1910 B, 1915 B, 1916 B, 1926 B, 1927 B
Morley, Edward Williams (1838–1923, American)—1879 P, 1881 P, 1887 P, 1892 P, 1905 P
Morley, Frank (1860–1937, American)—1899 Ma
Moro, Antonio *Lazzaro* (1687–1764, Italian)—1740 E
Morton, William Thomas Green (1819–1868, American)—1842 H, 1846 H
Morveau, Louis Bernard Guyton de (1737–1816, French)—1782 C, 1787 C (2 entries), 1790 C, 1794 C, 1798 C, 1799 C, 1800 H
Mosander, Carl Gustaf (1797–1858, Swedish)—1839 C, 1841 C, 1843 C
Moseley, Henry Gwyn Jeffreys (1887–1915, English)—1869 C, 1913 P
Muir, Thomas (1844–1934, Scottish)—1906–1923 Ma
Müller, Franz (Ferenc) Joseph (Baron de Reichenstein, 1740–1825, Austrian)—1783 C
Muller, Hermann Joseph (1890–1967, American)—1915 B, 1927 B
Müller, Johannes—*see* Regiomontanus, Johannes
Muller, Johannes Peter (1801–1858, German)—1840 B
Müller, Walther—1928 P
Münster, Sebastian (1489–1552, German)—1544 E
Murchison, Roderick Impey (1792–1871, Scottish)—1835 E, 1837 E, 1838 E, 1839 E
Murphy, J.J.—1876 E
Murphy, William Parry (b.1892, American)—1926 H
Murray, John (1841–1914, English)—1891 E, 1895 E
Musschenbroek, Petrus van (1692–1761, Dutch)—1745 P

N

Nansen, Fridtjof (1861–1930, Norwegian)—1878 E, 1888 B, 1888–

1889 E, 1893–1896 E, 1900–1906 E, 1902 E (2 entries), 1905 E, 1926 E
Napier, John (1550–1617, Scottish)—c.1580 Ma, 1614 Ma, 1624 Ma
Nasir al-Din al-Tūsī (Nasir Eddin, 1201–1274, Persian)—c.1250 Ma, 1259 A, c.1281 A/Met
Navier, Claude-Louis-Marie-Henri (1785–1836, French)—1823 P, 1845 P
Needham, John Turberville (1713–1781, English)—1740 B, 1767 B
Nemorarius—*see* Jordanus de Nemore
Nemore, Jordanus de—*see* Jordanus de Nemore
Nernst, Herman *Walther* (1864–1941, German)—1906 C
Netto, Eugen (1846–1919, German)—1878 Ma
Neumann, John von (1903–1957, Hungarian-American)—1928 Ma
Neumayr, Melchior (1845–1890, German)—1875 B, 1883 E
Newcomen, Thomas (1663–1729, English)—1769–1772 P, 1776 P
Newlands, John Alexander Reina (1837–1898, English)—1864 C, 1869 C
Newton, Isaac (1642–1727, English)—1328 P, 1644 P, 1656 P, 1664–1666 A/P, 1664–1666 Ma, 1664–1666 P, 1665–1666 P, 1668 A, 1668 Ma, 1669 Ma, 1670 Ma, 1671 A, 1671 Ma (2 entries), 1672 P, 1674 A, 1675 P, 1676 Ma (2 entries), 1679 A/P, 1683 E, 1685 A, 1685 A/Ma, 1687 A/P, 1687 C, 1687 E/A, 1687 Ma, 1689 S, 1691 Ma, 1693 Ma, 1694 Ma, 1701 P, 1704 C, 1704 Ma (2 entries), 1704 P (2 entries), 1704 S, 1705 A, 1707 Ma, 1709 S, 1711 Ma, 1712 A, 1712 Ma, 1716 P, 1717 P, 1728 S, 1734 Ma, 1734 P, 1735 Ma, 1736 Ma, 1736 P, 1736–1737 E, 1738 C, 1738 P, 1742 Ma, 1744 E, 1747 Met, 1748 Ma, 1754 Ma, 1758 Ma, 1759 A/P, 1759 P, 1775 P, 1781 S, 1784 A, 1786 P, 1787 A, 1788 A, 1789 C, 1801 C/Met, 1801 P, 1801–1804 P, 1802 C, 1810 P, 1812 Ma, 1827 A, 1827 P, 1843 A, 1851 E, 1872 P, 1911 P/A, 1916 P/A
Nicholas of Cusa—*see* Cusa, Nicholas
Nicholson, William (1753–1815, English)—1800 C
Nicod, Jean (1893–1924, French)—1930 Ma
Nicol, William (c.1768–1851, Scottish)—1829 E/P, 1831 E
Nicolaier, Arthur (1862–1934, German)—1884 H
Nicomachus of Gerasa (fl. c.100)—c.1240 Ma
Niepce, Joseph Nicéphore (1765–1833, French)—1824 C, 1827 C
Nieuwentijt, Bernard (1654–1718, Dutch)—1694 Ma
Nilson, Lars Fredrik (1840–1899, Swedish)—1879 C
Nobel, Alfred Bernhard (1833–1896, Swedish)—1866 C
Nöbeling, A. Georg (b.1907)—1928 Ma
Nobili, Leopoldo (1784–1835, Italian)—1826 C, 1826 P
Noddack, Ida Eva Tacke (b.1896, German)—1925 C
Noddack, Walter Karl (1893–1960, German)—1925 C
Noether, Amalie *(Emmy)* (1882–1935, German)—1907 Ma, 1910 Ma
Noguchi, Seisaku *Hideyo* (1876–1928, Japanese)—1913 H, 1918 H
Nollet, Jean Antoine (1700–1770, French)—1746 P, 1748 C, 1748 P
Norman, Robert (fl. c.1590, English)—1581 E/P, 1600 E/P
Nufer, Jakob (Swiss)—1500 H

O

Ockham, William of (c.1285–1349, English)—c.1290 A, c.1320 Ma, c.1330 B, c.1330 P, c.1330 S
Oersted, Hans Christian (1777–1851, Danish)—1812 P, 1820 P (4 entries), 1825 C, 1831 P, 1833 P
Ohm, Georg Simon (1789–1854, German)—1825 P, 1827 P, 1900 P
Ohm, Martin (1792–1872, German)—1822 Ma
Olbers, Heinrich Wilhelm Matthias (1758–1840, German)—1744 A, 1797 A, 1802 A, 1807 A, 1815 A, 1817 Ma, 1826 A
Oldenburg, Henry (c.1618–1677, English)—1676 Ma, 1677 Ma
Oliver, George (1841–1915, English)—1895 B
Olivi, Peter (1248–1298, French)—c.1280 P, c.1350 P
Omalius d'Halloy, Jean Baptiste Julien d' (1783–1875, Belgian)—1813 E, 1846 B
Onnes, Heike Kamerlingh—*see* Kamerlingh-Onnes, Heike
Oort, Jan Hendrik (b.1900, Dutch)—1927 A
Oresme, Nicole (c.1320–1382, French)—c.1350 P/Ma, c.1360 Ma (2 entries), c.1370 A, c.1370 S

Orfila, Mathieu Joseph Bonaventure (1787–1853, Spanish)—1814–1815 H
Ortelius (Oertel), Abraham (1527–1598, Flemish)—1570 E
Osborne, Thomas Burr (1859–1929, American)—1909 H (2 entries), 1914 H, 1918 H
Ostrogradski, Michel (1801–1861, Russian)—1828 P/Ma
Ostwald, Friedrich *Wilhelm* (1853–1932, Russian-German)—1894 C
Otho, Valentin (c.1550–1605, German)—1551 Ma, 1596 Ma
Oughtred, William (1575–1660, English)—c.1621 Ma, 1630 Ma, 1631 Ma, 1632 Ma
Owen, Richard (1804–1892, English)—1833–1840 B, 1848 B
Owen, Richard (1810–1897, American)—1857 E

P

Pacioli, Luca (c.1445–1517, Italian)—1494 Ma
Packe, Christopher (1686–1749, English)—1683 E, 1743 E
Palissy, Bernard (c.1514–1589, French)—1580 E
Pallas, Pierre Simon (1741–1811, Prussian)—1768–1774 S, 1777 E (2 entries)
Paracelsus, Theophrastus *Philippus Aureolus* Bombastus von Hohenheim (c.1493–1541, Swiss)—1530 H, c.1530 C, c.1530 H, 1540 C, 1597 C/H, 1648 C, 1661 C
Paré, Ambroise (1510–1590, French)—1545 H, 1563 H
Paris, Matthew (c.1200–1259, English)—c.1250 B, c.1250 E
Parkes, Alexander (1813–1890, English)—1865 C
Parkinson, James (1755–1824, English)—1804–1811 E, 1812 H, 1817 H
Parkinson, John (1567–1650, English)—1640 B
Parsons, William (Lord Rosse, 1800–1867, English-Irish)—1845 A
Pascal, Blaise (1623–1662, French)—1635–1648 S, 1636–1639 Ma, 1639–1640 Ma, 1640 Ma, 1642 Ma, 1646 Met/P, 1647 P, 1648 Met, c.1648 P, 1654 Ma (2 entries), 1657 Ma, 1658 Ma, 1659 Ma, 1663 P, 1665 Ma, 1673–1676 Ma
Pascal, Étienne (1588–1651, French)—1635–1648 S
Pasch, Moritz (1843–1930, German)—1882 Ma, 1894–1908 Ma, 1899 Ma
Pasteur, Louis (1822–1895, French)—1846–1848 C, 1856 B, 1862 B/H, 1865 B, 1867 H, 1871 B, 1879 C, 1879 H, 1880 B, 1880 H, 1882 H, 1885 C, 1885 H
Pauli, Wolfgang (1900–1958, Austrian-American)—1925 P
Pavlov, Ivan Petrovich (1849–1936, Russian)—1878 B, 1883 B, 1897 B, 1902 B, 1907 B
Peacock, George (1791–1858, English)—1812 Ma, 1833 Ma
Peano, Giuseppe (1858–1932, Italian)—1888 Ma, 1889 Ma, 1890 Ma, 1894–1908 Ma, 1899 Ma, 1901 Ma, 1910–1913 Ma, 1930 Ma
Pearson, Karl (1857–1936, English)—1893 Ma, 1900 Ma
Pecham, John (c.1230–1292, English)—1277–1279 P, 1604 P
Peirce, Benjamin (1809–1880, American)—1860–1869 Ma, 1870 Ma (3 entries), 1882 Ma
Peirce, Charles Sanders (1839–1914, American)—1870 Ma (2 entries), 1876 P, 1885 Ma
Peiresc, Nicolas Claude Fabri de (1580–1637, French)—1603 S, 1610 A
Pekelharing, Cornelius Adrianus (1848–1922, Dutch)—1905 H
Peligot, Eugène Melchoir (1811–1890, French)—1841 C
Pelletier, Pierre Joseph (1788–1842, French)—1817 B, 1818 C
Peltier, Jean Charles Athanase (1785–1845, French)—1834 P
Penck, Albrecht (1858–1945, German)—1903–1909 E, 1924 E
Peregrinus, Petrus—*see* Peter of Maricourt
Perkin, William Henry (1838–1907, English)—1856 C, 1868 C
Perraudin, Jean Pierre (1767–1858, Swiss)—1815 E, 1818 E, 1829 E
Perrault, Pierre (1611–1680, French)—1674 E
Perrin, Jean Baptiste (1870–1942, French)—1895 P, 1905 P, 1908 C, 1908 P, 1921 C
Peschel, Oscar (1826–1875, German)—1870 E
Peter of Maricourt (Pierre de Maricourt, Peter Peregrinus, Petrus Peregrinus, fl. c.1269, French)—1269 P
Peter Peregrinus—*see* Peter of Maricourt
Peter Philomena of Dacia (Petrus Dacus, Petrus Danus, Peter Nightingale, fl.1290–1300, Danish)—c.1240 Ma

Petit, Alexis Thérèse (1791–1820, French)—1819 C, 1912 C
Petrov, Vasily Vladimirovich (1761–1834, Russian)—1797–1812 C
Pfaff, Philipp (1716–1780, Prussian)—1756 H
Philoponus, Johannes—*see* John Philoponus
Piaget, Jean (1896–1980, Swiss)—1929 H
Piazzi, Giuseppe (1746–1826, Italian)—1801 A
Picard, Charles Émile (1856–1941, French)—1891–1896 Ma, 1893 Ma
Picard, Jean (1620–1682, French)—1641 A, 1666 A, 1667 A, 1684 E/A, 1687 A/P
A, 1666 A, 1667 A, 1684 E/A, 1687 A/P
Pickering, William Henry (1858–1938, American)—1898 A
Pictet, Raoul Pierre (1846–1929, Swiss)—1877 C
Pierre de la Ramée (Peter Ramus, Petrus Ramus, 1515–1572, French)—1543 P
Pilgrim, Ludwig (German)—1904 A, 1904 E
Pinel, Philippe (1745–1826, French)—1809 H
Pini, Pier Matteo (Italian)—1552 B
Planck, Max Karl Ernst Ludwig (1858–1947, German)—1896 P, 1900 P (2 entries), 1905 P, 1913 P, 1924 P, 1927 P
Plato (c.427–348/347 B.C., Greek)—1469 S, 1596 A
Playfair, John (1748–1819, Scottish)—1795 Ma, 1802 E (2 entries)
Plot, Robert (1640–1696, English)—1677 E
Plücker, Julius (1801–1868, German)—1828–1831 Ma, 1830 Ma (2 entries), 1831 Ma, 1834 Ma, 1837 Ma, 1839 Ma, 1868–1869 Ma
Poincaré, Jules Henri (1854–1912, French)—1881 Ma, 1886 Ma, 1890 P, 1892–1899 A/Ma, 1895 Ma, 1899 Ma, 1902 S, 1904 Ma, 1904 P, 1905 P, 1906 Ma
Poinsot, Louis (1777–1859, French)—1804 P
Poisson, Siméon Denis (1781–1840, French)—1813 P, 1837 Ma, 1839 P, 1926 P
Polo, Marco (1254–1324, Italian)—1296–1297 E, 1442 E
Ponce de Léon, Juan (c.1460–1521, Spanish)—1513 E
Poncelet, Jean Victor (1788–1867, French)—1822 Ma (2 entries), 1833 Ma
Porta, Giambattista Della (1535–1615, Italian)—1560 S, 1586 B, 1588 B/H, 1589 P, 1593 B/P, 1603 S
Post, Emil Leon (1897–1954, American)—1920–1921 Ma, c.1921 Ma
Poulsen, Valdemar (1869–1942, Danish)—1902 P
Pourtalès, Louis François de (1823–1880, Swiss-American)—1870 E, 1871 E
Powell, John Wesley (1834–1902, American)—1875 E
Prandtl, Ludwig (1875–1953, German)—1904 P, 1925 E
Prestwich, Joseph (1812–1896, English)—1875 E
Prévost, Jean Louis (1790–1850, Swiss)—1824 B
Priessnitz, Vincenz (1801–1851, German)—1829 H
Priestley, Joseph (1733–1804, English)—c.1770 C, 1770s C, 1771 C/B, 1771 C, 1772 C, 1772 P, 1774 C, 1774–1777 C, 1775 C (2 entries), 1777 H, 1781 C, 1783 C, 1800 C
Proust, Joseph Louis (1754–1826, French)—1792 C, 1797 C, 1799 C
Prout, William (1785–1850, English)—1816 C, 1823 B, 1827 C/H, 1886 C, 1892 C/Met, 1901 C, 1927 C
Ptolemy (Claudius Ptolemaeus, c.100–c.170 A.D., Greco-Egyptian)—1217 A, 1259 A, 1272 A, c.1281 A/Met, c.1290 A, c.1370 A, 1392 A, 1410 E, c.1420 E/A, 1436 E, 1514 A, 1543 A, 1551 A, 1588 A, 1603 A, 1604 P, 1609 A, 1621 A, 1627 A, 1632 A, 1647 A, 1718 A
Pyke—1892 P
Pylarini, Giacomo (1659–1718, Italian)—1715 H
Pythagoras of Samos (c.560–c.480 B.C., Greek)—1469 S, 1665 P, 1854 Ma, 1887 Ma

Q

Quetelet, Lambert *Adolphe* Jacques (1796–1874, Belgian)—1829 Ma
Qutb al-Dīn al-Shīrāzī (1236–1311, Persian)—c.1281 A/Met

R

Radó, Tibor (1895–1965, Hungarian)—1924 Ma
Ramsay, Andrew Crombie (1814–1891, Scottish)—1846 E, 1862 E
Ramsay, William (1852–1916, Scottish)—1892 C/Met, 1894 C/Met, 1895 C, 1898 C (3 entries)

Ramus, Peter—*see* Pierre de la Ramée
Rankin, William (1820–1872, English)—1859 P
Raspe, Rudolf Erich (1737–1794, German)—1785 C
Ratzel, Friedrich (1844–1904, German)—1882–1891 E
Raulin, Jules (1836–1896, French)—1869 B
Ray, John (1627–1705, English)—1686–1704 B, 1691 B, 1704 S
Rayleigh, Lord—*see* Strutt, John William
Réaumur, René Antoine Ferchault de (1683–1757, French)—1730 P, 1734–1742 B
Recorde, Robert (1510–1558, English)—1557 Ma
Redfield, William C. (1789–1857, American)—1831 Met, 1834 E, 1841 Met
Redi, Francesco (1626–1697/1698, Italian)—1684 B
Reed, Walter (1851–1902, American)—1900 H
Regiomontanus, Johannes (Johannes Müller, 1436–1476, German)—c.1250 Ma, 1464 Ma
Reich, Ferdinand (1799–1882, German)—1863 C
Reichenbach, Karl Ludwig (1788–1869, German)—1833 C
Reichenstein, Baron de—*see* Müller, Franz Joseph
Reinhold, Erasmus (1511–1553, German)—1551 A
Remsen, Ira (1846–1927, American)—1879 C
Renaldini, Carlo (1615–1698, Italian)—1694 P
Renard, Alphonse François (1842–1903, Belgian)—1891 E
Rennell, James (1742–1830, English)—1831 E, 1832 E
Rey, Jean (c.1582–c.1645, French)—1630 C
Reynolds, Osborne (1842–1912, Irish)—1883 P, 1894 P/E/Met
Rheticus, Georg Joachim (1514–1574, German)—1540 A, 1551 Ma, 1579 Ma, 1596 Ma
Ricci, Curbasto Gregorio (1853–1925, Italian)—1888 Ma
Riccioli, Giambattista (1598–1671, Italian)—1651 A (2 entries)
Richards, Theodore William (1868–1928, American)—1913 C, 1914 C
Richardson, Lewis Fry (1881–1953, English)—1920 Met, 1922 Met
Richer, Jean (1630–1696, French)—1672 A, 1672 P, 1687 E/A
Richter, Hieronymus Theodor (1824–1898, German)—1863 C
Richter, Jeremias Benjamin (1762–1807, German)—1791 C, 1792 C
Richthofen, Ferdinand Paul Wilhelm von (1833–1905, German)—1870 E
Ricketts, Howard Taylor (1871–1910, American)—1906 H, 1909 H
Ridley, Henry Nicholas (1855–1956, English)—1930 B
Riemann, Georg Friedrich Bernhard (1826–1866, German)—1851 Ma, 1854 Ma (3 entries), 1857 Ma, 1859 Ma, 1861 P/Ma, 1867 Ma, 1868 Ma, 1870 Ma, 1871 Ma (2 entries), 1877 Ma, 1881 Ma, 1884 Ma, 1900 Ma, 1902 Ma, 1916 P/A, 1918 Ma/P, 1922 Ma
Riggs, John—1844 H
Rinio, Benedetto (Italian)—1410 B
Río, Andrés Manuel del (1764–1849, Mexican)—1801 C
Ristoro d'Arezzo (c.1220–c.1282, Italian)—1282 E
Ritter, Carl (1779–1859, German)—1817–1859 E
Ritter, Johann Wilhelm (1776–1810, German)—1801 P
Robertson, Howard Percy (American)—1928 A
Roberval, Gilles Personne de (1602–1675, French)—1635–1648 S, 1643 A
Robinet, Jean Baptiste René (1735–1820, French)—1761–1768 B
Robins, Benjamin (1707–1751, English)—1735 Ma, 1747 P
Roche, Édouard Albert (1820–1883, French)—1848 A, 1873 A
Roentgen, Wilhelm Conrad—*see* Röntgen, Wilhelm
Rogers, Henry Darwin (1809–1866, American)—1842 E
Rogers, William Barton (1805–1881, American)—1842 E
Rolle, Michel (1652–1719, French)—1691 Ma
Romé de l'Isle, Jean Baptiste Louis (1736–1790, French)—1772 E/C
Römer, Olaus (Ole Christensen Roemer, 1644–1710, Danish)—1675 P/A
Röntgen (Roentgen), Wilhelm Conrad (1845–1923, German)—1895 H, 1895 P, 1896 P
Roscoe, Henry Enfield (1833–1915, English)—1869 C

Ross, James Clark (1800–1862, English)—1831 E, 1839–1843 E
Ross, John (1777–1856, English)—1818 E
Ross, Ronald (1857–1932, English)—1898 H, 1899 H
Ross, William Horace (1875–1947, American)—1907 C
Rouelle, Hilaire-Marin (1718–1779, French)—1773 B
Rousseau, Jean Jacques (1712–1778, Swiss-French)—1751–1772 S
Roux, Pierre Paul Émile (1853–1933, French)—1888–1891 H
Rowland, Henry Augustus (1848–1901, American)—1882 P, 1896 A
Rudolff (or Rudolf), Christoff (1500–1545, Austrian)—1530 Ma, 1585 Ma
Rue, Warren de la (1815–1889, English)—1860 A
Rueus, Franciscus—1566 S
Ruffini, Paolo (1765–1822, Italian)—1799 Ma, 1813 Ma, 1824 Ma
Rufinus (fl. 1250–1300, Italian)—c.1287 B
Rumford, Count—*see* Thompson, Benjamin
Runge, Friedlieb Ferdinand (1794–1867, German)—1834 C
Rush, Benjamin (1746–1813, American)—1812 H
Ibn Rushd—*see* Averroës
Russell, Alexander Smith (b.1888)—1913 C
Russell, Bertrand Arthur William (1872–1970, English)—1884 Ma, 1893–1903 Ma, 1902 Ma, 1903 Ma, 1903 P, 1905 Ma, 1906 Ma, 1908 Ma, 1910–1913 Ma, 1925 Ma, 1930 Ma
Russell, Henry Norris (1877–1957, American)—1905–1907 A, 1908 A, 1911 A, 1914 A
Rutherford, Daniel (1749–1819, Scottish)—1772 C (2 entries)
Rutherford, Ernest (1871–1937, New Zealander-English)—1902 C/P, 1902 P, 1906–1909 P, 1908 P, 1911 P/C, 1913 P, 1919 P (2 entries), 1921–1924 P, 1928 P

S

Sabine, Edward (1788–1883, English)—1825 Met, 1846 Met/E
Sabine, Wallace (1868–1919, American)—1900 P
Saccheri, Giovanni *Girolamo* (Gerolamo) (1667–1733, Italian)—c.1250 Ma, 1733 Ma, 1763 Ma, 1766 Ma
Sachs, Julius von (1832–1897, German)—1862 B
Sacrobosco, Johannes de—*see* John of Holywood
Sagredo, Francesco (Italian)—1613 Met
Saha, Meghnad (1894–1956, Indian)—1921 A
Saint-Hilaire, Geoffroy (1772–1844, French)—1796 B
Sala, Angelo (Angelus) (1576–1637, Italian)—1617 C
Salmon, George (1819–1904, Irish)—1849 Ma
Salomons, David Lionel (1851–1925, American)—1892 P
Sandwich, Earl of—see Montagu, Edward
Santorio, Santorio (Sanctorius, 1561–1636, Italian)—1603 H, 1614 H, 1626 H
Sarrus, Pierre Frédéric (1798–1861, French)—1819 Ma
Saussure, Horace Bénédict de (1740–1799, Swiss)—1779 E (2 entries), 1779–1796 E, 1783 Met, 1786 E, 1796 E
Saussure, Nicholas Théodore de (1767–1845, Swiss)—1804 B
Sauveur, Joseph (1653–1716, French)—1701 P
Savage, Thomas (1804–1880, American)—1847 B
Savart, Félix (1791–1841, French)—1820 P
Savary, Félix (1797–1841, French)—1827 A
Savery, Thomas (c.1650–1715, English)—1698 P
Scaliger, Julius Caesar (1484–1558, Italian)—1557 C
Scarperia, Jacopo Angelo de (Italian)—1410 E
Schäfer, Edward Albert Sharpey—*see* Sharpey-Schäfer, Edward Albert
Scheele, Carl Wilhelm (1742–1786, Swedish)—1769 C, 1771 C, 1772 C, 1774 C (3 entries), 1777 C/Met
Scheffer, Henric Theophil (1710–1759, Swedish)—1752 C
Scheiner, Christoph (1573–1650, German)—1612 A, 1613 A, 1619 B, 1630 A
Scheuchzer, Johann Jakob (1672–1733, Swiss)—1709 E, 1716 E, 1716–1718 E, 1726 E
Scheutz, Pehr Georg (1785–1873, Swedish)—1834 Ma, 1853 Ma
Schiaparelli, Giovanni Virginio (1835–1910, Italian)—1877 A, 1894 A, 1906 A
Schickard, Wilhelm (1592–1635, German)—1623 Ma

Schimper, Andreas Franz Wilhelm (1856–1901, German)—1898 E
Schimper, Karl (1803–1867, German)—1837 E
Schlafli, L. (Swiss)—1852 Ma
Schleiden, Jacob Matthias (1804–1881, German)—1838 B (2 entries)
Schmidt, Gerhard Carl Nathaniel (1865–1949, German)—1898 C
Schönbein, Christian Friedrich (1799–1868, Swiss)—1840 C
Schott, Gerhard (1866–1961, German)—1912 E
Schröder, Friedrich Wilhelm Karl *Ernst* (1841–1902, German)—1877 Ma, 1890–1905 Ma
Schrödinger, Erwin (1887–1961, Austrian)—1926 P (2 entries), 1927 P
Schultes, Johann (1595–1645, German)—1655 H
Schuster, Arthur (1851–1934, German-English)—1882 A, 1897–1898 Ma, 1897–1906 A/E/Met
Schwabe, Samuel Heinrich (1789–1875, German)—1843 A
Schwann, Theodor Ambrose Hubert (1810–1882, German)—1836 B (2 entries), 1837 B, 1838 B
Schwarzschild, Karl (1873–1916, German)—1916 A
Schweigger, Johann Salomo Christoph (1779–1857, German)—1820 P
Scopes, John Thomas (b.1901, American)—1925 S
Scot, Michael—*see* Michael Scot
Scott, K.J.—1922 H, 1923 H
Scott, Robert F. (1868–1912, English)—1910–1912 E
Sebastian de Elcano, Juan—*see* Elcano, Juan Sebastian de
Sechenov, Ivan Mikhaylovich (Michailovich) (1829–1905, Russian)—1865 B
Sedgwick, Adam (1785–1873, English)—c.1832 E, 1835 E, 1837 E, 1839 E, 1843 E
See, Thomas Jefferson Jackson (1866–1962, American)—1909 A
Seebeck, Thomas Johann (1770–1831, Estonian)—1822 P
Seeber, Ludwig August (1793–1855, German)—1831 Ma
Seeliger, Hugo von (1849–1924, German)—1892 A
Sefstrom, Nils Gabriel (1787–1845, Swedish)—1831 C
Seidel, Philipp Ludwig von (1821–1896, German)—1847 Ma
Semmelweis, Ignaz Philipp (1818–1865, Hungarian)—1847 H, 1861 H
Serturner, Friedrich Wilhelm Adam Ferdinand (1783–1841, German)—1805 C, 1817 C
Serveto, Miguel—*see* Servetus, Michael
Servetus, Michael (Miguel Serveto, 1511–1553, Spanish)—1553 B, 1555 B, 1559 B
Servois, Francois-Joseph (1767–1847, French)—1814 Ma
Shackleton, Ernest (1874–1922, English)—1907–1909 E
Shanks, William (1812–1882, English)—1853 Ma
Shapley, Harlow (1885–1972, American)—1912 A, c.1913 A, 1918 A, 1930 A
Sharp, Abraham (1651–1742, English)—1725 A
Sharpey-Schäfer, Edward Albert (1850–1935, English)—1895 B
Shaw, John C. (1845–1900, American)—1892 H
Shaw, William Napier (1854–1945, English)—1906 Met, 1925 Met, 1926–1931 Met
Sheffer, Henry Maurice (1883–1964, American)—1913 Ma
Shiga, Kiyoshi (1870–1957, Japanese)—1900 H
Shipley, P.G.—1921 H
Sickingen, Carl von—1782 C
Siemens, Ernst Werner von (1816–1892, German)—1839 C
Siemens, William (1823–1883, German)—1821 P, 1871 P
Simon, Théodore (1873–1961, French)—1905 S, 1914 S
Simpson, James Young (1811–1870, Scottish)—1847 H
Sitter, Willem de (1872–1934, Dutch)—1913 P/A, 1917 A, 1919 A, 1922 A, 1928 A
Slipher, Vesto Melvin (1875–1969, American)—1912 A, 1912–1917 A, 1929 A
Smale, Stephen—1904 Ma
Smeaton, John (1724–1792, English)—1752–1754 P, 1769–1772 P
Smith, Theobald (1859–1934, American)—1893 H
Smith, William (1769–1839, English)—1799 E, 1815 E, 1816 E, 1817 E, 1819 E
Smith, Willoughby—1873 P
Snell (Snel, Snellius), Willebrord (1580–1626, Dutch)—1621 P, 1637 P, 1657 P

Snow, John (1813–1858, English)—1847 H (2 entries), 1849 H, 1853 H, 1854 H, 1858 H
Soames, K.M.—1923 H, 1924 H
Sobrero, Ascanio (1812–1888, Italian)—1846 C
Soddy, Frederick (1877–1956, English)—1902 C/P, 1902 P, 1913 C (2 entries), 1917 C
Soldner, Johann Georg von (1776–1833, German)—1801 P, 1911 P/A
Sommerfeld, Arnold Johannes Wilhelm (1868–1951, German)—1915 P, 1915–1916 P, 1916 P, 1927 P
Sorby, Henry Clifton (1826–1908, English)—1853 E, 1858 E, 1859 E
Soubeiran, Eugène (1793–1858, French)—1831 C
Spallanzani, Lazzaro (1729–1799, Italian)—1765 C, 1767 B, 1779 B
Spencer, Herbert (1820–1903, English)—1852 B, 1858 S, 1862 S, 1876–1896 B, 1886 B
Spiegel (Spieghel), Adriaan van den (1578–1625, Italian)—1626 B
Sprengel, Christian Konrad (1750–1816, German)—1793 B
Sprengel, Hermann Johann Philipp (1834–1906, German)—1879 C
Spurzheim, Johann Christoph (John Caspar) (1776–1832, German)—1810–1819 B
Stahl, Georg Ernest (1660–1734, German)—1669 C, 1700 C, 1702 C, 1707 B
Stansbury, Howard (American)—1852 E, 1870–1879 E
Stark, Johannes (1874–1957, German)—1913 P
Starling, Ernest Henry (1866–1927, English)—1902 B, 1905 B
Staudt, Karl Georg Christian von (1798–1867, German)—1847 Ma, 1856–1860 Ma
Steenbock, Harry (1886–1967, American)—1919 H, 1924 H
Stefan, Josef (1835–1893, Austrian)—1879 P/A, 1883 P, 1900 P
Steiger, Otto—1895 Ma
Steiner, Jakob (1796–1863, Swiss-German)—1822 Ma, 1824 Ma, 1832 Ma, 1833 Ma, 1841 Ma, 1842 Ma
Steinitz, Ernst (1871–1928, German)—1910 Ma
Stelluti, Francesco (1577–1640, Italian)—1625 B
Steno, Nicolaus (Niels Stenson, 1638–1686, Danish)—1661 B, 1662 B, 1664 B, 1669 C/E, 1669 E (3 entries)
Stensen, Niels—*see* Steno, Nicolaus
Stenson, Niels—*see* Steno, Nicolaus
Stern, Théodore (1873–1961, French)—1914 S
Stevin, Simon (Stevinus, 1548–1620, Flemish)—1583 P, 1585 Ma, 1586 P
Stevinus—*see* Stevin, Simon
Stieltjes, Thomas Jan (1856–1894, French)—1894 Ma
Stifel, Michael (c.1487–1567, German)—1544 Ma
Stiles, Charles Wardell (1867–1941, American)—1902 H
Still, Andrew Taylor (1828–1917, American)—1874 H
Stirling, James (1692–1770, Scottish)—1730 Ma
Stokes, George Gabriel (1819–1903, English)—1823 P, 1845 P, 1847 Ma, 1849 P, 1850 P, 1854 Ma
Stolz, Otto (1842–1905, Austrian)—1886 Ma
Stoney, George *Johnstone* (1826–1911, Irish-English)—1874 P
Strachey, John (1671–1743, English)—1719 E
Stratton, Samuel Wesley (1861–1931, American)—1898 Ma
Strohmeyer, Friedrich (1776–1835, German)—1817 C
Stromeyer, M.—1819 A
Strutt, John (Robin) William (Lord Rayleigh, 1842–1919, English)—1877–1878 P, 1892 C, 1892 C/Met, 1894 C/Met, 1896 P, 1900 P, 1901 C
Strutt, Robert John (1875–1947, English)—1904 C
Study, Eduard (1862–1922, German)—1893 Ma
Sturgeon, William (1783–1850, English)—1825 P, 1837 P
Sturtevant, Alfred Henry (1891–1970, American)—1915 B
Suess, Eduard (1831–1914, Austrian)—1883–1901 E, 1885 E
Sumner, James Batcheller (1887–1955, American)—1926 C
Supan, Alexander (1847–1920, Austrian)—1898 Met
Sure, B.—1923 H
Sutton, Walter Stanborough (1877–1916, American)—1903 B
Svab, Anton von—1742 C

Sverdrup, Harald Ulrik (1889–1957, Norwegian)—1918–1925 E, 1927 E
Swammerdam, Jan (1637–1680, Dutch)—1658 B, 1667 B, 1682 B, 1737 B
Swan, Joseph (1828–1914, English)—1879 C
Swedenborg, Emanuel (1688–1772, Sweden)—1719 E, 1734 A
Swift, Jonathan (1667–1745, English)—1877 A
Sydenham, Thomas (1624–1689, English)—1676 H
Sylow, Peter *Ludvig* (Ludwig) Mejdell (1832–1918, Norwegian)—1872 Ma

T

Tacquet, Andreas (1612–1660, Belgian)—1651 Ma, 1656 Ma
Takamine, Jokichi (1854–1922, Japanese)—1901 B
Talbot, William Henry Fox (1800–1877, English)—1834 C, 1839 C
Tanner, W.F.—1925 H
Tartaglia, Niccolò (1499/1500–1557, Italian)—1535 Ma, 1537 P, 1539 P, 1545 Ma, 1546 Ma, 1770 Ma
Taylor, Brook (1685–1731, English)—1671 Ma, 1715–1717 Ma, 1742 Ma, 1797 Ma
Tchebycheff—*see* Chebyshev
Telesio, Bernardino (1509–1588, Italian)—1565–1586 B
Telles, Vicente Coelho de Seabra Silva (1764–1804, Portuguese)—1790 C
Tennant, Smithson (1761–1815, English)—1796 C, 1804 C
Tessan, Urbain Dortet de (1804–1879, French)—1811 E, 1844 E
Thenard, Louis Jacques (1777–1857, French)—1789 C, 1808 C, 1809 C (2 entries), 1810 C
Theodoric of Freiberg—*see* Dietrich von Freiberg
Theodosius, Franz Ulrich (1724–1802, German)—1759 P
Thomas Aquinas, Saint—*see* Aquinas, Saint Thomas
Thomas, Tracy Yerkes (b.1899, American)—1922 Ma
Thompson, Benjamin (Count Rumford, 1753–1814, American)—1797 E, 1798 P, 1799 P
Thompson, Leonard (1908–1935, Canadian)—1921 H
Thomson, George Paget (1892–1975, English)—1928 P
Thomson, James (1822–1892, English)—1849 P
Thomson, Joseph John (1856–1940, English)—1881 C/P, 1895 P, 1896 P, 1897 P, 1897 P/C, 1898 P, 1903 P, 1904 C/P, 1910–1911 C, 1911 P/C
Thomson, Thomas (1773–1852, Scottish)—1802 C, 1807 C, 1808 C (2 entries), 1813 C
Thomson, William (Lord Kelvin, 1824–1907, English)—1828 P/Ma, 1849 P, 1851 P (2 entries), 1852 C, 1877 C, 1898 C
Thomson, Charles *Wyville* (1830–1882, Irish)—1868 E, 1869 E
Timoni, Emmanuel (fl. 1741)—1715 H
Titian (1477–1576, Italian)—1543 B
Titius (Tietz), Johann Daniel (1729–1796, German)—1766 A
Tombaugh, Clyde William (b.1906, American)—1930 A
Tonelli, Leonida (1885–1946, Italian)—1922–1924 Ma
Torricelli, Evangelista (1608–1647, Italian)—1643 Met (2 entries), 1646 Met/P
Toscanelli dal Pozzo, Paolo (1397–1482, Italian)—1443–1472 A, c.1450 E
Tour, Charles Cagniard de la—*see* Cagniard de la Tour, Charles
Towneley, Richard (1629–1707, English)—1677–1704 Met
Townsend, John Sealy Edward (1868–1957, Irish-English)—1897 P
Travers, Morris William (1872–1961, English)—1898 C (3 entries)
Trembley, Abraham (1710–1784, Swiss)—1739 B
Treviranus, Gottfried Reinhold (1776–1837, German)—1800 B, 1802 B
Trimmer, Joshua (1795–1857, English)—c.1845 E
Trudeau, Edward Livingston (1848–1915, American)—1885 H
Trumpler, R.J. (1886–1956, Swiss-American)—1930 A
Tschermak von Seysenegg, Erich (1871–1962, Austrian)—1900 B
Tschirnhaus (Tschirnhausen), Ehrenfried Walther von (1651–1708, German)—1683 Ma, 1786 Ma
Tsiolkovsky, Konstantin Eduardovich (1857–1935, Russian)—1898 P
Turner, Edward (1796–1837, English)—1833 C
al-Tūsī, Nasir al-Din—*see* Nasir al-Din al-Tūsī

Twort, Frederick William (1877–1950, English)—1915 B
Tyndall, John (1820–1893, Irish)—1863 Met

U

Uhlenbeck, George Eugene (b.1900, Dutch-American)—1925 P
Ulloa y de la Torre Giral, Antonio de (1716–1795, Spanish)—1748 C
Unverdorben, Otto (1806–1873, German)—1826 C
Urbain, Georges (1872–1938, French)—1907 C
Uryson, Pavel Samuilovich (Paul Urysohn, 1898–1924, Russian)—1923 Ma, 1925 Ma
Ussher, James (1581–1656, Irish)—1650–1654 E, 1654 E

V

Valentine, Basil (Basilius Valentinus, Johann Thölde, fl. 1604, German)—1604 C
Vallée-Poussin, Charles-Jean-Gustave-Nicolas de la (1866–1962, Belgian)—1896 Ma
Vallisneri, Antonio (1661–1730, Italian)—1721 E
Valverde, Juan de (c.1520–c.1588, Spanish)—1556 B
Vandermonde, Alexandre Théophile (1735–1796, French)—1770 Ma, 1771 Ma
Van der Waals—*see* Waals, Johannes Diderik van der
Van't Hoff, Jacobus Hendricus (1852–1911, Dutch)—1874 C, 1877 C, 1884 C
Varen, Bernhard (Bernhardus Varenius, 1622–1650, German)—1650 E
Varolio, Costanzo (1543–1575, Italian)—1573 B
Vauquelin, Nicolas Louis (1763–1829, French)—1798 C (2 entries), 1808 B
Veblen, Oswald (1880–1960, American)—1906 Ma, 1907 Ma, 1910 Ma, 1922 Ma
Vedder, Edward Bright (1878–1952, American)—1912 H, 1925 H
Venetz, Ignatz (1788–1859, Swiss)—1818 E, 1829 E, 1835 E, 1836 E
Venn, John (1834–1923, English)—1881 Ma
Venturi, Giovanni Battista (1746–1822, Italian)—1797 P, 1892 P
Vesalius, Andreas (1514–1564, Flemish)—1537 H, 1539 B, 1543 B, 1555 B
Vespucci, Amerigo (1451–1512, Italian)—1502 E, 1507 E
Victoria, Alexandrina (Queen Victoria of Great Britain and Ireland, 1819–1901, English)—1853 H
Vieta—*see* Viète, François
Viète, François (Vieta, 1540–1603, French)—1579 Ma, c.1580 Ma, 1582 A, 1591 Ma, 1593 Ma, 1637 Ma, 1679 Ma, 1683 Ma
Vincent of Beauvais (c.1190–c.1264, French)—1250 S
Violle, Jules Louis Gabriel (1841–1923, French)—1882 C, 1890 C
Virchow, Rudolph Ludwig Carl (1821–1902, German)—1858 H
Vitello—*see* Witelo
Vives, Juan Luis (1492–1540, Spanish)—1531 S, 1538 B
Voegtlin, Carl (b.1879, Swiss)—1914 H, 1915 H
Vogel, Hermann Carl (1841–1907, German)—1889 A
Volta, Alessandro Giuseppe Antonio Anastasio (1745–1827, Italian)—1780 B, 1791 P, 1800 P, 1836 P
Voltaire, François Marie Arouet de (1694–1778, French)—1734 S, 1738 P, 1751–1772 S, 1772 Ma
Volterra, Vito (1860–1940, Italian)—1881 Ma, 1887 Ma
von Behring—*see* Behring, Emil Adolph von
von Buch—*see* Buch, Christian Leopold von
Vossius, Isaac (Isaac Voss, 1618–1689)—1663 E
Vries, Hugo Marie de (1848–1935, Dutch)—1894 B, 1900 B (2 entries), 1901–1903 B, 1907 B

W

Waage, Peter (1833–1900, Norwegian)—1803 C, 1867 C
Waals, Johannes Diderik van der (1837–1923, Dutch)—1873 C (2 entries)
Waldseemüller, Martin (1470–c.1522, German)—1507 E, 1522 A
Wallace, Alfred Russel (1823–1913, English)—1858 B, 1869 B, 1870 B, 1876 B
Wallingford (d.1213, English)—c.1210 E
Wallingford, Richard (c.1292–1335, English)—1326–1335 E
Wallis, John (1616–1703, English)—1655 Ma, 1685 Ma, 1693 Ma

Wantzel, Pierre Laurent (1814–1848, French)—1837 Ma
Ward, Joshua (1685–1761, English)—1740s C, 1806 C
Waring, Edward (c.1736–1798, English)—1770 Ma, 1776 Ma, 1909 Ma
Warren, John Collins (1778–1856, American)—1846 H
Wassermann, August von (1866–1925, German)—1906 H
Watson, John Broadus (1878–1958, American)—1928 H
Watson, Richard (1737–1816, English)—1781 C
Watson, Thomas (1792–1882, English)—1843 H
Watson, Thomas J. (1874–1956, American)—1924 Ma
Watson, William (1715–1787, English)—1746 P
Watt, James (1736–1819, Scottish)—1764 P, 1776 P
Weber, Heinrich (1842–1913, German)—1898–1899 Ma
Weber, Wilhelm Eduard (1804–1891, German)—1832 P/E, 1833 P, 1846 P
Wedderburn, Joseph Henry Maclagan (1882–1948, American)—1905 Ma
Wedgwood, Thomas (1771–1805, English)—1792 P, 1802 C
Wegener, Alfred Lothar (1880–1930, German)—1911 Met, 1912 E, 1915 E, 1922 Met, 1924 E
Weiditz, Hans—1530 B
Weierstrass, Karl Theodor Wilhelm (1815–1897, German)—1834 Ma, 1841 Ma, 1860–1869 Ma, 1861 Ma, 1863 Ma, c.1865 Ma, 1870 Ma, 1886 Ma, 1923 Ma
Weismann, August Friedrich Leopold (1834–1914, German)—1892 B
Wells, Horace (1815–1848, American)—1800 C/H, 1838 H, 1844 H
Wells, William Charles (1757–1817, American-English)—1813 B
Welsbach, Carl Auer von (1858–1929, Austrian)—1885 C (2 entries)
Wenzel, Carl Friedrich (1740–1793, German)—1777 C
Werner, Abraham Gottlob (1749–1817, German)—1774 E, 1787 E, 1789 E, 1791 E, 1795 E (2 entries), c.1796 E, 1802–1803 E, 1808 E, 1817 E
Werner, Johann (1468–1522, German)—1522 Ma
Wessel, Caspar (1745–1818, Norwegian)—1685 Ma, 1797 Ma, 1806 Ma
Western—1905 Ma, 1909 Ma
Weyl, Hermann (1885–1955, German-American)—1918 Ma, 1918 Ma/P
Wharton, Thomas (1614–1673, English)—1656 B
Wheatstone, Charles (1802–1875, English)—1831 P, 1834 P, 1837 P
Wheeler, George Alexander (b.1885, American)—1928 H
Whiston, William (1667–1752, English)—1696 E
White, Charles (1728–1813, English)—1769 H, 1773 H, 1784 H, 1790 H, 1799 B
White, Gilbert (1720–1793, English)—1789 B
Whitehead, Alfred North (1861–1947, English)—1903 Ma, 1910–1913 Ma, 1925 Ma, 1930 Ma
Whittlesey, Charles (1808–1866, American)—1868 E
Widman, Johannes (c.1462–c.1498, German)—1489 Ma, 1544 Ma
Wien, Wilhelm Carl Werner Otto Fritz Franz (1864–1928, German)—1896 P, 1900 P (2 entries)
Wiener, Norbert (1894–1964, American)—1930 Ma
Wilcke, Johan Carl (1732–1796, Swedish)—1759 P
Wilczynski, Ernest Julius (1876–1932, German-American)—1906 Ma
Wilhelmy, Ludwig Ferdinand (1812–1864, German)—1850 C
Wilkes, Charles—1846 B/E
Wilkins, John (1614–1672, English)—1638 A, 1640 A, 1640 P
Wilkinson, John (1728–1808, English)—1776 P
William of Moerbeke—*see* Moerbeke, William of
William of Ockham—*see* Ockham, William of
William of Saint Cloud (fl. c.1295, English-French)—1290 A, 1290 E
William of Saliceto (1210–1277, Italian)—c.1270 H, 1275 B/H
Williams, Daniel Hale (1856–1931, American)—1893 H
Williams, Robert Runnels (1886–1965, American)—1912 H
Willis, Thomas (1621–1675, English)—1659 Ma, 1664 B, 1670 H
Wilson, Charles Thomson Rees (1869–1959, Scottish)—1896 P
Wilson, Harold Albert (1874–1964, English-American)—1897 P

Wilson, William (1875–1965)—1915–1916 P
Winkler, Clemens Alexander (1838–1904, German)—1886 C
Winthrop, John (1714–1779, American)—1761 A
Witelo (Vitello, c.1230–c.1275, Silesian)—c.1270 P, 1604 P
Withering, William (1741–1799, English)—1775 H, 1776 H, 1785 H
Wittgenstein, Ludwig Josef Johann (1889–1951, Austrian)—1920–1921 Ma, 1921 Ma
Wittich, Paul (1555–1587, Silesian)—c.1580 Ma
Wöhler, Friedrich (1800–1882, German)—1823 C, 1827 C, 1828 C (2 entries), 1832 C
Wolf, Maximilian Franz Joseph Cornelius (1863–1932, German)—1891 A
Wolff, Caspar Friedrich (1734–1794, German)—1759 B
Wollaston, William Hyde (1766–1828, English)—1800 C, 1802 C, 1803 C, 1808 C, 1809 C
Woodward, John (1665–1728, English)—1695 E, 1704 S, 1728–1729 E
Worthen, Amos Henry (1813–1888, American)—1873 E
Wotton, William (1666–1727, English)—1694 S
Wren, Christopher (1632–1723, English)—1664 A, 1679 A/P, 1684 A, 1685 A
Wright, Almroth Edward (1861–1947, English)—1896 H
Wright, Edward (1561–1615, English)—1599 E
Wright, Orville (1871–1948, American)—1903 P
Wright, Thomas (1711–1786, English)—1750 A, 1755 A
Wright, Wilbur (1867–1912, American)—1903 P
Wundt, Wilhelm Max (1832–1920, German)—1872 H, 1879 B
Wurtz, Charles Adolphe (1817–1884, French)—1849 C, 1855 C, 1868–1878 C
Wurzer, Ferdinand—1830 B
Wüst, Georg Adolf (b.1890, German)—1922 E, 1925–1927 E, 1929 E
Wyman, Jeffries (1814–1874, American)—1847 B

Y

Yersin, Alexandre Emile Jean (1863–1943, French)—1894 H
Young, Charles Augustus (1834–1908, American)—1870 A
Young, John Wesley (1879–1932, American)—1907 Ma, 1910 Ma
Young, Thomas (1773–1829, English)—1801 H, 1801–1804 P, 1807 P (2 entries), 1809 P, 1818 P

Z

Zeeman, Pieter (1865–1943, Dutch)—1896 P, 1908 A, 1915 P
Zermelo, Ernst Friedrich Ferdinand (1871–1953, German)—1904 Ma, 1907 Ma, 1908 Ma, 1922 Ma
Zirm, Eduard Konrad (1863–1944, Austrian)—1905 H
Zöppritz, K.—1878 E
Zworykin, Vladimir Kosma (1889–1982, Russian-American)—1928 P

Subject Index

The indexing is done according to date and major subject heading, not page number. Major subject headings are identified as follows: A—Astronomy, B—Biology, C—Chemistry, E—Earth Sciences, H—Health Sciences, Ma—Mathematics, Met—Meteorology, P—Physics, S—Supplemental.

Abelian functions, 1832 Ma, 1857 Ma
Absolute value, 1913 Ma
Absolutism, 1903 P
Abyssal theory, oceanic, 1843 E
Acausality, 1919 S
Acetylene, 1836 C, 1860 C, 1895 C
Achromatic lens, 1733 A, 1757 A, 1827 B
Acids, 1611–1613 C, 1700 C, 1775 C, 1789 C, 1803 C; mineral, c.1210 C; muriatic, 1809 C, 1810 C
Acoustics, 1636 P, 1701 P, 1787 P, 1862 P, 1900 P
Acquired characteristics, 1794 B, 1801 B, 1809 B, 1868 B, 1875 B, 1886 B, 1892 B, 1930 B
Actinium, 1899 C; X, 1904–1905 C
Addison's disease, 1855 H
Adiabatic lapse rate, 1841 Met
Adrenaline, 1895 B, 1898 B, 1901 B
Affinity, chemical, 1718 C, 1775 C, 1777 C, 1792 C, 1801 C, 1884 C
Agglutinogens, 1927 H
Agriculture, 1306 B, 1748 B; fertilizers, 1842 C, 1855 C
Air: 1660 Met; Cavendish on, 1785 Met; composition, 1550 Met, 1665 Met, 1674 Met, 1777 C/Met, 1785 C/Met, 1801 Met/C, 1804 Met, 1892 C/Met, 1894 C/Met, 1895 C/Met; density, 1638 Met, c.1645 Met; masses, fronts, 1919 Met, 1928 Met; pollution, 1661 Met, 1925 Met; pollution control, 1863 Met; pressure, 1643 Met, 1646 Met, 1648 Met, 1654 Met, 1669 Met, 1869 Met; pump, 1645 P, 1662 C; Scheele on, 1777 Met; temperature, 1613 Met, 1787 Met, 1922 Met; da Vinci on, c.1500 S. *See also* Atmosphere
Airplane, 1903 P
Alchemy, c.1310 C, 1317 S, 1330 C, c.1530 C, c.1530 H, 1718 C
Alcohol, 1828 C, 1834 C, 1862 C
Alfonsine Tables, 1272 A, 1483 A, 1551 A, 1627 A
Algae, 1841 B
Algebra, 1202 Ma, c.1220 Ma, 1225 Ma, 1494 Ma, 1657 Ma, 1799 Ma, 1832–1833 Ma, 1844 Ma, 1898–1899 Ma; Boolean, 1847 Ma, 1916 Ma; Euler on, 1744 Ma, 1748 Ma; Lagrange on, 1798 Ma; linear associative, 1860s Ma, 1870 Ma; linear associative division, 1905 Ma; of matrices, 1857–1858 Ma; noncommutative, 1843 Ma; Peacock on, 1833 Ma; Vandermonde on, 1771 Ma; Viète on, c.1580 Ma, 1591 Ma. *See also* Diophantine analysis; Equations; Field theory; Group theory; Symbolism, mathematical
Algebraic forms, 1888 Ma, 1890 Ma
Algebraic invariance, 1845 Ma
Algebraic number(s), 1744 Ma, 1874 Ma; theory, 1897 Ma
Algol, 1782 A, 1889 A
Alkali, 1702 C, 1736 C
Almagest (Ptolemy), 1217 A, 1621 A
Alpha particles, 1906–1909 P, 1908 P
Aluminum, 1825 C, 1827 C, 1854 C, 1886 C
America, named, 1507 E
Amines, 1849 C
Amino acids, 1909 H
Ammonia, 1770s C, 1785 C, 1908 C; liquid, 1798 C
Ampère's Law, 1827 P
Amputation, 1563 H

Anastomoses, 1628 B
Anatomy: 1275 B/H, 1316 B, 1543 B, 1552 B, 1554 H/B, 1556 B, 1564 B, 1626 B; comparative, c.1500 S, 1555 B, 1760–1762 B, 1776 B, 1799 B, 1833–1840 B; pathological, 1761 H; plant, 1682 B; da Vinci's drawings, c.1513 B. *See also* Dissection; parts of the body
Andromeda nebula, galaxy, 1888 A, 1912 A, 1912–1917 A, 1923 A, 1925 A, 1929 A
Anemia, 1926 H
Anemometer, c.1450 Met, 1667 Met
Anesthesiology, 1266 H, 1858 H; carbon dioxide, 1824 H; chloroform, 1847 H, 1853 H; cocaine, 1884 H; ether, 1842 H, 1846 H, 1847 H; nitrous oxide, 1800 C/H, 1844 H
Aneurysms, 1785 H
Aniline, 1826 C
Animal heat, c.1778 B, 1779 B
Animals: electricity in, 1780 B, 1791 B; generation of, 1651 B. *See also* Classification, of animals
Anthrax, 1876 H, 1882 H
Antimatter, 1927 A
Antimony, 1604 C
Antisepsis, 1266 H, 1320 H, 1865 H, 1867 H, 1880 H
Appendicitis, 1812 H
Appleton Layer, 1924 Met
Arabic numerals. *See* Hindu-Arabic numerals
Archaeopteryx, 1861 B
Archetypes, 1795 B, 1796 B, 1848 B
Arc transmitter, 1902 P
Argand diagram, 1797 Ma, 1806 Ma
Argon, 1785 C, 1892 C, 1894 C
Arithmetic, 1202 Ma, c.1220 Ma, c.1240 Ma, 1491 Ma, 1494 Ma, 1544 Ma, 1817 Ma, 1832–1833 Ma, 1884 Ma, 1889 Ma, 1894–1908 Ma
Arithmetica (Diophantus), 1621 Ma
Arithmetical triangle, 1654 Ma
Artificial limbs, 1563 H
Artificial transmutation, 1919 P, 1921–1924 P
Aspirin, 1899 H
Asteroids, 1801 A, 1802 A, 1804 A, 1807 A, 1809 A, 1845 A, 1891 A; gaps among, 1857 A, 1866 A
Astigmatism, 1801 H, 1825 H
Astrea, 1845 A
Astrolabe, 1391 A, 1392 A
Astrology, c.1370 A, 1666 S
Astronomical tables, 1259 A, 1272 A, 1483 A, 1551 A, 1602 A, 1627 A
Asymptotic series, 1886 Ma
Atmosphere: circulation, 1699 E/Met, 1855 Met, 1856 Met, 1876–1880 Met, 1921 Met. *See also* Air
Atomic evolution, 1896 C
Atomic model, 1903 P, 1904 C/P, 1911 P/C, 1913 P, 1914 P, 1915 P, 1915–1916 P
Atomic number, 1913 P
Atomic structure, 1897 P/C, 1904 C/P, 1922 P
Atomic theory, c.1300 C, 1473 C, 1626 C, c.1640 C, 1758 C, 1890s P/C, 1919 P, 1921 C; and Aston, 1927 C; Berzelius, 1810–1820 C; Boyle, 1661 C, 1666 C; Dalton, 1801 Met/C, 1802 C, 1803 C/P, 1804 C, 1808 C; Einstein, 1905 P, 1919 P; Faraday, 1844 C; Higgins, 1789 C; Lémery, 1675 C; Maupertuis, 1751 B; Newton, 1704 C; Pauli, 1925 P; Prout, 1816 C; Strutt, 1901 C; J. Thomson, 1881 C/P; T. Thomson, 1807 C
Atomic volume curve, 1869 C
Atomic weights, 1803 C/P, 1814 C, 1816 C, 1818 C, 1820 C, 1826 C, 1869 C, 1886 C, 1892 C, 1910–1911 C
Aurora Borealis, 1733 Met, 1867 Ma
Avogadro's Hypothesis, 1811 C, 1814 C, 1860 C, 1864 C, 1865 C
Avogadro's number, 1865 C, 1908 C
Axiom of choice, 1904 Ma, 1907 Ma, 1908 Ma

Babcock test, 1890 C
Bacteria, 1683 B, 1692 B
Bacteriophage, 1915 B, 1917 B
Balance spring, 1658 P
Balduin's phosphorus, 1677 C
Balloon, 1783 Met
Balmer lines, 1885 P, 1913 P
Barium, 1808 C
Barometer, 1643 Met, 1645 P, 1665 Met, 1667 Met, 1704 Met; aneroid, 1702 Met; 3-liquid, 1685 Met
Barothermoscope, 1592 Met
Barycentric coordinates, 1827 Ma
Basalt, 1765 E, 1770 E, 1774 E, 1779 E, 1798 E, 1802–1803 E
Basedow's disease, 1840 H
Bathing, 1697 H
Bathymetry, ocean, 1854 E
Battery, 1800 P, 1836 P
Bauhin's valve, 1588 B
Beagle expedition, 1831–1836 E/B, 1836 E, 1842 E, 1844 E
Beaufort Scale, 1805 Met
Bees, 1609 B

Beet sugar, 1747 C
Bell-Magendie Law, 1811 B, 1822 B
Benzene, 1825 C, 1861 C; ring, 1865 C
Beriberi, 1890–1896 H, 1901 H, 1912 H
Berlin Academy of Sciences, 1700 S
Bernoulli numbers, polynomials, theorem, 1713 Ma
Bertrand's postulate, 1845 Ma, 1850 Ma
Beryllium, 1798 C, 1828 C
Bessel coefficients, 1824 Ma
Bessemer process, 1856 C
Beta functions, 1729 Ma
Beta particles, 1914 P
Betelgeuse, 1920 A
Betti numbers, 1871 Ma
Big Bang hypothesis, 1927 A
Binary forms, 1868 Ma, 1870 Ma, 1910 Ma
Binary system, c.1700 Ma
Binomial congruences, 1850 Ma
Binomial expansion, 1733 Ma
Binomial nomenclature, for plants, 1623 B, 1753 B, 1758 B
Binomial series, 1670 Ma, 1826 Ma
Binomial theorem, 1676 Ma, 1713 Ma, 1774 Ma
Biochemistry, 1915 B
Biology, term defined, 1800 B, 1802 B
Biot and Savart, Law of, 1820 P
Biquadratic reciprocity, 1825 Ma, 1832 Ma
Biquaternions, 1873–1876 Ma
Birds, c.1245 B, c.1250 B, 1555 B, 1781–1786 B, 1827–1839 B, 1880 B, 1920 B
Black body, 1860 P
Black Death, 1347–1350 H, 1348 H. *See also* Plague
Black holes, 1784 A, 1796 A, 1916 A
Bleeding. *See* Bloodletting
Blood, 1658 B, 1669 B, 1673 B, 1922 H/B; circulation, c.1260 B, 1553 B, 1556 B, 1559 B, 1603 B, 1616 B, 1628 B, 1651 B, 1660 B, 1663 B, 1668 B, 1733 B, 1756 B; coagulation, 1771 B; color, 1745 B, 1771 B, 1840 B/P; composition, 1745 B, 1773 B, 1830 B, 1837 B, 1920 B, 1922 B, 1927 H; Hewson on, 1774 B; Magnus on, 1837 B; Mayer on, 1840 B; Menghini on, 1745 B; pressure, c.1705 B, 1733 B; Rouelle on, 1773 B; transfusion, 1665 H, 1883 H; types, 1900 H
Bloodletting, 1347–1350 H, 1539 B
Bode's Law, 1766 A, 1772 A, 1781 A, 1801 A
Bohr atom, 1913 P, 1914 P
Bolometer, 1878 P
Bones, 1808 B; intermaxillary, 1784 B
Boolean algebra, 1847 Ma, 1870 Ma, 1877 Ma, 1916 Ma
Boric acid, 1702 C, 1778 C, 1779 C, 1808 C
Boron, 1808 C
Botany: descriptive, c.1250 B, 1544 B, 1576 B, 1675–1679 B, 1682 B, 1737 B, 1778 B, 1810 B, 1844 B, 1844–1847 B, 1875–1897 B; herbals, c.1287 B, 1410 B, 1500 H, 1530 B, 1542 B; observational, 1768–1771 B; plant motion, 1880 B; plant physiology, 1727 B; practical, 1306 B. *See also* Classification, of plants
Boulders, erratic, 1779 E, 1794 E, 1795 E, 1815 E, 1832 E, 1833 E
Boyle's Engine, 1662 C
Boyle's Law, 1662 C, 1676 C, 1687 C, 1738 C, 1834 C
Brain, 1573 B, 1745 H, 1810–1819 B, 1811 B, 1836 B, 1861 B, 1863 B, 1868 H, 1876 H/B, 1929 H
Brewster's Law, 1811 P
Bromine, 1825 C, 1826 C, 1920 B
Brownian motion, 1827 P, 1905 P, 1908 P
Brucia, 1891 A
Brückner period, 1887 E
Brunhes Epoch, 1906 E
Buffon needle problem, 1777 Ma
Butyl, 1864 C
Buys Ballot's Law, 1875 Met

Cadmium, 1817 C
Calcination, 1630 C, 1775 C, 1778 C, 1783 C
Calcite, 1669 P, 1678 P
Calcium, 1808 C
Calcium carbide, 1895 C
Calculators, 1623 Ma, 1642 Ma, 1673 Ma, 1888 Ma, 1905 Ma
Calculus, 1615 Ma, 1638 Ma, 1647 Ma, 1651 Ma, 1696 Ma; criticisms of, 1694 Ma, 1734 Ma; tensor, 1862 Ma; of variations, 1697 Ma, 1701 Ma, 1736 Ma, 1744 Ma, 1744 P, 1760 Ma, 1762 Ma, 1786 Ma, 1837 Ma, 1922–1924 Ma; and the following persons on: Agnesi, 1748 Ma; Ampère, 1806 Ma; Barrow, 1670 Ma; Bayes, 1736 Ma; J. Bernoulli, 1690 Ma; Bolzano, 1817 Ma; Cauchy, 1821 Ma, 1826 Ma; Euler, 1755 Ma, 1768–1770 Ma; Fourier, 1822 P/Ma; Green, 1828 Ma; Gregory, 1667 Ma, 1668 Ma; Hudde, 1658 Ma; Lagrange, 1797 Ma, 1799–1806 Ma, 1801 Ma; Landen, 1758 Ma; Leibniz, 1673–1676 Ma, 1677 Ma, 1684 Ma, 1686 Ma; Newton, 1664–1666 A/P, 1669 Ma, 1671 Ma, 1676 Ma, 1687 Ma, 1693 Ma,

1704 Ma, 1711 Ma, 1735 Ma, 1736 Ma; Pascal, 1658 Ma, 1659 Ma; Riemann, 1854 Ma; Stokes, 1854 Ma; Volterra, 1881 Ma
Calculus, fundamental theorem of, 1684 Ma
Calendar reform, 1576 A, 1582 A
Calorimeter, 1783 C; vapor, 1887 C
Cambrian System, 1835 E
Camera, 1827 C
Canal rays, 1886 P
Cancer, 1890–1891 H
Canyons, 1875 E
Capillaries, 1660 B, 1663 B
Carbolic acid, 1834 C, 1865 H
Carbon, 1785 C, 1828 C, 1858 C, 1861 C, 1874 C, 1900 C
Carbon cycle, 1779 B, 1804 B
Carbon dioxide, 1754 C, 1756 C, c.1770 C, 1778 C, 1779 B, 1824 H, 1837 B, 1896 E, 1899 E, c.1908 Met
Carbon monoxide, 1770s C; liquid, 1877 C
Carburetor, 1892 P
Cardinal numbers, 1884 Ma
Caries, 1728 H, 1778 H
Carnot cycle, 1824 P
Cartography. *See* Globes; Maps
Cassini division, 1675 A, 1866 A, 1888 A
Catalysis, 1740s C, 1806 C, 1816 C, 1823 C, 1836 C, 1894 C
Catastrophism, 1795 E, 1812 E, 1830–1833 E
Cathode rays, 1876 P, 1895 P, 1897 P
Cattle fever, 1893 H
Cauchy: integral test, 1837 Ma; integral theorem, 1825 Ma; sequences, 1872 Ma
Cauchy-Lipschitz Theorems, 1876 Ma, 1893 Ma
Cauchy-Riemann equations, 1851 Ma
Cauchy's Ratio Test, 1776 Ma
Causality, c.1300 S; Kant on, 1781 S
Cauterization, 1347–1350 H, 1545 H
Cayley's Theorem, 1854 Ma, 1870 Ma
Celestial mechanics, 1772 A, 1799–1825 A, 1825 A
Cell: as new term, 1665 B; nucleus of, 1831 B, 1838 B; theory, 1838 B, 1858 H
Cellular pathology, 1858 H
Celluloid, 1865 C, 1869 C
Celsius temperature scale, 1694 P, 1742 P, 1743 P
Census data, 1829 Ma
Centigrade temperature scale. *See* Celsius temperature scale
Centripetal force, 1659 P, 1673 P, 1679 A/P
Cepheid variables, 1912 A, c.1913 A, 1923 A, 1929 A
Cerebrospinal fluid, 1360 H
Ceres, 1801 A, 1809 A
Cerium, 1803 C
Cesarean operation, 1500 H
Cesium, 1860 C
CGS (measuring system), 1832 P
Challenger expedition, 1872–1876 E, 1895 E
Charles's Law, 1787 C, 1802 C, 1834 C
Chemical bond, 1927 C
Chemical compounds, 1919–1921 C
Chemical nomenclature, 1697 C, 1746 C, 1749 C, 1751 C, 1766 C, 1779 C, 1780 C, 1782 C, 1784 C, 1787 C, 1789 C, 1790 C, 1794 C, 1840 C, 1892 C
Chemistry: alkaloid, 1817 C; low-temperature, 1823 C, 1845 C, 1877 C, 1898 C, 1899 C, 1908 C, 1911 C; organic, 1806 C, 1810 C, 1828 C, 1832 C, 1837 C, 1840 C, 1858 C, 1861 C
Childbed fever. *See* Puerperal fever
Childbirth, 1563 H, 1751 H, 1853 H
Chi-square test, 1900 Ma
Chladni figures, 1787 P
Chlorine, 1774 C, 1800 H, 1810 C, 1811 C, 1819 C, 1823 C
Chloroform, 1831 C, 1847 H, 1853 H
Chlorophyll, 1817 B
Cholera, 1849 H, 1854 H, 1883 H; chicken, 1880 B
Chorda Tympani, 1843 B
Chromium, 1798 C
Chromosomes, 1903 B, 1904 B, 1915 B, 1926 B
Chronometers, 1765 P
Circumnavigation, 1522 E
Clairaut formula, 1743 E
Classification, of air masses, 1928 Met; of animals, 1551–1558 B, 1660 B, 1691 B, 1749 B, 1758 B, 1802 B, 1806 C, 1809 B, 1816–1822 B; of plants, c.1220 B, c.1250 B, 1554 B, 1565 B, 1583 B, 1623 B, 1640 B, 1660 B, 1686–1704 B, 1691 B, 1694 B, 1735 B, 1737 B, 1758 B, 1776 B, 1778 B, 1789 B, 1898 E; of chemicals, 1806 C; of climates, 1918 E; of clouds, 1803 Met, 1832 Met; of diseases, 1763 H, 1840 H; of earth's history, 1856 C; of foods, 1827 C/H; of insects, 1680 B, 1775 B; of minerals, 1546 E, 1774 E, 1782 C, 1784 C/E, 1814 E, 1817 E; of plant fossils, 1828 B; of rocks, 1843 E; of stars, 1918–1924 A; of teeth, 1778 H; of tissues, 1800 B
Climatology, 1733 B, 1784 E, 1887 E, 1899 E, 1918 E, 1919–1936 E, 1920 E.

See also Paleoclimatology
Clock, pendulum, 1656 P, 1672 P. *See also* Watches
Cloud chamber, 1896 P
Clouds, 1637 Met, 1803 Met, 1832 Met, 1889 Met
Cobalt, 1737–1738 C, 1819 A
Cocaine, 1884 H
Color, c.1300 P, 1665 P, 1772 P, 1810 P; blindness, 1777 H, 1794 H
Columbium, 1801 C, 1809 C
Combustion, c. 1500 S, 1665 C, 1669 C, 1674 C/Met, 1700 C, 1770 C, 1771 C, 1772 C, 1774 C, 1775 C, 1777 C/Met, 1778 C, 1783 C, 1789 C, 1797–1812 C, 1800 C
Comets, 1443–1472 A, 1540 A, 1577 A, 1578 A, 1620 A, 1623 A, 1664 A, 1694 E, 1705 A, 1758 A, 1797 A, 1809 A, 1815 A, 1818 A, 1927 A
Compass, 1269 P, 1581 E
Competition, among species, 1749 B, 1859 B
Complementarity Principle, 1928 P
Complex numbers, 1545 Ma, 1572 Ma, 1685 Ma, 1746 Ma, 1796 Ma, 1797 Ma, 1806 Ma, 1831 Ma, 1837 Ma, 1847 Ma, 1849 Ma, 1856–1860 Ma, 1863 Ma, 1870 Ma
Compton effect, 1923 P
Computers, 1822 Ma, 1833 Ma, 1834 Ma, 1853 Ma, 1880 Ma, 1890 Ma, 1895 Ma, 1898 Ma, 1924 Ma. *See also* Calculators
Conchology, 1862–1869 B
Condensation nucleii, 1875 Met, 1881 Met
Condenser, 1764 P
Conduction, heat, 1822 P
Conductivity: of gas, 1896 P, 1903 P; of metals, 1821 P, 1911 C
Conductors, 1732 P
Congruence, 1801 Ma
Conics, 1720 Ma
Conic sections, 1522 Ma, 1629 Ma, 1639–1640 Ma, 1667 Ma, 1668 Ma, 1679 Ma
Conjugate point, 1837 Ma
Consciousness, 1899 B
Conservation of angular momentum, 1664–1666 P
Conservation of energy, 1840 B/P, 1840 C, 1842 P, 1847 P, 1849 P, 1850 P, 1905 P
Conservation of matter, 1789 C, 1905 P
Conservation of momentum, 1644 P
Constant Composition, Law of, 1792 C, 1797 C, 1799 C, 1808 C
Continental drift, 1845 E, 1857 E, 1885 E, 1912 E, 1915 E
Continentality, 1845 Met
Continents, 1320 E
Continuity, 1344 Ma, 1806 Ma, 1854 Ma, 1861 Ma, 1873 Ma
Continuum Hypothesis, 1883 Ma, 1895–1897 Ma
Convection currents, ocean, 1797 E
Convergence: of series, 1669 Ma, 1673 Ma, 1703 Ma, 1705 Ma, 1776 Ma, 1812 Ma, 1821 Ma, 1826 Ma, 1837 Ma, 1873 Ma, 1879 Ma; conditional, 1833 Ma; uniform, 1847 Ma
Cooling, Newton's Law of, 1701 P
Coordinates: homogeneous, 1827 Ma, 1828–1831 Ma, 1830 Ma; line, 1830 Ma; polar, 1671 Ma, 1691 Ma, 1748 Ma
Copernican theory, 1514 A, 1540 A, 1543 A, 1543 P, 1551 A, 1576 A, 1578 A, 1584 A, 1588 A, 1605 A, 1609 A, 1610 A, 1613 A, 1616 S, 1632 A, 1633 S, 1647 A, 1651 A, 1690 A, 1835 S
Coral, 1734–1742 B, 1846 B, 1871 E; reefs, 1842 E
Coriolis effect, 1831 Met, 1835 Met, 1856 Met/E, 1857 Met, 1875 Met, 1905 E
Cornea transplant, 1905 H
Correlation coefficient, 1869 Ma, 1889 Ma
Correspondence Principle, 1923 P
Cosines, Law of, 1464 Ma
Cosmological constant, 1916 A, 1917 A
Cosmology, 1644 A, 1734 A, 1750 A, 1785 A, 1796 A, 1799–1825 A, 1845–1862 S, 1854 A, 1916 A, 1917 A, 1919 A, 1922 A, 1927 A
Cotes formula, 1714 Ma
Covariants, 1844 Ma, 1868 Ma, 1910 Ma
Creation: date of, 1650–1654 E, 1654 E, 1728 S; theories, 1566 S, 1754 B, 1778 E, 1804–1811 E
Cremona transformations, 1863 Ma
Creosote, 1833 C
Critique of Pure Reason (Kant), 1781 S
Crystallography, 1669 C/E, 1772 E/C, 1801 E, 1822 E, 1848 C, 1851 C
Cubic curves, 1704 Ma
Culturing, 1877 H
Curvature, 1781 Ma, 1827 Ma
Curve, space-filling, 1890 Ma, 1923 Ma
Cyanogen, 1815 C
Cyano radical, 1815 C
Cycloid, 1673 Ma, 1697 Ma
Cyclones, 1919 Met
Cyclotron, 1930 P

Daguerreotypes, 1824 C, 1839 C
D'Alembert's Principle, 1743 P

Daniell cell, 1836 P, 1837 P
Davisian System, 1889 E
Decalcification, 1745 H
Decameron (Boccaccio), 1353 H
Decimals, 1530 Ma, 1585 Ma
Dedekind cuts, 1833–1835 Ma, 1872 Ma
Deduction, 1644 S
Deferents, 1217 A, 1514 A, 1543 A
Deimos, 1877 A
De Moivre's Theorem, 1730 Ma, 1748 Ma
Dentistry, 1728 H, 1756 H, 1839 H, 1844 H, 1846 H. *See also* Teeth
Dentures, 1728 H
Denudation, of land, 1846 E
Denumerability, c.1840 Ma
Dephlogisticated air. *See* Oxygen
Depression, 1911 H, 1916 H
Derivative, 1754 Ma, 1758 Ma, 1821 Ma
Desargues's Theorem, 1648 Ma, 1825–1826 Ma
Descartes's Rule of Signs, 1637 Ma, 1707 Ma
Determinants, 1841 Ma, 1845 Ma, 1862 Ma, 1906–1923 Ma
Determinism, mathematical, 1812 P, 1820 Ma
Devonian System, 1837 E, 1839 E
Dew, 1781 C, 1783 C
Dewar flask, 1892 C, 1905 C
Dew point, 1751 Met
Diabetes, 1670 H, 1921 H
Diamagnetism, 1845 P, 1911 C
Diamonds, 1796 C, 1799 C
Didymium, 1841 C
Diesel engine, 1895 P
Dieting, 1731 H
Difference Engine. *See* Computers
Differentiation, 1684 Ma, 1695 Ma; partial, 1720 Ma
Diffraction, 1665 P, 1680 P; gratings, 1882 P
Diffusion, 1834 C
Digestion, 1648 B, 1833 B, 1836 B, 1843 H, 1878 B, 1897 B, 1902 B
Digitalis, 1775 H, 1785 H
Dimensions, 1822 P, 1878 Ma
Dinosaurs, 1884 B
Dione, 1684 A
Diophantine analysis, 1769–1770 Ma
Diphtheria, 1748 H, 1765 H, 1883 H, 1888–1891 H
Discours de la méthode (Descartes), 1637 S
Disease, 1858 H; classification of, 1763 H, 1840 H; descriptions of, 1275 H, 1348 H, 1530 H, 1554 H, 1572 H, 1623 H; nervous, 1892 H; theories of, c.1530 H, 1564 H, 1648 H, 1840 H, 1862 H, 1867 H, 1874 H, 1880 H; vitamin-deficiency, 1912 H. *See also* Medicine; Sanitation; Surgery; individual ailments
Disinfection, 1819 H
Disquisitiones arithmeticae (Gauss), 1801 Ma
Dissections, 1238 B, 1252 H, 1275 B, 1286 H, 1316 B, 1537 H, 1573 B
Distillation, 1500 C, 1512 C
Distribution, normal, 1733 Ma
Divergent series, 1843 Ma, 1882 Ma, 1894 Ma
Division, long, 1491 Ma
Doldrums, 1850 Met
Doppler Effect, 1842 P
Doppler-Fizeau Effect, 1848 A, 1868 A, 1912 A
Dropsy, 1775 H
Duality: logical, 1877 Ma; principle of, 1830 Ma, 1831 Ma
Dyes, 1856 C, 1878 C, 1907 H
Dynamite, 1866 C
Dynamo, 1878 P
Dysentery, 1900 H, 1917 H
Dysprosium, 1886 C

e, 1727–1728 Ma, c.1750 Ma, 1873 Ma
Ear, 1564 B, 1600–1601 B
Earth: age of, 1778 E; composition of, 1644 E, 1665 E; core of, 1866 E, 1916 E; crust of, 1756 E, 1759 E; history of, 1681 E, c.1693 E, 1695 E, 1696 E, 1740 E, 1748 E, 1749 E, 1762 E, 1778 E, 1785 E, 1787 E, 1791 E, 1795 E, 1865 C; mass of, 1774 E, 1798 E; motion of, 1748 A; orbit of, 1904 A; orbital variations of, 1843 A; radius of, 1684 E; revolution of, 1440 A, 1514 A, 1543 A, 1578 A, 1633 S, 1675 A; rotation of, c.1350 A, c.1370 A, 1440 A, 1543 A, 1600 A, 1640 E, 1687 A, 1835 Met, 1851 E; scale of, 1410 E, 1492 E; shape of, 1267–1268 E, c.1450 E, 1492 E, 1522 E, 1687 E, 1743 E, 1744 E, 1841 E; wobble of, 1752 E, 1891 E
Earthquakes, 1705 E, 1749 E, 1756 E, 1760 E, 1846 E, 1883 E, 1909 E. *See also* Seismology
Eclipse, c.1370 A, 1715 A, 1851 A, 1868 A/C, 1919 P/A
Ecliptic, obliquity of, 1290 A
EEG, 1929 H
Efficiency, 1752–1754 P
Ehrebungs Hypothesis, 1825 E
Ekman spiral, 1905 E
Elasticity, 1744 P

Elastic surfaces, 1816 P, 1821 P
Electric discharge speed, 1834 P
Electricity, 1747 P, 1759 P, 1800 P, 1828 P/Ma, 1832 P, 1882 P; atmospheric, 1753 Met; current converters, 1892 P; Franklin on, 1750 P, 1752 P; Gray on, 1729 P, 1732 P; Hauksbee on, 1705 P, 1706 P; and Leyden jar, 1745 P; Ohm on, 1825 P; one-fluid theory, 1746 P; two-fluid theory, 1734 P; von Guericke on, 1660 P, 1672 P. *See also* Electromagnetism
Electrocardiograph, 1903 H
Electrochemistry, 1807 C
Electrodynamometer, 1846 P
Electrolysis, 1800 C, 1803 C, 1833 C
Electromagnet, 1825 P
Electromagnetic induction, 1831 P, 1871 P
Electromagnetic rotation, 1821 P
Electromagnetic self-induction, 1832 P
Electromagnetic waves, 1888 P, 1892 P
Electromagnetism, 1820 P, 1821 P, 1827 P, 1828 P, 1831 P, 1833 P, 1855 P, 1865 P, 1873 Ma, 1885 P, 1886–1889 P, 1893 P
Electron(s), 1874 P, 1897 P, 1910 P, 1911 P/C, 1924 P, 1926 P, 1927 P, 1928 P; paths, 1915 P; shared, 1916 C; spin, 1925 P
Electroplating, 1839 C
Electroscope, 1748 P, 1791 P
Electrovalency, 1902 C
Elements, 1648 C, 1661 C, 1789 C
Elements (Euclid), 1220 Ma, c.1255 Ma, 1482 Ma, 1570 Ma
Ellipses, 1609 Ma
Elliptic functions, 1825 Ma, 1827 Ma, 1829 Ma
Elliptic integrals, 1716 Ma, 1761 Ma, 1825–1832 Ma, 1829 Ma
Embryology, 1559 B, 1621 E, 1651 B, 1673 B, 1759 B, 1828–1837 B, 1927 B
Emissivity, 1860 P
Emotions, 1872 B; anger, 1894 H
Emphysema, 1726 H
Enceladus, 1789 A
Encke's comet, 1818 A
Encyclopedias, c.1235 S, 1250 S, 1695 S, 1704 S, 1751–1772 S, 1754 Ma
Energy: activation, 1889 C, 1894 C; free, 1870–1879 C; solar, 1848 A, 1853 A, 1929 A. *See also* Conservation of energy
Engines, heat, 1824 P
Entropy, 1850 P, 1854 C, 1906 C
Enzymes, 1836 B, 1908 B, 1926 C
Ephemeris, 1767 A
Epicycles, 1217 A, 1514 A, 1543 A
Epigenesis, 1651 B, 1759 B
Epilepsy, c.1310 H, 1827 H, 1863 H
Epinephrine, *see* Adrenaline
Equals sign, 1557 Ma
Equation of continuity, 1775 P
Equation of state, 1834 C
Equations, 1591 Ma; cubic, 1225 Ma, 1515 Ma, 1535 Ma, 1545 Ma, 1546 Ma, 1572 Ma, 1591 Ma, 1683 Ma, 1704 Ma, 1720 Ma, 1770 Ma, 1853 Ma; difference, 1713 Ma, 1759–1792 Ma, 1850 Ma, 1900–1910 Ma; differential, 1671 Ma, 1692 Ma, 1736 Ma, 1739 Ma, 1744 Ma, 1774 Ma, 1820–1830 Ma, 1836 Ma, 1876 Ma, 1877 Ma, 1881 Ma, 1884 Ma, 1886 Ma, 1892–1899 A/Ma, 1893 Ma, 1908 Ma; higher-degree, 1799 Ma; linear, c.1220 Ma, 1225 Ma, 1494 Ma; partial differential, 1747 Ma; polynomial, 1225 Ma, 1629 Ma, 1658 Ma, 1679 Ma, 1683 Ma, 1720 Ma, 1770 Ma, 1853 Ma; quadratic, c.1220 Ma, 1225 Ma, 1494 Ma, 1591 Ma; quartic, 1544 Ma, 1591 Ma, 1720 Ma, 1770 Ma, 1853 Ma; quintic, 1770 Ma, 1786 Ma, 1813 Ma, 1824 Ma, 1830–1832 Ma, 1858 Ma; theory of, 1637 Ma, 1767 Ma, 1770 Ma. *See also* Algebra
Equinoxes, precession of, 1543 A, 1687 P, 1754 A
Equivalent Proportions, Law of, 1791 C
Erbium, 1843 C
Ergodic Hypothesis, 1877 P, 1890 P
Ergodicity, 1911 P
Erlanger Programm, 1872 Ma
Erosion, cycle of, 1889 E
Ether: medical, 1540 C, 1846 H, 1847 H; space as, 1643 A, 1675 P, 1678 P, 1690 P, 1704 P, 1716 P, 1744 A, 1801–1804 P, 1851 P, 1865 P, 1887 P, 1892 P, 1893 P, 1905 P
Ethylene, liquid, 1845 C
Euclid's parallel postulate, c.1250 Ma, 1733 Ma, 1763 Ma, 1766 Ma, 1781 S, 1786 Ma, 1795 Ma, 1816 Ma, 1823 Ma, 1826 Ma, 1873 Ma
Euler: buckling formula, 1744 P; identities, 1748 Ma; substitution, 1768–1770 Ma
Eulerian flow, 1761 E
Eulerian integrals, 1729 Ma, 1768–1770 Ma
Europium, 1901 C
Eustachian tube, valve, 1564 B
Evaporation, 1857 C/P
Evolution: catastrophic, 1770 B; Scopes trial, 1925 S; views of following persons on: Agassiz, 1860 B; Bagehot, 1873 B;

Bateson, 1894 B, 1914 B; Buffon, 1749 B; Cardano, 1550 B; Chambers, 1844 B; Cuvier, 1798 B; C. Darwin, 1844 B, 1858 B, 1859 B, 1868 B, 1871 B, 1872 B, 1875 B, 1900 B; E. Darwin, 1794 B; Fisher, 1930 B; Freke, 1851 B, 1861 B; Grant, 1813 B, 1826 B; Haldeman, 1843–1844 B; d'Halloy, 1846 B; Hooker, 1859 B; Huxley, 1859 B; Jenkin, 1867 B; Kammerer, c.1920 B; Lamarck, 1801 B, 1809 B, 1816–1822 B; de Maillet, 1748 B; Marsh, 1877 B; Matthew, 1831 B; Robinet, 1761–1768 B; Saint-Hilaire, 1796 B; Spencer, 1852 B, 1858 S, 1862 S, 1876–1896 B, 1886 B; de Vries, 1901–1903 B; Wallace, 1858 B, 1870 B, 1876 B; Wells, 1813 B
Excisions, 1769 H
Excluded middle, in logic, c.1320 Ma, 1923 Ma
Existence proofs, 1836 Ma
Existence theorems, 1820–1830 Ma, 1893 Ma
Expanding universe, 1848 A, 1922 A, 1928 A, 1929 A
Exponents, 1614 Ma; fractional, c.1360 Ma, 1655 Ma; imaginary, 1740 Ma; negative, 1655 Ma; notation for, 1637 Ma
Extinction, 1876 B, 1880 B
Extramission theory, c.1250 B
Extraterrestrial life, 1440 A, 1584 A, 1638 A, 1690 A, 1877 A, 1894 A, 1906 A
Eye, 1593 B, 1660 B, 1801 H, 1905 H. *See also* Vision
Eyeglasses, 1299 H, 1451 H, 1589 P, 1825 H

Fahrenheit temperature scale, 1714 P
Falcons, c.1245 B
Falling bodies, 1328 P, c.1350 P, c.1500 S, 1586 P, 1589–1592 P, 1604 P, 1609 P, 1638 P, 1784 P, 1850 P
Fallopian tubes, c.1560 B
Falsification, Principle of, c.1230 S
Farsightedness, 1299 H
Fats, 1823 C
Faulting, 1878 E
Fermat numbers, 1640 Ma, 1880 Ma, 1905 Ma, 1909 Ma
Fermat's Last Theorem, 1637 Ma, 1670 Ma, 1770 Ma, 1849 Ma
Fermat's Little Theorem, 1640 Ma, 1736 Ma, 1819 Ma
Fermat's Principle, 1657 P
Fermentation, 1680 B, 1856 B, 1897 B
Ferns, 1838 B, 1846–1864 B
Ferrel Cell, 1855 Met
Ferrel's Law, 1855 Met
Fertilization: human, 1779 B, 1824 B, 1875 B; of plants, 1740 B, 1763 B, 1823–1830 B, 1862 B, 1876 B
Fertilizers, 1842 C, 1855 C
Fetus, c.1276 B, 1604 B
Fibonacci sequence, 1220 Ma
Field theory, mathematics, 1830–1832 Ma, 1879 Ma, 1881 Ma, 1893 Ma, 1910 Ma; physics, 1918 Ma/P
Film, photographic, 1887 C
Fingerprints, 1885 B
Finite differences, 1715–1717 Ma
Fish, 1788–1804 B, 1828 B; fossil, 1833–1844 E
Fitzgerald contractions, 1892 P
Fixed air. *See* Carbon dioxide
Fjords, 1870 E
Flame test, 1758 C
Flight: airplane, 1903 P; balloon, 1783 Met; mechanics of, 1809–1854 P
Fluid dynamics, 1738 P, 1744 P, 1755 P, 1761 P, 1775 P, 1788 P/E, 1797 P, 1823 P, 1845 P
Fluid pressure, 1583 P, c.1648 P, 1661 P, 1663 P
Fluorine, 1886 C
Fluxions, 1664–1666 A/P, 1669 Ma, 1693 Ma, 1704 Ma, 1734 Ma, 1735 Ma, 1736 Ma, 1754 Ma, 1797 Ma
Folding, 1878 E
Foods: classification of, 1827 C/H; milk, 1890 C, 1905 H; preservation of, 1765 C, 1810 C; pumpkin, Indian corn, 1542 B. *See also* Nutrition; Vitamins
Foot and mouth disease, 1898 H
Forceps, 1696 H
Force(s): Knight on, 1748 P; line of, 1821 P, 1855 P
Formalism, mathematical, 1904 Ma, 1912 Ma, 1918 Ma, 1925 Ma
Fossils, 1282 E, 1808 E; Agassiz on, 1833–1844 E; Bonnet on, 1770 B; classification of plant, 1828 B; Cuvier on, 1812 E; Guettard on, 1757 E, 1765 E; Hooke on, 1705 E; Lister on, 1671 E, 1678 E; North American, 1832 E; Parkinson on, 1804–1811 E; Plot on, 1677 E; Scheuchzer on, 1709 E, 1716 E, 1726 E; Smith on, 1799 E, 1815 E, 1816 E, 1817 E; Steno on, 1669 E; da Vinci on, c.1500 S; von Buch on, 1830 E; Woodward on, 1695 E. 1728–1729 E

Foucault pendulum, 1851 E
"Foundation of the General Theory of Relativity, The" (Einstein), 1916 P/A
Four-color problem, 1852 Ma
Fourier analysis, 1807 Ma, 1822 Ma, 1897–1898 Ma, 1930 E
Fourier series, 1728 Ma, 1873 Ma
Foxglove, 1785 H
Fractions, c.1340 Ma, 1822 Ma, 1887 Ma; continued, 1572 Ma, 1658 Ma, 1737 Ma, 1850 Ma, 1879 Ma
Fraunhofer lines, 1814 A, 1859 A, 1870 A
French Academy of Sciences, 1666 S, 1793 S
Friction, 1798 P
Frogs, 1668 B, 1780 B, 1791 B
Fruit flies, 1910 B, 1926 B, 1927 B
Fuel cell, 1839 C
Fuels, liquid, 1910 C
Functionals, 1887 Ma
Functions, 1806 Ma; of a complex variable, 1825 Ma, 1851 Ma; continuous, 1817 Ma, 1834 Ma; Dirichlet, 1837 Ma; elliptic, 1825 Ma, 1827 Ma, 1829 Ma; Euler on, 1748 Ma; Hankel on, 1870 Ma; hyperelliptic, 1832 Ma; Mathieu on, 1868 Ma; of a real variable, 1821 Ma; theory of, 1822 Ma, 1886 Ma, 1902 Ma
Fundamental Theorem: of Algebra, 1799 Ma, 1816 Ma; of Arithmetic, 1801 Ma; of the Calculus, 1684 Ma
Furnace, electric, 1892 C

Gadolinite, 1794 C
Galaxies. *See* Stars
Gallium, 1869 C, 1874 C, 1875 C
Galois group, 1830–1832 Ma
Galois imaginaries, 1830–1832 Ma
Galois Theory, 1830–1832 Ma, 1846 Ma
Galvanometer, 1820 P, 1826 P
Game theory, 1921 Ma, 1928 Ma
Gamma functions, 1729 Ma, 1768–1770 Ma
Gangrene, 1790 H
Gas(es), 1624 C, 1648 C, 1754 C, 1766 C, 1770s C, 1803 C; laws, 1662 C, 1676 C, 1687 C, 1699 C, 1738 C, 1787 C, 1801 C, 1802 C, 1803 C, 1808 C, 1809 C, 1834 C, 1873 C; liquefied, 1882 C; mantle, 1885 C. *See also* Kinetic theory of gases
Gastric juices, 1823 B, 1833 B
Gaussian complex integers, 1825 Ma
Gaussian distribution, 1809 Ma
Gay-Lussac Law, 1809 C, 1811 C
Geiger counter, 1908 P, 1928 P
Generators, 1672 P, 1831 P, 1878 P
Genes, 1865 B, 1900 B, 1909 B, 1915 B, 1926 B
Genetics: Bateson on, 1902 B; Boveri on, 1889 B; Bridges on, 1916 B; Fisher on, 1930 B; Galton on, 1869 B; Johannsen on, c.1903 B; Mendel on, 1865 B, 1900 B; Morgan on, 1910 B, 1916 B, 1926 B; Muller on, 1927 B; Sutton on, 1903 B; de Vries on, 1901–1903 B, 1907 B; Weismann on, 1892 B. *See also* Chromosomes; Genes
Geognosy, 1808 E
Geography, 1410 E, 1817–1859 E; descriptive, 1296–1297 E, 1442 E, 1502 E, 1507 E, 1805–1834 Met/E, 1855 E; explorations, 1296–1297 E, 1416–1434 E, 1442 E, 1502 E, 1522 E; historical, 1877 E; speculative, c.1450 E; text, 1650 E. *See also* Globes; Maps
Geography (Ptolemy), 1410 E
Geoid, 1735–1744 E, 1736–1737 E, 1744 E
Geological column, 1787 E, c.1796 E
Geological epochs, 1830 E
Geological mapping, 1683 E, 1716–1718 E, 1743 E, 1746 E, 1815 E, 1817 E, 1819 E, 1826 E, 1840 E
Geomagnetism, 1492 E, 1581 E, 1600 E/P, 1609 E, 1635 E, 1698–1700 E, 1804 Met, 1831 E, 1832 E/P, 1839–1843 E
Geometry, 1220 Ma, c.1250 Ma, c.1255 Ma, 1604 Ma, 1673 Ma, 1704 Ma, 1795 Ma, 1817 Ma, 1833 Ma, 1837 Ma, 1854 Ma, 1856–1860 Ma, 1871 Ma, 1872 Ma; duality in, 1825–1826 Ma; foundations of, 1832–1833 Ma, 1854 Ma, 1882 Ma, 1894–1908 Ma, 1899 Ma; parallel displacement in, 1917 Ma; Solomon's seal in, 1849 Ma; various kinds of: analytical, c.1350 Ma, c.1360 Ma, 1629 Ma, 1637 Ma, 1668 Ma, 1679 Ma, 1685 Ma, 1731 Ma, 1803 Ma, 1856–1860 Ma; of curves, 1887–1896 Ma; differential, 1731 Ma, 1807 Ma; elliptic, 1854 Ma, 1868 Ma; Euclidean, 1570 Ma, 1687 Ma, 1781 S, 1796 Ma, 1854 Ma, 1900 Ma; finite projective, 1906 Ma; hyperbolic, 1823 Ma, 1826 Ma, 1829 Ma, 1832 Ma, 1854 Ma, 1868 Ma; line, 1868–1869 Ma; *n*-dimensional, 1843 Ma, 1844 Ma; non-Euclidean, 1733 Ma, 1763 Ma, 1766 Ma, 1786 Ma, 1816 Ma, 1823 Ma, 1826 Ma, 1829 Ma, 1832 Ma, 1854 Ma, 1868 Ma, 1870 Ma, 1871 Ma, 1916 P; non-Riemannian,

1918 Ma; projective, 1636–1639 Ma, 1639 Ma, 1648 Ma, 1685 Ma, 1799 Ma, 1803 Ma, 1822 Ma, 1825–1826 Ma, 1830 Ma, 1832 Ma, 1834 Ma, 1847 Ma, 1859 Ma, 1873 Ma, 1907 Ma, 1910 Ma, 1928 Ma; projective differential, 1906 Ma; spherical, 1854 Ma
Geriatrics, 1724 H
Germanium, 1869 C, 1886 C
Germ plasm, 1892 B
Germ theory, of disease, 1546 H, 1840 H, 1862 H/B, 1864 H/B, 1867 H, 1880 H
Glacial: epochs, 1899 E; history, 1815 E, 1818 E, 1829 E, 1834 E, 1835 E, 1836 E; periods, 1903–1909 E. *See also* Ice ages; Interglacial periods; Paleoclimatology
Glaciology, c.1250 E, 1716 E, 1760 E, 1779 E, 1787 E, 1794 E, 1795 E, 1840 E. *See also* Glacial
Glands, 1662 B; suprarenal, 1564 B
Glass eyes, 1579 H
Globes, 1492 E, 1541 E. *See also* Maps
Glucose, 1812 C
Goldbach conjecture, 1742 Ma
Gondwanaland, 1885 E
Gorillas, 1847 B
Gough map, 1325–1330 E
Graham's Law, 1834 C
Gram, 1790 P
Graphs, representational, c.1350 Ma, c.1360 Ma
Granite, 1795 E
Gravitation, 1679 A/P, 1781 A, 1801 P, 1827 A; Adams on, 1925 A; Borelli on, 1666 A; centers of, 1565 Ma; Clairaut on, 1743 E; Copernicus on, 1543 P; Descartes on, 1644 A; Einstein on, 1911 A, 1916 P, 1919 P; Gilbert on, 1600 E/P; and Halley, 1705 A; Hooke on, 1657 P, 1662–1666 P, 1664 A, 1674 A, 1684 A; Huygens on, 1669 P; Kepler on, 1609 A/P, 1621 A, 1643 A; Laplace on, 1799–1825 A; Newton on, 1664–1666 A/P, 1685 Ma, 1687 A, 1704 P; Richer on, 1672 P
Gravitational constant, 1798 P
Gravitational potential, 1773 P
Greenhouse effect, 1863 Met, c.1908 Met
Green's Theorem, 1828 Ma
Greenwich Observatory, 1675 A, 1712 A
Gregory's series, 1668 Ma
Group representations, 1896–1903 Ma
Group theory, 1848 Ma, 1851 C, 1854 Ma, 1870 Ma, 1872 Ma, 1878 Ma, 1884 Ma, 1888–1890 Ma, 1893 Ma; permutation, 1770 Ma, 1815 Ma. *See also* Galois theory, Group representations
Gulf Stream, 1513 E, 1770 E, 1825 Met, 1831 E, 1846 Met, 1885 E
Gunter's Scales, 1620 Ma, c.1621 Ma
Gutenberg discontinuity, 1914 E
Gyroscope, 1852 P

Hadley cell, 1735 Met, 1855 Met
Hafnium, 1923 C
Hall effect, 1879 P
Halley's comet, 1443–1472 A, 1540 A, 1705 A, 1758 A
Hall-Héroult process, 1886 C
Hansen's bacillus, 1873 H
Hauksbee machine, 1706 P
Heart, 1543 B, 1553 B, 1554 B, 1555 B, 1559 B, 1616 B, 1628 B, 1664 B, 1871 B, 1883 B; surgery, 1893 H
Heat(s), 1738 P, 1783 C, 1789 C, 1798 P, 1799 P, 1807 P, 1819 C, 1837 P, 1847 P, 1849 P, 1850 P; conduction, 1822 P, 1861 P; latent, 1762 C; mechanical equivalent of, 1843 P, 1847 P; specific, 1760 C, 1779 C, 1819 C
Hébé, 1845 A
Heine-Borel Theorem, 1872 Ma, 1923 Ma
Heliocentricity, 1514 A, 1540 A, 1543 A, 1584 A, 1632 A, 1647 A. *See also* Earth, revolution of; Sun
Helium, 1868 C, 1895 C, 1906–1909 P, 1908 P; liquid, 1908 C
Henry Draper Catalogue (Cannon), 1918–1924 A, 1925–1936 A
Henry's Law, 1803 C, 1808 C
Herbals, c.1287 B, 1410 B, 1500 H, 1530 B, 1542 B
Heredity: germ plasm theory of, 1892 B; Mendel on, 1865 B. *See also* Genetics
Hermitian forms, 1854–1857 Ma
Hertzsprung gap, 1922 A
Hertzsprung-Russell diagram, 1905–1907 A, 1908 A, 1914 A
Hess's Law, 1840 C
Hexagon, 1640 Ma
Hilbert Basis Theorem, 1890 Ma
Hindu-Arabic numerals, 1220 Ma, c.1220 Ma, 1299 Ma, 1494 Ma, c.1500 Ma
Hirondelle expedition, 1885 E
Histoire naturelle (Buffon), 1749 B, 1749 E, 1749–1788 B/E, 1753–1767 B, 1777 Ma, 1778 E, 1781–1786 B, 1788–1804 B
Hodgkin's disease, 1832 H
Holmium, 1886 C
Homeopathy, 1810 H
Homotopy group, 1895 Ma
Hooke's Law, 1678 P
Hookworms, 1902 H

Hormones, 1902 B, 1905 B
Horse latitudes, 1850 Met
Horticulture, 1875 B
Hospitals, hygiene in, 1773 H
Hubble's Law, 1929 A
Humboldt current, 1811 E
Humidity, relative, 1751 Met. *See also* Hygrometer
Huronian System, 1855 E
Hurricanes, 1835 Met, 1871 Met
Hydrocephalus, c.1270 H
Hydrochloric acid, 1611–1613 C, 1775 C, 1810 C
Hydrodynamics, 1760 P, 1883 P
Hydrofluoric acid, 1771 C
Hydrogen, 1671 C, 1766 C, 1781 C, 1783 C, 1789 C, 1816 C, 1885 P; atom, 1926 P; liquid, 1898 C; model, 1913 P; solid, 1899 C
Hydrogen bromide, 1823 C
Hydrogen chloride, 1775 C
Hydrogeology, 1802 E
Hydrolysis, 1850 C
Hydropathy, 1829 H
Hydrostatics, 1583 P, 1586 P, 1612 P, c.1648 P
Hygiene, in hospitals, 1773 H
Hygrometer, c.1450 Met, 1657 Met, 1665 Met, 1687 Met, 1773 Met, 1774 Met, 1783 Met; wet and dry bulb, 1800 Met
Hypercomplex number systems, 1863 Ma, 1870 Ma
Hypergeometric series, 1812 Ma
Hyperion, 1848 A
Hypernumbers, 1870 Ma
Hypnotism, 1774 H, 1841 H, 1893 H, 1895 H
Hysteria, 1880–1882 H, 1893 H, 1895 H

i, c.1750 Ma
Iapetus, 1671 A
Iatrochemistry, c.1530 H, 1669 C
Ice ages, 1832 E, 1837 E, 1839 E, 1840 E, 1842 E, c.1845 E, 1847 E, 1852 E, 1852–1855 E, 1863 E, 1864 E, 1867 E, 1870 E, 1873 E, 1874 E, 1875 E, 1876 E; cause of, 1920 E; mapped, 1894 E; sea level in, 1841 E. *See also* Interglacial periods
Iceland spar. *See* Calcite
Iconoscope, 1928 P
Icosahedron, 1884 Ma
Ideal numbers, 1846 Ma
Ideas, 1739 B, 1749 H, 1754 B
Ileocecal valve, 1588 B
Imaginary numbers. *See* Complex numbers
Immunochemistry, 1907 C
Impetus theory, c.1280 P, c.1350 A, c.1350 P, 1463 P
Impredicative definitions, 1906 Ma
Incandescent lamp, 1879 C
Indium, 1863 C
Induction: electrical, 1732 P; electromagnetic, 1831 P, 1832 P; magnetic, 1269 P; mathematical, 1321 Ma; as scientific method, 1620 S
Inertia, c.1350 A, 1613 P, 1644 P, 1664–1666 A/P, 1687 P, 1872 P
Infinite classes, 1638 Ma, c.1840 Ma, 1850 Ma, 1884 Ma
Infinite descent, 1659 Ma
Infinite point set, 1879 Ma
Infinite series, 1668 Ma, 1669 Ma, 1812 Ma
Infinitesimals, 1676 Ma, 1755 Ma, 1817 Ma
Infinities, 1344 Ma, 1604 Ma, 1655 Ma, 1656 Ma, 1874 Ma, 1883 Ma, 1887 Ma. *See also* Transfinite numbers
Inflammable air. *See* Hydrogen
Ingrafting, inoculation method, 1717 H
Inoculation: smallpox, 1715 H, 1717 H, 1721 H; typhoid fever, 1896 H. *See also* Vaccination
Insects, 1669 B, 1682 B, 1734–1742 B, 1737 B; anatomy of, c.1610 B; classification of, 1680 B, 1775 B
Insulators, 1732 P
Insulin, 1921 H
Integers, 1886 Ma
Integrals: elliptic, 1716 Ma, 1761 Ma, 1825–1832 Ma, 1829 Ma; hyperelliptic, 1832 Ma; Lebesgue, 1902 Ma; Riemann, 1854 Ma; ultraelliptic, 1888 Ma
Integral sign, 1686 Ma
Integration, 1558 Ma, 1565 Ma, 1615 Ma, 1635 Ma, 1647 Ma, 1655 Ma, 1684 Ma, 1884 Ma; Lebesgue, 1902 Ma; Riemann, 1854 Ma
Intelligence testing, 1905 S, 1914 S
Intercellular space, in plants, 1806 B
Interference: as light phenomenon, 1665 P; Principle of, 1801–1804 P
Interferometer, 1881 P
Interglacial periods, 1873 E, 1874 E
International Polar Year, 1882–1883 Met/E
Intuition, 1795 S
Intuitionism, 1907 Ma, 1912 Ma, 1918 Ma, 1925 Ma
Invariants, algebraic, 1845 Ma, 1907 Ma, 1910 Ma
Inversion, 1824 Ma
Iodine, 1811 C, 1814 C
Ionium, 1907 C

Ionization, 1884–1887 C
IQ, 1914 S
Iridium, 1804 C
Iron, 1762 C
Irradiation, 1923 H, 1924 H
Irrationals, 1220 Ma, 1544 Ma, 1833–1835 Ma, 1860s Ma, 1872 Ma, 1886 Ma
Irritability, 1752 B
Isarithmic map, 1698–1700 E
Islands, 1825 E; volcanic, 1844 E
Isobaths, 1737 E
Isobutylene, 1825 C
Isohyets, 1898 Met
Isomers, 1814 C, 1823 C, 1825 C, 1828 C, 1830 C, 1858 C; optical, 1846–1848 C
Isomorphism, Law of, 1820 C
Isoperimetric figures, 1701 Ma
Isoperimetry, 1841 Ma, 1842 Ma
Isopropyl, 1862 C
Isotherms, 1845 Met
Isotopes, 1886 C, 1907 C, 1910–1911 C, 1913 C

Jacobians, 1841 Ma
Jacquard attachment, 1805 Ma, 1833 Ma
Jordan Curve Theorem, 1887 Ma, 1911 Ma
Joule's Law, 1841 P
Joule-Thomson effect, 1852 C, 1877 C, 1898 C
Juno, 1804 A
Jupiter, 1664 A, 1799–1825 A; satellites of, 1610 A, 1668 A, 1675 P, 1787 A, 1808 A, 1891 A, 1892 A

Kaleidoscope, 1816 P
Karlsruhe Congress, 1860 C
Keeler Gap, 1888 A
Kelvin temperature scale, 1851 P
Kennelly-Heaviside layer, 1902 Met
Kepler's laws, 1609 A, 1619 A, 1664–1666 A/P, 1679 A/P, 1687 P, 1927 A
Kidneys, 1564 B, 1920 B
Kinetic theory, 1827 P; of gases, 1738 C, 1851 C/P, 1857 C/P, 1858 C/P, 1859 C/P, 1860s C, 1896 P
Klein bottle, 1882 Ma
Knot theory, 1927–1928 Ma
Knowledge: Comte on, 1840–1842 S; Kant on, 1781 S; scientific, c.1450 S; Spencer on, 1862 S
Koenigsberg bridge problem, 1735 Ma
Krypton, 1898 C

Lagrange's Theorem, 1770 Ma
Lagrangian flow, 1788 P
Lakes, 1862 E
Lanthanum, 1839 C
Laplace coefficients, 1785 Ma
Laplace's equation, 1760 P, 1813 P
Laplace tidal equations, 1776 E
Laplace transform, 1812 Ma
Larynx, 1600–1601 B
Latitude, 1290 E, 1436 E, 1524 E, 1568 E, 1698–1700 E
Lattice algebra, 1928 Ma
Lattice theory, 1897 Ma
Laudable pus, 1266 H
Laurentian System, 1855 E
Lead, 1913 C, 1914 C; poisoning, 1745 H
Least Action, Principle of, 1744 P, 1833 P
Least Constraint, Principle of, 1829 P
Least squares, 1794 Ma, 1806 Ma, 1809 Ma, 1823 Ma
Lebesgue integration, 1902 Ma, 1904 Ma
Legendre conditions, 1786 Ma
Legendre polynomials, 1784 Ma
Leibniz-Newton dispute, 1712 Ma
Lemniscate, 1694 Ma
Lens, achromatic, 1733 A, 1757 A, 1827 B
Lenz's Law, 1834 P
Leprosy, 1873 H
Leucocytes, 1922 B
Lever, Law of, c.1220 P
Leyden jar, 1745 P
L'Hospital's Rule, 1696 Ma
Life, deep sea, 1868 E, 1869 E
Light: Brewster's Law, 1811 P; Einstein on, 1905 P, 1916 P; Faraday on, 1845 P, 1846 P; Fermat on, 1657 P; Fresnel on, 1818 P; Grimaldi on, 1665 P; Goethe on, 1810 P; Grosseteste on, c.1220 P, c.1231–1235 P; Hooke on, 1665 P; Huygens on, 1678 P, 1680 P, 1690 P; incandescent, 1879 C; Leibniz on, 1682 P; lime-, 1825 C; Newton on, 1665–1666 P, 1672 P, 1675 P, 1704 P; polarization of, 1678 P, 1717 P, 1808 P; pressure, 1899 P; red shift, 1917 A; refraction of, 1690 P, 1809 P, 1880 P; speed of, 1637 P, 1675 P, 1849 P, 1865 P, 1878 P, 1879 P, 1887 P, 1905 P, 1913 P; von Soldner on, 1801 P; Young on, 1801–1804 P. *See also* Optics; Rainbow; Refraction; Wave theory
Lightning, 1750 Met, 1752 Met
Lightning rod, 1750 Met
Limelight, 1825 C
Limits, theory of, 1655 Ma, 1754 Ma, 1821 Ma
Line curve, 1830 Ma
Liquid drop model, 1930 P
Lithium, 1817 C, 1818 C, 1834 C, 1855 C
Loess, 1870 E

Logarithms, 1544 Ma, c.1580 Ma, 1614 Ma, 1620 Ma, 1624 Ma, 1627 Ma, 1727–1728 Ma
Logic, c.1320 Ma; Brouwer on, 1908 Ma; Daniell on, 1916 Ma; Dedekind on, 1888 Ma; De Morgan on, 1847 Ma, 1860 Ma, 1874 Ma; Frege on, 1879 Ma; Gödel on, 1930 Ma; Hegel on, 1812–1816 Ma; Leibniz on, 1666 Ma; Peirce on, 1870 Ma; Post on, c.1921 Ma; Russell on, 1902 Ma, 1903 Ma, 1905 Ma, 1910–1913 Ma; symbolic, 1666 Ma, 1854 Ma, 1890–1905 Ma, 1894–1908 Ma, 1901 Ma; and truth tables, 1920–1921 Ma
Logical connectives, 1913 Ma
Longitude, 1436 E, 1524 E, 1568 E, 1635 E, 1668 A, 1675 A, 1735–1744 E, 1736–1737 E, 1753 E, 1765 E
Lorentz-Lorenz relations, 1880 P
Lorentz transformations, 1892 P, 1904 P
Lungs, 1920 B
Lutetium, 1907 C
Lux concept, c.1220 P
Lymphatic vessels, 1787 B
Lysozyme, 1922 H/B

Mach's Principle, 1872 P
Maclaurin series, 1715–1717 Ma, 1742 Ma
Magic Lyre, 1831 P
Magnesia alba, 1754 C
Magnesium, 1808 C, 1831 C
Magnetic declination, 1622 P
Magnetic dip, 1581 P, 1600 E/P, 1609 E
Magnetic induction, 1269 P
Magnetic poles: north, 1831 E; south, 1907–1909 E
Magnetic reversals, 1906 E
Magnetic variations, 1698–1700 E, 1701 E
Magnetism, 1269 P, 1581 E/P, 1845 P, 1896 P
Magnus effect, 1835 P
Malaria, 1880 H, 1898 H, 1899 H
Manganese, 1774 C
Manifolds, 1924 Ma
Manometer, 1661 P
Maps, c.1250 E, 1325–1330 E, 1436 E, c.1450 E, 1500 E, 1507 E, 1524 E; atlases, 1570 E, 1594 E, 1837–1848 E; projections for, 1524 E, 1568 E, 1599 E. *See also* Geological mapping; Globes
Marble, 1779 E/C
Mariotte's Law, 1676 C
Mars, 1600 A, 1672 A, 1877 A, 1894 A, 1906 A
Mass, 1687 P, 1891 P, 1911 P
Mass Action, Law of, 1803 C, 1867 C
Mastectomy, 1718 H, 1890–1891 H
Masurium, 1925 C
Mathematical induction, 1321 Ma
Mathematical expectation, 1657 Ma
Mathieu functions, 1868 Ma
Matrices, 1857–1858 Ma, 1862 Ma, 1925 P
Matrix mechanics, 1925 P
Matter, 1748 C, 1758 C, 1812 P, 1870 P. *See also* Conservation of matter
Maud expedition, 1918–1925 E, 1927 E
Maupertuis's principle, 1744 P
Mean free path length, 1858 C/P
Measles, 1676 H, 1917 H
Mechanics, c.1280 P, 1328 P, c.1330 P, c.1350 P/Ma, 1537 P, 1539 P, 1585 P, 1589–1592 P, 1638 P, 1641 P, 1656 P, 1659 P, 1736 P, 1743 P, 1744 P, 1765 P, 1784 P, 1788 P, 1803 P, 1822 P; celestial, 1799–1825 A, 1825 A, 1892–1899 A. *See also* Falling bodies, Impetus theory, projectile
Medicine: clinical, 1834 H; diagnostic, 1348 H, c.1450 H, 1761 H; experimental, 1865 H; homeopathy, 1810 H; hydropathy, 1829 H; inoculation, vaccination, 1715 H, 1717 H, 1721 H, 1796 H, 1798 H, 1882 H, 1896 H; regulations in, 1238 B; treatments, 1266 H, c.1270 H, 1320 H, 1347–1350 H, c.1350 H, 1360 H, 1539 B, 1545 H, 1563 H, 1695 H, 1910 H, 1912 H, 1921 H. *See also* Anesthesiology; Disease; Medicines; individual ailments, specialties
Medicines, c.1350 A, 1500 C, 1530 B, c.1530 H, 1535 H, 1588 H, 1775 H, 1928 H
Meiosis, 1892 B
Mendel's Laws, 1902 B
Mental age, 1905 S, 1914 S
Mental illness, 1801 H, 1812 H
Mental processes, 1538 B
Mephitic air. *See* Nitrogen
Mercator projection, 1568 E, 1599 E
Mercator's series, 1668 Ma
Mercury (metal), c.1530 H, 1669 C, 1911 C
Mercury (planet), 1859 A/C, 1916 A, 1919 P/A; transit of, 1631 A, 1661 A, 1677 A
Mesmerism. *See* Hypnotism
Mesothorium, 1907 C
Messier catalog, 1771 A, 1784 A
Metabolism, 1614 H, 1838 B
Metallurgy, 1540 C, 1556 E. *See also* Metals
Metals, c.1260 E, 1630 C, 1753 E; and alchemy, c.1310 C; defining, c.1550 C; luster of, 1751 C; rusting of, 1700 C;

structure of, 1927 P. *See also* Metallurgy; individual metals
Meteorites, 1772 A, 1790 A, 1794 A, 1803 A; composition of, 1817 C, 1819 A
Meteorological observations, 1337–1344 Met, 1657 Met, 1663 Met, 1723 Met, 1728 Met, 1771 Met, 1780 Met, 1888–1889 Met, 1918–1925 E
Meter, 1790 P, 1791 P, 1892–1893 P
Metric system, 1790 P, 1795 P
Michelson-Morley experiment, 1887 P, 1892 P, 1905 P
Microbes, 1837 B
Micrographia (Hooke), 1665 B, 1665 C/Met
Micrometer, 1641 A, 1666 A, 1667 A
Microorganisms, 1677 B, 1767 B, 1871 B
Microphone, 1877 P
Microscope, 1609 B, 1625 B, 1665 B, 1668 B; achromatic, 1827 B
Migration, bird, c.1250 B
Milk, 1890 C, 1905 H
Milk leg, 1784 H
Milky Way, 1610 A, 1750 A, 1785 A, 1906 A, 1918 A, 1927 A, 1930 A
Mimas, 1789 A
Mind, 1538 B, 1781 S, 1890 B, 1899 B; blank, 1690 B
Mineral acids, 1210 C
Minerals, c.1260 E, 1781–1786 E, 1858 E, 1859 E. *See also* Classification, of minerals
Mineral salts, c.1350 H
Mineral veins, 1753 E, 1791 E
Minimum principles, 1640 P
Minus sign, 1489 Ma, 1557 Ma
Mirrors, c.1270 P
Mixing length concept, 1925 Met
Möbius band, 1858 Ma
Moho discontinuity, 1909 E
Molar heat capacities, 1819 C, 1912 C
Molecules, 1811 C, 1846–1848 C, 1860 C, 1905 P, 1921 C
Molybdenum, 1781 C
Momentum: conservation of, 1644 P, 1656 P, 1687 P; conservation of angular, 1664–1666 A/P, 1735 Met
Monsoons, 1686 Met
Moon, 1543 A, 1638 A, 1647 A, 1651 A, 1687 A/P, 1788 A; distance from earth, 1610 A, 1751–1753 A, 1878 A; librations of, 1637 A; mapped, 1750 A; origin of, 1873 A, 1878 A, 1879 A, 1882 E/A, 1909 A; photographed, 1839–1840 A
Moraines, 1779 E, 1787 E
Morphine, 1805 C, 1817 C
Mortality tables, 1693 Ma
Motion, 1328 P, c.1330 P, 1537 P, 1539 P, 1687 P, 1786 P. *See also* Impetus theory, mechanics
Motor, alternating-current, 1885 P
Mountains, c.1260 E, 1282 E, 1546 E, 1557 E, 1561 E, 1669 E, 1734 E, 1740 E, 1756 E, 1759 E, 1777 E, 1779–1796 E, 1796 E, 1842 E; formation of, 1830 E, 1847 E, 1852 E, 1880 E, 1887 E
Multiple Proportions, Law of, 1804 C, 1808 C, 1810–1820 C
Multiplication, 1494 Ma; sign, 1698 Ma
Muriatic acid, 1809 C, 1810 C
Muscles, c.1310 B
Mutations, 1900 B, 1901–1903 B, 1907 B, 1927 B, 1930 B

n!, 1730 Ma
Nasal secretions, 1670 B
Natural numbers, 1796 Ma, 1887 Ma, 1889 Ma
Natural selection, 1813 B, 1844 B, 1859 B, 1875 B, 1914 B, 1930 B
Navier-Stokes equations, 1823 P, 1845 P
Navigation, c.1300 E, c.1420 A/E, 1492 E, 1522 E, 1568 E, 1599 E, 1668 A, 1753 A, 1765 E, 1770 E, 1802 A
N-dimensional space, 1844 Ma
Nearsightedness, 1451 H
Nebulae. *See* Stars
Nebular Hypothesis, 1796 A
Negative numbers, 1225 Ma, 1545 Ma, 1591 Ma, 1629 Ma, 1655 Ma, 1657 Ma, 1796 Ma, 1797 Ma, 1822 Ma, 1831 Ma, 1887 Ma
Neighborhoods, 1914 Ma
Neodymia, 1885 C
Neon, 1898 C, 1910–1911 C
Neptune, 1772 A, 1821 A, 1845 A, 1846 A
Neptunism, 1787 E, 1791 E, 1795 E
Nernst heat theorem, 1906 C
Nervous diseases, 1892 H
Nervous system, 1664 B, 1811 B, 1888 B
Neuritis, 1912 H
Neurology, 1749 H, 1864 H
Neutralization, 1700 C
New Atlantis, The (F. Bacon), 1627 S
Newton-Leibniz dispute, 1712 Ma
Newton's identities, 1707 Ma
Newton's Laws of Motion, 1687 P, 1786 P
New World, 1500 E, 1502 E, 1507 E
Niacin, 1928 H
Nickel, 1751 C
Nicod's Criterion, 1930 Ma
Nicol Prism, 1829 E/P
Niobium. *See* Columbium

Nitrogen, 1665 Met, 1772 C, 1777 C, 1785 C, 1892 C, 1894 C/Met; liquid, 1877 C
Nitrogen trichloride, 1813 C
Nitroglycerin, 1846 C
Nitrous oxide, 1800 C
Noah's flood, 1282 E, 1681 E, 1694 E, 1695 E, 1696 E, 1705 E, 1709 E, 1726 E, 1770 B, 1823 E, 1829 E, 1833 E
Nobili's Rings, 1826 C
Nomenclature. *See* Binomial nomenclature; Chemical nomenclature
Normal distribution, 1733 Ma
Northern lights, 1733 Met
North Pole, magnetic, 1831 E
Nose, 1670 B
Notation, differential, 1812 Ma
Novas, 1572 A, 1573 A, 1604 A, 1606 A, 1834–1837 A, 1866 A, 1892 A. *See also* Stars; Supernovas
Novum organum (F. Bacon), 1620 S
Nuclear atom, 1911 P/C
Nuclear structure, 1930 P
Nucleus, 1911 P/C, 1919 P
Number couples, 1831 Ma, 1837 Ma, 1860s Ma
Number field, 1879 Ma
Numbers, geometry of, 1844 Ma, 1896 Ma
Number systems, c.1700 Ma
Number theory, 1621 Ma, 1630–1665 Ma, 1637 Ma, 1640 Ma, 1770 Ma, 1849 Ma; Dirichlet on, 1839 Ma; Eisenstein on, 1847 Ma, 1850 Ma; Euler on, 1732 Ma, 1736 Ma, c.1750 Ma, 1761 Ma; Fermat on, 1670 Ma; Gauss on, 1801 Ma; Goldbach on, 1742 Ma; Lagrange on, 1773 Ma; Legendre on, 1798 Ma, 1830 Ma; Sarrus on, 1819 Ma
Numerals. *See* Hindu-Arabic numerals; Roman numerals
Numerical stability, 1929 Ma
Nutation, of earth's axis, 1748 A
Nutrition, 1898 H, 1905 H, 1907 H, 1909 H, 1915 H, 1918 H, 1921 H. *See also* Food; Vitamins

Observational errors, theory of, 1823 Ma
Observatories, astronomical, 1259 A, c.1420 A/E, 1675 A
Obstetrics, 1696 H
Ocean: abyssal theory of, 1843 E; basins, 1882 E; bathymetry, 1854 E; circulation, 1663 E, 1787 E, 1797 E, 1895 E, 1922 E, 1923 E, 1929 E; currents, 1699 E, 1811 E, 1832 E, 1834 E, 1847 E, 1856 E, 1878 E, 1905 E, 1926 E; depth measurements, 1521 E; floor, 1818 E; layering, 1911 E; sediment mapping, 1870 E, 1891 E; temperature, 1751 E, 1868 E, 1869 E; waves, 1776 E, 1802 E, 1889 E. *See also* Gulf Stream; Life, deep sea; Oceanography; Sea Water; Tides
Oceanography, 1725 E, 1855 E, 1875 E, 1900–1906 E, 1902 E, 1912 E, 1926 E, 1927 E; observational, 1661 E, 1698–1700 E, 1839–1843 E, 1853 E, 1857–1858 E, 1868 E, 1872–1876 E, 1877 E, 1885 E, 1918–1925 E, 1925–1927 E; theoretical, 1738 P/E. *See also* Ocean
Ockham's razor, c.1290 A, c.1330 S
Octaves, Law of, 1864 C
Ohm's Law, 1825 P, 1827 P, 1900 P
Olbers's comet, 1815 A
Olbers's Paradox, 1744 A, 1826 A
On the Origin of Species (Darwin), 1858 B, 1859 B
On the Revolutions of the Heavenly Spheres (Copernicus), 1514 A, 1543 A, 1616 S
Operators: logical, 1844 Ma; mathematical, 1814 Ma
Ophthalmoscope, 1851 H, 1864 H
Opticks (Newton), 1704 C, 1704 Ma, 1704 P, 1717 P
Optics, c.1220 P, c.1231–1235 P, c.1270 P, 1277–1279 P, 1589 P, 1604 P, 1637 P, 1738 P, 1744 P, 1828 P. *See also* Light; Rainbow
Organic compounds, 1832 C
Orion nebula, 1610 A, 1880 A
Osmium, 1804 C
Osmosis, 1748 C, 1826 C
Osteopathy, 1874 H
Ostrogradski's Theorem, 1828 Ma
Overthrusting, 1884 E
Ovum, mammalian, 1827 B
Oxygen, 1665 Met, 1674 C, 1771 C, 1774 C, 1774–1777 C, 1775 C, 1777 C, 1778 C, 1779 B, 1781 C, 1783 C, 1785 C/Met, 1789 C, 1892 C; isotopes, 1929 C; liquid, 1877 C
Oxymuriatic acid. *See* Chlorine
Ozone, 1840 C

Paleoclimatology, 1840 E, 1846 E, 1868 E, 1870–1879 E, 1878 E, 1903–1909 E, 1904 E, 1907–1909 E, 1912–1913 E, 1924 E, 1930 E. *See also* Glacial history, Ice ages
Paleogeology, 1719 E, 1752 E
Paleontology, 1808 E, 1812 E, 1847–1894 E, 1883 E
Paleozoic System, 1843 E
Palladium, 1803 C
Pallas, 1802 A, 1803 C, 1809 A

Palsy, 1817 H
Paradoxes, 1902 Ma, 1905 Ma, 1906 Ma, 1908 Ma
Parallax, 1572 A, 1588 A, 1728 A, 1750 A, 1838 A, 1839 A, 1911 A, 1914 A
Parallel lines, 1604 Ma
Paralysis, c.1530 H, 1745 H
Paramagnetism, 1845 P
Paramour Pink expedition, 1698–1700 E
Parkinson's disease, 1817 H
Partial Pressures, Law of, 1793 Met, 1801 Met/C
Particle counter, 1908 P
Particle-wave duality, 1905 P, 1924 P
Pascal's Principle, c.1648 P
Pascal's Theorem, 1640 Ma
Pasteurization, 1856 B, 1885 C
Pauli exclusion principle, 1925 P
Pellagra, 1912 H, 1914 H, 1915 H, 1916 H, 1925 H, 1928 H
Peltier effect, 1834 P
Pendulum, c.1583 P, 1589–1592 P, 1603 H, 1609 P, 1656 P, 1659 P, 1672 P, 1673 P; Foucault, 1851 E
Penicillin, 1928 H
Pepsin, 1836 B
Perception: center in brain, 1876 H/B; sense, 1754 B; Voltaire on, 1738 P
Percussion, 1761 H
Perfume, 1868 C
Periodicity, 1748 Ma, 1834 Ma; double, 1800 Ma, 1825 Ma, 1829 Ma
Periodic table, 1829 C, 1862 C, 1864 C, 1869 C, 1870–1875 C, 1904 C/P, 1913 P
Peritonitis, 1848 H
Permutation(s), 1321 Ma; groups, 1770 Ma, 1815 Ma, 1858 Ma, 1870 Ma, 1872 Ma
Pharmacopoeias, 1535 H. *See also* Medicines
Phase rule, 1876–1878 C
Phenol, 1834 C
Philosopher's stone, c.1310 C
Philosophical Transactions of the Royal Society, 1665 S
Phlogisticated air. *See* Nitrogen
Phlogiston, 1669 C, 1700 C, 1750 C, 1751 C, 1766 C, 1771 C, 1772 C, 1775 C, 1777 C, 1778 C, 1781 C, 1783 C, 1786 C, 1797–1812 C, 1800 C
Phobos, 1877 A
Phoebe, 1898 A
Phonograph, 1877 P
Phosphorus, 1677 C, 1769 C, 1772 C, 1909 H, 1918 H
Photoconductivity, 1873 P
Photoelectric effect, 1902 P, 1905 P, 1915 P
Photoelectricity, 1887 P, 1898 P, 1902 P
Photography: color, 1869 C; daguerreotypes, 1824 C, 1839 C; dry collodion process in, 1856–1857 C; film for, 1887 C; first image, 1827 C; Wedgwood on, 1802 C
Photosynthesis, 1779 B, 1804 B, 1837 B, 1862 B
Photovoltaic effect, 1839 P
Phrenology, 1810–1819 B
Physiognomy: human, 1586 B; plant, 1588 B
Physiology: human, 1840 B; plant, 1727 B
Pi (π), 1593 Ma, 1596 Ma, 1706 Ma, c.1750 Ma, 1770 Ma, 1853 Ma, 1882 Ma
Pistil, 1682 B
Plague, 1347–1350 H, 1348 H, 1353 H, 1646 H, 1894 H, 1897 H
Planck's Law, 1896 P, 1900 P
Planes, inclined, c.1220 P, 1589–1592 P, 1609 P
Planetary motions, orbits, 1217 A, 1272 A, c.1281 A, c.1350 A, 1514 A. 1543 A, 1551 A, 1596 A, 1600 A, 1605 A, 1609 A, 1619 A, 1627 A, 1632 A, 1644 A, 1666 A, 1674 A, 1679 A/P, 1684 A, 1685 A, 1787 A, 1809 A, 1916 A
Planetary positions, 1392 A, 1483 A, 1588 A, 1627 A
Planimeter, 1814 Ma
Plant dispersal, 1906 B, 1930 B
Plants. *See* Botany
Platinum, 1557 C, 1748 C, 1752 C, 1782 C, 1800 C, 1803 C
Platonic Academy, 1469 S
Pleistocene period, 1846 E; glaciation, 1903–1909 E
Plücker's equations, 1839 Ma
Plus sign, 1489 Ma, 1557 Ma
Pluto, 1906 A, 1930 A
Plutonism, 1795 E
Pneumatic trough, 1766 C, 1770s C
Poincaré conjecture, 1904 Ma
Poisson distribution, 1837 Ma
Poisson equation, 1813 P, 1839 P
Polar coordinates, 1671 Ma, 1691 Ma, 1748 Ma
Polar front, 1906 Met
Polarization, of light, 1678 P, 1717 P, 1808 P
Pollination, 1793 B
Pollution, 1388 H; air, 1661 Met, 1925 Met; control, 1863 Met
Polonium, 1898 C
Polyhedra, 1639 Ma, 1752 Ma
Polyps, 1739 B

Poncelet-Steiner Theorem, 1822 Ma, 1833 Ma
Pons Varolii, 1573 B
Population, 1798 B
Portolano charts, c.1300 E
Positivism, 1830–1842 S
Postmortems, 1286 H
Potassium, 1807 C
Potential theory, 1839 P
Power law, c.1545 Ma
Pragmatism, 1876 P
Praseodymia, 1885 C
Precession of equinoxes, 1543 A, 1687 P, 1754 A
Precipitation, 1751 Met. *See also* Rain; Snowfall
Preformation theory, 1651 B, 1759 B
Pregnancy, 1774 B
Prehistoric man, 1863 E
Pressure, mapped, 1869 Met. *See also* Air, Atmosphere, Meteorology
Prime mover, concept of, c.1270 S, c.1350 P
Prime numbers, 1761 Ma, 1831 Ma, 1846 Ma; theorem of, 1850 Ma, 1896 Ma
Principia mathematica (Newton), 1665–1666 A/P, 1685 Ma, 1687 A, 1687 C, 1687 E, 1709 S, 1759 A/P
Printing, 1289 S, 1440 S, c.1455 S, 1476 S, 1482 Ma, 1482–1515 S, 1483 A
Probability, 1321 Ma, c.1545 Ma, 1654 Ma, 1657 Ma, 1665 Ma, 1713 Ma, 1718 Ma, 1763 Ma, 1772 Ma, 1777 Ma, 1812 Ma, 1820 Ma
Projectile motion, 1537 P, 1589–1592 P, 1638 P, 1747 P, 1835 P
Projections, stereographic, c.1220 E/Ma
Prosthaphaeresis, c.1580 Ma, 1593 Ma
Protactinium, 1917 C
Proton, 1919 P
Protozoa, 1676 B, 1677 B
Prout's Hypothesis, 1816 C, 1892 C, 1901 C, 1927 C
Prussian Tables, 1551 A, 1627 A
Psychiatry, 1901 H. *See also* Mental illness, Psychoanalysis
Psychical research, 1882 B
Psychoanalysis, 1880–1882 H, 1893 H, 1895 H. *See also* Psychology
Psychology, 1538 B, 1749 H, 1890 B; child, 1928 H, 1929 H; criminal, 1876 B; experimental, 1872 H, 1879 B
Pterodactyl, 1871 B
Ptolemaic theory, 1217 A, c.1290 A, 1514 A, 1588 A, 1632 A
Puerperal fever, 1751 H, 1843 H, 1847 H, 1861 H, 1879 H
Pulse, c.1450 H, 1603 H, 1707 H
Pulsilogium, 1603 H
Purity, of substances, 1823 C
Pus, laudable, 1266 H
Pyrometer, 1892 P

Quadratic forms, 1769–1770 Ma, 1801 Ma; binary, 1773 Ma; differential, 1827 Ma, 1861 Ma; ternary, 1831 Ma
Quadratic reciprocity, 1744 Ma, 1795 Ma, 1798 Ma, 1801 Ma
Quadrupeds, 1753–1767 B
Quantification: in biochemistry, 1915 B; chemical, 1617 C, 1792 C
Quantitative methods, in chemistry, 1769 C
Quantum spin number, 1925 P
Quantum theory, 1900 P, 1905 P, 1913 P, 1923 P, 1925 P, 1926 P, 1927 P
Quaternions, 1843 Ma, 1844 Ma, 1853 Ma, 1862 Ma, 1866 Ma, 1870 Ma

Rabies, 1885 H
Radiation, 1804 P, 1859 P, 1878 P, 1879 P/A, 1900 P; cosmic, 1911 Met/P; infrared, 1800 P; ultraviolet, 1801 P
Radicals, 1834 C, 1837 C, 1843 C; benzoyl, 1832 C; in chemical reactions, 1815 C, 1828 C, 1834 C
Radioactivity, 1896 C, 1896 P, 1897 P, 1898 C, 1899 C/P, 1901–1906 E, 1902 P, 1913 C, 1928 P
Radiometer, 1873–1876 Met
Radiothorium, 1905 C
Radio waves, 1886–1889 P
Radium, 1898 C, 1900 C, 1902 P/C, 1904 C, 1907 C, 1910 C
Radon, 1900 C, 1924 P
Rain, 1677–1704 Met, 1751 Met, 1911 Met, 1922 Met; gauge, 1639 Met
Rainbow, c.1235 Met/P, c.1281 A/Met, 1304 Met/P
Rationals, 1860–1869 Ma
Ratio test, 1776 Ma, 1821 Ma
Rayleigh's radiation formula, 1900 P
Reactions, 1894 C; classified, 1877 C; rates, 1777 C, 1899 C
Reals, 1872 Ma
Réaumur temperature scale, 1730 P
Red shift, 1848 A, 1917 A, 1925 A, 1929 A
Reefs, 1818 E, 1842 E
Reflection, Law of, c.1270 P
Reflexes, conditioned, 1907 B
Refraction: 1690 P, 1809 P; conical, 1828 P; double, 1669 P; law of, c.1231–

1235 P, c.1270 P, 1621 P, 1637 P, 1657 P; of lenses, 1843 P
Refrigeration, 1877 C
Relativity, 1904 P, 1905 P, 1908 P, 1911 P/A, 1916 A, 1916 P, 1919 P, 1922 A, 1925 A, 1927 P
Religion and science: 1924 S; Aquinas, c.1270 S, 1879 S; Aristotle, 1215 S, 1277 S; Bruno, 1584 A, 1600 S; Buckland, 1820 E; Cotton, 1652 S; Catholic Index, 1835 S; Galileo, 1616 S, 1633 S; Newton, 1728 S; Pope Gregory IX, 1231 S; Rueus, 1566 S; Servetus, 1553 B; Wilkins, 1640 A
Reproduction: by aphids, 1740 B; sexual, of plants, 1682 B, 1694 B
Respiration, c.1500 S, 1667 B, 1674 C/Met, 1727 B, 1771 B, 1778 C, 1837 B
Reversibility, 1803 C
Reynolds' Number, 1883 P
Reynolds' Stresses, 1894 P/E/Met
Rhea, 1672 A
Rhenium, 1925 C
Rhodium, 1803 C
Richardson Number, 1920 Met
Rickets, 1650 H, 1912 H, 1921 H, 1923 H, 1924 H
Riemann-Christoffel tensor, 1861 Ma
Riemannian spaces, 1867 Ma
Riemann integral, 1854 Ma, 1867 Ma, 1881 Ma, 1884 Ma
Riemann's Hypothesis, 1859 Ma
Riemann surface, 1851 Ma, 1877 Ma
Rivers, c.1360 E, 1580 E, 1674 E, 1889 E, 1890 E; graded, 1875 E; subterranean, c.1646 E
Roche limit, 1848 A
Rock crystal, 1603 E
Rockets, 1898 P, 1920 P, 1926 P
Rocks, 1859 E; origins of, 1858 E; silicious, 1862 E. *See also* Stones
Rocky Mountain spotted fever, 1906 H, 1909 H
Rolle's Theorem, 1691 Ma
Roman numerals, 1202 Ma, 1299 Ma, c.1500 Ma
Root test, 1821 Ma
Royal Society of London, 1660 S, 1665 S, 1666 S
Rudolphine Tables, 1627 A
Rusting, 1700 C
Ruthenium, 1844 C

Saccharin, 1879 C
Salts, 1736 C
Salt wells, 1695 H
Samarium, 1879 C
Sanitation, 1388 H
Sassolite, 1779 C
Satellite orbits, 1880 A
Saturn: motion of, 1787 A; rings of, 1610 A, 1612 A, 1655 A, 1659 A, 1675 A, 1785 A, 1799–1825 A, 1848 A, 1850 A, 1857 A, 1859 A, 1866 A, 1888 A, 1895 A; satellites of, 1655 A, 1659 A, 1671 A, 1672 A, 1684 A, 1789 A, 1848 A, 1898 A; shape of, 1610 A, 1612 A, 1655 A, 1659 A
Scandium, 1869 C, 1879 C
Scarlet fever, 1736 H, 1748 H, 1925 H
Schizophrenia, 1911 H
Schrodinger wave equation, 1926 P
Schwann cells, 1838 B
Schwarzschild singularity, 1916 A
Science: history of, 1694 S; philosophy of, 1709 S
Science and religion. *See* Religion and science
Science fiction, 1634 S
Scientific methods, 1531 S, 1608 S, 1648 S; F. Bacon on, 1605 S, 1620 S, 1627 S; R. Bacon on, 1267–1268 S; Descartes on, 1637 S, 1644 S; Galileo on, 1623 S; Goethe on, 1795 S; Grosseteste on, c.1230 S; Ockham on, c.1330
Scientific societies, 1560 S, 1603 S, 1657 S, 1666 S, 1667 Met, 1700 S, 1793 S
Scopes trial, 1925 S
Scurvy, 1623 H, 1720 H, 1753 H, 1912 H
Sea ice, 1616 E, 1902 E, 1905 E
Sea level, 1841 E, 1868 E
Sea water: composition of, 1820 E, 1865 E; salinity of, 1578 E, 1673 E, c.1693 E, 1715 E, 1740 E, 1902 E
Sedimentary cycle, 1894 E
Seismograph, 1880 E
Seismology, 1760 E, 1880 E. *See also* Earthquakes
Selenium, 1818 C, 1873 P
Semen, 1677 H, 1779 B
Sense perception, 1754 B
Senses, 1754 B
Sensibility, 1752 B
Sensory organs, 1609 B
Series: Bernoulli, 1696 Ma. *See also* Convergence, Divergent series, Hypergeometric series, Infinite series
Set theory, 1874 Ma, 1883 Ma, 1902 Ma, 1904 Ma, 1906 Ma, 1908 Ma, 1914 Ma, 1922 Ma
Sex, 1897–1928 B
Sexual reproduction, plants, 1682 B, 1694 B
Shells, 1678 E

Sight. *See* Vision
Signs, Descartes's Rule of, 1637 Ma, 1707 Ma
Silicates, 1862 E
Silicon, 1823 C, 1824 C, 1854 C
Silkworm disease, 1865 B
Silkworms, 1669 B
Silurian System, 1835 E, 1838 E
Silver cyanate, 1823 C
Silver fulminate, 1823 C
Sines: law of, c.1250 Ma, 1464 Ma; table of, 1534 Ma
Sirius, 1844 A, 1868 A, 1916 A; Sirius B, 1862 A, 1914 A, 1916 P/A, 1925 A
Skin diseases, 1572 H
Slate cleavage, 1853 E
Slide rule, 1620 Ma, c.1621 Ma; circular, 1630 Ma, 1632 Ma
Smallpox inoculation, vaccination, 1715 H, 1717 H, 1721 H, 1796 H, 1798 H
Snell's Law of Refraction, 1621 P, 1637 P, 1657 P
Snowfall, 1751 Met
Soda water, c.1770 C
Sodium, 1807 C, 1913 P
Soil, 1881 B
Solar atmosphere, 1862 A
Solar chemistry, 1887 A
Solar composition, 1859 A, 1862 A, 1868 A
Solar eclipse, 1715 A, 1919 A
Solar energy, 1848 A, 1853 A, 1929 A
Solar flares, 1859 A
Solar prominences, 1860 A, 1868 A
Solar rotation, 1853–1861 A, 1891 A, 1906 A
Solar spectrum, 1814 A, 1844 A, 1859 A, 1868 A, 1870 A, 1876 A, 1896 A
Solar storms, 1908 A
Solar system: origin, evolution of, 1734 A, 1785 A, 1796 A, 1799–1825 A; scale of, 1672 A. *See also* Copernican theory; Planetary motions, orbits, positions; Sun; individual planets
Solar temperature, 1879 A
Solomon's seal, 1849 Ma
Solvability by radicals, of equations, 1770 Ma, 1771 Ma, 1799 Ma, 1813 Ma, 1824 Ma, 1830–1832 Ma
Soul, 1707 B, 1869 B
Sound, c.1230 P, 1701 P, 1877–1878 P; Doppler effect, 1842 P
Space, 1872 P, 1916 A. *See also* Ether
Space-time, 1870 P, 1907 P, 1908 P, 1916 P
Species: fixity of, 1798 B; origin of, 1751 B
Spectral analysis, 1873 Ma
Spectral lines, 1916 P
Spectrograph, mass, 1919 C
Spectroheliograph, 1890 A
Spectroscopy, 1802 C, 1814 A, 1834 C, 1853 C, 1854 C, 1859 A/C
Speech center, in brain, 1745 H, 1836 H, 1861 B
Sperm, 1677 H, 1824 B, 1875 B
Spontaneous generation, 1648 B, 1684 B, 1740 B, 1767 B, 1856 B, 1862 B
Springs, 1580 E, 1721 E
Squaring the circle, 1882 Ma
Stamen, 1682 B
Standard deviation, 1893 Ma
Standard error, 1889 Ma
Stark Effect, 1913 P
Stars: aberration of, 1728 A; atlas of, 1687 A; binary, 1781 A, 1782 A, 1827 A, 1833 A, 1834–1837 A, 1844 A, 1895–1935 A, 1913 A; catalogs of, 1602 A, 1627 A, 1661 A, 1676–1678 A, 1678 A, 1712 A, 1725 A, 1727–1747 A, 1771 A, 1774 A, 1798 A, 1801 A, 1818 A, 1833 A, 1862 A; classified, 1918–1924 A; colors, 1922 A; and Copernican theory, 1543 A; and dark clouds, 1930 A; diameter of, 1920 A; distances of, 1543 A, 1576 A, 1744 A, 1785 A, 1838 A, 1839 A, 1912 A, 1913 A, 1923 A; evolution of, 1755 A, 1916 A, 1919 A; galaxies, 1610 A, 1750 A, 1755 A, 1784 A, 1785 A, 1905 A, 1912 A, 1918 A, 1923 A, 1924–1935 A, 1927 A, 1929 A; giant, 1905–1907 A; identification of, 1603 A; motions of, 1718 A; nebulae, 1610 A, 1612 A, 1755 A, 1771 A, 1783 A, 1784 A, 1785 A, 1796 A, 1801 A, 1828 A, 1834–1837 A, 1845 A, 1864 A, 1880 A, 1888 A, 1912 A, 1912–1917 A, 1925 A; pulsating, c.1913 A; Southern Hemisphere, 1676–1678 A, 1678 A, 1834–1837 A; spectra of, 1872 A, 1901 A, 1918–1924 A, 1921 A, 1925–1936 A; streams, 1904 A; variable, 1667 A, 1908 A; white dwarf, 1914 A. *See also* Novas; Supernovas
Statical equilibrium, 1583 P
Statics, c.1220 P, 1583 P
Statistics, 1588 E, 1662 Ma, 1693 Ma, 1713 Ma, 1733 Ma, 1749 Ma, 1763 Ma, 1829 Ma, 1845–1850 Ma, 1889 Ma
Steam engine, 1698 P, 1769–1772 P, 1776 P
Steam pump, 1698 P
Steel, 1785 C, 1856 C
Stefan-Boltzmann Law, 1879 P, 1883 P, 1900 P
Stellar aberration, 1728 A

Steno, duct of, 1661 B
Stereochemistry, 1846–1848 C, 1874 C
Stereographic projections, c.1220 E/Ma
Stethoscope, 1816 H, 1819 H
Stirling's approximation, 1730 Ma
Stirling series, 1730 Ma
Stokes's Law, 1850 P
Stomach acid, 1823 B
Stones, c.1260 E, 1502 E, 1546 E, 1779–1796 E. *See also* Boulders; Rocks
Storms, 1831 Met, 1841 Met, 1871 Met, 1919 Met
Straightedge and compass constructions, 1220 Ma, 1796 Ma, 1801 Ma, 1822 Ma, 1833 Ma, 1837 Ma, 1882 Ma, 1899 Ma
Stratigraphy, 1282 E, 1669 E, 1695 E, 1719 E, 1756 E, 1762 E, 1799 E, 1808 E, 1811 E, 1813 E, 1815 E, 1816 E, 1817 E, c.1832 E, 1838 E
Stratosphere, 1902 Met, 1922 Met
Strontium, 1808 C, 1834 C
Strychnine, 1818 C
Subsidence, 1846 E
Substitution: law of, 1834 C; reactions, 1820 C
Sulfur, 1669 C, 1808 C, 1809 C
Sulfur dioxide, 1775 C
Sulfuric acid, 1611–1613 C
Summation sign, c.1750 Ma
Sun, 1605 A, 1609 A, 1783 A, 1870 A, 1892 A, 1929 A; age of, 1853 A; corona of, 1851 A, 1882 A, 1898 A; distance from earth, 1672 A, 1684 A, 1716 A, 1761 A, 1769 A, 1824 A; -spots, 1612 A, 1613 A, 1843 A, 1845 A, 1853–1861 Ma, 1874 A, 1906 A, 1908 A. *See also* Heliocentricity; Solar entries
Superconductivity, 1911 C
Supernova, 1572 A, 1573 A, 1602 A. *See also* Novas; Stars
Superphosphates, 1842 C
Supersaturation, 1875 Met
Suppuration, 1266 H, 1320 H, 1867 H
Surfaces, theory of, 1887–1896 Ma
Surgery, 1275 B/H, c.1290 H, 1296 H, 1890 H; antiseptic, 1266 H, 1320 H, 1865 H, 1867 H; Cesarean, 1500 H; heart, 1893 H; mastectomy, 1718 H, 1890–1891 H; texts, 1252 H, 1360 H, 1655 H; transplants, 1905 H. *See also* Anesthesiology; Dissection
Surveying, 1522 A, 1533 Ma
Survival of the fittest, 1852 B
Sylow's Theorem, 1872 Ma
Symbolism, chemical, 1787 C, 1789 C, 1802 C, 1808 C, 1813 C, 1814 C, 1818 C, 1833 C, 1834 C, 1843 C, 1858 C, 1861 C, 1864 C
Symbolism, mathematical, 1489 Ma, 1545 Ma, 1557 Ma, c.1580 Ma, 1591 Ma, 1637 Ma, 1655 Ma, 1698 Ma, 1706 Ma, 1727–1728 Ma, 1730 Ma, c.1750 Ma, 1873 Ma, 1888 Ma
Syphilis, 1530 H, 1906 H, 1910 H, 1912 H, 1913 H

Tangents, law of, 1579 Ma
Tantalum, 1802 C, 1809 C
Tautochrone, 1690 Ma
Taylor series, 1671 Ma, 1715–1717 Ma
Teeth, 1563 B, 1756 H, 1771 H, 1778 H, 1788 H, 1838 H. *See also* Caries; Dentistry
Telegraph, 1833 P, 1837 P, 1877 P; wireless, 1895 P, 1902 P
Telephone, 1876 P
Telescope, 1589 P, 1608 A, 1609 A, 1610 A, 1611 A, 1630 A, 1671 A, 1897 A; achromatic lens, 1733 A, 1757 A; cross hairs in, 1641 A, 1666 A, 1667 A; reflecting, 1663 A, 1668 A, 1671 A
Tellurium, 1783 C
Temperature, 1626 H, 1792 P; critical, 1822 P, 1869 C, 1873 C; isotherms, 1845 Met
Temperature scales, 1777 Met; Celsius, 1694 P, 1742 P, 1743 P; critical, 1822 P, 1869 C; Fahrenheit, 1714 P; Kelvin, 1851 P; Reaumur, 1730 P
Tensor analysis, 1862 Ma, 1869 Ma, 1888 Ma
Ternary quadratics, 1798 Ma
Territorial drive, 1920 B
Tetanus, 1884 H, 1892 H
Tethys, 1684 A
Thallium, 1861 C, 1862 C
Theodolite, 1522 A
Theorema aureum (Gauss), 1795 Ma, 1801 Ma
Theory of the Earth (Hutton), 1785 E, 1795 E, 1802 E
Thermodynamics, 1824 P, 1859 P, 1873 C; chemical, 1876–1878 C; laws of, 1850 P, 1851 P, 1854 C, 1859 C/P, 1877 P, 1906 C
Thermoelectricity, 1822 P
Thermometer, 1592 Met, 1613 Met, 1657 Met, 1665 Met, 1667 Met, 1702 Met; platinum resistance, 1871 P; wet bulb, 1755 Met. *See also* Temperature scales
Thermometry, 1885 P
Thermos bottle, 1892 C
Thermoscope, 1626 H
Theta functions, hyperelliptic, 1877 Ma

Thiamin, 1912 H
Thin slices method, 1831 E, 1859 E
Thoracic duct, 1564 B
Thorium, 1829 C, 1897 P, 1898 C; X, 1902 C
Three-body problem, 1772 Ma
Thyroxine, 1914 B
Tides, c.1210 E, 1326–1335 E, c.1350 E, 1578 E, 1640 E, 1663 E, 1683 E, 1687 P, 1845 E
Timepieces. *See* Clocks; Watches
Tissues, classified, 1800 B
Titan, 1655 A
Titanium, 1795 C, 1825 C, 1910 C
Tobacco mosaic disease, 1892 H, 1898 H
Topology, 1639 Ma, 1735 Ma, 1833 Ma, 1847 Ma, 1852 Ma, 1857 Ma, 1858 Ma, 1863 Ma, 1871 Ma, 1887 Ma, 1887 Ma, 1895 Ma, 1899 Ma, 1904 Ma, 1906 Ma, 1911 Ma, 1914 Ma, 1923 Ma, 1924 Ma, 1925 Ma, 1927–1928 Ma, 1928 Ma
Torque, 1804 P
Torsion, 1777 P, 1784 P
Torsion balance, 1777 P
Toxicology, 1814–1815 H
Traité élémentaire de chimie (Lavoisier), 1789 C
Transcendental numbers, 1744 Ma, 1844 Ma, 1873 Ma, 1882 Ma
Transfinite numbers, 1895–1897 Ma, 1897 Ma, 1907 Ma
Transfinite set theory, 1883 Ma, 1891 Ma
Tree rings, c.1500 S, 1919–1936 E
Triads, element, 1829 C
Triangulation: of polyhedra, 1863 Ma; technique, 1533 Ma
Triassic formations, 1898 E
Trigonometry, 1220 Ma, c.1250 Ma, 1464 Ma, 1533 Ma, 1591 Ma; spherical, 1635 Ma, 1853 Ma, 1893 Ma; tables, 1551 Ma, 1579 Ma, 1596 Ma
Trilobites, 1757 E
Troposphere, 1902 Met
Truth function, 1879 Ma
Truth tables, 1920–1921 B
Truth values, 1885 Ma, 1920–1921 Ma
Tschirnhaus substitution, 1786 Ma
Tuberculosis, 1882 H, 1885 H
Tungsten, 1783 C, 1785 C
Typhoid fever, 1659 H, 1880 H, 1896 H

Ultraviolet radiation, 1801 P
Umbilical cord, 1604 B
Uncertainty Principle, 1927 P
Unconscious, 1912 B
Unified field theory, 1918 P, 1929 P
Uniformitarianism, 1785 E, 1795 E, 1802 E, 1830–1833 E, 1863 E
Universe, 1916 A, 1927 A
Uplifting, of land, 1810 E
Upwelling, of water, 1844 E
Uranium, 1789 C, 1841 C, 1896 P, 1897 P, 1904 C; X_1, 1900 C; X_2, 1913 C; Y, 1911 C; Z, 1921 C
Uranus, 1772 A, 1781 A, 1789 C, 1821 A, 1845 A, 1906 A
Urea, 1828 C
Uterus, 1774 B

Vaccination, 1796 H, 1798 H, 1882 H. *See also* Inoculation
Vacuum, 1600 A, 1621 A, 1643 P, 1645 P, 1646 P, 1647 P, 1654 P, 1657 P, 1672 P
Valency, 1852 C, 1902 C
Valleys, 1786 E, 1795 E, 1802 E, 1862 E, 1890 E
Vanadium, 1801 C, 1831 C, 1869 C
Van der Waals equation, 1873 C
Vector(s), 1843 Ma, 1844 Ma; analysis, 1881–1884 Ma, 1893 Ma; curl, 1873 Ma
Veins, 1603 B
Venn diagrams, 1881 Ma
Venus, 1610 A, 1632 A, 1757 A; transit of, 1639 A, 1716 A, 1761 A, 1768–1774 S, 1769 A
Vertebrates, 1848 B
Vesta, 1807 A
Viruses, 1898 B, 1892 B
Viscous boundary layer, 1904 P
Vision, c.1250 B, 1268 B, 1277–1279 P, c.1330 B, 1593 B, 1604 P, 1619 B, 1637 P; color, 1794 H, 1807 P. *See also* Astigmatism; Eye; Eyeglasses
Vitalism, 1828 C, 1833 B, 1860 C
Vitamins, 1890–1896 H, 1906 H, 1907 H, 1912 H, 1913 H, 1914 H, 1915 H, 1919 H, 1921 H, 1922 H, 1923 H, 1924 H, 1928 H. *See also* Nutrition
Volcanoes, c.1260 E, 1546 E, 1748 E, 1749 E, 1752 E, 1760 E, 1765 E, 1775 E, 1776 E, 1789 E, 1795 E, 1798 E, 1802–1803 E, 1805 E, 1809 E, 1815 E, 1825 E, 1839 E, 1844 E, 1846 E, 1906 E; and climate, 1784 E; Tertiary, 1888 E
Voltaic cell, pile, 1800 C, 1800 P, 1836 P
Vortex: motion, 1858 P; theory, 1644 A, 1659 P, 1669 P
Vulcan, 1859 A/C
Vulcanism, 1787 E, 1795 E
Vulcanization, 1839 C

Waring's problem, 1770 Ma, 1909 Ma
Watches, 1502 P, 1658 P. *See also* Clocks
Water: 1843 C; composition of, 1781 C, 1783 C, 1811 C; effects of, 1802 E; purification of, 1800 H, 1919 H;

subterranean, 1872 E; vapor, 1666 Met, 1793 Met, 1802 Met, 1881 Met
Waterproofing, 1823 C
Wave mechanics, 1926 P, 1927 P
Wave theory, 1665 P, 1675 P, 1678 P, 1680 P, 1690 P, 1704 P, 1801–1804 P, 1809 P, 1818 P, 1819 P, 1849 P
Weather: impact, 1733 B; mapping, 1855 Met, 1863 Met; prediction, 1337–1344 Met, 1921 Met, 1922 Met. *See also* Climatology; Meteorological observations; Precipitation; Wind
Weathervane, c.1450 Met
Weierstrass-Bolzano Theorem, c.1865 Ma
Well-ordered sets, 1883 Ma
Well-ordering Theorem, 1904 Ma
Wharton's duct, 1656 B
Wien's Law, 1896 P, 1900 P
Winds, 1747 Met, 1857 Met, 1875 Met; global surface, 1850 Met; mapped, 1869 Met; measuring, c.1450 Met, 1805 Met; ocean, 1847 E; resultant, 1771 Met; in storms, 1831 Met; subterranean, c.1260 E; trade, 1684 Met, 1686 Met, 1699 Met, 1735 Met, 1831 E, 1835 Met
Worms, 1881 B
Wurtz Reaction, 1855 C

Xenon, 1898 C
X rays, 1895 H, 1895 P, 1912 P

Yeast, 1680 B, 1836 B
Yellow fever, 1623 H, 1881 H, 1900 H, 1918 H, 1919 H
Ytterbia, 1878 C
Yttria, 1794 C

Zeeman effect, 1896 P, 1908 A, 1915 P
Zermelo-Fraenkel system, 1922 Ma
Zermelo's postulate. *See* Axiom of choice
Zeta function, 1737 Ma, 1748 Ma, 1859 Ma
Zinc, c.1530 C, 1742 C, 1746 C, 1854 B, 1869 B
Zirconium, 1824 C
Zoophytes, 1846 B